Nanobiophotonics

Edited by
Gabriel Popescu

New York Chicago San Francisco
Lisbon London Madrid Mexico City
Milan New Delhi San Juan
Seoul Singapore Sydney Toronto

The McGraw-Hill Companies

Library of Congress Cataloging-in-Publication Data

Nanobiophotonics / [edited by] Gabriel Popescu.
p. ; cm.
Includes bibliographical references and index.
Summary: "How can we realize super resolution medical imaging using nanotechnology and optical technology. This introductory tutorial provides readers sufficient knowledge to understand this new branch of multidisciplinary study, nanobiophotonics, and further allow them to review most resent research results in the biomedical application"—Provided by publisher.
ISBN 978-0-07-173701-2 (alk. paper)
1. Imaging systems in medicine. 2. Nanophotonics. 3. Photobiology.
4. Ultrastructure (Biology) I. Popescu, Gabriel, date.
[DNLM: 1. Nanotechnology—methods. 2. Biosensing Techniques—methods.
3. Diagnostic Imaging—methods. 4. Optical Phenomena. QT 36.5 N1823 2010]
R857.N34N328 2010
616.07′54—dc22 2010020595

Nanobiophotonics

1 2 3 4 5 6 7 8 9 0 DOC/DOC 1 9 8 7 6 5 4 3 2 1 0

ISBN 978-0-07-173701-2
MHID 0-07-173701-4

The pages within this book were printed on acid-free paper.

Sponsoring Editor
Michael Penn

Acquisitions Coordinator
Michael Mulcahy

Editorial Supervisor
David E. Fogarty

Project Manager
Shruti Vasishta
Glyph International

Copy Editors
Patti Scott and
Carol A. Loomis

Proofreader
Linda Leggio

Indexer
Robert Swanson

Production Supervisor
Pamela A. Pelton

Composition
Glyph International

Art Director, Cover
Jeff Weeks

Contents

Contributors

Catherine A. Best, Ph.D. *Department of Medical Cell and Structural Biology, College of Medicine, University of Illinois at Urbana–Champaign, Urbana, IL* (CHAP. 15)

Joerg Bewersdorf, Ph.D. *Department of Cell Biology, Yale University School of Medicine, New Haven, CT* (CHAP. 16)

Rohit Bhargava, Ph.D. *Department of Bioengineering, Micro and Nanotechnology Laboratory and Beckman Institute for Advanced Science and Technology, University of Illinois at Urbana–Champaign, Urbana, IL* (CHAP. 10)

Stephen A. Boppart, M.D., Ph.D. *Departments of Electrical and Computer Engineering, Bioengineering, and Medicine, Biophotonics Imaging Laboratory, Beckman Institute for Advanced Science and Technology, University of Illinois at Urbana–Champaign, Urbana, IL* (CHAPS. 8 AND 11)

Nicholas X. Fang, Ph.D. *Department of Mechanical Science and Engineering, University of Illinois at Urbana–Champaign, Urbana, IL* (CHAP. 9)

Kin Hung Fung, Ph.D. *Department of Mechanical Science and Engineering, University of Illinois at Urbana–Champaign, Urbana, IL* (CHAP. 9)

Travis J. Gould, Ph.D. *Department of Cell Biology, Yale University School of Medicine, New Haven, CT* (CHAP. 16)

Renu John, Ph.D. *Biophotonics Imaging Laboratory, Beckman Institute for Advanced Science and Technology, University of Illinois at Urbana–Champaign, Urbana, IL* (CHAP. 11)

Manuel F. Juette, Dipl.-Phys. *Department of Cell Biology, Yale University School of Medicine, New Haven, CT*
Department of Biophysical Chemistry, University of Heidelberg, Heidelberg, GER
Department of New Materials and Biosystems, Max Planck Institute for Metals Research, Stuttgart, GER (CHAP. 16)

Logan Liu, Ph.D. *Department of Electrical and Computer Engineering, University of Illinois at Urbana–Champaign, Urbana, IL* (CHAPS. 3 AND 14)

Marina Marjanovic, Ph.D. *Strategic Initiative on Imaging, Beckman Institute for Advanced Science and Technology, University of Illinois at Urbana–Champaign, Urbana, IL* (CHAPS. 1 AND 4)

Monal R. Mehta, M.S. *Laboratory for Photonics Research of Bio/Nano Environments (PROBE), Department of Mechanical Science and Engineering, University of Illinois at Urbana–Champaign, Urbana, IL* (CHAP. 12)

Gabriel Popescu, Ph.D. *Quantitative Light Imaging Laboratory, Department of Electrical and Computer Engineering, Beckman Institute for Advanced Science and Technology, University of Illinois at Urbana–Champaign, Urbana, IL* (CHAPS. 2 AND 5)

Raghu Ambekar Ramachandra Rao, M.S. *Laboratory for Photonics Research of Bio/Nano Environments (PROBE), Department of Electrical and Computer Engineering, University of Illinois at Urbana–Champaign, Urbana, IL* (CHAPS. 6 AND 12)

S. Sayegh, M.D., Ph.D. *The Eye Center, Urbana, IL* (CHAP. 7)

Utkarsh Sharma, Ph.D. *Biophotonics Imaging Laboratory, Beckman Institute for Advanced Science and Technology, University of Illinois at Urbana–Champaign, Urbana, IL* (CHAP. 8)

Maxim Sukharev, Ph.D. *Department of Applied Sciences and Mathematics, Arizona State University at the Polytechnic Campus, Mesa, AZ* (CHAP. 13)

Krishnarao Tangella, M.D. *Department of Pathology, College of Medicine, University of Illinois at Urbana–Champaign and Christie Clinic, Urbana, IL* (CHAPS. 1 AND 4)

Kimani C. Toussaint, Jr., Ph.D. *Laboratory for Photonics Research of Bio/Nano Environments (PROBE), Department of Mechanical Science and Engineering, Affiliate, Departments of Electrical and Computer Engineering, and Bioengineering, Affiliate, Beckman Institute for Advanced Science and Technology University of Illinois at Urbana–Champaign, Urbana, IL* (CHAPS. 6 AND 12)

Michael J. Walsh, Ph.D. *Department of Bioengineering, Micro and Nanotechnology Laboratory and Beckman Institute for Advanced Science and Technology, University of Illinois at Urbana–Champaign, Urbana, IL* (CHAP. 10)

Preface

Nanobiophotonics is designed to serve as a desktop reference for the field at the boundary between nanotechnology, photonics, and biomedicine. Although this interdisciplinary topic is of current, specialized research interest, the authors contributed preparatory material, which makes the book accessible to both students and scholars outside the field. To this end, the book consists of 16 chapters, grouped in three different parts, which reflect the progression from introductory to specialized topics. The Introduction covers the basics of cancer cell biology, electromagnetic fields and nano-optics. In the Review of Methods, the reader is exposed to a number of techniques of broad interest: tissue pathology, light scattering, second-harmonic generation, vision restoration, low-coherence interferometry, plasmonics, and metamaterials. Finally, Part III Current Research Areas cover a range of active research directions: infrared spectroscopic imaging, coherence imaging, second-harmonic imaging, plasmonics and plasmon resonance energy transfer, nanoscale red blood cell fluctuations, super-resolution microscopy, and spatial light interference microscopy.

The authors organized the 2009 Nanobiophotonics Summer School at University of Illinois at Urbana–Champaign, during the period of June 1 - June 12, 2009. The school was sponsored by the Network for Computational Nanotechnology (NCN), which is funded by the National Science Foundation.

Nanobiophotonics would not have been possible without the support from NCN and nanoHub, National Science Foundation, Beckman Institute for Advanced Science and Technology, Materials Computing Center, and Department of Electrical and Computer Engineering. Crucial in organizing the summer school were Umberto Ravaioli, Nahil Sobh, and Julie McCartney.

About the Editor

Gabriel Popescu is an assistant professor in the Department of Electrical and Computer Engineering and a full-time faculty member with the Beckman Institute for Advanced Science and Technology at the University of Illinois at Urbana–Champaign. He and his colleagues started the Nanobiophotonics Summer School at the Beckman Institute in 2009. The school was sponsored by the Network for Computational Nanotechnology (NCN), which is funded by the National Science Foundation.

PART I

Introduction

CHAPTER 1

Biology of the Cancer Cell

Marina Marjanovic

Beckman Institute for Advanced Science and Technology
University of Illinois at Urbana–Champaign

Krishnarao Tangella

Department of Pathology
College of Medicine
University of Illinois at Urbana–Champaign and Christie Clinic

1.1 Cell—A Basic Unit of Life

The cell is the simplest structure capable of performing all the activities of life. Due to its small size, the discovery and study of cells was only possible after the invention of the microscope. Robert Hooke, an English scientist, was first to describe and name cells in 1665, when he observed a slice of cork (bark from an oak tree), using a set of lenses that was able to magnify objects 30 times. Hooke thought that the "tiny boxes" or "cells," were only characteristic of cork and never realized the significance of his discovery. Anthony van Leeuwenhoek was the first to discover microorganisms. He constructed a microscope that could magnify 300 times, which revealed a microbial world in a drop of pond water. He also observed the blood cells and sperm cells of animals. In 1839, almost two centuries after the discoveries of Hooke and Leeuwenhoek; Mathias Schleiden and Theodor Schwann, two German scientists, concluded that all living things are composed of cells. This generalization forms the basis of the *Cell Theory*.

The light microscope is still a basic tool for studying cells, although its resolution is limited by the shortest wavelength of light (about 200 nm). Techniques such as phase-contrast, fluorescence, or confocal microscopy, as well as different types of staining, can further increase the contrast between the cellular structures and thus improve their visibility. However, most subcellular structures, or *organelles*, are not visible with a light microscope. The development of the electron microscope in the 1950s has revealed the complex structural organization of cells. Instead of light, a beam of electrons is passed through the specimen; this allows resolution of about 2 nm, 100 times greater than the light microscope. Two basic types of electron microscopes provide different but complementary sets of images. In a transmission electron microscope (TEM) an electron beam is passed through a thin section of samples stained with heavy metals, which gives the details of the cell's internal structure. For scanning electron microscopy (SEM), an electron beam scans the surface of the specimen coated with a gold film and the excited electrons form a three-dimensional image on the screen. One disadvantage of electron microscopy is that the methods used to prepare samples kill the cells. Also, these methods may introduce artifacts that do not exist in the living cells. Although light microscopy offers fewer details of the cellular structure, it has the huge advantage of allowing the study of live cells.

1.2 Cell Cycle

One of the main properties of the cell is the ability to reproduce itself. Cell division, or *mitosis*, is a process in which the chromosomes of the nucleus divide and separate with great precision so that two sets of exactly the same number and type of chromosomes are formed and each set is passed to a new (daughter) cell. Mitosis is the basis for the growth, development, and repair in multicellular organisms. It is an integral part of the cell cycle, the life of the cell from its origin in the division of a parent cell until its own division.

The cell cycle (Fig. 1.1) includes the *mitotic (M) phase*—in which the chromosomes and cytoplasm are divided, and the *interphase*—during which most of the cell's growth, metabolic activities, and chromosome replication occur. The interphase usually lasts 90% of the cell cycle and includes the G_1 *(first gap) phase*, the *S phase*, and the G_2 *(second gap) phase*. During all three phases, the cell grows by synthesizing proteins and producing organelles. During the S phase, chromosomes are duplicated. To recreate a structure as unique and complex as a cell, all information must be transferred. This means that the exact duplication and equal division of the genetic material (DNA) that contains the cell's genetic information is necessary. Copying and dividing the tens of thousands of genes in a cell involves a precise and tightly controlled process. Most of the cell's DNA is located in the nucleus, and it is associated with specific proteins into material called *chromatin*. As a cell prepares to divide, entangled chromatin becomes organized into separate structures or chromosomes. Each eukaryotic species has a characteristic number of chromosomes. For example,

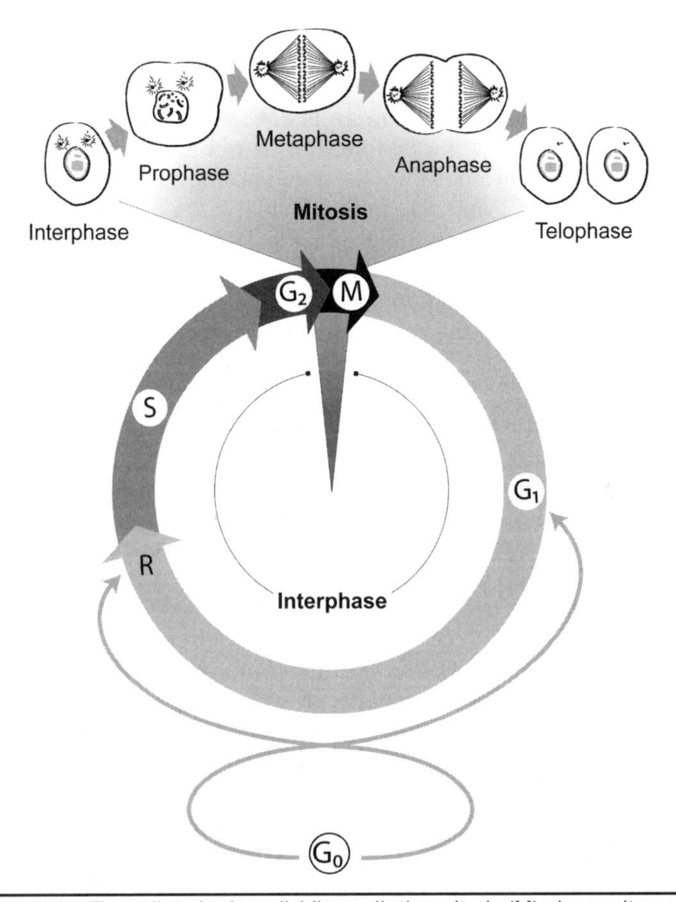

FIGURE 1.1 The cell cycle. In a dividing cell, the mitotic (M) phase alternates with the interphase. The first part of the interphase, called the *first gap phase* (G_1), and the following period of active DNA synthesis (S phase) each occupy 30% to 40% of the cycle. The *second gap phase* (G_2), during which there is no DNA synthesis but other preparations for cell division take place, takes only 10% to 20% of the total cell cycle. Mitosis (M) is the shortest phase of the cell cycle. At the *restriction point* (R), the cell can exit the cycle and remain in the nondividing (G_0) state for a longer or indefinite period of time.

humans have 46 chromosomes. They are visible only when coiled and condensed in a dividing cell. Thus, during a cell cycle a cell grows (G_1), continues to grow as it copies its chromosomes (S), grows more as it prepares for cell division (G_2), and divides (M). The daughter cells may then repeat the cycle.

The phase of the cell cycle between two divisions is referred to as the interphase. Although mitosis is a dynamic continuum of events, it is conventionally described in subphases: prophase, metaphase, anaphase, and telophase. Each subphase has characteristic morphological and biochemical events that can be used to identify the status of the cell cycle as well as the frequency of the cell divisions. Some of these events are schematically presented in Fig. 1.1. During the interphase there are only a few microscopically recognizable structures in the nucleus, since the DNA molecules are too uncoiled and extended. During the prophase, changes occur in both the nucleus and the cytoplasm. In the nucleus, the chromatin fibers become tightly coiled, condensing into chromosomes so they become visible with a light microscope in the living cell and stainable in fixed material as continuous threadlike structures. Since the DNA duplication already occurred during the S phase, each chromosome appears as two joined strands (two identical DNA molecules).

In the cytoplasm, the microtubules form a mitotic spindle that would assume the central position in the cell after the nuclear membrane fragments. Chromosomes are binding to the mitotic spindle, which is essential for organizing and equally dividing chromosomes between daughter cells. The metaphase begins with the gathering of chromosomes in the middle of the dividing cell. They line up in the imaginary plane called the *metaphase plate* so that the two strands of each chromosome are facing the opposite sides of the mitotic spindle. The anaphase begins when the chromosome strands separate and start migration along the microtubules toward the opposite sides of the cell (or poles). By the end of the anaphase, the two opposite poles of the cell have equivalent and complete sets of chromosomes. In the telophase, fragments of the nuclear membrane, preserved in the cytoplasm, begin to reform around the chromosomes gathered at the poles. The chromosomes begin to uncoil and lose most of their stainability. With these events, mitosis, the division of the nucleus, is now complete. While the reconstruction of new nuclei takes place, constriction of cytoplasm starts in the middle of the cell, between two sets of chromosomes. *Cytokinesis*, the division of the cytoplasm, usually takes place parallel to the last phases of the nuclear division or shortly after it.

The studies with cell cultures have helped us understand the events during cell proliferation and growth, but it is important to note that the constant doubling time is rarely found in normal adult tissue. Cells may remain for weeks or months (sometimes indefinitely) in the interphase and occasionally undergo one or more division cycles.

1.3 Control of the Cell Cycle

Normal growth, development, and maintenance of tissues depend on the proper control of timing and rate of cell division. The rate of cell division depends on the type of the cell. For example, human skin cells divide frequently to replace dead or damaged cells. On the other hand, very specialized cells, such as nerve or muscle cells, do not divide at all. However, the cell cycle is subject to adjustment by both external and internal cues. At certain critical points in the cell cycle, called *checkpoints*, "a go/no-go decision" is made. Human cells generally have built-in stop signals that keep the cell cycle at the checkpoints until the "go" signal is released. Checkpoints register both the internal completion of the events necessary for the next phase as well as the external signals outside the cell. Important checkpoints occur during the G_1, G_2, and M phases, but it seems that the one during the G_1 phase is the most important. It is often called the *restriction point (R)*, and takes place just before DNA synthesis (S phase). If all systems are "go" at this point, which means that all the internal and external factors are favorable, the cell proceeds to copy its DNA. Then the cell divides. Alternatively, the cell may exit the cell cycle at the restriction point and switch to a nondividing state called the *G_0 phase*. Most cells of the human body are actually in the G_0 phase. The most specialized cells, such as nerve and muscle cells, will never reenter the cell cycle. Other cells, such as liver cells, can be "called back" to the cell cycle by certain internal and external cue factors, such as growth factors released during injury.

Control mechanisms of cell division have been studied through the use of cell culture. Cells isolated from plants or animals are grown in glass containers supplied with a nutrient solution called a *growth medium*. Normal cells will multiply in a cell culture only until they form a single layer of cells, as seen in Fig. 1.2. External factors, such as the availability of nutrients and growth factors, limit the density of cell population. However, if some cells are removed, those near to the gap will divide until the gap is filled with cells. This is called *density-dependent inhibition* of cell division, and it is related to diminishing supplies of essential nutrients and growth factors. Most cells also exhibit *anchorage dependence*. For normal growth and division, cells must attach to the surface, such as the glass wall in the cell cultures or the extracellular matrix of the tissue. Both density-dependent inhibition and anchorage dependence ensure the optimal density of the cells for the normal function of the tissue.

1.4 Biology of a Cancer Cell

Cancer cells escape from the body's normal control mechanisms. When grown in a cell culture, cancer cells do not exhibit density-dependent inhibition and may continue to divide indefinitely, even when the growth factors are depleted from the culture (Fig. 1.2). It is logical to assume that they either produce their own growth factors or do not need them. If they are continually provided with nutrients, cancer cells in the culture can divide indefinitely, rather than stop dividing after 20 to 50 divisions, as in the case of normal mammalian cells. Some cancer cell lines, such as the famous HeLa cells, have been kept dividing for research for the past 40 to 50 years.

The abnormal dividing behavior of cancer cells can have serious consequences when it occurs in the tissue. When a normal cell is

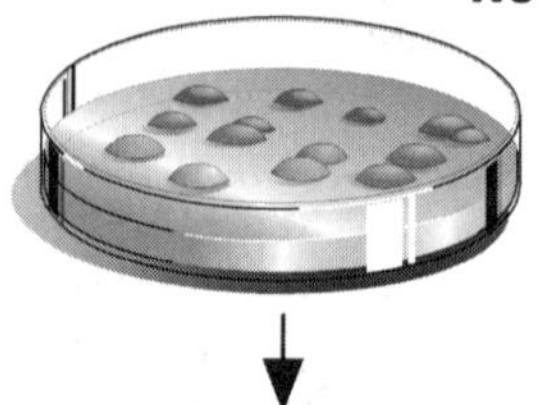

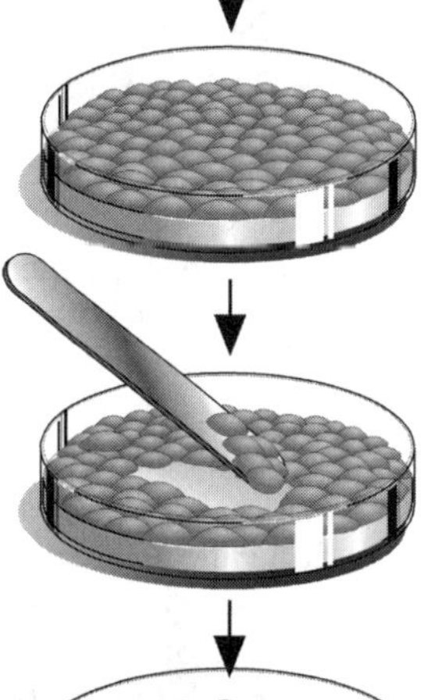

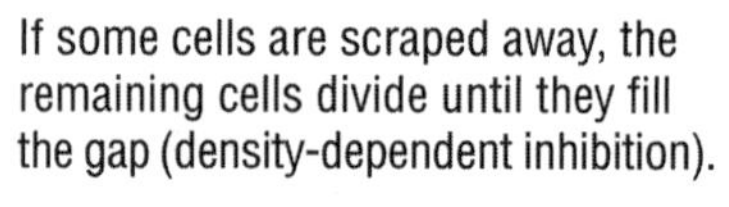

FIGURE 1.2 Effect of external factors on cell division in normal and cancer cells. The availability of nutrients, growth factors, and surface for attachment limits the number of cell divisions. Cancer cells, however, exhibit neither density-dependent inhibition nor anchorage dependence.

transformed or converted to a cancer cell, the body's immune system usually destroys it. However, if the cancer cell escapes destruction, it may proliferate to form a *tumor*, a mass of cancer cells within a tissue. Although cancer cells arise from normal cells, they lose their normal form and function in the body and their only goal is to divide. *Benign tumors* remain at their original site and can be removed by surgery. They are usually slow-growing, encapsulated, and well-differentiated. The majority of benign tumors are not life threatening, unless they grow enough to press on vital organs or tissues and prevent their normal function. *Malignant tumors* are the ones that cause cancer as they invade and disrupt the function of one or more organs. This spreading of cancer cells to distant locations from the original site is called *metastasis* and usually has fatal consequences. Malignant tumors are fast-growing, without a capsule, and poorly differentiated. In addition to uncontrolled proliferation, malignant tumor cells may also have abnormal metabolism and unusual numbers of chromosomes; this causes the loss of normal cell functions.

Due to the changes in the cell membrane, cancer cells may lose their attachments to other cells and the extracellular matrix. If the cancer cell separates from the original tumor, it can initiate secondary tumors in the body through a complex sequence of events called the *invasion-metastasis cascade* (Fig. 1.3). In the initial step, cancer cells penetrate the basement membrane of the lymphatic or blood vessel wall and enter the circulation (*intravasation*). They can then travel by

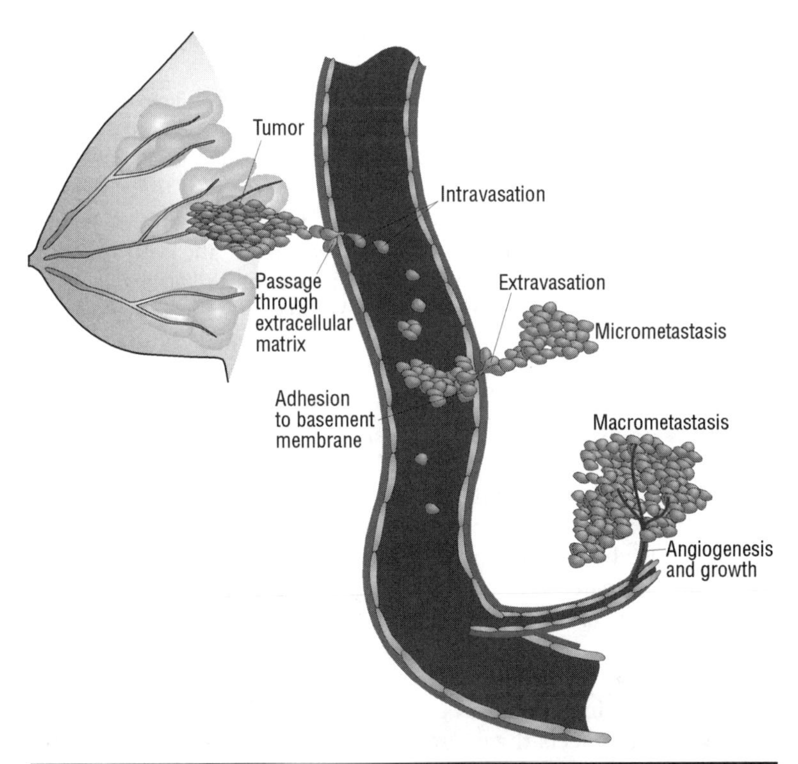

FIGURE 1.3 The metastasis of a breast cancer. The cells of the malignant tumor detach and enter the circulation, spreading to the other parts of the body. The metastatic tumors have the ability to stimulate the growth of new blood vessels (angiogenesis) that supplies them with necessary nutrients.

blood or lymph circulation to distant sites in the body. Invasion of the tissues and organs occurs when the cancer cells move again through the epithelial tissue of the vessel wall (*extravasation*) and form small, dormant groups within the healthy tissue (*micrometastases*). In the last step, some of these micrometastases may colonize the tissue and cause a formation of secondary tumors (*macrometastases*). The probability for a single cancer cell to successfully complete this series of events is very small; this explains why not all cancer cells leaving the primary tumor form macrometastases.

In order to survive, all live cells including cancer cells depend on the continuous supply of nutrients and oxygen, as well as the removal of metabolic waste and carbon dioxide by blood circulation. Cells have to be ≥ 0.2 mm away from the nearest blood vessel to ensure sufficient oxygen diffusion. Because of this, the capillary network in all tissues branch extensively and come into close contact with all cells (directly or no more than several cell diameters away). It has been shown that cancer cells also prefer to grow around the blood vessels. However, with the fast cell proliferation during the formation of tumors, the distance between the cancer cells and existing blood capillaries increases. This can cause lack of oxygen (*hypoxia*) and death in cancer cells. At a certain stage, tumors have the ability to stimulate the development of new blood vessels and assemble their own blood supply (*angiogenesis*).

The growth of tumors is followed by gradual morphological changes in the cells. Transformed cells lose not only their specific functions but also their appearance (Fig. 1.4). At the initial stages, transformed cells are not much different from the surrounding healthy cells of the same tissue, but their number is *increasing.* This stage is usually indicated as *hyperplasia*. At the next stage, or *metaplasia,* cancer cells are still not very different from normal cells, but they appear in tissue where they are usually not found, replacing the cells characteristic for that tissue. Metaplasia is the most common in the transition zones, between two different types of epithelial tissues (e.g., between the cervix and uterus) and can be used to identify early phases of malignancy. With further cell reproduction, cancer cells lose their normal appearance. Cytological changes at the stage of *dysplasia* (Fig. 1.4) feature an increased diversity in nuclear size and shape and an increased number of cells in mitosis. Since the cancer cells proliferate rather quickly, cells do not have time to grow between two divisions, and the ratio of nucleus versus cytoplasm size increases. The nucleus also shows an increased affinity for stains and appears darker in cytological preparations. In dysplasia, the loss of cell uniformity affects the overall loss of normal tissue architecture. However, cells are not yet leaving their site of origin and invading the neighboring tissue. This stage is usually considered as a premalignant change. All the abnormal growths that are easily visible are named *neoplasia* ("new growth"). However, that includes both benign and

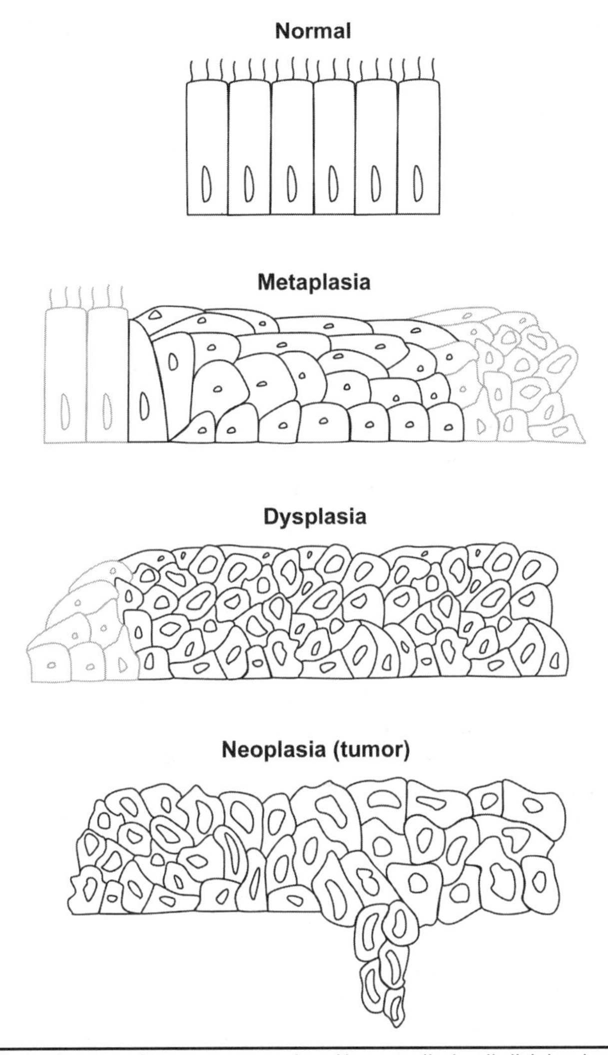

FIGURE 1.4 Stages of tumor progression. Uncontrolled cell division leads to gradual morphological and functional changes of the cells, and finally to metastasis.

more invasive malignant growths. At this stage, cells are able to separate from the primary tumor and invade adjacent and distant tissues causing metastasis (Fig. 1.3).

Although cancer cells can significantly differ from the cells they originated from, in most cases they preserve some features of their normal precursors. These features allow pathologists to determine the tissue of origin and tumor classification. Only in a small number of cases do cancer cells lose all features of their normal precursor cells and become completely dedifferentiated (*anaplasia*).

Tumors can potentially form from any type of cell in the human body, but most tumors develop from *epithelial tissues*. Epithelial tissues are layers of specialized cells that line the external and internal surfaces of the body, including the surfaces of all organs and cavities. They serve as protective shields and are often exposed to harsh mechanical and chemical conditions. Epithelial cells have a high-proliferation capacity to replace damaged and dead cells, which increases the probability of loss of control over the cell cycle.

The incidence of nonepithelial cancer is much lower. The major type of malignant nonepithelial tumors called *sarcomas*, develop from different cells of connective tissues, such as adipocytes (adipose or fat tissue), osteoblasts (bone), fibroblasts (skin, tendons), and so on. The second type are cancers that develop from circulating blood and lymph cells. Among these cancers are different types of leukemia and lymphoma. And the third group of nonepithelial tumors consists of tumors of the nervous tissue (gliomas, neuroblastomas, etc.).

1.5 Molecular Biology of a Cancer Cell

With the development of molecular biology, we have started to understand why a normal cell would transform into a cancer cell. Certain normal cellular genes, called *proto-oncogenes*, the code for the proteins that regulate the cell cycle. They regulate cell division and differentiation in the presence of mitogenic stimuli. Their structural and functional changes (*mutations*) can convert them into *oncogenes*, cancer-causing genes, which are able to promote cell growth in the absence of mitogenic stimuli. These genetic changes can be caused by environmental factors, such as chemical carcinogens or X-rays. They can also stem from random, spontaneous changes of the genetic material or insertion of viral oncogenes into the human DNA. There are three types of genetic changes that convert proto-oncogenes into oncogenes: (1) movement of DNA within the genome, (2) amplification of a proto-oncogene, and (3) mutation of a proto-oncogene. Often cancer cells contain chromosomes with abnormal structures. If the chromosomes break and then reconnect incorrectly, some genes can be moved from one chromosome to another. A proto-oncogene can be translocated to a new place on a chromosome that is involved in an active transcription; this can cause an excessive production of normal growth-stimulating proteins. The amplification of a proto-oncogene has a similar effect by increasing the number of proto-oncogene copies in the cell. The third possibility is that mutation changes the product of the proto-oncogene so it becomes more active or more resistant to inhibition.

Besides growth-stimulating proteins, cells also produce tumor-suppressor proteins that normally inhibit cell division and prevent uncontrolled tissue growth. Mutations that decrease the activity of these tumor-suppressor proteins can contribute to the transformation of a normal cell into a cancer cell.

It has been shown that a sequence of mutations, rather than one mutation, is needed to gradually produce a cancer cell. These changes usually include the presence of at least one active oncogene and the loss of several tumor-suppressor genes. The fact that an accumulation of genetic changes is required to produce a cancer cell can explain why the probability of developing cancer increases with age. It also explains predispositions of some families to certain types of cancers. If members of a family are inheriting one or more oncogenes or mutated tumor-suppressor genes, they need to accumulate fewer genetic changes during life to develop cancer. A lot of effort today is devoted to detecting inherited cancer genes so that a predisposition to certain cancers can be detected early in life. For example, breast cancer is the second most common cancer in the United States. Each year about 200,000 new cases are registered and about 40,000 women die. Researchers have identified evidence for strong inherited predispositions. Mutations in two tumor-suppressor genes, BRCA1 and BRCA2, significantly increase the probability of developing breast cancer. A woman who inherits a mutant BRCA1 gene has a 60% probability of developing breast cancer, compared with only a 2% probability for a woman with a normal gene.

1.6 Summary

In summary, development of cancer (carcinogenesis) is a complex, multistep process at both genetic and histological levels in which genetic changes cause the uncontrolled proliferation of affected cells.

Suggested Reading

Junqueira, L. C., and J. Carneiro, *Basic Histology: Text and Atlas*, McGraw-Hill, New York, 2005.

Weinberg, Robert T., *The Biology of Cancer*, Garland Science, Taylor & Francis Group, New York, 2007.

CHAPTER 2

Review of Electromagnetic Fields

Gabriel Popescu

Quantitative Light Imaging Laboratory
Department of Electrical and Computer Engineering
Beckman Institute for Advanced Science and Technology
University of Illinois at Urbana–Champaign

2.1 Maxwell's Equations

In this section we will review the main features of Maxwell's equations in differential forms. We discuss these equations in different representations: space-time ($\mathbf{r}$, t), space-frequency ($\mathbf{r}$, ω), and wave vector frequency ($\mathbf{k}$, ω).

2.1.1 Maxwell's Equations in the Space-Time Representation

The first Maxwell equation, known as *Faraday's induction law,* establishes that a varying magnetic field produces a rotating electric field.

$$\nabla \times \mathbf{E}(\mathbf{r}, t) = -\frac{d\mathbf{B}(\mathbf{r}, t)}{dt} \tag{2.1}$$

In Eq. (2.1), $\mathbf{E}(\mathbf{r}, t)$ is the instantaneous *electric field* (in V/m) and $\mathbf{B}(\mathbf{r}, t)$ is the instantaneous *magnetic induction* or *magnetic flux density* (in Wb/m^2). *Ampère's circuital law* (with Maxwell's correction) states that a magnetic field can be generated both by existing currents (the original Ampère's law) and by varying electric fields.

$$\nabla \times \mathbf{H}(\mathbf{r}, t) = \frac{d\mathbf{D}(\mathbf{r}, t)}{dt} + \mathbf{J}_S(\mathbf{r}, t) + \mathbf{J}_C(\mathbf{r}, t) \tag{2.2}$$

In Eq. (2.2), $\mathbf{H}(\mathbf{r}, t)$ is the instantaneous *magnetic field intensity* (in A/m); $\mathbf{D}(\mathbf{r}, t)$ is the instantaneous *electric induction* or *flux density* (in C/m^2); $\mathbf{J}_s(\mathbf{r}, t)$ is the instantaneous source current density (in A/m^2); and $\mathbf{J}_c(\mathbf{r}, t)$ is the instantaneous conduction current density (in A/m^2).

Gauss's law for the electric and magnetic fields is:

$$\nabla \cdot \mathbf{D}(\mathbf{r}, t) - \rho(\mathbf{r}, t) \tag{2.3}$$

where $\rho(\mathbf{r}, t)$ is the volume charge density (C/m^3), and

$$\nabla \cdot \mathbf{B}(\mathbf{r}, t) = 0 \tag{2.4}$$

Equation (2.4) establishes the *nonexistence of magnetic charge*. In addition to Gauss's equations, the following *constitutive relations* apply:

$$\begin{aligned} \mathbf{D} &= \varepsilon \mathbf{E}, \qquad \varepsilon = \varepsilon_0 \cdot \varepsilon_r \\ \mathbf{B} &= \mu \mathbf{H}, \qquad \mu = \mu_0 \cdot \mu_r \\ \mathbf{J}_C &= \sigma \mathbf{E} \end{aligned} \tag{2.5}$$

in Eq. (2.5), ε is the *dielectric permittivity*, μ is the *magnetic permeability*, and σ is the *conductivity* associated with the material. Generally, these material quantities, ε, μ, and σ, are *inhomogeneous* ($\mathbf{r}$-dependent) and *anisotropic* (direction-dependent tensors). They can also be field-dependent for *nonlinear* media. For linear, homogeneous, and isotropic media, ε, μ, and σ are simple scalars. The *relative* dielectric permittivity ε_r and magnetic permeability μ_r are defined with respect to the vacuum values $\varepsilon_0 = 10^{-9}/36\pi$ F/m and $\mu_0 = 4\pi \cdot 10^{-7}$ N/A^2 (or Henry/m).

2.1.2 Boundary Conditions

In order to solve Maxwell's equations for fields propagating between two media, boundary conditions for both the electric and magnetic fields are necessary. There are four equations that describe the relationship between the tangent components of **E** and **H**, and the normal components of **D** and **B** in the two media (Fig. 2.1), as follows:

$$\begin{aligned} \hat{\mathbf{n}} \times (\mathbf{E}_1 - \mathbf{E}_2) &= 0 \\ \hat{\mathbf{n}} \times (\mathbf{H}_1 - \mathbf{H}_2) &= \mathbf{j}_S \\ \hat{\mathbf{n}} \cdot (\mathbf{D}_1 - \mathbf{D}_2) &= \rho_S \\ \hat{\mathbf{n}} \cdot (\mathbf{B}_1 - \mathbf{B}_2) &= 0 \end{aligned} \tag{2.6}$$

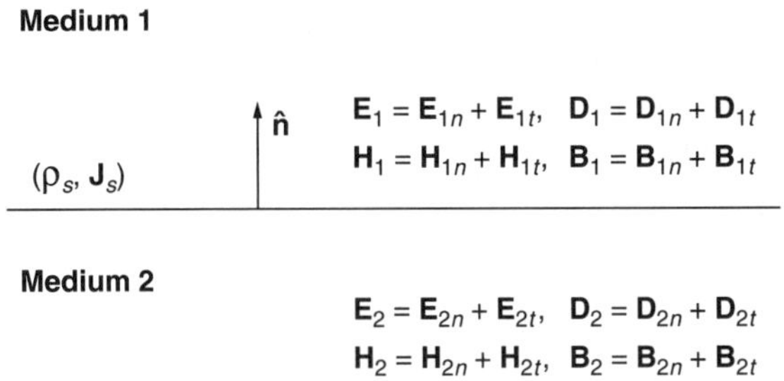

Medium 1

$\hat{\mathbf{n}}$ $\mathbf{E}_1 = \mathbf{E}_{1n} + \mathbf{E}_{1t}$, $\mathbf{D}_1 = \mathbf{D}_{1n} + \mathbf{D}_{1t}$

$(\rho_s, \mathbf{J}_s)$ $\mathbf{H}_1 = \mathbf{H}_{1n} + \mathbf{H}_{1t}$, $\mathbf{B}_1 = \mathbf{B}_{1n} + \mathbf{B}_{1t}$

Medium 2

$\mathbf{E}_2 = \mathbf{E}_{2n} + \mathbf{E}_{2t}$, $\mathbf{D}_2 = \mathbf{D}_{2n} + \mathbf{D}_{2t}$

$\mathbf{H}_2 = \mathbf{H}_{2n} + \mathbf{H}_{2t}$, $\mathbf{B}_2 = \mathbf{B}_{2n} + \mathbf{B}_{2t}$

FIGURE 2.1 Interface between two media.

In Eq. (2.6), $\hat{\mathbf{n}}$ is the unit vector normal to the interface (Fig. 2.1), $\mathbf{J}_s$ is the surface current, and ρ_s is the surface charge density. Note that in the absence of free charge and currents, the boundary conditions indicate that the *tangential fields* (**E**, **H**) and normal inductions (**D**, **B**) are conserved across the interface.

2.1.3 Maxwell's Equations in the Space-Frequency Representation (r, ω)

For situations that involve broad band fields, the *spatial behavior of each temporal frequency* ω is often of interest. In order to obtain the Maxwell equation in the space-frequency representation, we use the *differentiation property* of Fourier transforms,

$$\frac{d}{dt}\mathbf{F}(t) \rightarrow i\omega \cdot \mathbf{F}(\omega) \tag{2.7}$$

In Eq. (2.7), **F** denotes the Fourier transform operator (in time). $\mathbf{F}(t)$ is an arbitrary time-dependent vector, and $\mathbf{F}(\omega)$ is its Fourier transform. Note that we will denote the Fourier transform of a function by the same symbol, with the understanding that $\mathbf{F}(t)$ and $\mathbf{F}(\omega)$ are two distinct functions.

Taking the Fourier transform of Eqs. (2.1 to 2.4) and using the differentiation property in Eq. (2.7), we can rewrite the Maxwell equations in the (**r**, ω) representation.

$$\begin{aligned}
&\nabla \times \mathbf{E}(\mathbf{r}, \omega) = -i\omega \mathbf{B}(\mathbf{r}, \omega) \\
&\nabla \times \mathbf{H}(\mathbf{r}, \omega) = i\omega \mathbf{D} + \mathbf{J}_S(\mathbf{r}, \omega) + \mathbf{J}_C(\mathbf{r}, \omega) \\
&\nabla \times \mathbf{B}(\mathbf{r}, \omega) = 0 \\
&\nabla \times \mathbf{D}(\mathbf{r}, \omega) = \rho(\mathbf{r}, \omega)
\end{aligned} \tag{2.8}$$

Similarly, the constitutive relations take the form

$$\begin{aligned}
&\mathbf{D}(\mathbf{r}, \omega) = \varepsilon \mathbf{E}(\mathbf{r}, \omega), \qquad \varepsilon = \varepsilon_0 \cdot \varepsilon_r, \qquad \mathbf{D} = \varepsilon_0 \mathbf{E} + \mathbf{P} \\
&\mathbf{B}(\mathbf{r}, \omega) = \mu \mathbf{H}(\mathbf{r}, \omega), \qquad \mu = \mu_0 \cdot \mu_r, \qquad \mathbf{B} = \mu_0 \mathbf{H} + \mathbf{M} \\
&\mathbf{J}_C(\mathbf{r}, \omega) = \sigma \cdot \mathbf{E}(\mathbf{r}, \omega)
\end{aligned} \tag{2.9}$$

Note that even for homogeneous media, ε is a function of frequency; this establishes the *dispersion relation* associated with the medium. The boundary conditions remain in the same form; that is, they apply for each individual frequency.

$$
\begin{aligned}
&\hat{\mathbf{n}}\cdot\left[\mathbf{E}_1(\mathbf{r},\omega)-\mathbf{E}_2(\mathbf{r},\omega)\right]=0\\
&\hat{\mathbf{n}}\cdot\left[\mathbf{H}_1(\mathbf{r},\omega)-\mathbf{H}_2(\mathbf{r},\omega)\right]=\mathbf{J}_S\\
&\hat{\mathbf{n}}\cdot(\mathbf{B}_1-\mathbf{B}_2)=0\\
&\hat{\mathbf{n}}\cdot(\mathbf{D}_1-\mathbf{D}_2)=\rho_s
\end{aligned}
\tag{2.10}
$$

2.1.4 The Helmholtz Equation

By eliminating **H**, **B**, and **D** from Eqs. (2.1 to 2.5), we obtain an equation only in $\mathbf{E}(\mathbf{r}, t)$, which is referred to as the *wave equation.* An analogous equation, called the *Helmholtz equation,* can be obtained in the $(\mathbf{r}, \omega)$ representation, as follows: First, for simplicity, let us consider linear, isotropic, and charge-free media. In this case, Eq. (2.8) simplifies to

$$\nabla\times\mathbf{E}(\mathbf{r},\omega)=-i\omega\mu\mathbf{H}(\mathbf{r},\omega) \tag{2.11a}$$

$$\nabla\times\mathbf{H}(\mathbf{r},\omega)=(\sigma+i\omega\varepsilon)\mathbf{E}(\mathbf{r},\omega) \tag{2.11b}$$

$$\nabla\cdot\mathbf{D}=0 \tag{2.11c}$$

$$\nabla\cdot\mathbf{B}=0 \tag{2.11d}$$

Applying the curl operator to Eq. (2.11*a*), we obtain

$$\nabla\times\nabla\times\mathbf{E}=-i\omega\nabla\times\mu\mathbf{H} \tag{2.12}$$

Using the identity $\nabla\times\nabla\times\mathbf{E}=\nabla(\nabla\mathbf{E})-\nabla^2\mathbf{E}$ and the fact that $\nabla\mathbf{E}=0$ for charge-free media, we obtain the Helmholtz equation

$$\nabla^2\mathbf{E}(\mathbf{r},\omega)+k^2\mathbf{E}(\mathbf{r},\omega)=0 \tag{2.13a}$$

$$k^2(\omega)=\omega^2\mu\varepsilon(\omega)-i\omega\mu\sigma(\omega) \tag{2.13b}$$

Note that in Eq. (2.13*a*), the Laplace operator applied to vector **E** is defined as $\nabla^2\mathbf{E}=\nabla^2E_x\cdot\hat{\mathbf{x}}+\nabla^2E_y\cdot\hat{\mathbf{y}}+\nabla^2E_z\cdot\hat{\mathbf{z}}$. Equation (2.13*b*) establishes the frequency dependence of the modulus of the wave vector $\mathbf{k}(\omega)$, that is, *the dispersion relation of the medium.* While typical biological structures are characterized by the constant μ, the permittivity ε and conductivity σ can have a strong dependence on frequency.

In rectangular coordinates, the vector, Eq. (2.13*a*) breaks down into three scalar equations.

$$\begin{aligned} \nabla^2 E_x(\mathbf{r}, \omega) + k_x^2 E_x^2(\mathbf{r}, \omega) &= 0 \\ \nabla^2 E_y(\mathbf{r}, \omega) + k_y^2 E_y^2(\mathbf{r}, \omega) &= 0 \\ \nabla^2 E_z(\mathbf{r}, \omega) + k_z^2 E_z^2(\mathbf{r}, \omega) &= 0 \end{aligned} \tag{2.14}$$

Using the separation of variables method, that is, assuming solutions of the form $X(x, \omega) \cdot Y(y, \omega) \cdot Z(z, \omega)$, we obtain the *plane wave solution* of the form

$$E(\mathbf{r}, \omega) = A \cdot e^{i\mathbf{k}\cdot\mathbf{r}} \tag{2.15a}$$

$$k^2 = k_x^2 + k_y^2 + k_z^2 = \omega^2\mu\varepsilon - i\mu\varepsilon\omega\sigma \tag{2.15b}$$

Decomposing the wave vector into its real and imaginary parts,

$$k(\omega) = k'(\omega) + ik''(\omega) \tag{2.16}$$

we can rewrite Eq. (2.15*a*) as

$$E(\mathbf{r}, \omega) = A \cdot e^{-k''(\omega)\cdot r} \cdot e^{ik'(\omega)\cdot r} \tag{2.17}$$

It is apparent in Eq. (2.17) that the real and imaginary parts of k capture the refraction (phase term) and absorption/gain (amplitude term) of the medium.

Note that in lossless media, $k = \omega\sqrt{\varepsilon\mu} = \omega\sqrt{\varepsilon_0\mu_0} \cdot \sqrt{\varepsilon_r\mu_r} = (\omega/c) \cdot \sqrt{\varepsilon_r\mu_r}$. In dielectric media at optical frequencies, $\mu_r \simeq 1$ and $\sqrt{\varepsilon_r}$ equal the *refractive index*; thus

$$k(\omega) = n(\omega)\frac{\omega}{c} = n(\omega)k_0 \tag{2.18}$$

In Eq. (2.18), $k_0 = \omega/c$ is the wave vector in a vacuum.

2.1.5 Maxwell's Equations in the (k, ω) Representation

Often we deal with optical fields characterized by broad *angular spectra*. These fields can be decomposed by wave vector **k** of different directions. The modulus of each **k**-vector is defined by the *dispersion relation* (Eq. 2.13*b*) that depends on the material properties (ε, μ, and σ) and optical frequency ω. Thus, the normal representation of the fields in this case is in the **k**-ω space. Note that a differentiation property similar to Eq. (2.7) holds for the ∇ operator, as follows:

$$\begin{aligned} \nabla \times \mathbf{F}(\mathbf{r}) &\rightarrow i\mathbf{k} \times \mathbf{F}(\mathbf{k}) \\ \nabla \cdot \mathbf{F}(\mathbf{r}) &\rightarrow i\mathbf{k} \cdot \mathbf{F}(\mathbf{k}) \end{aligned} \tag{2.19}$$

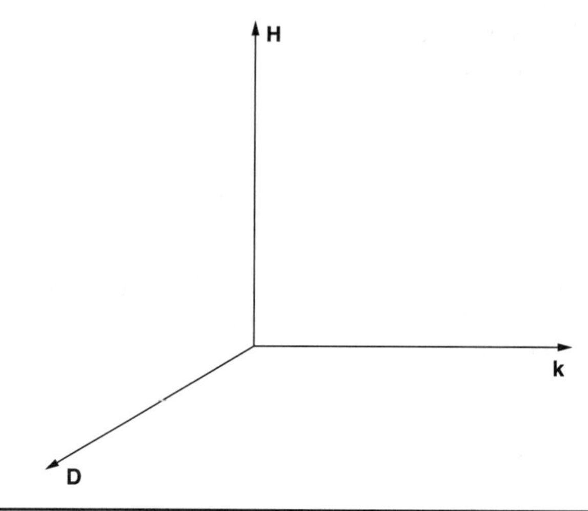

FIGURE 2.2 Mutually orthogonal vectors.

Thus, the (**k**, ω) representation of Maxwell's equations is obtained by the spatial Fourier transformation of Eq. (2.11). For media with no free charge (ρ = 0) or currents (**J** = 0), these equations are

$$\mathbf{k}\times\mathbf{E}(\mathbf{k},\omega) = -\omega\mathbf{B}(\mathbf{k},\omega) \tag{2.20a}$$

$$\mathbf{k}\times\mathbf{H}(\mathbf{k},\omega) = \omega\mathbf{D}(\mathbf{k},\omega) \tag{2.20b}$$

$$\mathbf{k}\cdot\mathbf{D} = 0 \tag{2.20c}$$

$$\mathbf{k}\cdot\mathbf{B} = 0 \tag{2.20d}$$

Equation (2.20) describes the propagation of the frequency component ω and plane wave of wave vector **k**. Right away, Eqs. (2.20*c* and *d*) establish that $\mathbf{k}\perp\mathbf{B}$ and $\mathbf{k}\perp\mathbf{D}$. Generally, μ is a scalar—that is, **B**||**H**—but ε is a tensor—that is, **D** is not necessarily parallel to **E**. Thus, from Eq. (2.20*b*) we see that $\mathbf{D}\perp\mathbf{H}$, such that

$$\mathbf{k}\perp\mathbf{D} \quad \text{and} \quad \mathbf{k}\perp\mathbf{H} \quad \text{and} \quad \mathbf{H}\perp\mathbf{D} \tag{2.21}$$

Equations (2.20*a–c*) show that **H**, **D**, and **k** are mutually orthogonal vectors (Fig. 2.2). For isotropic media, **D**||**E**, such that **H**, **E**, and **k** are also mutually orthogonal.

The characteristic impedance η of the medium is defined as

$$\eta = \frac{\omega\mu}{k} \tag{2.22}$$

Using Eq. (2.15*b*) to express **k** for media with nonzero conductivity, we obtain

$$\eta = \left(\frac{\mu}{\varepsilon - i(\sigma/\varepsilon)} \right)^{1/2} \tag{2.23}$$

For nonconducting media (electric insulators), $\eta = \sqrt{\mu/\varepsilon}$. For isotropic media, η connects the moduli of **E** and **H** directly [from Eq. (2.19*b*)],

$$E = \eta \cdot H \tag{2.24}$$

2.1.6 Phase, Group, and Energy Velocity

Let us consider the electric field associated with a light beam propagating along the +*z* direction with an average wave vector <**k**>. As illustrated in Fig. 2.3, the (real) temporal signal is characterized by a slow modulation (*envelope*), due to the superposition of different frequencies, and a fast sinusoidal modulation (*carrier*), at the average frequency <ω>.

$$E(z,t) = A(z,t) \cdot e^{-i(<\omega>t - <k>z)} \tag{2.25}$$

Thus the phase delay of the field is

$$\phi(z,t) = \langle\omega\rangle t - \langle k\rangle z \tag{2.26}$$

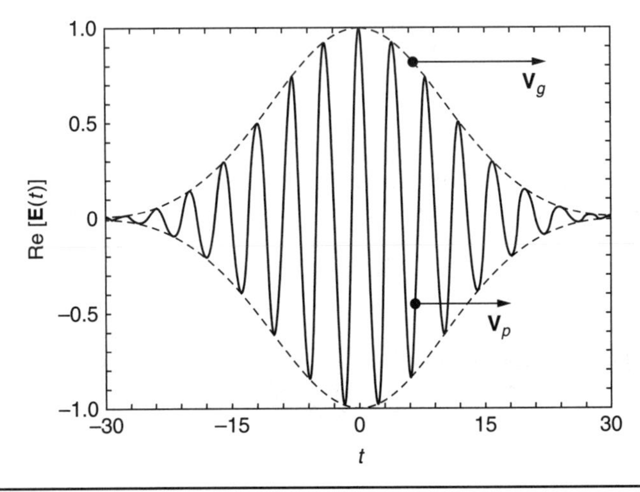

FIGURE 2.3 Group and phase velocity.

The *phase velocity* is associated with the advancement of wave fronts, for which ϕ = constant. From differentiating Eq. (2.26), we obtain

$$\begin{aligned} d\phi(z,t) &= \langle\omega\rangle dt - \langle k\rangle dz \\ &= 0 \end{aligned} \tag{2.27}$$

Thus, the phase velocity is

$$\begin{aligned} v_\phi &= \frac{dz}{dt} = \frac{\langle\omega\rangle}{\langle k\rangle} \\ &= \frac{c}{n} \end{aligned} \tag{2.28}$$

Equation (2.28) represents the zeroth-order approximation in the expansion.

$$\begin{aligned} \frac{d\omega(k)}{dk} &= \frac{d}{dk}\left(\frac{c}{n}\cdot k\right) \\ &= \frac{c}{n} - \frac{ck}{n^2}\cdot\frac{dn}{dk} = v_\phi n\left(1 - \frac{\omega}{n}\cdot\frac{dn}{d\omega}\right) \end{aligned} \tag{2.29}$$

Equation (2.29) defines the group velocity $\mathbf{v}_g$ at which the envelope of the light signal propagates. Thus, the group and phase velocities are equal in nondispersive media, that is, when $dn/d\omega = 0$.

In the following, we prove that the group velocity in fact defines the velocity at which the electromagnetic energy flows; in other words we show that the group and energy velocity are the same. By definition, the energy velocity is the ratio between the Poynting vector **S** and electromagnetic volume density U.

$$\mathbf{v}_e = \frac{1}{U}\cdot\mathbf{S} \tag{2.30a}$$

$$\mathbf{S} = \mathbf{E}\times\mathbf{H} \tag{2.30b}$$

$$U = \frac{1}{2}\mathbf{E}\cdot\mathbf{D} + \frac{1}{2}\mathbf{H}\cdot\mathbf{B} \tag{2.30c}$$

We now differentiate the first two Maxwell equations in the ($\mathbf{k}$, ω) representation, that is, Eqs. (2.20*a* and *b*), as follows:

$$\begin{aligned} \delta\mathbf{k}\times\mathbf{E} + \mathbf{k}\times\delta\mathbf{E} &= \delta\omega\mu\mathbf{H} + \omega\mu\delta\mathbf{H} \\ \delta\mathbf{k}\times\mathbf{H} + \mathbf{k}\times\delta\mathbf{H} &= -\delta\omega\varepsilon\mathbf{E} - \omega\varepsilon\delta\mathbf{E} \end{aligned} \tag{2.31}$$

Note that the meaning of $\delta\mathbf{k}$ and $\delta\omega$ is that of a spread in **k**-vectors (spatial frequency spectrum) and optical frequency (optical spectrum). Let us dot-multiply Eq. (2.31*a*) by **H** and Eq. (2.31*b*) by **E**, so that we obtain

$$\begin{aligned} \delta\mathbf{k}\cdot(\mathbf{E}\times\mathbf{H}) + \mathbf{k}(\mathbf{H}\times\delta\mathbf{E}) &= \mu\delta\omega\mathbf{H}\cdot\mathbf{H} + \omega\mu\mathbf{H}\cdot\delta\mathbf{H} \\ \delta\mathbf{k}\cdot(\mathbf{E}\times\mathbf{H}) + \mathbf{k}(\delta\mathbf{H}\times\mathbf{E}) &= -\delta\omega(\mathbf{E}\cdot\varepsilon\mathbf{E}) - \omega(\mathbf{E}\cdot\varepsilon\delta\mathbf{E}) \end{aligned} \tag{2.32}$$

Adding these two equations, we obtain

$$\delta\mathbf{k}\cdot(\mathbf{E}\times\mathbf{H})=\frac{1}{2}\delta\omega\left[\mathbf{E}\cdot\varepsilon\mathbf{E}+\mu\mathbf{H}\cdot\mathbf{H}\right] \tag{2.33}$$

where we used the fact that the products $\delta\mathbf{H}(\omega\mu\mathbf{H}-\mathbf{k}\times\mathbf{E})$ and $\delta\mathbf{E}(\varepsilon\cdot\mathbf{E}\omega+\mathbf{k}\cdot\mathbf{H})$ vanish. Finally, using Eq. (2.30*a*), we find

$$\begin{aligned}\mathbf{v}_e &= \frac{\delta\mathbf{k}}{\delta\omega} \\ &= v_g\end{aligned} \tag{2.34}$$

Equation (2.34) establishes the anticipated result that the electromagnetic energy flows at group velocity. It is apparent from Eq. (2.29) that v_g can exceed the speed of light in vacuum c, in special circumstances where $dn/d\omega < 0$ (anomalous dispersion). However, this does not pose a conflict with the postulate of the relativity theory that states that the *signal velocity,* at which information can be transmitted via electromagnetic fields is bounded by c.

2.1.7 The Fresnel Equations

Let us consider the problem of light propagation at the interface between two media (Fig. 2.4). In the following, we will derive the expressions for the field reflection and transmission coefficients.

In Fig. 2.4, the plane xz is typically referred to as the plane of incidence (plane of the paper), that is, the plane defined by the incident wave vector and normal ($\hat{\mathbf{n}}$) at the interface.

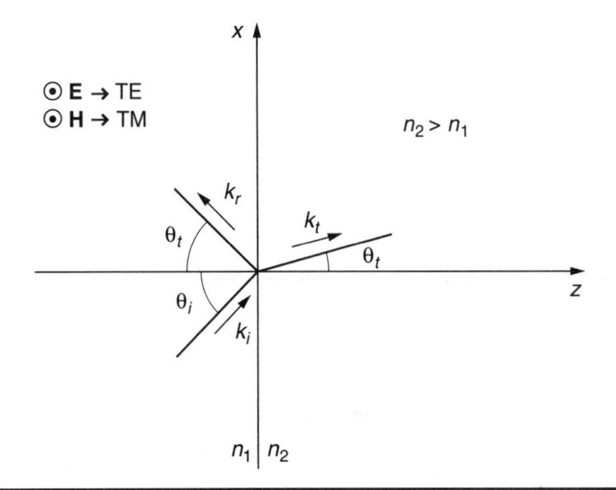

Figure 2.4 Two media of refractive index n_1 and n_2 separated by the xy plane. The subscripts i, t, and r refer to incident, transmitted, and reflected light, respectively.

Recall the tangential components of the fields and normal components of inductions are conserved in the absence of surface currents and charges [Eq. (2.6)].

$$\hat{\mathbf{n}}\times(\mathbf{E}_1-\mathbf{E}_2)=0 \tag{2.35a}$$

$$\hat{\mathbf{n}}\times(\mathbf{H}_1-\mathbf{H}_2)=0 \tag{2.35b}$$

$$\hat{\mathbf{n}}\cdot(\mathbf{D}_1-\mathbf{D}_2)=0 \tag{2.35c}$$

$$\hat{\mathbf{n}}\cdot(\mathbf{B}_1-\mathbf{B}_2)=0 \tag{2.35d}$$

For $\rho = 0$ and $\mathbf{J} = 0$, Maxwell's equation in the **k**-ω representation [Eq. (2.19)] yields

$$\mathbf{H}=\frac{1}{\omega\mu}\mathbf{k}\times\mathbf{E} \tag{2.36a}$$

$$\mathbf{E}=-\frac{1}{\omega\varepsilon}\mathbf{k}\times\mathbf{H} \tag{2.36b}$$

Expanding the cross products in Eq. (2.36), we find that the problem breaks into two independent cases (modes): (1) *transverse electric (TE) mode,* when **E** is perpendicular to the plane of incidence ($\mathbf{E}\,||\,y$) and (2) *transverse magnetic (TM) mode,* when $\mathbf{H}\,||\,y$. We discuss these cases separately.

TE Mode (E||y)

If $\mathbf{E}\,||\,y$, the boundary conditions for the tangent **E**-field and normal **B**-field [Eqs. (2.35*a* and *d*)] are

$$E_{yi}+E_{yr}=E_{yt} \tag{2.37a}$$

$$H_{zi}+H_{zr}=H_{zt} \tag{2.37b}$$

Using Eq. (2.36*b*) to express Eq. (2.37*b*) in terms of E_y components, we can rewrite the system of equations [Eq. (2.37)] as

$$\begin{gathered} k_{xi}\cdot E_{yi}+k_{xr}\cdot E_{yr}=k_{xt}\cdot E_{yt} \\ E_{yi}+E_{yr}=E_{yt} \end{gathered} \tag{2.38}$$

Since Eq. (2.38) must hold for any incident field E_{yi}, we obtain

$$k_{xi}=k_{xr}=k_{xt} \tag{2.39}$$

Expressing the wave vector components in terms of angles with respect to the normal, the result in Eq. (2.39) establishes a very basic result, known as *Snell's law*.

$$n_1 \sin\theta_i = n_1 \sin\theta_r = n_2 \sin\theta_t \tag{2.40}$$

where we used wavenumber k in a medium of refractive index n relates to the wavenumber in vacuum as $k = nk_0$. Note that Eq. (2.40) implies $k_{zi} = -k_{zn}$.

In order to obtain the *field reflection coefficient*, we now use the continuity of tangent **H** components.

$$H_{xi} + H_{xr} = H_{xt} \tag{2.41}$$

This can be expressed in terms of **E** components [via Eq. (2.36*b*)].

$$k_{zi}E_{yi} + k_{zr}E_{yr} = k_{zt}E_{yt} \tag{2.42}$$

Finally, combining Eqs. (2.42) and (2.37*a*) to solve for the **E**-field transmission and reflection coefficients, we obtain

$$\begin{aligned} r_{TE} &= \frac{E_{yr}}{E_{yi}} = \frac{k_{zi} - k_{zt}}{k_{zi} + k_{zt}} \\ t_{TE} &= \frac{E_{yt}}{E_{yi}} = \frac{2k_{zt}}{k_{zi} + k_{zt}} \end{aligned} \tag{2.43}$$

TM Mode (H||y)

Using the analog equations to the TE mode [Eq. (2.37)], the conservation of H_y components and normal **D** components, we find that k_x = constant for TM, as expected, since Snell's law holds in this case as well. In order to obtain the field reflection and transmission coefficients, we use the conservation of both the tangent field components E_x and H_y.

$$\begin{aligned} H_{yi} + H_{yr} &= H_{yt} \\ \frac{k_{zi}}{n_1^2} \cdot H_{yi} + \frac{k_{zr}}{n_1^2} \cdot H_{yr} &= \frac{k_{zt}}{n_2^2} \cdot H_{yt} \end{aligned} \tag{2.44}$$

The $1/n_{1,2}^2$ factor occurs due to the $1/\varepsilon$ factor in Eq. (2.36*b*) ($\varepsilon = n^2$). Thus the **H**-field reflection and transmission coefficients for the TM mode are

$$\begin{aligned} r_{TM} &= \frac{H_{yr}}{H_{yi}} = \frac{k_{zi}/h_1^2 - k_{zt}h_2^2}{k_{zi}/h_1^2 + k_{zt}h_2^2} \\ t_{TM} &= \frac{H_{yt}}{H_{yi}} = \frac{2k_{zt}}{k_{zi}/h_1^2 + k_{zt}h_2^2} \end{aligned} \tag{2.45}$$

We kept r_{TM} and t_{TM} in terms of the **H**-fields to emphasize the symmetry with respect to the TE case. Of course, the quantities can be further expressed in terms of the **E**-fields via $\mathbf{H} = \mathbf{k} \times \mathbf{E}/\omega\mu$. Note that conservation of energy is satisfied in both cases.

$$\begin{aligned} |k_{TE}|^2 + |t_{TE}|^2 &= 1 \\ |k_{TM}|^2 + |t_{TM}|^2 &= 1 \end{aligned} \tag{2.46}$$

Taken together, Eqs. (2.43) and (2.45), referred to as the *Fresnel equations*, provide the reflected and transmitted fields for an arbitrary incident field. Because of the polarization dependence of the reflection and refraction coefficients, polarization properties of light can be modified via reflection and refraction. We discuss two particular cases that follow from the Fresnel equations, where the transmission or reflection coefficient vanishes.

2.1.8 Total Internal Reflection

Setting $t_{TE} = t_{TM} = 0$ yields the same condition for "no transmission" in both TE and TM modes; that is, $k_{zt} = 0$. Thus,

$$\begin{aligned} k_{zt} &= \sqrt{k_t^2 - k_{xt}^2} \\ &= \sqrt{n_2^2 \cdot k_0^2 - n_1^2 \cdot k_0^2 \cdot \sin^2\theta_i} \\ &= 0 \end{aligned} \tag{2.47}$$

where we used the property that k_x = constant [Eq. (2.38)]; that is, $k_{xt} = k_{xi}$. Thus, the transmission vanishes for

$$\theta_c = \sin^{-1}\left(\frac{n_2}{n_1}\right) \tag{2.48}$$

The angle θ_c is referred to as the *critical angle,* at which total *internal reflection* takes place. Note that total internal reflection can occur for both TE and TM polarizations, the only restriction being that $n_2 < n_1$. For angles of incidence that are larger than the critical angle, $\theta_i > \theta_c$ the field reflection coefficient becomes

$$\begin{aligned} r_{TE} &= \frac{k_{zi} - i|k_{zt}|}{k_{zi} + i|k_{zt}|} \\ &= \frac{e^{-i\phi_{TE}}}{e^{i\phi_{TE}}} = e^{-i2\phi_{TE}} \end{aligned} \tag{2.49}$$

Where $\phi_{TE} = \tan^{-1}(|k_{zt}|/k_{zi})$. Thus, for $\theta_i > \theta_c$ the reflection coefficient is purely imaginary—that is, the power is 100% reflected—but the reflected field is shifted in phase by $2\phi_{TM}$. Similarly, for the TM mode we obtain

$$r_{TM} = e^{-i2\phi_{TM}}$$
$$\phi_{TM} = \tan^{-1}\left(\frac{|k_{zt}|}{k_{zi}}\right)\cdot\frac{n_1^2}{n_2^2} \tag{2.50}$$

Since the ϕ_{TM} and ϕ_{TE} have different values, total internal reflection can be used to change the polarization state of optical fields. This basic property has been used in the past to make polarization wave plates.

Note that the transmitted plane wave has the form

$$\begin{aligned} E_t &= E_0 \cdot e^{ik_{zt}\cdot z} \\ &= E_0 \cdot e^{-|k_{zt}|\cdot z} \end{aligned} \tag{2.51}$$

Equation (2.51) indicates that the field in medium 2 is decaying exponentially (Fig. 2.5). Thus the field is significantly attenuated over a distance on the order of $1/k_{zt}$, that is, the field does not propagate, or is *evanescent*.

2.1.9 Transmission at the Brewster Angle

Another particular case of Fresnel equations is when the reflection coefficient vanishes.

For the TE mode, we have

$$r_{TE} = 0 \Rightarrow k_{zi} = k_{zt} \Rightarrow n_1 = n_2 \tag{2.52}$$

Thus, for TE polarization, the only way to obtain 100% transmission through an interface is in the trivial case when there is no refractive

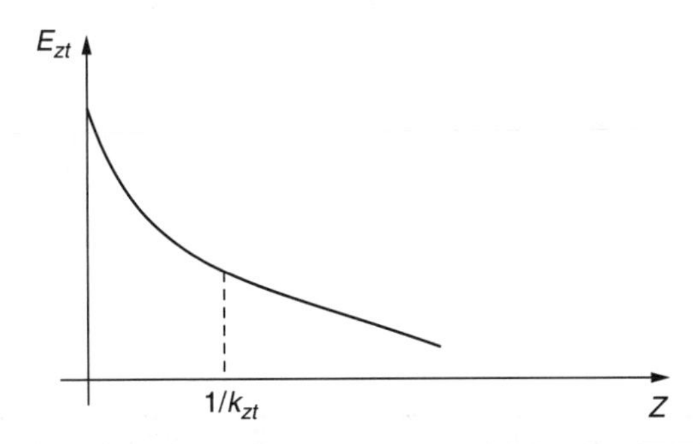

Figure 2.5 Decaying evanescent field.

index contrast between the two media (or when there is no interface at all). However, for the TM mode the situation is very different.

$$r_{TM} = 0 \Rightarrow \frac{k_{iz}}{n_1^2} = \frac{k_{tz}}{n_2^2} \Rightarrow n_2 \cos\theta_i = n_1 \cos\theta_t \tag{2.53}$$

The condition is satisfied simultaneously with Snell's law, such that we have

$$n_1 \sin\theta_i = n_2 \sin\theta_t \tag{2.54a}$$

$$n_2 \cos\theta_i = n_1 \cos\theta_t \tag{2.54b}$$

Multiplying Eqs. (2.54*a* and *b*) side by side, we obtain

$$\sin 2\theta_i = \sin 2\theta_t \Rightarrow \theta_i + \theta_t = \frac{\pi}{2} \tag{2.55}$$

Finally, the angle of incidence at which $r_{TM} = 0$, referred to as the *Brewster angle,* is defined by combining Eq. (2.55) and Eq. (2.54*a*).

$$\tan\theta_B = \frac{n_2}{n_1} \tag{2.56}$$

Thus, unlike with total internal reflection where the *transmission* can vanish for both polarizations, the *reflection* can only vanish in the TM mode. Physically, the absence of reflection at the Brewster angle for TM polarization can be understood by the absence of radiation by an (induced) dipole along its axis (Fig. 2.6). The concept of *induced dipoles* followed by reradiation is essential for the Lorentz model of light-matter interaction, as detailed in the next section.

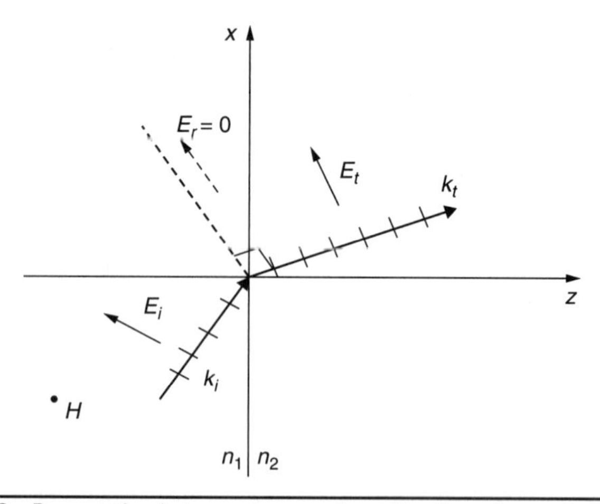

FIGURE 2.6 Propagation between two media at the Brewster angle (TM polarization only).

2.2 The Lorentz Model of Light-Matter Interaction

2.2.1 From Microscopic to Macroscopic Response

In this section we review the main concepts in basic atom-field interactions. In particular, we present the Lorentz model, a prequantum–mechanics description and its asymptotic case for metals known as the Drude model. Despite its simplicity, the Lorentz model explains much of classical optics via a physical picture borrowed from mechanics. Lorentz developed his model of light-atom interaction in the late 19th century. The starting point is the "mass on a spring" description of electrons connected to nuclei. Thus, the incident electric field induces displacement to the electron that is under the influence of a spring–like restoring force due to the nucleus. The equation of motion for the electron can be expressed as

$$\frac{d^2x(t)}{dt^2}+\gamma\frac{dx(t)}{dt}+\omega_0^2x(t)=\frac{-e}{m}E(t) \tag{2.57}$$

where γ is the damping constant, ω_0 is the resonant frequency, e is the electronic charge, m is the mass of the electron, and E is the incident field. Note that γ and ω_0 are characteristic of the material, the first describing the energy dissipation property of the medium and the second the ability of the medium to store energy. Since Eq. (2.57) is a linear differential equation, Fourier transforming both sides of the equation provides straightforward access to the frequency-domain solution. Using the Fourier property of the differential operator,

$$\frac{d^n}{dt^n}\rightarrow(i\omega)^n \tag{2.58}$$

we obtain Eq. (2.58) in the frequency domain as:

$$-\omega^2x(\omega)+i\gamma\omega x(\omega)+\omega_0^2x(\omega)=-\frac{e}{m}E(\omega) \tag{2.59}$$

Thus, from Eq. (2.59), we readily find the solution for the charge displacement in the frequency domain as

$$x(\omega)=\frac{(e/m)E(\omega)}{\omega^2-i\gamma\omega-\omega_0^2} \tag{2.60}$$

In order to obtain the time domain solution $x(t)$, we need to Fourier-transform Eq. (2.60) back to the time domain. However, here we explore further the frequency domain solution. The induced dipole moment due to the charge displacement $x(\omega)$ is, by definition,

$$\mathbf{p}(\omega)=-e\cdot\mathbf{x}(\omega) \tag{2.61}$$

So far, in Eqs. (2.59 to 2.61) we obtained *microscopic* quantities, that is, the atomic-level response of the system. The *macroscopic* behavior of the medium is obtained by evaluating the induced polarization **P**, which captures the contribution of all dipole moments within a certain volume.

$$\mathbf{P} = N\langle \mathbf{p} \rangle \tag{2.62}$$

In Eq. (2.62), N is the volume concentration of dipoles (in m^{-3}) and the angular brackets denote the ensemble average. Assuming that all induced dipoles are parallel within the volume, we obtain

$$P(\omega) = \frac{Ne^2}{m} \cdot \frac{E(\omega)}{\omega_0^2 + i\gamma\omega - \omega^2} \tag{2.63}$$

Note that, generally, each atom has multiple resonances, or *dipole-active modes,* such that Eq. (2.63) can be generalized to

$$P(\omega) = \frac{Ne^2}{m} \cdot \sum_i \frac{\beta_i E(\omega)}{\omega_{0i}^2 - \omega^2 + i\gamma_i\omega} \tag{2.64}$$

where the summation is over all modes i, characterized by different resonant frequencies ω_{0i} and damping constants γ_i. The weight β_i is called the *oscillator strength* and has the quantum mechanical meaning of a *transition strength.* With this in mind, we reverse to the single normal mode description, which captures the origin of absorption and refraction of materials. The induced polarization only captures the contribution of the medium itself; that is, it excludes the vacuum contribution. Thus,

$$\mathbf{P} = \varepsilon_0 \chi \mathbf{E} \tag{2.65a}$$

$$\chi = \varepsilon_r - 1 = n^2 - 1 \tag{2.65b}$$

In Eq. (2.65*a*), χ is called the *dielectric susceptibility*, which is generally a *tensor* quantity. However, for isotropic media, we obtain the complex scalar permittivity $\varepsilon_r = \varepsilon_r''(\omega) + i\varepsilon_r''(\omega)$.

$$\varepsilon_r(\omega) - 1 = \frac{Ne^2}{m\varepsilon_0} \cdot \frac{1}{\omega_0^2 - \omega^2 + i\gamma\omega} \tag{2.66}$$

The prefactor $Ne^2/m\varepsilon_0$ in Eq. (2.66) has units of frequency squared, ω_p^2, and ω_p is referred to as the *plasma frequency*. The physical meaning of this frequency is discussed a bit later. From Eq. (2.66), we readily obtain the real and imaginary parts of ε_r.

$$
\begin{aligned}
\varepsilon_i'(\omega) &= 1+\omega_p^2\frac{\omega_0^2-\omega^2}{(\omega_0^2-\omega^2)+\gamma^2\omega^2} \\
\varepsilon_r'(\omega) &= \omega_p^2\frac{\gamma\omega}{(\omega_0^2-\omega^2)^2+\gamma^2\omega^2}
\end{aligned}
\tag{2.67}
$$

Figure 2.7 illustrates the main features of ε' and ε'' versus frequency. In order to gain further physical insight into Eq. (2.67), we discuss three different frequency regions.

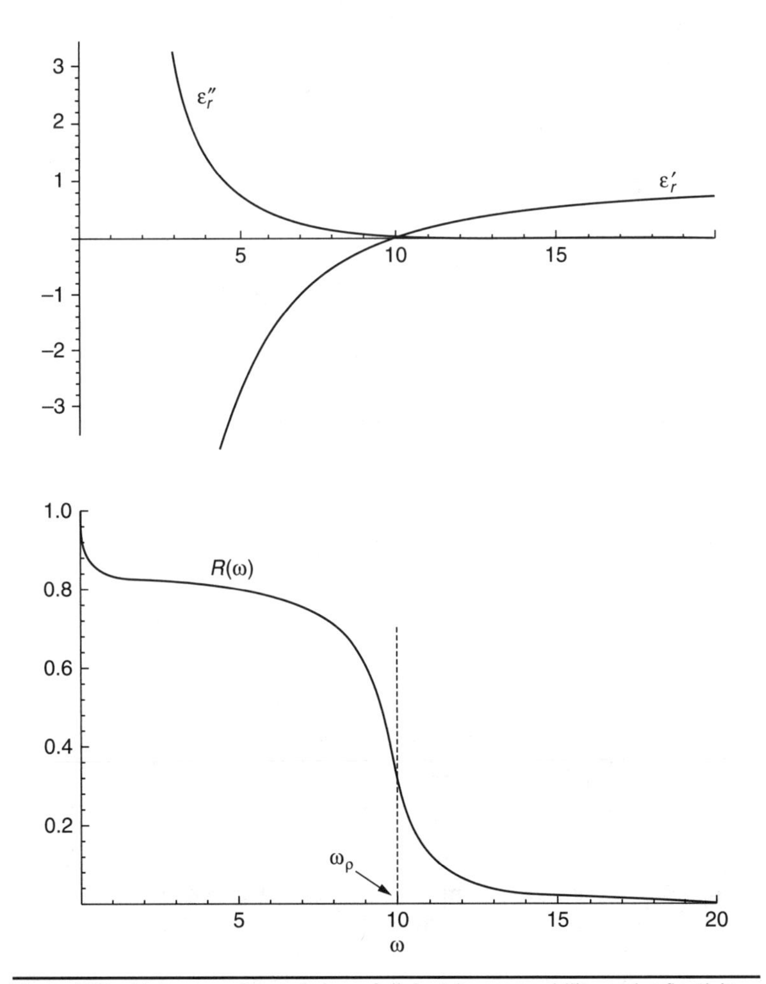

FIGURE 2.7 Frequency dependence of dielectric permeability and reflectivity around the plasma frequency ($\omega_p = 10$, $\gamma = 1$).

2.2.2 Response below the Resonance $\omega << \omega_0$

In this case, Eq. (2.67) simplifies to

$$\begin{aligned}\varepsilon_r'(\omega) &\simeq 1+\frac{\omega_p^2}{\omega_0^2}\cdot\frac{1}{1-2(\omega/\omega_0)^2}\\ \varepsilon_r''(\omega) &\simeq \frac{\omega_p^2\cdot\gamma\omega}{\omega_0^4}\cdot\frac{1}{1-2(\omega/\omega_0)^2}\end{aligned} \tag{2.68}$$

Since $\gamma\omega << \omega_0^2$, then $\varepsilon' > \varepsilon''$; this indicates that absorption is negligible, or equivalently, that below the resonance the material is *transparent*. Further note that $d\varepsilon'/d\omega > 0$; this by definition, defines a region of *normal dispersion*. In fact, it can be seen that expanding further the denominator in Eq. (2.68), we obtain $\varepsilon_r'(\omega) \propto \omega^2$ and $\varepsilon_r''(\omega) \propto \omega^3$. This explains why, for transparent materials, the refractive index dependence on frequency is often approximated by a quadratic function. Of course, as we approach resonance, this dependence becomes more complicated.

2.2.3 Response at the Resonance, $\omega \simeq \omega_0$

For frequencies that are comparable to ω_0, Eq. (2.67) is well approximated by

$$\begin{aligned}\varepsilon_r'(\omega) &\simeq 1+\frac{\omega_p^2}{2\omega_0}\cdot\frac{\omega-\omega_0}{1+\left(\frac{\omega-\omega_0}{\gamma/2}\right)^2}\\ \varepsilon_r''(\omega) &\simeq \frac{\omega_p^2}{2\omega_0}\cdot\frac{\gamma/2}{1+\left(\frac{\omega-\omega_0}{\gamma/2}\right)^2}\end{aligned} \tag{2.69}$$

It can be seen that under these conditions, the absorption is significant, $\omega-\omega_0 < \gamma$, and the absorption line has a characteristic shape, called, not surprisingly, the *Lorentzian line*. This shape is characterized by a central frequency ω_0 and a full-width half maximum of γ. While ω_0 has a clear physical significance of the frequency at which the system "resonates" or absorbs strongly, the meaning of γ is somewhat more subtle. The damping constant γ represents the average frequency at which electrons collide inelastically with atoms, and so induce loss of energy. Thus, $\gamma = 1/\langle\tau_{col}\rangle$, where τ_{col} is the average time between collisions. Finally, around resonance, $d\varepsilon'/d\omega < 0$; this defines *anomalous dispersion*.

2.2.4 Response above the Resonance, $\omega >> \omega_0$

Well above the resonance, the following equation applies:

$$\varepsilon_r'(\omega) \simeq 1+\omega_p^2 \cdot \frac{1}{\omega^2+\gamma^2}$$
$$\varepsilon_r''(\omega) \simeq 1+\omega_p^2 \cdot \frac{\gamma/\omega}{\omega^2+\gamma^2} \tag{2.70}$$

Again, the absorption becomes less significant, as expected in a frequency range away from the resonance. Similarly, the dispersion is normal again; $d\varepsilon'/d\omega > 0$.

In sum, the Lorentz oscillatory model provides great insight into the classical light-matter interaction. In the following section, we will investigate the particular situation of metals, when the charge moves freely within the material.

2.3 Drude Model of Light-Metals Interaction

The optical properties of metals were first modeled by Drude in the context of conductivity. In highly conductive materials, the restoring force in Eq. (2.57), $m\omega_0^2 x$, vanishes; this establishes that the charge can move freely. Under these conditions, we obtain *Drude's model*, in which Eq. (2.67) reduces to ($\omega_0 \to 0$).

$$\varepsilon_r'(\omega) = 1-\frac{\omega_p^2}{\omega^2+\gamma^2}$$
$$\varepsilon_r''(\omega) = \frac{\gamma}{\omega} \cdot \frac{\omega_p^2}{\omega^2+\gamma^2} \tag{2.71}$$

Typically $\gamma = 1/\langle \tau_{col} \rangle << \omega$; that is, the frequency of collisions is much lower than that of optical frequencies. In this high frequency limit, $\varepsilon_r'(\omega) \simeq 1-\omega_p^2/\omega^2$ and $\varepsilon_r''(\omega) \simeq \omega_p^2 \gamma/\omega^3$. From the Fresnel equations, we can derive the power reflection coefficient. Thus, for normal incidence, the intensity-based reflectivity is

$$R(\omega) = \left| \frac{n-1}{n+1} \right|^2 \tag{2.72}$$

where n is the (complex) refractive index. Since $n = \sqrt{\varepsilon_r}$, Eq. (2.72) becomes

$$R(\omega) = \left| \frac{\sqrt{\varepsilon_r' + i\varepsilon_r''} - 1}{\sqrt{\varepsilon_r' + i\varepsilon_r''} + 1} \right|^2 \tag{2.73}$$

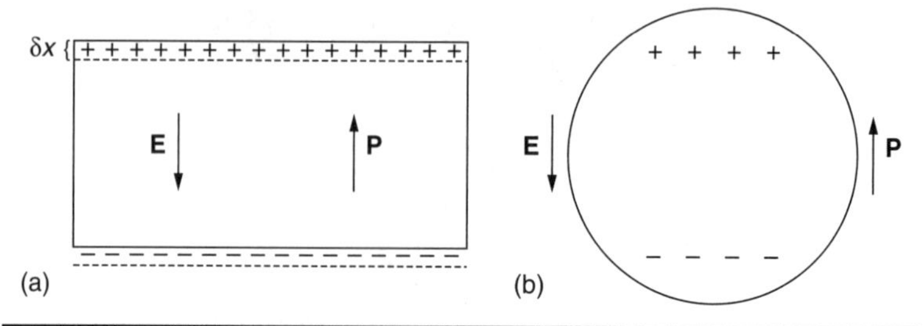

FIGURE 2.8 Polarization induced in a thin layer of metal (*a*) and a nanoparticle (*b*) at plasma frequency.

Figure 2.7 illustrates the frequency dependence of ε', ε'', n', n'', and R for various values of ω_p and γ.

Interestingly, at $\omega = \omega_p$, $\varepsilon_r'(\omega)$ vanishes. In order to gain a physical understanding of the plasma frequency ω_p, let us consider a thin film of metal (Fig. 2.8*a*). The applied electric field induces a polarization $\mathbf{P} = \varepsilon_0 \chi \mathbf{E}$, with $\chi = \varepsilon_r - 1$. Tuning the frequency of the incident field to the plasma frequency, $\varepsilon_r = 0$ $\chi = -1$ and $\mathbf{P} = -\varepsilon_0 \mathbf{E}$. On the other hand, the induced polarization is the total charge times the displacement per unit volume,

$$P = -N \cdot e\delta x \tag{2.74}$$

Therefore, the electric field is

$$\begin{aligned} E &= -\frac{P}{\varepsilon_0} \\ &= \frac{Ne}{\varepsilon_0} \cdot \delta x \end{aligned} \tag{2.75}$$

If we construct the electric force due to the charge displacement, $F = -eE$, we obtain

$$\begin{aligned} F &= -\frac{Ne^2}{\varepsilon_0} \cdot \delta x \\ &= -k_e \cdot \delta x \end{aligned} \tag{2.76}$$

In Eq. (2.76), we define k_e as the *spring* constant of the restoring force. Thus, by definition, the system is characterized by a resonant frequency, $\omega_p = \sqrt{k_e/m}$. This is the plasma frequency associated with the thin film.

$$\omega_p = \sqrt{\frac{Ne^2}{m\varepsilon_0}} \tag{2.77}$$

From Maxwell's equations, we have that $\nabla \times \mathbf{H} = \partial D/\partial t = \varepsilon_0 \partial E/\partial t + \partial P/\partial t$. At plasma frequency, $P = -\varepsilon_0 E$, and thus, the magnetic field vanishes. This indicates that there is no bulk propagation of the electromagnetic field. The quantum of energy associated with these charge oscillations at plasma frequency is $\hbar\omega_p$, and the respective quantum particle is called *plasmon*. Light scattering on plasma oscillations, including in nanoparticles (Fig. 2.8*b*), can be described as a *photon-plasmon interaction*.

Suggested Reading

Bass, M., and Optical Society of America, *Handbook of Optics*, McGraw-Hill, New York, 1995.

Born, M., and E. Wolf, *Principles of Optics: Electromagnetic Theory of Propagation, Interference and Diffraction of Light*, Cambridge University Press, Cambridge, New York, 1999.

Bracewell, R. N., *The Fourier Transform and Its Applications*, McGraw-Hill, Boston, 2000.

Feynman, R. P., R. B. Leighton, and M. L. Sands, *The Feynman Lectures on Physics*, Pearson/Addison-Wesley, San Francisco, 2006.

Jackson, J. D., *Classical Electrodynamics*, Wiley, New York, 1999.

Landau, L. D., E. M. Lifshits, and L. P. Pitaevskii, *Electrodynamics of Continuous Media*, Pergamon, Oxford and New York, 1984.

CHAPTER 3

Introduction to Nanophotonics

Logan Liu
Department of Electrical and Computer Engineering
University of Illinois at Urbana–Champaign

3.1 Overview

An area in nanoscience called *nanophotonics* is defined by the National Academy of Science as "the science and engineering of light matter interactions that take place on wavelength and subwavelength scales where the physical, chemical, or structural nature of natural or artificial nanostructured matter controls the interactions." Nanophotonics is an emerging and quickly growing field that incorporates engineering, physics, and chemistry. Nanophotonics first starts with the subwavelength optical visualization of sample surfaces using near-field scanning optical microscopy. Recently, the principles of nanophotonics have also been extended into new active research areas such as the luminescent semiconductor quantum dots and metallic plasmonic nanoparticles. As a matter of fact, nanophotonics may be also found in many natural structures such as the colorful peacock feathers and butterfly wings that contain nanoscale array structures reflecting specific wavelengths of light. The materials used in nanophotonics are quite diverse; however, only the materials involved in the discussed aspects will be addressed. In general, most of the nanoscale materials discussed here are nanoparticles. These particles range in size from 2 to 100 nm in diameter. They are made up of many different substances such as metals, semiconductors, and metallic oxides. They are aggregates of atoms in a relatively crystalline form. Several different metals are used in nanophotonics. Gold and silver are both used to make nanoplasmonic structures that act as optical resonators and near-field light condensers.

Compound semiconductor nanoparticles are used for their quantum confinement and luminescent properties. The introduction will start with laying out the fundamental physical laws behind nanophotonics, especially the photon–electron interaction at nanoscale. In the following the quantum nanophotonic structures are discussed including semiconductor quantum dots and nanoplasmonic particles. At the end, the modeling and simulation methods of nanophotonic devices, particularly nanoplasmonic structures, will be introduced.

3.2 Foundation of Nanophotonics

The dimensions of nanophotonic devices are much smaller than the optical wavelength, so the light-matter interactions in nanophotonic devices involve the photon–electron interactions within the quantum confinement region. In the following, the analogy between the nanoscale physics of photons and electrons will be drawn to manifest the significance of photon–electron interactions.

The periodic electron movements within nanoscale boundaries can lead to an electromagnetic wave, and if the movements are at optical frequencies, photons are emitted. On the other hand, the electrons within nanoscale boundaries may be moved upon light excitations, and the movements can follow the same frequency as light. As a matter of fact, the wavelengths of light and electrons have similar definitions. The wavelength of a light wave is defined as $\lambda = h/p = c/\nu$, and the wavelength of an electron wave is defined as $\lambda = h/p = h/mv$, where c is the speed of light, ν is the optical frequency, m is the electron mass, and v is the electron velocity. In addition, basic equations describing propagation of photons in dielectrics have some similarities to those for propagation of electrons in crystals. The electromagnetic wave equation is

$$\frac{-h^2/4\pi^2}{2m}[\nabla \cdot \nabla + V(r)]\psi(r) = E_n \psi(r)$$

and Schrödinger's electronic wave equation is $\nabla \cdot \nabla E(r) = \varepsilon k_0^2 E(r)$. The free-space solutions to the above two equations share the same format as $E = E^0(e^{ik \cdot r - \omega t} + e^{-ik \cdot r + \omega t})$ and $\psi = c(e^{ik \cdot r - \omega t} + e^{-ik \cdot r + \omega t})$.

The propagations of light and electrons within bounded structures such as nanophotonic devices can be analyzed in a similar fashion. The propagation of light is affected by the dielectric property of the boundary material, and the propagation of electrons is affected by the Coulomb potential of the boundaries. As shown in Fig. 3.1, photons can tunnel through classically forbidden zones into the boundary as the exponentially decaying evanescent wave with an imaginary k vector. Electron wave function also tunnels into the potential boundary and decays exponentially in forbidden zones.

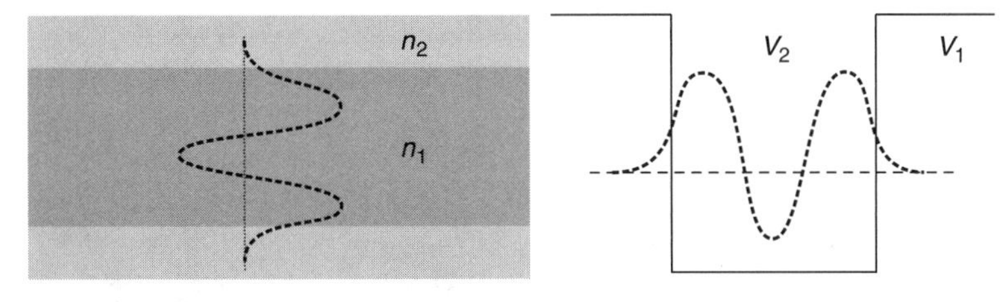

FIGURE 3.1 Bounded optical wave function (left) and electronic wave functions in potential well structures.

Confinement of light results in field variations similar to the confinement of electrons in a potential well when $V_1 > V_2$. For light, the analog of a potential well is a region of a high refractive index bounded by a region of a lower refractive index, that is, $n_1 > n_2$. The dispersion relationship of electrons and photons is nevertheless different, as shown in Fig. 3.2. For photons, energy $E = pc = (h/2\pi)kc$ is in a linear relationship with wave vector k while for electrons the energy$E = p^2/2m = (h/2\pi)^2 k^2/2m$ is in a quadratic relationship with wave vector k.

3.3 Quantum Confinement in Nanophotonics

Quantum-confined materials refer to structures that are constrained to nanoscale lengths in one, two, or all three dimensions, as shown in Fig. 3.3. The length along which there is quantum confinement must be smaller than the de Broglie wavelength of electrons for thermal energies in the medium. For effective quantum confinement, one or more dimensions are usually less than 10 nm. Structures that are quantum-confined show a strong effect on their optical properties. Artificially created structures with quantum confinement on one, two, or three dimensions are called *quantum wells, quantum wires,* and *quantum dots,* respectively. Nanoscale confinement in one dimension results in a quantum well. The quantization of energy into discrete levels has applications for fabrication of new nanoscale solid-state lasers. Two or more quantum wells side by side give rise to a multiple quantum well (MQM) structure.

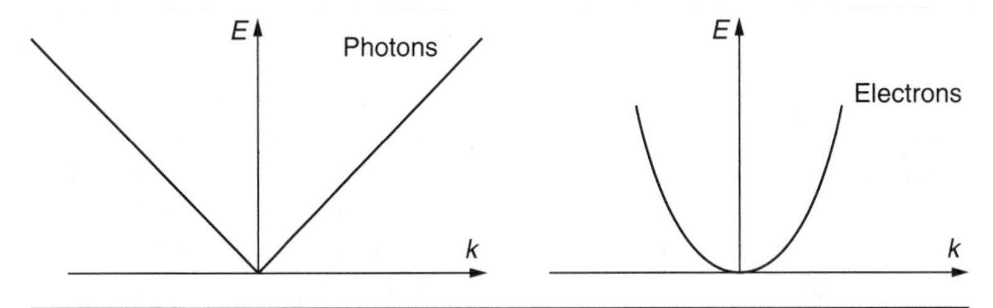

FIGURE 3.2 Dispersion characteristics of photons (left) and electrons (right).

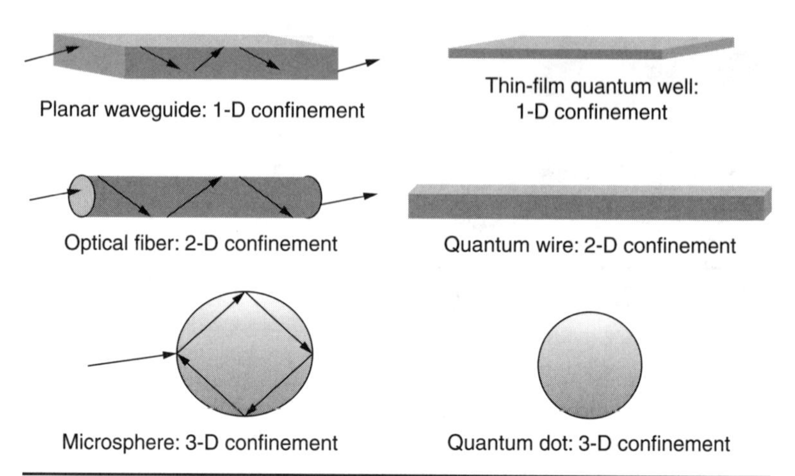

FIGURE 3.3 Confinement of light in microphotonic devices and electrons in nanophotonic devices.

In general, confinement produces a blue shift of the bandgap. Location of discrete energy levels depends on the size and nature of confinement. This implies an increase of optical transition probability. This happens any time the energy levels are squeezed into a narrow range, resulting in an increase of energy density. The oscillator strengths increase as the confinement increases from bulk to quantum well to quantum wire to quantum dot. Semiconductor quantum dots as three-dimensional quantum confinement nanostructures have attracted significant attention. The confined electron wave functions have the resonance energy defined by the following equation based on the model of "a particle in a box."

$$E_{n_1,n_2,n_3} = E_C + \frac{h^2}{8m_e^*}\left(\frac{n_1^2}{L_x^2} + \frac{n_2^2}{L_y^2} + \frac{n_3^2}{L_z^2}\right)$$

If the quantum dots are made of direct bandgap semiconductor material, the photon emission due to the electron-hole pair recombination process is possible. Specifically, II–VI and III–V direct bandgap semiconductors are the most common ones under investigation. These substances exhibit an interesting property when in the size range under 10 nm in diameter that is comparable to the material Bohr radius. Once the particles are smaller than this threshold size, the color of fluorescence is determined by the size, not only the material, as shown in Fig. 3.4. As these nanoparticles decrease in size, the color shifts toward the blue end of the visible spectrum owing to the higher resonance energy. Different substances have different spectrums, so indium phosphate which fluoresces red in bulk form will shift with the particle size down to the yellow/green region of the spectrum. If a substance fluoresces yellow in bulk form, it is possible

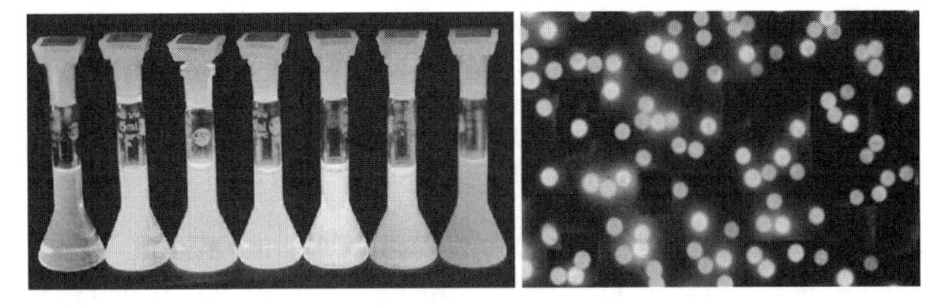

FIGURE 3.4 Fluorescent photographs of CdSe quantum dots solutions (left) and quantum dots stained microparticles (right).

that the smaller particles may even reach the blue/violet end of the spectrum. The semiconductor quantum dots such as Cadmium Selenide (CdSe) nanoparticles have been used in many biomedical applications as fluorescence labels.

3.4 Plasmonics

Plasmon is defined as the "collective and periodic free electron movement in the metallic structures under the electromagnetic excitation." As depicted in Fig. 3.5, the free electrons in the metal conduction band are not bounded to the nucleus, so they can move in the applied alternating electromagnetic field within the physical boundary of metal materials, i.e., an alternating electron current. Due to the electron-lattice collision and the wave dispersion in metal, the free electron movements are subject to the resistance, and electron kinetic energy

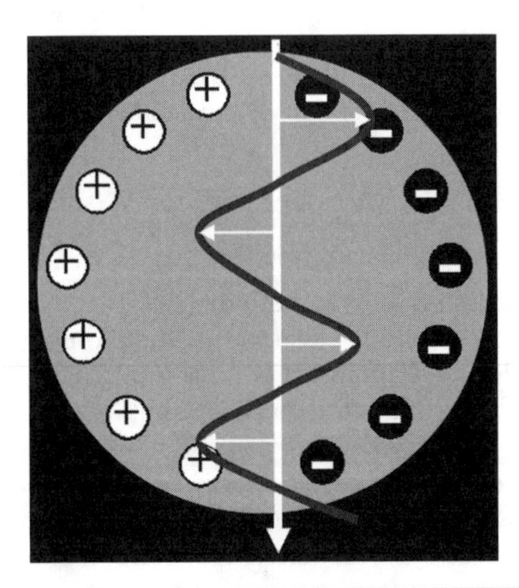

FIGURE 3.5 Schematic diagram of plasmon resonance in a spherical metallic nanoparticle. The free electrons oscillate with the undulating electromagnetic wave.

will be dissipated. As in many other mechanical movements, there exist the plasmon resonance modes with minimal energy damping. If the plasmon resonance modes fall into the optical frequency range, the alternating free electrons form an optical dipole to reemit light. A detailed discussion on nanoplasmonics can be found in Chapter 4.

3.4.1 Optical Field Enhancement and Concentration

Since the electrons in nanophotonic structures can be manipulated spatially and temporarily by an optical field with minimal energy damping, the time-averaged local electron density in nanophotonic structures becomes controllable by intentionally designing the nanostructure geometries. For instance, in nanoplasmonic structures, the electron density can be artificially tuned to be concentrated at the location with very small surface areas such as sharp tips, corners, and contact points, so the reemitted local electromagnetic field strength at these positions becomes enhanced for orders of magnitude in comparison with the excitation field strength.

The enhanced electromagnetic field in nanoplasmonic structures provides a powerful excitation source in molecular optical spectroscopy. Raman scattering and fluorescence emission can be enhanced for several orders of magnitudes, since the light emission intensities in the electronic energy level transitions are proportional to the excitation power. Surface-enhanced Raman scattering (SERS) was discovered in the 1970s[1], and later became a powerful bioanalytical tool; many theoretical works on SERS effects were reported [2–22]. Although the complete theoretical understanding of the SERS phenomenon is still at large, consensus has been reached that the local electromagnetic field enhancement on nanoplasmonic structures plays a major role in SERS. The radiating dipole moment of the nanoplasmonic structures can be described as $p = \varepsilon_m \alpha E_{\text{inc}} e^{-i\omega t}$, where p is the dipole moment, α is the polarizability defined previously, and E_{inc} is the incident electric field. The scattering field from the nanoplasmonic structures is defined as

$$E_{\text{sca}} = \frac{e^{ikr}}{-ikr} \frac{ik^3}{4\pi\varepsilon_m} \hat{r} \times (\hat{r} \times \vec{p})$$

Incident radiation interacts with both the metal to create a surface plasmon and the target molecule where the variations in the vibrational levels of the molecule result in a photon of unidentical frequency being returned to the metal and scattered. The combination of incident reradiation and the vibrational energy of the molecule results in significantly increased scattering power observed as surface-enhanced Raman scattering. The Raman scattering enhancement factor ρ is proportional to the fourth order of the local field amplitude enhancement.

$$\rho(r,\omega) = \left| \frac{E_{\text{inc+sca}}(r,\omega)}{E_{\text{inc}}(r,\omega)} \right|^4$$

In addition to the electromagnetic enhancement mechanism which is active during the SERS process, there is a chemical enhancement mechanism that contributes considerably to the observed SERS signal. The chemical enhancement mechanism involves the incident radiation striking the nanoplasmonic SERS substrate, resulting in a photon being excited within the metal to a higher energy level. From this excited state, a charge transfer process to a vibrational level of the same energy within the target molecule takes place. Variations in vibrational energy states occur resulting in the transfer of a photon of different frequency being passed back to the metallic energy levels and returned to the ground state of the metal.

3.5 Nanophotonic Characterization with Near-Field Optics

Near-field scanning optical microscopy (NSOM) is a microscopic technique for nanostructure investigation that breaks the far-field resolution limit by directly detecting the near-field emission or scattering. This is done by placing the detector very close (lots less than one wavelength, typically a few nanometers) to the specimen surface. This allows for surface optical inspection with a high spatial, spectral, and temporal resolving power. One can use NSOM to characterize the near-field scattering intensity from the nanophotonic device surface and establish the two-dimensional optical field distributions. NSOM can be operated in both an aperture mode and a nonaperture mode. As illustrated in Fig. 3.6, the tips used in the apertureless mode are very sharp and do not have a metal coating.

The aperture-mode NSOM system resembles an atomic force microscopy system. The primary components of an NSOM setup include the light source, feedback mechanism, scanning tip, detector, and piezoelectric sample stage. One can use the same laser used to excite the SERS signal as the light source in NSOM measurements. The laser light will be focused into an optical fiber through a polarizer, a beam splitter, and a coupler. The polarizer and the beam splitter

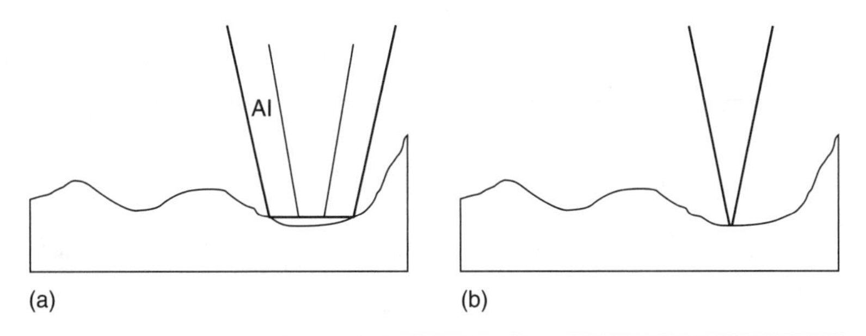

Figure 3.6 Configurations of the NSOM tip in an (*a*) aperture mode and (*b*) nonaperture mode.

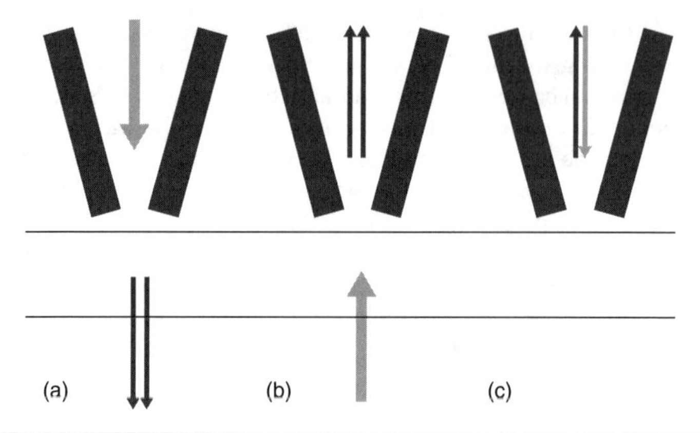

Figure 3.7 Illustrations of three operational methods in an aperture-mode NSOM. (*a*) Near-field excitation, far-field collection; (*b*) Far-field excitation, near-field collection; (*c*) Near-field excitation, near-field collection.

serve to remove stray light from the returning reflected light. The scanning tip is an etched optical fiber coated with metal except at the tip. An avalanche photo diode (APD) is usually used to pick up the scattering light. For the same system configurations, there are still many different ways of near-field excitation and detection. Figure 3.7 illustrates three possible methods that can be applied in the nanophotonic characterization experiments: near-field illumination, far-field collection; far-field illumination, near-field collection; and near-field illumination, near-field collection.

3.6 Computation and Simulation in Nanophotonics

The above analysis of nanophotonic devices comprises analytical approaches that can be applied well to nanophotonic structures with regular geometries such as spheres and cubes. However, the analytical approach is typically hampered by oversimplifying assumptions and limited by irregular geometries. In addition, at the nanoscale, many material properties are different from those in the bulk scale. Numerical computation and simulation are important to analyzing and designing of nanophotonic devices in arbitrary shapes. The numerical simulations in nanophotonics include frequency-domain and time-domain simulations. The frequency-domain simulations generally produce steady-state solutions and the time-domain simulations can produce transient solutions. These two simulations have their own advantages and disadvantages which are discussed in great detail next. As an example, the analytical approach on the nanoplasmonic particle scattering problem will be presented based on the Mie scattering theory, which will be followed by the frequency- and time-domain simulations of nanoplasmonic structures.

3.6.1 Mie Scattering Theory

The plasmon resonance wavelength of metallic nanoparticles is represented as optical absorption or scattering peak wavelength in far-field measurements. The optical cross-sectional absorbance of the single Au nanosphere may be calculated using the Mie scattering theory with known refractive indices of Au and the surrounding medium as well as the diameter of the Au nanosphere. In fact, the Mie scattering theory may be used to calculate the scattering cross section for any nanoparticles. The refractive index of Au is a complex value, dramatically varying with incident wavelength in the optical range. The complex refractive index N may be calculated from the complex dielectric constant ε as

$$N = \sqrt{\varepsilon}$$

According to Mie scattering theory, the optical coefficients in single spherical gold particle scattering are given by

$$Q_{\text{sca}} = \sum_{n=1}^{\infty} \frac{2}{x^2}(2n+1)(|a_n|^2 + |b_n|^2)$$

$$Q_{\text{ext}} = \sum_{n=1}^{\infty} \frac{2}{x^2}(2n+1)\,\text{Re}(a_n + b_n)$$

$$Q_{\text{abs}} = Q_{\text{ext}} - Q_{\text{sca}}$$

where Q_{sca}, Q_{ext}, and Q_{abs} are the efficiencies of scattering, extinction, and absorptions, and a_n and b_n represent the partial wave coefficients, defined as

$$a_n = \frac{N_r \psi_n(N_r x)\psi'_n(x) - \psi_n(x)\psi'_n(N_r x)}{N_r \psi_n(N_r x)\xi'_n(x) - \xi_n(x)\xi'_n(N_r x)}$$

$$b_n = \frac{\psi_n(N_r x)\psi'_n(x) - N_r \psi_n(x)\psi'_n(N_r x)}{\psi_n(N_r x)\xi'_n(x) - N_r \xi_n(x)\psi'_n(N_r x)}$$

Here ψ_n are the Riccati-Bessel functions for the first-order spherical Bessel function, defined as

$$\psi_n(x) = x j_n(x)$$

ξ_n is the Riccati-Bessel function for the first-order spherical Hankel function, defined as

$$\xi_n(x) = x h_n^{(1)}(x)$$

and N_r is the ratio of the refractive index of the Au nanosphere N_{Au} to the refractive index of the medium N_m.

The single nanoparticle optical parameters can be correlated to the bulk solution parameter by

$$\sigma_{i,\lambda} = Q_{i,\lambda} A \qquad \mu_{i,\lambda} = \rho\sigma_{i,\lambda}$$
$$i = \text{sca}, \quad \text{ext}, \quad \text{abs}$$

Where A and σ_{abs} are, respectively, geometrical and optical cross sections, μ is the optical coefficient, and ρ is the particle density.

The absorption of the colloid suspension can be calculated by only considering isotropic scattering. Using Beer's law for homogeneous suspensions, the relative total transmittance T is exponentially proportional to the absorption coefficient $\mu_{\text{abs},\lambda}$, and the relative total absorbance R may then be calculated as

$$T_\lambda = \exp(-\mu_{\text{abs},\lambda} L) \qquad R_\lambda = 1 - T_\lambda$$

If another layer forms a shell surrounding the Au nanospheres, which is the case of molecule binding to the Au nanosphere, the partial wave coefficients are given by

$$a_n = \frac{\psi_n(x_2)[\psi_n'(N_2x_2) - A_n\chi_n'(N_2x_2)] - N_2\psi'(x_2)[\psi_n(N_2x_2) - A_n\chi_n(m_2x_2)]}{\xi_n(x_2)[\psi_n'(N_2x_2) - A_n\chi_n'(N_2x_2)] - N_2\xi_n'(x_2)[\psi_n(N_2x_2) - A_n\chi_n(m_2x_2)]}$$

$$b_n = \frac{N_2\psi_n(x_2)[\psi_n'(N_2x_2) - B_n\chi_n'(N_2x_2)] - \psi_n'(x_2)[\psi_n(N_2x_2) - B_n\chi_n(N_2x_2)]}{N_2\xi_n(x_2)[\psi_n'(x_2) - B_n\chi_n'(N_2x_2)] - \xi_n'(x_2)[\psi_n(N_2x_2) - B_n\chi_n(N_2x_2)]}$$

Here χ_n is a second-order spherical Neumann function; A_n and B_n are defined as

$$A_n = \frac{N_2\psi_n(N_2x_1)\psi_n'(N_1x_1) - N_1\psi_n'(N_2x_1)\psi_n(N_1x_1)}{N_2\chi_n(N_2x_1)\psi_n'(N_1x_1) - N_1\chi_n'(N_2x_1)\psi_n(N_1x_1)}$$

$$B_n = \frac{N_2\psi_n'(N_2x_1)\psi_n(N_1x_1) - N_1\psi_n(N_2x_1)\psi_n'(N_1x_1)}{N_2\chi_n'(N_2x_1)\psi_n(N_1x_1) - N_1\chi_n(N_2x_1)\psi_n'(N_1x_1)}$$

$$N_1 = \frac{N_c}{N_m} \qquad N_2 = \frac{N_s}{N_m} \qquad x_1 = \frac{2\pi N_m a_c}{\lambda} \qquad x_2 = \frac{2\pi N_m a_{cs}}{\lambda}$$

Here N_c and N_s are refractive indices of the Au core and dielectric shell for coated nanospheres; a_c and a_{cs} are the radii of the core and whole particles.

3.6.2 Maxwell Equations

The surface plasmon wave on metallic thin film or any arbitrary nanoplasmonic structures can be calculated using the electrodynamic theories or the Maxwell equations. The aforementioned Mie scattering formulas were derived from the Maxwell equations in the case of spherical coordination and harmonic waves. The general forms of Maxwell equations are as follows:

$$\nabla \cdot D = \rho, \qquad \nabla \cdot B = 0$$

$$\nabla \times H = J + \frac{\partial}{\partial t} D, \qquad \nabla \times E = -\frac{\partial}{\partial t} B$$

$$D = \varepsilon_0 E + P, \qquad H = \mu_0^{-1} B - M$$

$$J = \sigma E, \qquad B = \mu\mu_0 H, \qquad P = \varepsilon_0 \chi E$$

Here D is the electric displacement, E is the electric field, B is the magnetic field, H is the magnetic induction, ρ is the charge density, J is the current density, P is the electric polarization, M is the magnetization, and σ, μ, and χ are the conductivity, permittivity, and electric susceptibility, respectively.

The surface plasmon wave at a planar metallic thin film can be solved by using the above Maxwell equations. Let's assume the surface plasmon wave as a conventional planar electromagnetic wave in transverse electric mode

$$E_1 = E_{x1}\hat{x} \exp[-k_x \cdot x + i(k_z z - \omega t)]$$

$$E_2 = -E_{x2}\hat{x} \exp[k_x' \cdot x + i(k_z z - \omega t)]$$

$$H_1 = H_{y1}\hat{y} \exp[-k_x \cdot x + i(k_z z - \omega t)]$$

$$H_2 = H_{y1}\hat{y} \exp[-k_x' \cdot x + i(k_y z - \omega t)]$$

Here subscripts 1 and 2 represent the parameters in the dielectric material and metal, respectively, and x is defined as the direction perpendicular to the interface. According to the Maxwell equations, the propagating wave vector k is

$$\nabla \times E = -\frac{\partial B}{\partial t} = -\mu \frac{\partial H}{\partial t}$$

We can then obtain

$$(ik_z\hat{z} - k_x\hat{x}) \times (E_{x1}\hat{x}) = i\omega\mu H_{y1}\hat{y}$$

$$(ik_z\hat{z} - k_x'\hat{x}) \times (-E_{x2}\hat{x}) = i\omega\mu H_{y1}\hat{y}$$

from which we know $E_{x1} = E_{x2}$ and thus the electric field has no component perpendicular to the interface of the metal and dielectric media, for example, air if only the electric field also has a component along the surface. The electric field can be rewritten as

$$E_1 = (E_{x1}\hat{x} + E_{z1}\hat{z}) \exp[-k_x x + i(k_z z - \omega t)]$$

$$E_2 = (-E_{x2}\hat{x} + E_{z1}\hat{z}) \exp[-k_x' x + i(k_z z - \omega t)]$$

$$\Rightarrow ik_z E_{x1} + k_x E_{z1} = -ik_z E_{x2} + k_x' E_{z1}$$

The boundary condition without a free surface charge ($D_1 = D_2$) is $\varepsilon_0\varepsilon_1 E_{x1} + \varepsilon_0\varepsilon_2 E_{x2} = 0$. On the other hand, the magnetic field satisfies

$$\nabla \times H_1 = \frac{\partial D_1}{\partial t} = \varepsilon_0\varepsilon_1 \frac{\partial E_1}{\partial t}$$

$$\nabla \times H_2 = \frac{\partial D_2}{\partial t} = \varepsilon_0\varepsilon_2 \frac{\partial E_2}{\partial t}$$

$$\Rightarrow$$

$$-ik_z H_{y1} = -i\omega\varepsilon_0\varepsilon_1 E_{x1}$$

$$-ik_x H_{y1} - -i\omega\varepsilon_0\varepsilon_1 E_{z1}$$

$$-ik_z H_{y1} = -i\omega\varepsilon_0\varepsilon_2 E_{x2}$$

$$-ik_x' H_{y1} = -i\omega\varepsilon_0\varepsilon_2 E_{z1}$$

$$\Rightarrow \quad \frac{k_x}{k_x'} = \frac{\varepsilon_1}{\varepsilon_2} \qquad E_{x1} = E_{x2} = 0$$

If we take the curl at both sides of the Maxwell equation, we can obtain

$$\nabla \times \nabla \times H_1 = (ik_z\hat{z} - k_x\hat{x}) \times (ik_z\hat{z} - k_x\hat{x}) \times (H_{y1}\hat{y})$$
$$= -(k_z^2 - k_x^2) H_{y1}\hat{y}$$

$$\nabla \times \frac{\partial D}{\partial t} = \frac{\partial}{\partial t}\varepsilon_0\varepsilon_1 \nabla \times E_1 = \frac{\partial}{\partial t}\varepsilon_0\varepsilon_1(-\mu_0 \frac{\partial H_1}{\partial t}) = -\varepsilon_0\varepsilon_1\mu_0(i\omega)^2 H_{y1}\hat{y}$$

$$\Rightarrow -k_z^2 + k_x^2 = -\varepsilon_0\varepsilon_1\mu_0\omega^2$$

Similarly, we can obtain $-k_z^2 + k_x'^2 = -\varepsilon_0\varepsilon_2\mu_0\omega^2$. By summarizing the above relationships, the propagation wave vectors perpendicular and along the interface can be described as

$$k_x = \sqrt{\frac{-\varepsilon_0\varepsilon_1\mu_0\omega^2}{1+\varepsilon_2}}$$

$$k_z = \sqrt{\frac{\varepsilon_0\varepsilon_1\varepsilon_2\mu_0\omega^2}{1+\varepsilon_2}}$$

Since the dielectric constant of metal is complex, k_x and k_z are also complex. The decay direction and propagation direction are in the *xz* plane but not necessarily along *x* and *z*. There is no electric component perpendicular to the interface ($E_x = 0$), and all E ($E = E_z$) and H ($H = H_y$) fields are along the surface of the so-called surface plasmon wave.

3.6.3 Finite Element Frequency-Domain Nanophotonic Simulation

Analytical solutions usually only exist for conformal geometries when Maxwell equations are used. For nanophotonic devices with arbitrary geometries, the numerical simulation methods are applied. Introduced here is the finite-element electromagnetic simulation method which has been extensively applied in the nanophotonic modeling work. The finite element method (FEM) is a method for solving partial differential equations (PDEs) for arbitrary geometric systems. This method requires the discretion of the complex geometric domains into smaller subregions or cells in conformal geometries such as rectangles or triangles. On each cell the function is approximated by a standard characteristic form. This method can be applied to a wide range of physical and engineering problems as long as it can be expressed as PDEs. The details related to electromagnetic simulation using FEM can be found elsewhere[3]; here only the general formulations are discussed. For an electromagnetic wave in a time-harmonic form

$$E(x,y,z,t) = E(x,y,z)\exp(j\omega t)$$

$$H(x,y,z,t) = H(x,y,z)\exp(j\omega t)$$

the Maxwell equations can be turned into

$$\nabla \times (\mu^{-1}\nabla \times E) - \omega^2 \varepsilon_c E = 0$$

$$\nabla \times (\varepsilon^{-1}\nabla \times H) - \omega^2 \mu H = 0$$

In the above form, the time term is eliminated, and thus the equation can be directly applied to the discrete FEM cells. There are many boundary conditions in the electromagnetic FEM simulations. For nanoplasmonic structures, usually the low-reflecting and matched boundary conditions are used. They are defined as

$$\sqrt{\frac{\mu_0\mu_r}{\varepsilon_c}}\, n \times H + E - (n \cdot E)n = 2E_0 - 2(n \cdot E_0)n + 2\sqrt{\frac{\mu_0\mu_r}{\varepsilon_c}}\, n \times H_0$$

and

$$n \times (\nabla \times E) - j\beta[E - (n \cdot E)n] = -2j\beta[E_0 - (n \cdot E_0)n] - 2j\beta\sqrt{\frac{\mu_0\mu_r}{\varepsilon}}\, n \times H_0$$

respectively, where β is the propagation constant and E_0 and H_0 are the sources of electric and magnetic fields, respectively. The above formulas are applicable for three-dimensional simulations. For two-dimensional or plane wave simulation, the wave equation and boundary conditions can be simplified as

$$\nabla \cdot \nabla E_z - \varepsilon k_0^2 E_z = 0 \qquad \text{for transverse electric (TE) waves}$$

$$-\nabla \cdot (\varepsilon \nabla H_z) - k_0^2 H_z = 0 \qquad \text{for transverse magnetic (TM) waves}$$

The low-reflecting boundary condition becomes

$$n\times\sqrt{\mu}H+\sqrt{\varepsilon}E_z=2\sqrt{\varepsilon}E_{0z} \qquad \text{(TE waves)}$$

$$n\times\sqrt{\varepsilon}E-\sqrt{\mu}H_z=-2\sqrt{\mu}H_{0z} \qquad \text{(TM waves)}$$

And the matched boundary condition becomes

$$n\times(\nabla\times E_z)-j\beta E_z=-2j\beta E_{0z} \qquad \text{(TE waves)}$$

$$n\times(\nabla\times H_z)-j\beta H_z=-2j\beta H_{0z} \qquad \text{(TM waves)}$$

All the simulation processes including geometry definitions, FEM cell mesh generation, and equation solving can be integrated into one program, and the results can be exported for further analysis. Presented here are some simulation results using FEM frequency-domain analysis. Figure 3.8 shows the simulation results of electric field amplitude enhancement under 785-nm light excitation with TM polarization.

In another example, a nanoplasmonic particle with irregular geometries is analyzed using FEM simulation. Figure 3.9 shows the local field enhancement results for a gold nanocrescent particle. The outer radius of the nanocrescent is 200 nm, and an electromagnetic plane wave is normally incidental at a free-space wavelength of 785 nm with TM polarization. By varying the tip sharpness (edge in three dimensions), the maximum field enhancement can vary significantly. The numerical simulation is a useful tool to seek and verify the optimal designs of nanophotonic devices.

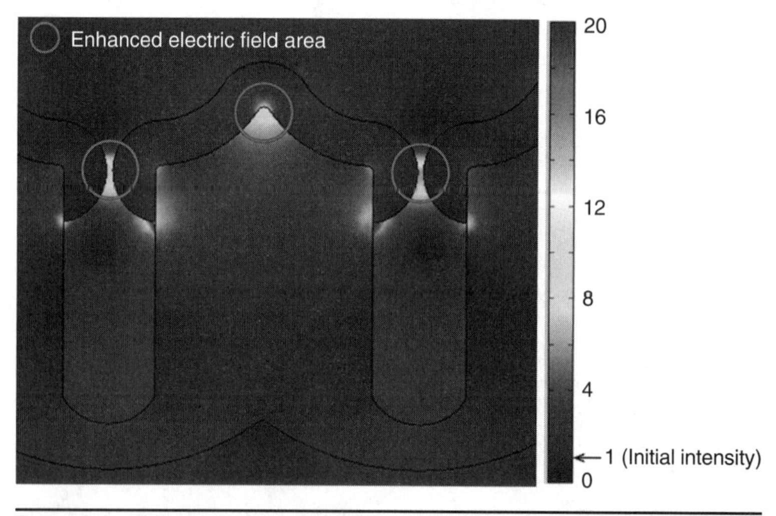

FIGURE 3.8 FEM simulation results of electric field enhancement of a thin gold layer deposition on a nanohole array (cross-sectional view). The nanohole size is 10 nm. The unit of enhancement shown in the intensity scale bar is the decibel.

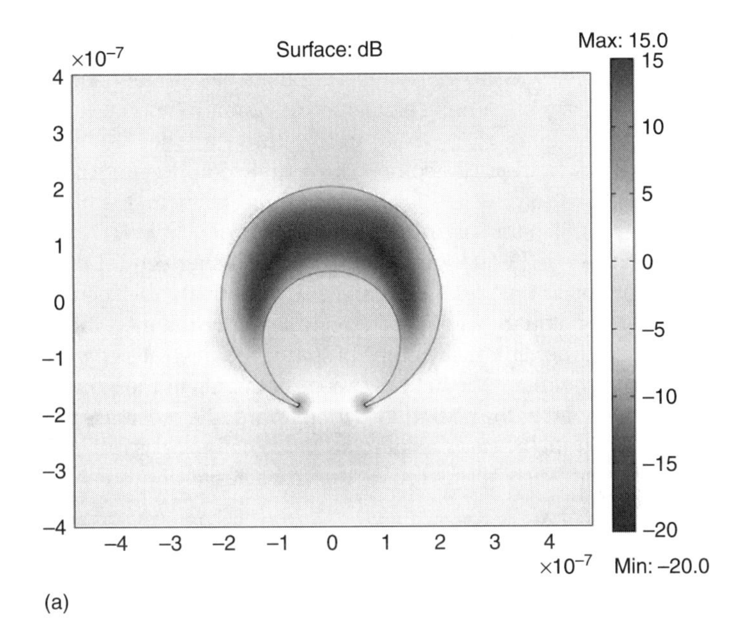

(a)

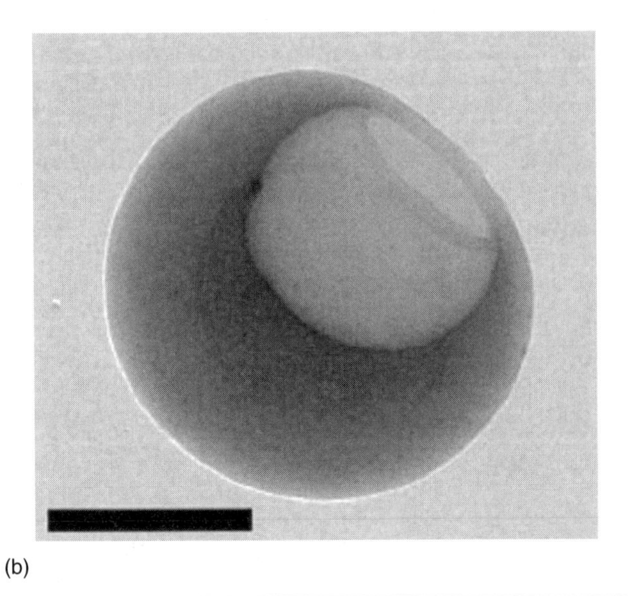

(b)

Figure 3.9 Electric field amplitude distribution map (top) and transmission electron microscopy image for a nanocrescent device. The local field enhancement near the tip is about 15 times higher than the incident field amplitude.

3.6.4 Finite Difference Time Domain (FDTD)

For the more accurate and general time-domain electromagnetic simulation, one can carry out a finite element time-domain (FDTD) simulation which has the ability to generate unstructured meshes (for the most accurate representation of geometry), accurate solutions of the partial differential equations with higher-order finite element methods, and massive parallelism for solving very large problems. In FDTD, the time-dependent Maxwell equations are discretized using central-difference approximations to the space and time partial derivatives. The resulting finite difference equations are solved in a leapfrog manner: the electric field vector components in a volume of space are solved at a given instant in time; then the magnetic field vector components in the same spatial volume are solved at the next instant in time; and the process is repeated over and over until the steady-state electromagnetic field is fully evolved. Let's rewrite the Maxwell curl equations in Cartesian coordinates as

$$\frac{\partial H_x}{\partial t} = \frac{1}{\mu}\left(\frac{\partial E_y}{\partial z} - \frac{\partial E_z}{\partial y}\right) \qquad \frac{\partial E_x}{\partial t} = \frac{1}{\varepsilon}\left(\frac{\partial H_z}{\partial y} - \frac{\partial H_y}{\partial z} - \sigma E_x\right)$$

$$\frac{\partial H_y}{\partial t} = \frac{1}{\mu}\left(\frac{\partial E_z}{\partial x} - \frac{\partial E_z}{\partial z}\right) \qquad \frac{\partial E_y}{\partial t} = \frac{1}{\varepsilon}\left(\frac{\partial H_x}{\partial z} - \frac{\partial H_z}{\partial x} - \sigma E_y\right)$$

$$\frac{\partial H_z}{\partial t} = \frac{1}{\mu}\left(\frac{\partial E_x}{\partial y} - \frac{\partial E_y}{\partial x}\right) \qquad \frac{\partial E_z}{\partial t} = \frac{1}{\varepsilon}\left(\frac{\partial H_y}{\partial x} - \frac{\partial H_x}{\partial y} - \sigma E_z\right)$$

If we define the discrete mesh grid (i, j, k) in Cartesian coordinates, the above partial differential equations can be transformed to the following finite difference equations. Here we list only the first two sets of equations; others can be deduced in similar fashion.

$$\begin{aligned} &E_x^{n+1}(i+\tfrac{1}{2},j,k) \\ &=[1-\sigma(i+\tfrac{1}{2},j,k)/\varepsilon(i+\tfrac{1}{2},j,k)]E_x^n(i+\tfrac{1}{2},j,k) \\ &\quad +\Delta t/\varepsilon(i+\tfrac{1}{2},j,k)\delta[H_z^{n+\frac{1}{2}}(i+\tfrac{1}{2},j+\tfrac{1}{2},k)-H_z^{n+\frac{1}{2}}(i+\tfrac{1}{2},j-\tfrac{1}{2},k) \\ &\quad +H_y^{n+\frac{1}{2}}(i+\tfrac{1}{2},j,k-\tfrac{1}{2})+H_y^{n+\frac{1}{2}}(i+\tfrac{1}{2},j,k-\tfrac{1}{2})-H_y^{n+\frac{1}{2}}(i+\tfrac{1}{2},j,k+\tfrac{1}{2})] \end{aligned}$$

$$\begin{aligned} H_x^{n+\frac{1}{2}}(i,j+\tfrac{1}{2},k+\tfrac{1}{2}) &= H_x^{n-\frac{1}{2}}(i,j+\tfrac{1}{2},k+\tfrac{1}{2})+\Delta t/\mu\delta[E_y^n(i,j+\tfrac{1}{2},k+1) \\ &\quad -E_y^n(i,j+\tfrac{1}{2},k)+E_z^n(i,j,k+\tfrac{1}{2})-E_z^n(i,j+1,k+\tfrac{1}{2})] \end{aligned}$$

Here, n is the time index, Δt is the time increment, and δ is the space increment.

One may use commercially available FDTD simulation software to numerically solve the above coupled equation in three dimensions, and one can also repeat the simulation for different meshing grids,

FIGURE 3.10 Electric field amplitude distribution of a silver nanocavity produced from an FDTD simulation analysis.

Cartesian grids, or triangle grids. Using state-of-the-art supercomputing facilities, one may be able to set the time increment to subfemtosecond scales and the space increment to subnanometer scales. Such small spatial/temporal resolutions demand extreme computing power but will be able to provide us with more accurate results, which is critical to the understanding of electromagnetic enhancement mechanism at the sub-10 nm nanophotonic devices.

As the examples show the FDTD simulation capability, Fig. 3.10 shows the enhancement in a silver nanocavity. The nanocavity has a width of 4 nm and a depth of 18 nm. The excitation wavelength is 582 nm, and the electric field enhancement is 50 times.

References

1. Fleischm, M., Hendra, P. J., and Mcquilla, A. J. "Raman-Spectra of Pyridine Adsorbed at a Silver Electrode." *Chemical Physics Letters* **26**, 163–166 (1974).
2. Kirtley, R., Jha, S. S., and Tsang, J. C. "Surface Plasmon Model of Surface Enhanced Raman Scattering." *Solid State Communications* **35**, 509–512 (1980).
3 Lee, T. K. and Birman, J. L. "Molecule Adsorbed on Plane Metal Surface: Coupled System Eigenstates," *Physical Review B* **22**, 5953–5960 (1980).
4. Lee, T. K., and Birman, J. L. "Quantum Theory of Enhanced Raman-Scattering by Molecules on Metals—Surface-Plasmon Mechanism for Plane Metal Surface." *Physical Review B* **22**, 5961–5966 (1980).
5. Arya, K., and Zeyher, R. "Theory of Surface-Enhanced Raman Scattering from Molecules Adsorbed at Metal Surfaces." *Physical Review B* **24**, 1852–1865 (1981).

6. Gorobei, N. N., Ipatova, I. P., and Subashiev, A. V. "Electromagnetic Theory of Surface-Enhanced Raman Scattering." *JETP Letters* **34**, 149–152 (1981).
7. Lee, T. K., and Birman, J. L. "A Note on a Quantum Theory of Surface-Enhanced Raman Scattering." *Journal of Raman Spectroscopy* **10**, 140–144 (1981).
8. Persson, B. N. J. "On the Theory of Surface-Enhanced Raman Scattering." *Vacuum* **31**, 601 (1981).
9. Persson, B. N. J. "On the Theory of Surface-Enhanced Raman Scattering." *Chemical Physics Letters* **82**, 561–565 (1981).
10. Arunkumar, K. A., and Bradley, E. B. "Theory of Surface-Enhanced Raman Scattering." *Journal of Chemical Physics* **78**, 2882–2888 (1983).
11. Philpott, M. R. "Quantum Theory of Surface-Enhanced Raman Scattering and Its Relation to Experiment." *Abstracts of Papers of the American Chemical Society* **185**, 25-Coll (1983).
12. Ueba, H. "Theory of Charge-Transfer Excitation in Surface-Enhanced Raman Scattering." *Surface Science* **131**, 347–366 (1983).
13. Arya, K., and Zeyher, R. "Theory of Surface-Enhanced Raman Scattering." *Topics in Applied Physics* **54**, 419–462 (1984).
14. Brodskii, A. M., and Daikhin, L. I. "Theory of Surface-Enhanced Raman Scattering at the Metal Electrolyte Interface." *Soviet Electrochemistry* **20**, 955–961 (1984).
15. Lombardi, J. R., Birke, R. L., Lu, T. H., and Xu, J. "Charge-Transfer Theory of Surface-Enhanced Raman Spectroscopy—Herzberg-Teller Contributions." *Journal of Chemical Physics* **84**, 4174–4180 (1986).
16. Rojas, R., and Claro, F. "Theory of Surface-Enhanced Raman Scattering in Colloids." *Journal of Chemical Physics* **98**, 998–1006 (1993).
17. GarciaVidal, F. J., and Pendry, J. B. "Collective Theory for Surface-Enhanced Raman Scattering." *Physical Review Letters* **77**, 1163–1166 (1996).
18. Kambhampati, P., Child, C. M., Foster, M. C., and Campion, A. "On the Chemical Mechanism of Surface-Enhanced Raman Scattering: Experiment and Theory." *Journal of Chemical Physics* **108**, 5013–5026 (1998).
19. Otto, A. "Theory of First Layer and Single Molecule Surface Enhanced Raman Scattering (SERS)." *Physica Status Solidi a-Applied Research* **188**, 1455–1470 (2001).
20. Giese, B., and McNaughton, D. "Surface-Enhanced Raman Spectroscopic and Density Functional Theory Study of Adenine Adsorption to Silver Surfaces." *Journal of Physical Chemistry B* **106**, 101–112 (2002).
21. Janesko, B. G., and Scuseria, G. E. "Surface-Enhanced Raman Optical Activity of Molecules on Orientationally Averaged Substrates: Theory of Electromagnetic Effects." *Journal of Chemical Physics* **125**, (2006).
22. Pustovit, V. N., and Shahbazyan, T. V. "Microscopic Theory of Surface-Enhanced Raman Scattering in Noble-Metal Nanoparticles." *Physical Review B* **73**, (2006).
23. Jin, J. *The Finite Element Method in Electromagnetics*, John Wiley & Sons, New York, 1993.

PART II

Review of Methods

CHAPTER 4

Tissue Pathology: A Clinical Perspective

Krishnarao Tangella

Department of Pathology
College of Medicine
University of Illinois at Urbana–Champaign and Christie Clinic

Marina Marjanovic

Beckman Institute for Advanced Science and Technology
University of Illinois at Urbana–Champaign

4.1 Introduction

Pathologists are physicians who are integral and essential parts of patients' diagnostic workup. Due to the nature of their work, they can be a valuable resource for research projects. The purpose of this chapter is to give the reader a brief overview of a pathologist's role and the processes that occur in a medical laboratory that lead to a final diagnosis.

When patients find a lump or bump lesion, they go to a primary care physician or a surgeon for further workup. The physician then performs various diagnostic tests to work up the patient. These diagnostic tests, depending on the clinical situation, may require tests such as CT scan, MRI, PET scan, X-rays, and blood tests [1]. On the basis of the diagnostic teing, further workup, such as surgical intervention, may be performed. The surgery performed could either be a diagnostic procedure such as a diagnostic biopsy, or it could be a total curative therapy such as a complete resection of the tumor. The surgeon, depending on the clinical scenario, also utilizes fine-needle aspiration as a modality in the initial workup to diagnose a lump or a bump process. During the fine-needle aspiration, a syringe is used to procure individual cells and a small cluster of cells with the aspiration technique, which a pathologist examines under the microscope to give a diagnosis of benignity or malignancy [2]. After the surgical specimen arrives in the pathology department, the specimen is processed, and after a thorough evidence-based process, a pathology report is issued. On the basis of the pathology report, further management of the patient is pursued. In cases of malignancy, a multidisciplinary approach using modalities that include but are not limited to, further surgeries, chemotherapy, and radiation is utilized to treat the patient [3].

4.2 Tissue Preparation

4.2.1 Accessioning the Specimen

The pathology specimen is sent to a histology laboratory for further processing. The histology laboratory is charged with the receipt, accessioning, grossing, processing, embedding, sectioning, and staining of the various specimens received from physician offices and hospital operating rooms. These specimens include tissue samples, various body fluids and smears for cytology, cells and tissue collected via fine-needle aspiration, CT-guided core needle biopsies, and assorted other items that are hard to fit into the previous classifications. Tissue may be submitted as the result of a major surgical procedure or as the result of a very small biopsy. In general, the majority of the tissues removed from the body are submitted to histology for processing and examination. It is important to recognize that histology is a labor- and time-intensive process. Whereas the rest of the clinical laboratories where blood specimens are processed have experienced a great decrease in turnaround times for specimens due to extensive automation, the histology laboratory is relatively somewhat lagging behind in automation. The bottom line is that tissue specimens take time for proper fixation to occur, embedding is still largely a manual procedure, and sectioning of the tissue is much more of an art than it is a science. In most cases, a specimen that is received will routinely take two working days before results can be expected. There is new microwave technology for rapid-tissue processing that is beginning to become more readily available and more affordable. However, many smaller laboratories do not have such an instrument. There are also instruments that will automatically embed the tissue for sectioning, but again, these are not yet routinely available.

The following will be a general description of what happens to a specimen from the time of receipt at the histology laboratory to the time it is turned over to a pathologist for diagnostic interpretation. More complete information may be found in a variety of textbooks that have been written on the subject [4, 5, 6]. Some of the more common

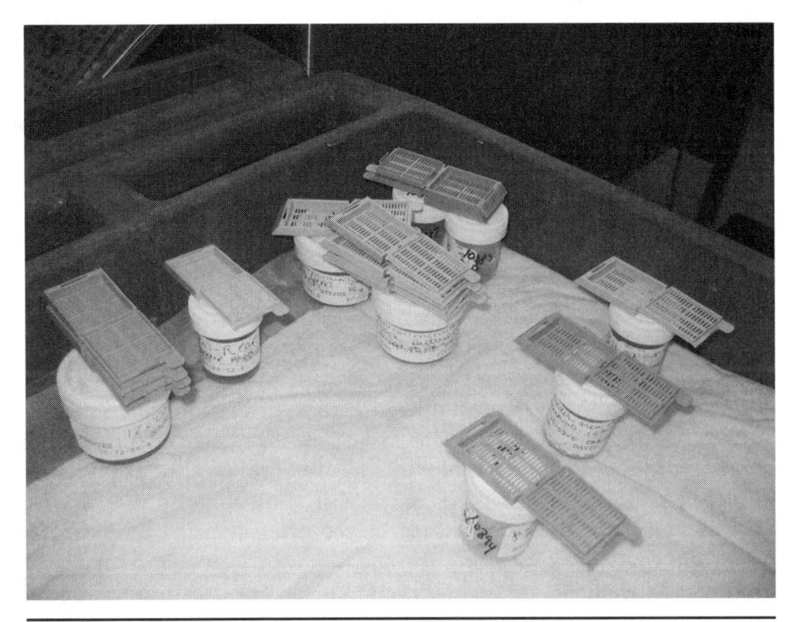

Figure 4.1 Small biopsy specimens with prelabeled cassettes.

problems that occur throughout the process will be mentioned to enhance the understanding of the reader.

After the specimen is removed from the patient, the specimen is either delivered fresh or submitted in a fixative. A fixative is a solution that helps preserve the tissue specimens for a prolonged duration. The specimen is then sent to a histology laboratory. On receipt of the specimen in a histology laboratory, it will routinely be entered into the laboratory computer information system, which assigns a unique specimen number (Fig. 4.1). It is also very common for a worksheet to be created for use by the histology technicians as they move the specimen through its subsequent steps. It is during this process that demographics are entered for the patient, and it is here that most mistakes are discovered. These mistakes commonly include

1. Incomplete specimen requisitions
2. Nonspecific or incorrect specimen site information
3. Missing specimen site
4. Specimen and requisition information that do not match
5. No unique identifying number
6. Specimens that are completely unlabeled
7. Improper preservative
8. Wrong patient demographic information

Errors of this sort do not mean that the specimen will be discarded; most specimens are irreplaceable. However, delays may occur while the errors are being corrected, and it is often necessary to return the specimen to the submitting office or operating room that was the source of the specimen to fix the inaccuracies. As with everything else in the laboratory, accurate specimen identification is of prime importance.

4.2.2 Grossing in the Specimen

After assigning a unique specimen number, the specimen is ready to be "grossed" [6].The gross examination is a macroscopic examination of the specimen, performed by a pathologist or a pathology assistant (Figs. 4.2 and 4.3). The first step is to again verify that the specimen is correctly labeled, the requisition is properly filled out, and the specimen and requisition match. Typically, the specimen is physically described; dimensions, weight, color, unusual lesions or markings, distance of the lesion from the resection margin, and so on. Representative tissue sections are taken from the specimen using a set of predetermined protocols that are evidence based in their creation (Fig. 4.4). Thin sections are cut to help with the tissue fixation process [7]. It is common for the submitting physician to indicate the specimen orientation by "tagging" a portion of the specimen to designate its original orientation prior to removal, for example, "suture at 12:00 o'clock position." Specimens are commonly inked to help find the margins microscopically or to designate a portion of the specimen that is of particular interest. Sections of the specimen are

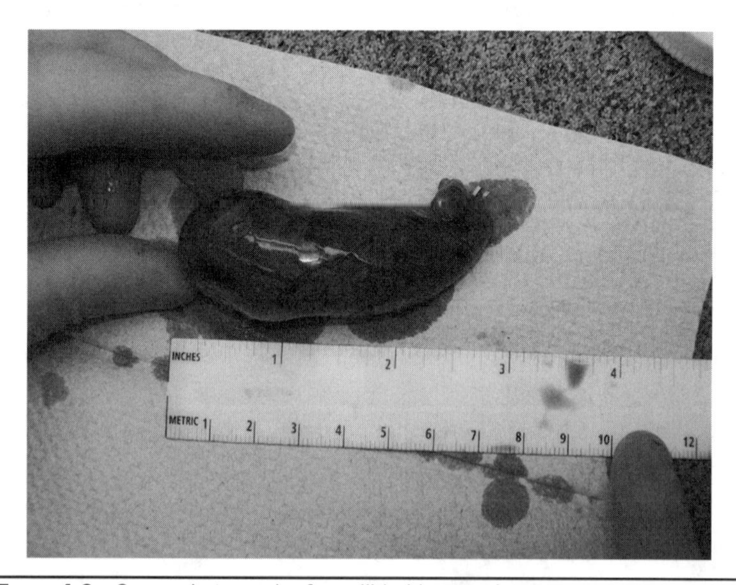

Figure 4.2 Gross photograph of a gallbladder specimen.

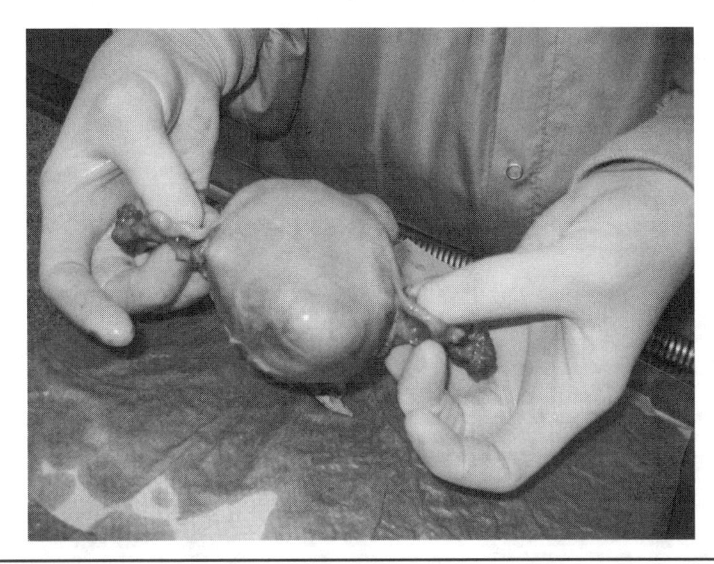

FIGURE 4.3 Gross photograph of a uterus with bilateral ovaries and fallopian tube specimen. The uterus shows grossly bosselated appearance due to fibroids.

FIGURE 4.4 Cassettes showing section of dissected tissue sections.

placed into tissue cassettes. The origin of the sections is also annotated in the gross description. Very small specimens are usually placed between sponges or in bags to prevent them from escaping the cassette during subsequent tissue processing (Fig. 4.5). They may also be marked with a dye to help the histology tech find extremely

FIGURE 4.5 Cassette showing small biopsy submitted between sponges.

small specimens. After the cassettes are loaded and closed, they are placed into a 10% neutral-buffered formalin for fixation prior to actual processing. The pathologist or pathology assistant dictates the "gross description," which becomes part of the final pathology report. An example of the gross description is: "Received is specimen number 1234, labeled Jane Doe, core needle biopsy, left breast. The name and specimen ID number match. Received in formalin are three yellow-white core needle biopsies, measuring 1.5, 1.3, and 1.2 cm in length, and 0.1 cm in diameter. They are submitted in their entirety between sponges in one cassette" [7].

4.2.3 Frozen Sections

A special mention about frozen sections: Frozen sections are commonly submitted by surgery to obtain an immediate impression regarding malignancy versus benignity or to determine if the margins of the specimen are free from disease [7]. The College of American Pathology dictates that a simple frozen section be completed and the results relayed to the surgeon within 20 minutes. To achieve this, the specimen is submitted without fixative ("fresh") to histology. After the appropriate demographics and labeling are verified, the specimen is grossly examined to determine the appropriate tissue for freezing. It is then quick-frozen using a cryostat (Fig. 4.6), which is basically a microtome in a freezer (Fig. 4.7), and a section is cut for quick-staining using hematoxylin and eosin. It is then submitted to the pathologist for microscopic examination and the pathologist calls the results to the submitting surgeon as soon as possible [7].

Figure 4.6 Frozen block containing the specimen during a frozen section.

Figure 4.7 Microtome showing a frozen block in process of cutting.

Errors that can occur in the grossing process include

1. Cassettes are given the wrong number, which does not match the specimen number.
2. Cassettes are overfilled with the tissue sections. This results in histology techs having a difficult time cutting the specimen.

If the cassette does not allow for some margin around the tissue prior to cutting the tissue on the microtome, the specimen is much more difficult to cut.

3. Duplicate cassette numbers are assigned.
4. Cassette numbers may be accidentally skipped.
5. Bony tissue is not properly decalcified prior to tissue processing.

Fixation of the specimen is an important part of tissue processing. Generally, a tissue fixes in 10% formalin at an approximate rate of 1 to 2 mm per hour. However, some tissue, especially fatty tissue, can be expected to take much longer. Cutting thin sections at the grossing table (3 to 4 mm in thickness) can help with this problem, but fatty specimens may routinely require additional time in formalin for fixation to occur. Unfixed, fatty tissue is almost impossible to section at the microtome.

4.2.4 Specimen in Tissue Processor

The next step is to transfer the tissue cassettes to the tissue processor [5]. As mentioned earlier, microwave tissue processing is becoming more commonly available and affordable but is still not the standard in the typical histology lab. During classic tissue processing, the tissue is automatically passed through a sequential series of steps, moving from 10% formalin, to 95% alcohol, to 100% alcohol, to xylene, to paraffin. To facilitate the process of tissue infiltration, the processor often alternates between increased pressure and vacuum. The process itself is fairly lengthy, taking from 10 to 12 hours. Because of this, it is usual for the histology lab to have a cutoff time for processing specimens to allow enough time for the tissue to complete the process. Tissue specimens received after this time will routinely be delayed until the following day for processing. At the end of processing, the tissue is ready to be embedded in paraffin (Fig. 4.8).

4.2.5 Embedding of Tissue

Embedding tissue in paraffin is another art, and experience is of great value [5]. The tissue must be properly oriented when it is embedded in paraffin, or the microscopic interpretation of the subsequent stained slide may be extremely difficult or even impossible. There are entire books that have been written to address the proper embedding of tissue. Such detailed description is beyond the scope of this book. However, skin biopsies tend to present many problems. If a skin lesion is embedded lesion side down, the lesion itself may be cut away as the paraffin block is "faced" at the microtome. For this reason,

FIGURE 4.8 Tissue blocks after paraffin embedding the tissue.

shave biopsies of skin are typically embedded on edge to allow the pathologist to view all the layers of skin. Punch biopsies of skin are typically bisected off-center along the long axis and embedded with the cut side down, again allowing all the layers of skin to be visualized. The pathologist or pathology assistant will often help the histology tech by leaving specific instructions (e.g., embed on edge) in order to ensure proper embedding.

Another potential embedding problem concerns multiple specimens embedded within the same block. For example, it is common for core needle biopsies to be embedded together. It is important for each of the cores to be embedded within the same plane of the block or risk either a slide with a missing core or a slide where one core is completely cut away in order to expose the "hidden" core. Both scenarios are unacceptable, and such a block should be melted and the tissue reembedded.

The actual embedding process is usually manual in nature. The cassette containing the tissue is removed from the processor and opened, and the tissue removed. Melted paraffin is dispensed into a tissue mold, and the tissue itself is manually oriented within the paraffin as it begins to cool. The original cassette is used to top the mold, which is then filled with more paraffin and allowed to cool in a refrigerator. Once it is completely cooled, it is ready to be sectioned using the microtome.

Again, poor embedding can create problems, including

1. Allowing the paraffin to get too hot, making it much more difficult to cut
2. Accidentally embedding a small piece of tissue from one cassette into the incorrect block
3. Not orienting the tissue properly
4. Inability to determine proper orientation of tissue
5. Not embedding all tissue within the same plane
6. Tissue fragmenting as it is removed from the cassette prior to embedding
7. Not being able to find the tissue that was originally placed into the cassette; missing pieces (Some tissue may not survive the tissue processor, in spite of efforts to protect it.)

4.2.6 Microtome Sectioning of Paraffin Blocks

After the tissue has been processed, embedded, and cooled, it is finally ready for the microtome (Fig. 4.9), where diagnostic slides are prepared for staining. Most histology labs try to cut the tissue at 3 to 4 microns in thickness (Fig. 4.10), although thinner and thicker sections may be required at times depending on the clinical scenario. Cutting thin tissue sections may be the most demanding job in the laboratory. It is a frustrating job to learn, a difficult job to master,

FIGURE 4.9 A microtome.

Figure 4.10 A ribbon of tissue block containing the biopsy specimen.

and so repetitive in nature that great attention to detail is essential. It is also very physically demanding, with a high potential for repetitive motion injuries, even with the use of proper engineering controls, like motorized microtomes, ergonomic chairs, and proper positioning of ancillary tools [5]. Although difficult, it is also absolutely vital to provide the pathologist the best technical-quality slides, which help in securing a proper diagnosis for the submitting physician.

The first step in the sectioning process is to reexpose the tissue within the paraffin block, using a process called *facing*. After all, the tissue is embedded in paraffin, and the paraffin must be first cut away to expose the tissue if a section is to be achieved. Once the tissue is faced, a thin section is cut for mounting on a glass slide. These slides are numbered in the same way as the original cassettes. Special coated glass slides are commonly used, especially if special stains or immunohistochemical stains are a possibility. The coated or charged slides will more tightly bind the tissue to the glass slide. It is also important to cut a section of tissue that displays a representative section of the entire piece of tissue. The physician may also request levels through the tissue, with each subsequent level deeper within the tissue than the previous level.

The tissue is mounted onto the glass slide by first floating it on a warm water bath (Fig. 4.11), and then picking it up with the labeled slide. The slides are then allowed to dry in a warm oven for 15 to 30 minutes prior to staining [5].

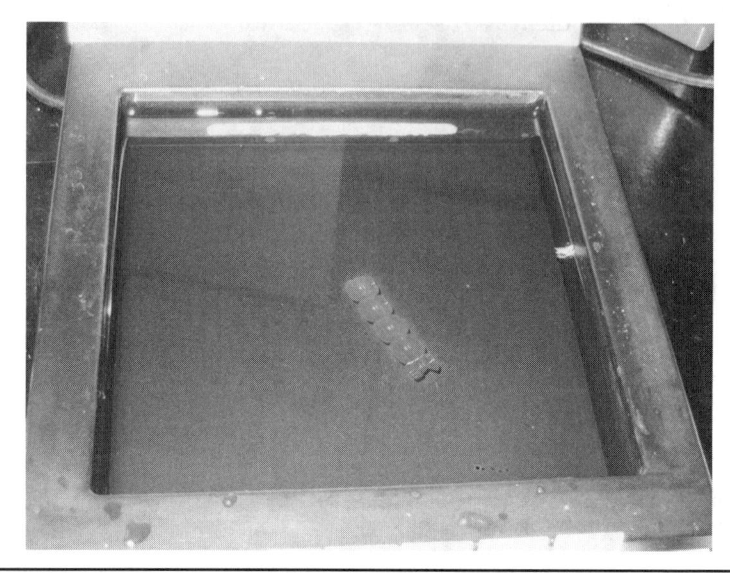

Figure 4.11 Warm water bath showing a floating ribbon prior to pickup on a slide.

Problems associated with microtomy include

1. Improper facing of the block
2. Incomplete sections
3. Difficulty in sectioning (Bony tissue or fatty tissue is much more difficult to cut.)
4. Labeling the slide incorrectly
5. Using a microtome blade that is nicked, and so creates uneven and incomplete sections
6. Folding and wrinkling within the tissue section
7. Tissue that is cut too thick
8. Tissue from more than one block on the same slide (*floaters*)
9. Dirty slides/or fingerprints

4.2.7 Tissue Staining

The routine stain of the histology lab is the hematoxylin and eosin (H&E) stain (Figs. 4.12 and 4.13). The staining of the slides can be manual, semiautomated, or completely automated depending on the laboratory. A typical slide stainer will run the slides first through a deparaffinization process, then through hematoxylin and eosin stains, and finally through a dehydration process. At the end of staining, the slides are ready to be cover-slipped. Manually cover-slipping several slides is a very time-consuming process. Fortunately, automatic cover-slip

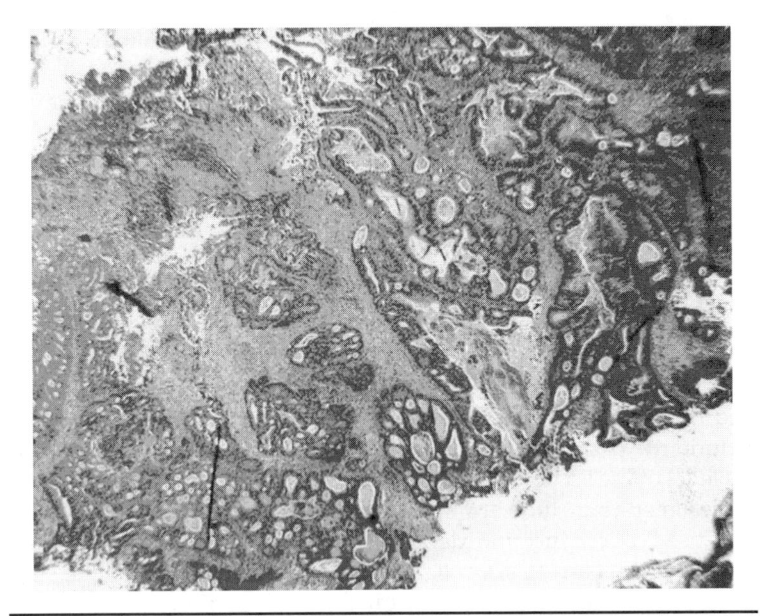

FIGURE 4.12 Microscopic photograph showing an infiltrating adenocarcinoma of colon stained with an H&E stain.

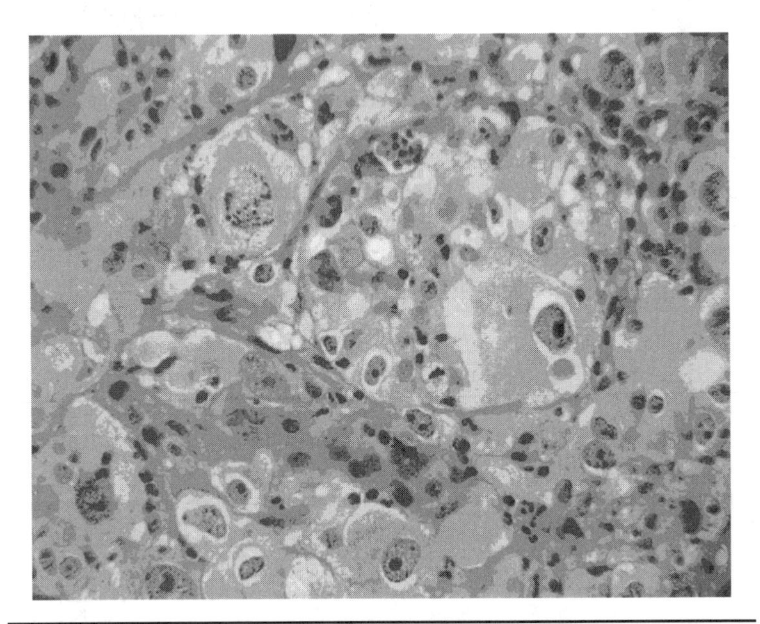

FIGURE 4.13 Microscopic photograph showing an infiltrating non-small-cell carcinoma of lung stained with an H&E stain.

machines are available, which use either a plastic cover-slip or a glass cover-slip.

Problems associated with staining and cover-slipping include

1. Slides not completely dehydrated/or dirty dehydration solvents
2. Floaters in the staining process attaching to other slides
3. Stains and reagents not changed on a regular schedule
4. Sloppy cover-slipping—too much mounting media
5. Dirty cover-slips/or fingerprints

Most laboratories complete the process by affixing an identifying label to each slide to better identify it. These labels are usually generated by the laboratory's computer information system and typically include the patient name, the specimen number, the block number, the level of the section, and the name of the laboratory. They can also be designed to include any special treatment or stain that may have been used [Decalcification (decal), Periodic acid-Schiff (PAS), Gomori-Grocott methenamine silver (GMS), etc.] [4]. As with every other step in the process, mistakes may occur in the labeling of slides. Because of the potential danger in misidentifying the final slides, technologies are being developed and sold that will help reduce the problem. There are now automatic cassette-labeling systems, automatic slide-labeling systems, and bar code–driven systems that demand that the cassette bar code match the slide bar code. These systems can be interfaced to the laboratory computer system to greatly reduce the potential for human error.

Before the completed slides are presented to the pathologist for examination, they are usually paired with a transcription of the original dictation from the time of initial grossing. They will often also include the original requisition, all of the patient demographics, a listing of previous specimens from the same patient, and the final results from those specimens. The pathologist will frequently request an earlier case for comparison with the current case if needed. Armed with this information, and requesting more as needed, the final diagnosis is reached and the report is issued.

However, the histology laboratory is still not finished with the patient. Any tissue that is unused from the time of grossing is held for a predetermined amount of time—from one week to one month—in case there is any need to revisit the specimen or additional tissue must be submitted for examination. The slides and blocks are stored a specified amount of time as determined by the Clinical Laboratory Improvement Amendments (CLIA) Act which states:

Sec. 493.1105 Standard: Retention requirements

(ii) Pathology test reports for at least 10 years after the date of reporting.

(7) Slide, block, and tissue retention—

(i) Slides.

(A) Retain cytology slide preparations for at least 5 years from the date of examination.

(B) Retain histopathology slides for at least 10 years from the date of examination.

(ii) Blocks. Retain pathology specimen blocks for at least 2 years from the date of examination.

(iii) Tissue. Preserve remnants of tissue for pathology examination until a diagnosis is made on the specimen.

(b) If the laboratory ceases operation, the laboratory must make provisions to ensure that all records and, as applicable, slides, blocks, and tissue are maintained and available for the time frames specified in this section.

A laboratory can choose to store the slides and blocks for longer periods of time. Usually, the diagnostic reports become a permanent part of the patient's medical record. These slides, blocks, and records may become a valuable potential source for future medical research.

The question now becomes, how does fluid for cytology differ from tissue for histology? The difference is mainly in the initial preparation of the specimen. Since the focus in cytology changes from tissue to individual cells and groups of cells, it becomes necessary to concentrate the cells by first spinning the fluid in a centrifuge. Most of the time the red blood cells that are in the fluid are destroyed by means of a special fixative that both fixes the cells of interest and destroys the red blood cells. It is from this spun sediment that a cell block is prepared. The supernatant is poured off, a few drops of plasma are added to the sediment, and the whole thing is clotted using a drop or two of thrombin. This cell-filled thrombin clot is poured into a tissue cassette, usually between sponges or histology paper, and placed into 10% formalin until the tissue processor is started. The cell block is then embedded, cut, stained, and labeled just like the histology blocks.

4.2.8 Cytology Specimens

Other techniques in cytology may include the preparation of slides with a monolayer of cells for examination [1]. A portion of the fluid is spun down and decanted as above. It is then placed on an instrument that makes and stains a monolayer slide, which can be examined by the pathologist or cytologist. This same technique is now employed in place of the traditional Pap smear, yielding a higher detection rate of clinically significant results, while reducing the incidence of false negative reports. This Pap smear technique is also now used coupled with computerized optical programs that can screen pap slides with great accuracy, marking any suspicious cells for pathologist review.

Finally, there is still a place for the old-fashioned cellular smear wherein the cells are directly smeared on a glass slide [1]. This is commonly utilized during the fine-needle aspiration procedure to determine the adequacy of the specimen. On receipt in the histology or cytology lab, the specimen is often stained with the Pap technique and examined along with the rest of the case. As with the histology cases, the final slides are matched up with pertinent information—demographics, requisitions, histories—prior to submission to the pathologist.

After examining the slides, whether histology or cytology or both, it may become necessary for the pathologist to order additional slides and/or stains before the final diagnosis is given. It is not uncommon to request that additional levels of the tissue be cut and stained for examination, often in a search for clean margins or due to embedding difficulties. There are times when the case needs to be referred to another pathologist at another location, so recut slides are frequently ordered to send out. There are times when follow-up information that demands additional testing of some kind is received.

4.2.9 Special Stains

Finally, there are other stains beside the H&E. Most histology laboratories offer a variety of histochemical stains, often referred to as special stains. With these stains, additional information is able to be gleaned that may help to nail down a difficult diagnosis or to confirm what is suspected by the pathologist. Some of the more commonly ordered stains include

1. GMS stain for fungal organisms
2. Acid fast stain (AFB) for acid-fast bacillus such as mycobacterium, cryptosporidium parvum, isospora, and cyclospora cysts, and hooklets of cysticerci
3. Giemsa stain for mast cells and microorganisms, such as Giardia or *Helicobacter pylori*
4. PAS stain for staining basement membranes, fungal organism, glycogen, and a variety of inclusions in certain clinical scenarios
5. Reticulin stain for highlighting reticulin fibers
6. Prussian blue stain to detect iron deposits
7. Trichrome stains collagen blue
8. Congo Red stain to stain amyloid deposits [1, 4]

Special stains are done by both manual methods and by using newer special stain instruments that are fully automated. Positive control slides are stained at the same time the patient slide is prepared to ensure proper staining results.

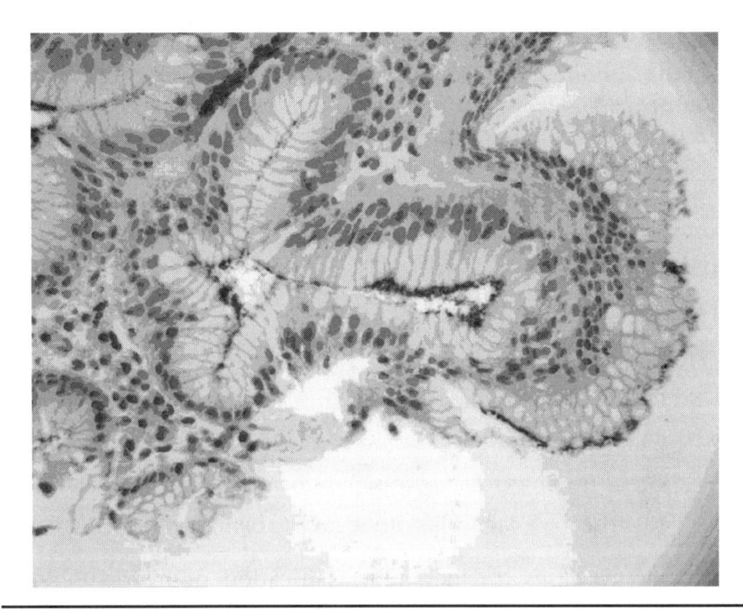

FIGURE 4.14 Microscopic photograph showing a gastric biopsy showing positive staining with a *Helicobacter pylori* organism on an immunohistochemical stain.

Immunohistochemical (IHC) staining (Fig. 4.14) is an integral part of pathology workup [8]. IHC staining has a variety of uses in today's pathology lab. Based on antigen-antibody reactions, it allows the pathologist to identify various tumor antigens to determine the nature of the tumor, aggressiveness of the tumor, prognostic information, and suitability for some drug treatment protocols (predictive markers). There are literally hundreds of antibodies available to provide information to both the pathologist and the submitting physician. The staining protocol for IHC testing is very complex. Because of this, instruments have been developed that are essentially programmable robots to properly dispense the reagents and incubate the reagents for the proper time. Both positive and negative controls are utilized to ensure the proper functioning of the IHC system. IHC stains are only to be interpreted by a qualified pathologist. They are a powerful tool in today's histology lab.

The "product" that is ultimately produced by the pathology department is the pathology report. The histology laboratory plays an instrumental part in the development of this report and only succeeds by utilizing a team approach, where clear communication is the key.

In the pathologist's office, the slides are reviewed by a pathologist under a microscope and a final diagnosis report is issued [1]. There are a variety of tools the pathologist uses to make a diagnosis.

The tools utilized to make a diagnosis may include, but are not limited to, immunohistochemical stains, histochemical stains, examination of the slides under polarized light, molecular or cytogenetic findings of the specimen, and consultation with other pathologists on diagnostically challenging cases.

4.3 Pathology Report

The pathology report delineates

1. Patient demographics
2. Origin of the tissue sample
3. Clinical history
4. Gross findings (macroscopic) of the specimen
5. Microscopic diagnosis with regard to benignity or malignancy

In benign cases, further subclassification of histopathology findings is reported. In malignant cases, based on reporting protocols [9], a pathology report may include, but is not limited to, the type of the cancer, tissue origin of the cancer, margins of the surgical resection, grade of the tumor, vascular invasion, depth of invasion, lymph node status if a lymph node dissection is performed, size of the cancer, prognostic factors such as mitotic count and receptor status, and predictive factors such as the absence or presence of a receptor to a targeted chemotherapy [9].

4.4 Prognostic and Predictive Markers in Pathology

A prognostic biomarker helps in determining the overall outcome for a patient regardless of therapy. There are many prognostic markers depending on the type and origin of the cancer. Examples of prognostic markers include the grade of the tumor, size of the tumor, receptor status, vascular invasion, stage of the tumor, and so on.

A predictive biomarker helps in determining the efficacy of and response to a particular therapeutic intervention. There are a number of predictive biomarkers that assist in the therapy of malignancies. New predictive markers are discovered through research and development every day. These help in improving both the morbidity and mortality in patients. Some of the examples include

1. HER2 (c-erbB-2 onco protein) receptor: HER2 is an 185kD transmembrane growth factor receptor protein. HER2 overexpression in breast cancer is associated with a poorer prognosis and survival. It has been found that an overexpression

of the HER2 receptor may benefit from Herceptin® therapy, particularly in patients with metastatic breast cancer.[10,11]

2. CD117 (c-kit) receptor: C-kit is a proto-oncogene that encodes a transmembrane tyrosine kinase receptor. It is overexpressed in tumor cells from a variety of cancers, mainly gastrointestinal stromal tumors (GIST) [12] and a subset of myeloid leukemia [13]. Gleevec therapy can be helpful in patients whose tumors express CD117 mutations.
3. EGFR receptor: Epidermal Growth Factor Receptor (EGFR) is a 170kD cell-surface receptor. The patients whose colon cancer overexpresses EGFR may be treated with cetuximab therapy [14, 15, 16]. It is also helpful in certain non-small-cell carcinomas of pulmonary origins [17] where erlotinib therapy can be utilized.
4. Estrogen receptors: Estrogen receptor positive breast cancer can be treated with Tamoxifen therapy [10, 11].

4.5 Opportunities for Biophotonics at the Nanoscale

As mentioned, the tissue slides and blocks are valuable sources for research. The pathologist makes the diagnosis by observing the slides under a microscope, which is a time-consuming process. Newer methodologies can be utilized to analyze the slides in an automated manner; this can aid the pathologist by saving time and improving the diagnostic accuracy.

Current procedures in cancer diagnosis involve the analysis of a thin slice of *stained tissue* by a trained pathologist. The staining process accounts for a significant portion of the time and cost involved in the procedure. Understanding the *nanoscale architecture of unstained tissues* and how it deteriorates with cancer will help develop more sensitive tools for screening, early diagnosis, and higher rates of successful treatments. Light has the unique capability of harvesting information from the tissue without physical contact. In particular, interferometry provides a powerful means of extracting optical thickness information with *nanometer accuracy* while preserving the diffraction-limited transverse resolution. We developed a new type of microscopy, Spatial Light Interference Microscopy (SLIM), which combines microscopy and interferometry, and thus renders nanoscale architectural information from unstained tissue slices. Preliminary results on skin and prostate biopsies indicate that this new approach is feasible and can become a significant clinical tool that will aid pathologists in diagnosing a broad range of cancers.

This research has the potential to transform cancer pathology in all clinical areas. Currently, the staining procedure costs a significant portion of the total tissue biopsy preparation. The SLIM optical method provides clinical information without staining the specimens; these will bring a tremendous reduction in cost. Tissue-staining processing takes approximately 1 hour from the biopsy to the pathologist's desk. Our procedure can reduce or eliminate this delay time. Such an imaging method can be implemented in fiber optics and applied *in vivo* to exposed tissue directly or to inner organs via needle delivery.

We have recently developed SLIM as a new optical method, capable of measuring optical path-length changes of 0.3 nm spatially (i.e., point-to-point change) and 0.03 nm temporally (i.e., frame-to-frame change) [18]. SLIM combines two classic ideas in light imaging: Zernike's phase contrast microscopy and Gabor's holography. The resulting topographic accuracy is comparable to that of atomic force microscopy (AFM), while the acquisition speed is 1000 times higher. Figure 4.15 shows SLIM and AFM measurements on a carbon film (color bar in nanometers). We used SLIM to image 4-μm thick tissue biopsy slices for prostate, skin, and breast cancer. From each tissue block, two successive slices were cut: one slice was stained with H&E and the second, unstained, was imaged by SLIM. An illustrative example is shown in Fig. 4.16, where the SLIM optical path-length map is shown in Fig. 4.16*a* (color bar in nanometers), the H&E stain in Fig. 4.16*b*, and the root mean square (rms) of the path-length fluctuations in Fig. 4.16*c* (color bar in nanometers). The SLIM images covered >1 cm^2, with a transverse resolution of 350 nm (in panel *a*, the arrows point to well-resolved individual cells). Remarkably, the rms path length associated with the unstained tissues correlates very well with the H&E image. The red and blue regions were drawn by a board-certified pathologist to indicate the cancer and normal area, respectively. It can be seen that the rms of the optical path length is significantly larger in the cancer region, suggestive of a more "disordered" tissue. The histogram of rms path length (Fig. 4.16*d*) captures this trend very clearly, as the curve broadens significantly for cancer tissue. This is a strong evidence that the *nanoscale distribution* of optical path lengths is an excellent intrinsic marker for cancer.

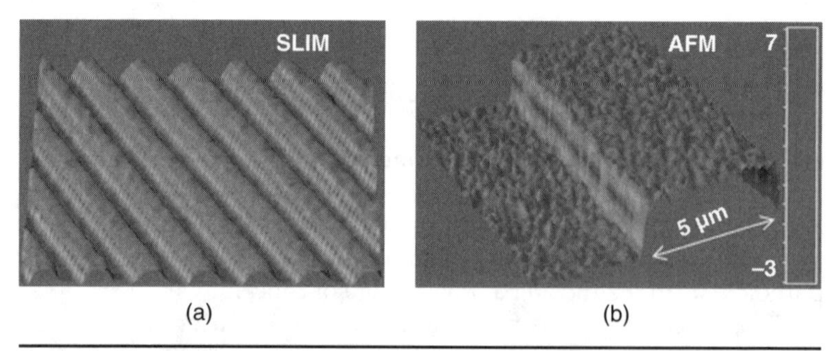

Figure 4.15 SLIM measurement of (*a*) a nanometer-thick carbon film as compared with (*b*) the AFM measurement.

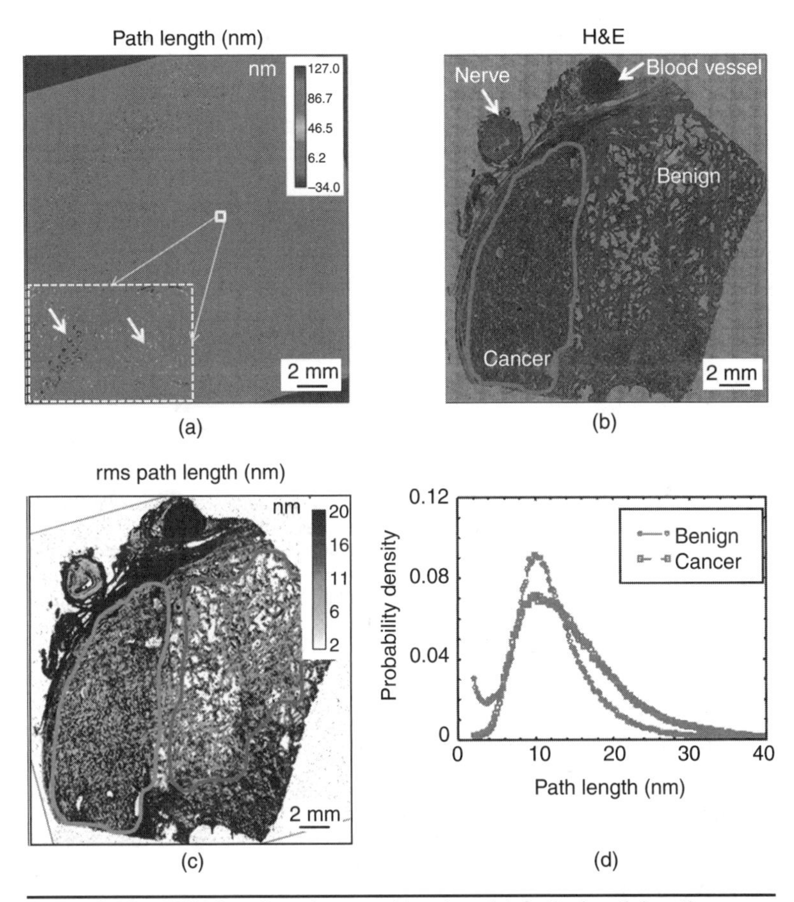

FIGURE 4.16 SLIM imaging of cancer biopsies: (*a*) Optical path-length map, (*b*) H&E stain, and (*c*) rms path length and path-length distributions for cancer and normal, as indicated.

4.6 Summary

In summary, pathologists are helpful in arriving at a final diagnosis. The specimen processing in a histology laboratory is a complex and labor-intensive process. Rapid developments are being made in molecular tumor pathology. These newer techniques help facilitate the patient's treatment using newer targeted therapies. Based on a pathologist report, further interventions to alleviate the patient's overall health are instituted. There is a greater need for thorough and timely pathology reports, which are essential in instituting the targeted therapies in patients with malignancies. Newer optical techniques can assist the pathologist not only by saving time but also by helping with the accuracy of the diagnosis. Detecting subtle, that is, nanoscale, tissue changes that occur early in cancer development may provide the future key for a cure. Such translational research can potentially transform the practice of pathology.

Acknowledgments

I am grateful for the images shown in Figs. 4.15 and 4.16, which measured by Zhuo Wang at Prof. Gabriel Popescu's Quantitative Light Imaging Laboratory, UIUC. I acknowledge Drs. Karthik Vaideeswaran and Anil Gopinath's valuable insights during the preparation of this chapter. I am obliged to Chuck McNeil for reviewing the manuscript and helping with the pictures.

References

1. V. Kumar, A. K. Abbas, J. Aster, and N. Fausto, *Robbins & Cotran Pathologic Basis of Disease*, 8th ed., Elsevier Health Sciences, 2009.
2. D. Chhieng, D. Jhala, N. Jhala et al. "Endoscopic Ultrasound-Guided Fine-Needle Aspiration Biopsy: A Study of 103 Cases," *Cancer* 96: 232–239, 2002.
3. M. D. Abeloff, J. O. Armitage, J. E. Niederhuber, M. B. Kastan, and G. McKenna, *Clinical Oncology*, 4th ed., Elsevier Health Sciences, 2008.
4. J. A. Kiernan, *Histological and Histochemical Methods: Theory and Practice*, 4th ed., Cold Spring Harbor Laboratory Press, Cold Spring Harbor, NY, 2008.
5. J. D. Bancroft and M. Gamble, *Theory and Practice of Histological Techniques*, 6th ed., Elsevier, Philadelphia, 2007.
6. *Advanced Laboratory Methods in Histology and Pathology*, Armed Forces Institute of Pathology, Washington, DC, 1994.
7. S. C. Lester, *Manual of Surgical Pathology*, 2nd ed., W. B. Saunders Co., Philadelphia, 2005.
8. H. Hibshoosh and R. Lattes, "Immunohistochemical and Molecular Genetic Approaches to Soft-Tissue Tumor Diagnosis: A Primer," *Seminars in Oncology* 24(5): 515–525, 1997.
9. J. R. Srigley, T. McGowan, A. Maclean, M. Raby, J. Ross, S. Kramer, and C. Sawka, "Standardized Synoptic Cancer Pathology Reporting: A Population-Based Approach," *J. Surg. Oncol.* 99(8): 517–524, 2009.
10. J. S. Ross, E. A. Slodkowska, W. F. Symmans, L. Pusztai, P. M. Ravdin, and G. N. Hortobagyi, "The HER-2 Receptor and Breast Cancer: Ten Years of Targeted Anti-HER-2 Therapy and Personalized Medicine," *Oncologist*, 14(4): 320–368, 2009.
11. F. Rastelli and S. Crispino, "Factors Predictive of Response to Hormone Therapy in Breast Cancer," *Tumori*, 94(3): 370–383, 2008.
12. M. Miettinen and J. Lasota, "Gastrointestinal Stromal Tumors: Review on Morphology, Molecular Pathology, Prognosis, and Differential Diagnosis," *Arch. Pathol. Lab. Med.* 130(10): 1466–1478, 2006.
13. H. de Lavallade, J. F. Apperley, J. S. Khorashad, D. Milojkovic, A. G. Reid, M. Bua, R. Szydlo, E. Olavarria, J. Kaeda, J. M. Goldman, and D. Marin, "Imatinib for Newly Diagnosed Patients with Chronic Myeloid Leukemia: Incidence of Sustained Responses in an Intention-to-Treat Analysis," *J. Clin. Oncol.* 26(20): 3358–3363, 2008.
14. S. Siena, A. Sartore-Bianchi, F. Di Nicolantonio, J. Balfour, and A. Bardelli, "Biomarkers Predicting Clinical Outcome of Epidermal Growth Factor Receptor-Targeted Therapy in Metastatic Colorectal Cancer," *J. Natl. Cancer Inst.* 101(19): 1308–1324, 2009.
15. J. Wils "Adjuvant Treatment of Colon Cancer: Past, Present, and Future," *J. Chemother.* 19(2): 115–122, 2007.

16. T. P. Plesec and J. L. Hunt, "KRAS Mutation Testing in Colorectal Cancer," *Adv. Anat. Pathol.* 16(4): 196–203, 2009.
17. T. K. Owonikoko, S. Y. Sun, and S. S. Ramalingam, "The Role of Cetuximab in the Management of Non-Small-Cell Lung Cancer," *Clin. Lung Cancer*, 10(4): 230–238, 2009.
18. Z. Wang, L. J. Millet, H. Ding, M. Mir, S. Unarunotai, J. A. Rogers, M. U. Gillette, and G. Popescu, "Spatial Light Interference Microscopy (SLIM)," *Nature Methods*, under review.

CHAPTER 5

Light Scattering in Inhomogeneous Media

Gabriel Popescu

Quantitative Light Imaging Laboratory
Department of Electrical and Computer Engineering
Beckman Institute for Advanced Science and Technology
University of Illinois at Urbana–Champaign

5.1 Elastic (Static) Light Scattering

5.1.1 Scattering Properties of One Dielectric Particle

Particle is a generic term used in the context of light scattering; it refers to a localized region in space that has optical properties different from those of the surrounding medium. The purpose of this chapter is to introduce some of the basic concepts of wave scattering by a single dielectric particle and by distribution of such particles. Rather than presenting a comprehensive description of scattering phenomena, we will restrict the presentation and only define the main scattering parameters, reviewing the scattering models as they apply to specific situations that are encountered in practice.

Let us consider a linearly polarized electromagnetic plane wave,

$$\mathbf{E}_i(\mathbf{r}) = \mathbf{E}_0 e^{i \cdot \mathbf{kr}} \tag{5.1}$$

where $\mathbf{k} = \omega/c$ is the wave number and c is the wave velocity in the medium. The time variation of the field is included in the amplitude $\mathbf{E}_0$. Let the wave be the incident on a particle with a complex dielectric constant given by

$$\varepsilon_r(\mathbf{r}) = \frac{\varepsilon(\mathbf{r})}{\varepsilon_0} = \mathrm{Re}\left[\varepsilon_r(\mathbf{r})\right] + i\,\mathrm{Im}\left[\varepsilon_r(\mathbf{r})\right] \tag{5.2}$$

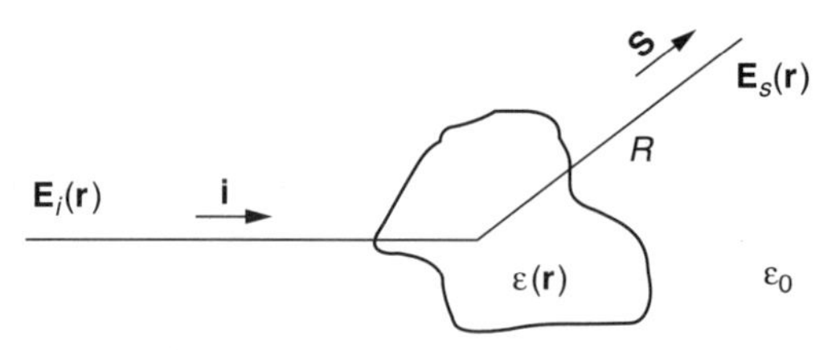

FIGURE 5.1 Light scattering by an arbitrary particle.

The complex part of $\varepsilon_r(\mathbf{r})$ defines the loss in the particle, while its position dependence includes the general case of *an* inhomogeneous particle. Regarding Fig. 5.1, the field at the distance R from a reference point inside the particle represents the superposition of the incident field $\mathbf{E}_i$ and the scattered field $\mathbf{E}_s$. Conventional scattering theory deals with the electromagnetic waves in the far field of the scatterers. Usually, the far field is defined by the condition $R > d^2/\lambda$, with d the dimension of the particle; in this region, $\mathbf{E}_s$ approaches a spherical wave [1].

$$\mathbf{E}_s(\mathbf{r}) = \mathbf{E}_0 \frac{e^{i\cdot kR}}{R} f(\mathbf{s},\mathbf{i}) \tag{5.3}$$

Here, $f(\mathbf{s},\mathbf{i})$ represents the amplitude, phase, and polarization of the scattered wave in the far field and along the direction $\mathbf{s}$ when the particle is illuminated by a plane wave from direction $\mathbf{i}$. Equation 5.3 represents a propagating spherical wave which originates inside the particle; its magnitude depends on the characteristics of the scattering event through the function $f(\mathbf{s},\mathbf{i})$.

The efficiency of a scattering process in terms of the energy balance is measured by the differential scattering cross section, which is defined as

$$\sigma_d(\mathbf{s},\mathbf{i}) = \lim_{R\to\infty}\left[R^2 \frac{S_s}{S_i}\right] = |f(\mathbf{s},\mathbf{i})|^2 \tag{5.4}$$

S_s and S_i are the magnitudes of the Poynting vectors corresponding to a scattered and an incident wave, respectively.

$$\mathbf{S}_i = \frac{1}{2\eta}|\mathbf{E}_i|^2 \mathbf{i}, \quad \text{and} \quad \mathbf{S}_s = \frac{1}{2\eta}|\mathbf{E}_s|^2 \mathbf{s} \tag{5.5}$$

In Eq. (5.5), η is the impedance of the surrounding medium, defined as the square root of the ratio between the magnetic and electric permeability; $\eta = \sqrt{\mu/\varepsilon}$. The *total cross section of scattering* σ has units of area and can be interpreted as a measure of the scattering probability. The ratio between the scattering differential and total cross section is called the *phase function* and defines the angular distribution of the scattered light.

$$p(\mathbf{s},\mathbf{i}) = 4\pi\frac{\sigma_d(\mathbf{s},\mathbf{i})}{\sigma} = 4\pi\frac{|f(\mathbf{s},\mathbf{i})|^2}{\sigma} \tag{5.6}$$

As one can see, the phase function $p(\mathbf{s},\mathbf{i})$ describes the anisotropy of a scattering process, and it is therefore extensively used in multiple-scattering theory.

In some applications, it is of interest to have a measure of how strongly the particle scatters light in the direction opposite to the incident one. One can define a *backscattering (radar) cross section* σ_b, which relates to the differential cross section σ_d as

$$\sigma_b = 4\pi\sigma_d(-\mathbf{i},\mathbf{i}) \tag{5.7}$$

The total observed power at all the angles surrounding the particle is described by the scattering cross section σ_s.

$$\sigma_s = \int_{4\pi} \sigma_d \, d\Omega = \frac{\sigma}{4\pi}\int_{4\pi} p(\mathbf{s},\mathbf{i})\, d\Omega \tag{5.8}$$

Note that $\sigma_s = \sigma$ only in the absence of absorption. In general, the ratio between the scattering and the total cross section, also called *albedo,* is smaller than unity. The quantity that measures the total power, which is absorbed by the particle is the absorption cross section, and it relates to the total cross section (extinction) and scattering cross section by

$$\sigma = \sigma_s + \sigma_a \tag{5.9}$$

Equation 5.9 merely states the law of energy conservation as applied to an individual scattering process.

So far, we have presented the main descriptors of the scattering process that involves a particle with arbitrary size and shape; note that Eq. (5.8) is very general and states that if the detailed phase function is known, then the energy balance of the scattering process can be established. In the following, we will discuss how this general theory applies to scatterers with specific properties, such as size, shape, and refractive index.

5.1.2 Rayleigh Particles

When the scattering particle is much smaller in size than the wavelength of light, the field inside the sphere can be approximated as uniform and the radiation pattern approaches that of a simple dipole.

$$f(\mathbf{s},\mathbf{i}) = \frac{k^2}{4\pi}\left[\frac{3(\varepsilon_r - 1)}{\varepsilon_r + 2}\right] V\left[-\mathbf{s}\times(\mathbf{s}\times\mathbf{i})\right] \tag{5.10}$$

In Eq. (5.10), V is the volume of the particle and **s** and **i** are the unit vectors associated with the incident and scattered directions, respectively. Under these conditions, the scattering process is commonly referred to as the *Rayleigh regime,* and by Eq. (5.8), the scattering cross section is found to be [1]:

$$\sigma_s = \pi a^2 \frac{8(ka)^4}{3} \left| \frac{\varepsilon_r - 1}{\varepsilon_r + 2} \right|^2 \tag{5.11}$$

It is worth noting the dependence of the Rayleigh scattering cross section with the fourth power of the optical frequency and with the sixth power of the particle radius a.

When the total phase delay associated with the scattering process is very small, the scattering lies in the so-called Rayleigh-Debye scattering regime. Note that the restriction is now not only in terms of the size but also involves the dielectric constant ε_r of the particle.

$$(\varepsilon_r - 1)kd << 1 \tag{5.12}$$

where d is the dimension of the particle. In this case, the electric field inside the particle is approximated to the incident field $\mathbf{E}_i$ and the scattering function is no longer isotropic.

$$f(\mathbf{s},\mathbf{i}) = \frac{k^2}{4\pi}[-\mathbf{s}\times(\mathbf{s}\times\mathbf{i})]\cdot \int_{4\pi} [\varepsilon_r(\mathbf{r}') - 1]\cdot e^{i\cdot \mathbf{q}\cdot \mathbf{r}'} dV' \tag{5.13}$$

As can be seen, the scattering function is proportional to the Fourier transform of the function $\varepsilon_r(\mathbf{r}') - 1$, evaluated at the wave vector $\mathbf{q} = k(\mathbf{s} - \mathbf{i})$, with $|\mathbf{q}| = 2k\sin(\theta/2)$, and θ the scattering angle. The Fourier transform relationship in Eq. (5.13) represents the first Born approximation, that is, a weakly scattering approximation.

For a spherical particle of radius a, Eq. (5.13) becomes

$$\begin{aligned} f(\mathbf{s},\mathbf{i}) &= \frac{k^2}{2\pi}[-\mathbf{s}\times(\mathbf{s}\times\mathbf{i})](\varepsilon_r - 1)VF(\theta) \\ F(\theta) &= \frac{3}{q^2 \cdot a^3}[\sin(qa) - qa\cos(qa)] \end{aligned} \tag{5.14}$$

It is worth noting that the scattering function associated with the Rayleigh-Debye regime factor into a dipole and an anisotropic contribution. The fact that the phase function depends only on the size of the scattering center makes this approximation extremely appealing for applications where a simple analytical form is needed for the scattering amplitude.

5.1.3 Mie Theory

The exact solution of the scattering of a plane electromagnetic wave by a homogeneous and isotropic sphere was obtained by Mie in 1908. Using Hertz potentials and continuity equations for the tangential fields at the boundary, the exact expression for the scattering cross section can be obtained [1].

$$\sigma_s = \pi a^2 \frac{2}{\alpha^2} \sum_{n=1}^{\infty} (2n+1)(|a_n|^2 + |b_n|^2) \tag{5.15}$$

where the quantities a_n and b_n are defined as

$$\begin{aligned} a_n &= \frac{\psi_n(\alpha)\psi_n'(\beta) - m\psi_n(\beta)\psi_n'(\alpha)}{\zeta_n(\alpha)\psi_n'(\beta) - m\psi_n(\beta)\zeta_n'(\alpha)} \\ b_n &= \frac{m\psi_n(\alpha)\psi_n'(\beta) - \psi_n(\beta)\psi_n'(\alpha)}{m\zeta_n(\alpha)\psi_n'(\beta) - \psi_n(\beta)\zeta_n'(\alpha)} \end{aligned} \tag{5.16}$$

The parameters a_n and b_n, referred to as *Mie coefficients*, are expressed through the special functions ψ and ζ. The summation in Eq. (5.15) can only be evaluated numerically, but, with the computing power available today, that became a rather facile task. In Eq. (5.16), $\alpha = ka$, and $\beta = kma$, with m the refractive index of the particle relative to the environment. The dependence of the scattering efficiency on the size parameter ka is depicted in Fig. 5.2. One can see that, for large-size parameters, the curve becomes more rippled, which is an indication of the resonant modes that develop in large spheres. In addition, note that in the macroscopic limit, that is, for infinitely large spheres, the asymptotic value of the scattering efficiency is 2. While limited only

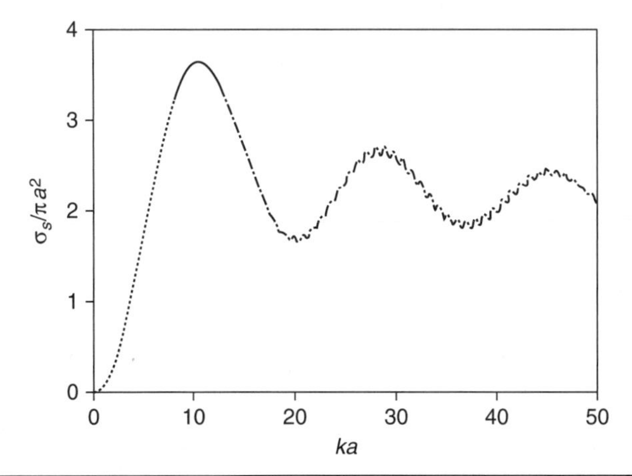

Figure 5.2 Scattering efficiency versus size parameter for a spherical particle.

to spherical shapes, the Mie formalism brings important physical insight into the scattering interaction in general. Other analytical models, accounting for scattering on particles of different shapes, are available as well for specific situations.

5.1.4 Single-Scattering Approximation for a Volume Distribution of Dielectric Particles

In practical situations, one rarely encounters isolated scattering particles. Thus, for a collection of particles, the parameters introduced in the previous section must be corroborated with the information about the whole ensemble. The situation is depicted in Fig. 5.3, where a volume V containing a random distribution of particles is illuminated by a continuous wave (c.w.) beam with an angular distribution $\Omega(\mathbf{i})$. The practical task is to find the total power received at the detector D that has a cross-section A_D and an angular acceptance of $\Omega(\mathbf{s})$. Considering that the scattering medium is in the far field of both the source and the detector, the total power detected is given by [2]

$$P_D = P_0 \int_V \frac{\lambda^2 \boldsymbol{\Omega}(\mathbf{i}) \cdot \boldsymbol{\Omega}(\mathbf{s})}{(4\pi)^3 R_1^2 R_2^2} \cdot N \sigma_{bi}(s,i)\, dV \tag{5.17}$$

$$\sigma_{bi}(\mathbf{s},\mathbf{i}) = 4\pi \left| f(\mathbf{s},\mathbf{i}) \right|^2 = \sigma \cdot p(\mathbf{s},\mathbf{i}) \tag{5.18}$$

In Eq. (5.17), P_0 is the incident power, N the volume density of particles, and λ is the wavelength of the electromagnetic field. R_1 and R_2 are the distances from the scattering element to the source and receiver, respectively, as depicted in Fig. 5.3. The efficiency of scattering from the incident direction **i** to the scattered direction **s** is quantified by the bistatic cross section σ_{bi}. Equation 5.17 is of high importance in many

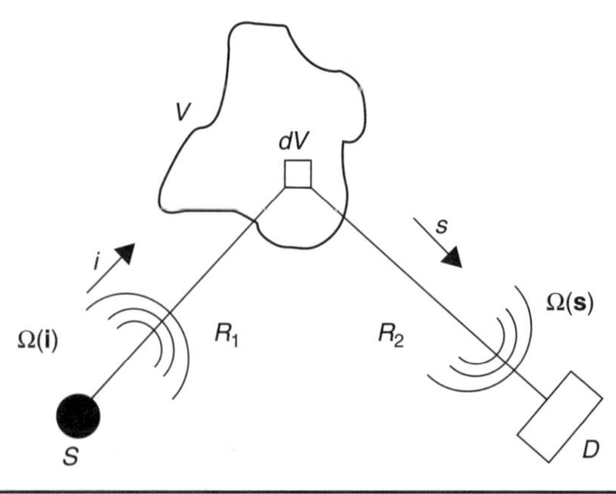

Figure 5.3 Scattering geometry, where the incident and scattered light are not co-directional (bistatic geometry).

applications that involve a point-source and point-detection geometry; it provides the total scattered power collected at a certain angle as a function of the properties of the scattering medium, divergence of the incident beam, and collection angle at the detector.

A particular case of this geometry, pertinent to the work presented in the following chapters, is obtained when one makes use of optical fibers to both send the light into and collect it from an inhomogeneous medium. In this case, Eq. (5.17) has a particular form which applies for $\mathbf{s} = -\mathbf{i}$. This geometry is referred to as the *monostatic radar (backscattering) configuration,* and the total power collected is given by

$$P_D^b = P_0 \int_V \frac{\lambda^2 |\Omega(\mathbf{i})|^2 N\sigma_b}{(4\pi)^3 R^4} dV \tag{5.19}$$

In this particular case, the bistatic cross section becomes the backscattering cross section σ_b and the geometry is characterized by only one distance $R_1 = R_2 = R$. The monostatic geometry that uses a point-like source and a point-like detector is very common in measurements on tissues, where the backscattering light is of interest.

Equation 5.19 is an idealization of the physical phenomenon, as it does not contain any information about the light attenuation during the multiple-scattering process. For situations where the field undergoes multiple-scattering events, this correction turns out to be quite important.

The attenuation due to the scattering of a plane wave during the propagation through a scattering medium can be quantified by a global parameter characterizing the medium, that is, its scattering mean free path.

$$l_s = \frac{1}{N\sigma} \tag{5.20}$$

Thus, after a propagation distance L, the remaining, unscattered power is given simply by the Lambert-Beer law of attenuation.

$$P = P_0 e^{-L/l_s} \tag{5.21}$$

Using Eqs. (5.19 and 5.21), we can write a more accurate result for the power received in the backscattering geometry, which includes the attenuation due to successive scattering events.

$$P_D^b = P_0 \int_V \frac{\lambda^2 |\Omega(\mathbf{i})|^2 N\sigma_b}{(4\pi)^3 R^4} \cdot e^{-\frac{2r}{L_S}} dV \tag{5.22}$$

Equation 5.22 represents the so-called first-order multiple-scattering approximation, since it takes into account the attenuation due to the scattering process but neglects the component of the light that is scattered into the scattering direction $\mathbf{s}$. Several approaches to account for higher-order multiple scattering will be presented in Section 5.2.

5.2 Quasi-Elastic (Dynamic) Light Scattering

In the previous sections, we discussed the scattering phenomenon in general, by describing the so-called direct problem for certain situations associated with scattering by a single particle or by collections of particles. In general, solving the direct scattering problem means to fully quantify the scattering process once the scattering medium is known. In Section 5.1.2, we did not take into account the fact that the scattering centers may change their properties in time. In other words, we tacitly assumed that the detector is slow enough to integrate over the time fluctuations of the scattered light. In the following, we will quantify these fluctuations in order to infer information about the scatterers. This is an aspect of solving the *inverse scattering problem*, that is, using the scattered light to infer information about the (dynamic) scattering medium.

Time fluctuations of the scattered field arise from temporal modifications of the scattering potential $\varepsilon_r(\mathbf{r}) - 1$. This may be due to time fluctuations of the dielectric constant, spatial coordinates, or both. In this section, we will discuss the situations where the only time variable is the coordinate of the scattering centers. While these displacements can be due to intrinsic (internal) or extrinsic (external) forces, we will restrict the discussion to intrinsic forces, where the scattering system is considered isolated from any external influence.

Dynamic systems such as solids, liquid crystals, gels, solutions of biological macromolecules, simple molecular fluids, electrolyte solutions, dispersions of microorganisms, membrane vesicles, and colloidal suspensions have been investigated using laser light for quite some time. The principle of the technique, referred to as *dynamic light scattering,* relies on quantifying, in a time-resolved manner, the displacements of the particulates dispersed in a solution. The mean square displacement is retrieved from the analysis of the temporal fluctuations of the light scattered by the system and can be used to obtain information about the structural properties of the dispersed medium.

The schematic of a typical dynamic light-scattering experiment is shown in Fig. 5.4. The incident monochromatic light is scattered by the dynamic system, and the fluctuating scattered light is detected and subsequently analyzed by an autocorrelator. In order to evaluate this intensity autocorrelation function, the mechanical behavior of the dynamic scatterers first has to be modeled. One of the most common types of particle motion encountered in practice is Brownian motion, which typically describes the evolution of colloids under thermal excitation. Within this model, the displacement of the particles $\Delta R(t)$ is a result of numerous independent interactions with the surrounding fluid. Thus, the probability of displacement within

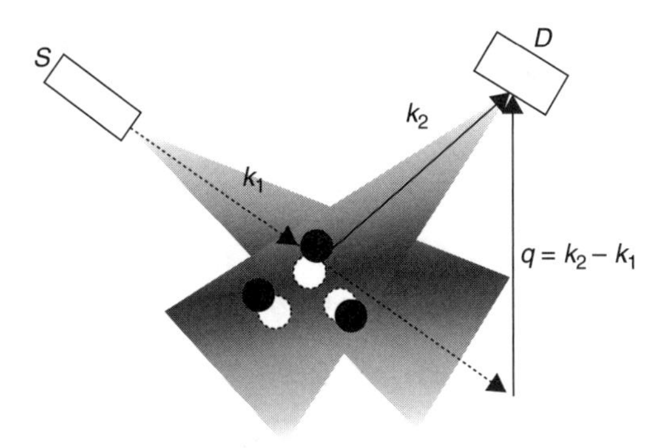

FIGURE 5.4 Generic dynamic light-scattering geometry.

an elementary volume d^3R around R has a Gaussian (normal) shape [3].

$$G(R,t) = \left[\frac{2\pi}{3}\cdot\left\langle \Delta R^2(t)\right\rangle\right]^{-3/2} \exp\left[-\frac{3R^2}{2\left\langle \Delta R^2(t)\right\rangle}\right] \tag{5.23}$$

where $G(R, t)$ is normalized such that $\int G(R,t)d^3R = 1$. The width of the distribution relates to the particle mean square displacement $\langle \Delta R^2(t)\rangle$. Due to the displacement of each individual particle, the optical field undergoes phase changes during the scattering process, which results in intensity fluctuations of the scattered light. Neglecting the amplitude fluctuations of the scattered light, the factor that quantifies the light fluctuations has the form

$$F(q,t) = \langle \exp\ \{iq[r_j(t) - r_j(0)]\}\rangle_j \tag{5.24}$$

Equation 5.24, $F(q,t)$ is referred to as the *intermediate scattering function* and represents an average of the phase factors over the ensemble of particles of coordinates r_j. The scattering vector (or *momentum transfer*) q relates to the scattering angle θ by $q = (4\pi/\lambda)\sin\theta$. Using the probability density $G(R, t)$ of Eq. (5.23) to calculate this average, one obtains the final expression for the intermediate scattering function [4].

$$F(q,t) = \exp\left[-\frac{q^2\langle \Delta R^2(t)\rangle}{6}\right] \tag{5.25}$$

For Brownian particles, the mean square displacement $\langle \Delta R^2(t)\rangle$ can be related directly to the diffusion coefficient of particles in the suspension.

$$\left\langle \Delta R^2(t)\right\rangle = 6Dt \tag{5.26}$$

Using the Stokes-Einstein equation, D relates to the size a of the particle (assumed spherical) and the viscosity η of the surrounding liquid.

$$D = \frac{k_B T}{6\pi\eta a} \tag{5.27}$$

where k_B is the Boltzmann constant. The field autocorrelation function of the scattered field is given simply by

$$\Gamma(\tau) = I_0 F(q,\tau) \tag{5.28}$$

where I_0 is the average-scattered intensity. Thus the normalized first-order autocorrelation function $\Gamma(\tau)/\Gamma(0)$ becomes

$$g^{(1)} = \exp\left[-\frac{q^2\langle\Delta R^2(t)\rangle}{6}\right] \tag{5.29}$$

In practice, the intensity rather than the field autocorrelation function is usually the measurable quantity. For radiation that is fully coherent both spatially and temporally, the second- and first-order autocorrelations are connected through a simple relation [3].

$$g^{(2)}(\tau) = 1 + \left|g^{(1)}(\tau)\right|^2 \tag{5.30}$$

Equation 5.30, which relates the measurable quantity $g^{(2)}(\tau)$ to the field autocorrelation function $g^{(1)}(\tau)$ is referred to as the *Siegert relation*.

In conclusion, by measuring the intensity autocorrelation function of the scattered light, one can retrieve information about the motion of the particles [Eqs. (5.29 and 5.30)], which can be further related to the properties of the liquid or size of the particles [Eqs. (5.26 and 5.27)]. New advances of the dynamic light-scattering technique allow enlarging its applicability toward investigating structural properties of various systems, including polydisperse colloidal suspensions, glassy systems, and viscoelastic fluids [5].

5.3 Multiple Scattering

In Section 5.2, we introduced the basic description of the single-scattering process, which applies to sparse distribution of particles. For the single-scattering regime to be applicable, the average dimension of the scattering medium must be smaller than the *mean free path* l_s. However, in many situations this is not the case, and the light travels through the medium over distances much longer than l_s and encounters many scattering events during the propagation. It is obvious that, for this case, the simple treatment presented earlier does not apply and the complexity of the problem is tremendously increased. Actually, a full vector solution to the field propagation in dense media is not available; scalar theories give results in an analytical form only for special, intermediate cases of multiple-scattering regimes. Nevertheless, important simplifications can be introduced whenever the scattering medium is very dense and the electromagnetic waves can be assumed to travel randomly through the medium.

5.3.1 Elements of Radiative Transport Theory

The transport description in the case of a highly scattering medium is based on a photon random walk picture that is similar to that used in various other areas of physics, such as solid-state physics and nuclear reactor physics [6]. The wave correlations that may survive this heavy scattering regime are completely neglected, while each individual scattering event is considered an independent process.

Using the energy- (photon-) conservation principle, the final goal of the transport theory is to find the time-dependent photon distribution at each point in the scattering medium. We will make use of the following definitions and notations:

- $N(r,t)$: photon density [m^{-3}]
- $\nu\mu_s = \nu/l_s$: frequency of interaction [s^{-1}]
- $\mu = \mu_a + \mu_s$: attenuation coefficient [m]
- $\Phi\,(r,\,t) = \nu N(\mathbf{r},\,t)$: photon flux [$m^{-2}s^{-1}$]
- $n(\boldsymbol{r},\,\boldsymbol{\Omega},\,t)$: density of photons [$m^{-3}sr^{-1}$] having the direction of propagation within the solid angle interval [Ω, $\Omega + d\Omega$]
- $\varphi\,(\mathbf{r},\,\boldsymbol{\Omega},\,t)R = \nu n(\mathrm{r},\,\boldsymbol{\Omega},\,t)$: angular photon flux [$m^{-2}s^{-1}sr^{-1}$]; *scalar quantity*
- $\mathbf{J}(\mathrm{r},\,\boldsymbol{\Omega},\,\mathrm{t}) = \boldsymbol{\Omega}\varphi(\mathbf{r},\,\boldsymbol{\Omega},\,t)$: angular current density [$m^{-2}s^{-1}sr^{-1}$]; *vector quantity*
- $\mathbf{J}(\mathbf{r},\,t) = \int_{4\pi} \mathbf{j}(\mathbf{r},t,\Omega)\,d\Omega$ photon current density [$m^{-2}s^{-1}$]

In order to write the transport equation, we first identify all the contributions that affect the photon balance within a specific volume V, as depicted in Fig. 5.5. Thus, we obtain the following transport equation in its integral form:

$$\int_V \left\{ \frac{dn}{dt} + v\Omega\nabla n + v\mu n(\mathbf{r},\boldsymbol{\Omega},t) - \int_{4\pi} [v'\mu_s(\Omega' \rightarrow \Omega) n(\mathbf{r},\boldsymbol{\Omega}',t)]\, d\Omega' - s(\mathbf{r},\Omega,t) \right\} d^3r\, d\Omega = 0 \qquad (5.31)$$

The first term denotes the rate of change in the energy density, the second term represents the total leakage out of the boundary, the third term is the loss due to scattering, and the fourth term accounts for the photons that are scattered into the direction $\boldsymbol{\Omega}$, while the

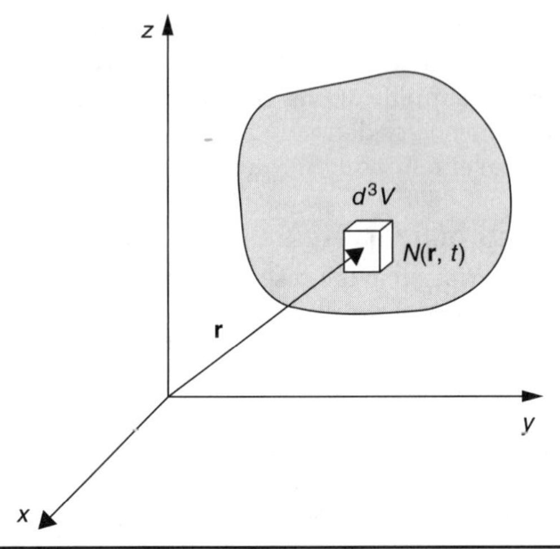

FIGURE 5.5 Elementary volume in a strongly scattering medium.

last term represents the possible source of photons inside the volume. Using the fact that the volume V was arbitrarily chosen, the integral of Eq. (5.31) is identically 0 if and only if the integrand vanishes. Thus, the differential form of the transport equation is obtained as

$$\frac{dn}{dt} + \nu\Omega\nabla n + \nu\mu n(\mathbf{r},\Omega,t)$$
$$- \int_{4\pi} [\nu'\mu_s(\Omega' \to \Omega) n(r,\Omega',t)]\, d\Omega' - s(\mathbf{r},\Omega,t) = 0 \tag{5.32}$$

The same equation can be expressed in terms of the angular flux $\varphi(\mathbf{r}, \mathbf{\Omega}, \mathrm{t}) = \nu \cdot n(\mathbf{r}, \mathbf{\Omega}, \mathrm{t})$.

$$\frac{1}{\nu}\frac{d\phi}{dt} + \Omega\nabla\phi + \mu\phi(\mathbf{r},\Omega,t)$$
$$- \int_{4\pi} [\mu_s(\mathbf{\Omega}' \to \mathbf{\Omega})\phi(\mathbf{r},\mathbf{\Omega}',t)]\, d\Omega' - s(\mathbf{r},\mathbf{\Omega},t) = 0 \tag{5.33}$$

The initial condition for this equation is $\varphi(\mathbf{r}, \mathbf{\Omega}, 0) = \varphi_0(\mathbf{r}, \mathbf{\Omega})$, while the boundary condition is $\varphi(\mathbf{r}_s, \mathbf{\Omega}, \mathrm{t})_{rs\in S} = 0$ if $\mathbf{\Omega} e_s < 0$. The transport equation, although just an approximation of the real physical problem, can be solved in most cases only numerically. As we will see in the next section, further simplifications to this equation are capable of providing solutions in closed forms and will, therefore, receive special attention.

5.3.2 Diffusion Approximation of the Transport Theory

Equation 5.33 can be integrated with respect to the solid angle $\mathbf{\Omega}$, such that the result can be expressed in terms of the photon flux Φ and the current density $\mathbf{J}$ as a continuity equation.

$$\frac{1}{v}\frac{d\Phi}{dt}+\nabla\mathbf{J}(\mathbf{r},t)+\mu\Phi(\mathbf{r},t)=\mu_s\Phi(\mathbf{r},t)+s(\mathbf{r},t) \tag{5.34}$$

As can be seen, Eq. (5.34) contains two independent variables Φ and $\mathbf{J}$.

$$\begin{aligned}\Phi(\mathbf{r},t)&=\int_{4\pi}\phi(\mathbf{r},\Omega,t)\,d\Omega\\ J(\mathbf{r},t)&=\int_{4\pi}\Omega\phi(\mathbf{r},\Omega,t)\,d\Omega\end{aligned} \tag{5.35}$$

Therefore, an additional assumption is needed to make the computations tractable. In most cases, one can assume that the angular flux ϕ is only linearly anisotropic.

$$\phi(\mathbf{r},\Omega,t)=\frac{1}{4\pi}\Phi(\mathbf{r},t)+\frac{3}{4\pi}J(\mathbf{r},t)\cdot\Omega \tag{5.36}$$

Equation 5.36 represents the first-order Legendre expansion of the angular flux and is the key approximation that allows obtaining the diffusion equation. It is also referred to as the $\mathbf{P}_1$ approximation, since it represents a Legendre polynomial P_l, for $l = 1$. In addition, the photon source is considered to be isotropic and the photon current to vary slowly with time: $\frac{1}{|\mathbf{J}|}\frac{\partial\mathbf{J}}{\partial t} << \nu\mu_t$. Under these conditions, the equation that relates $\mathbf{J}$ and Φ can be derived and is known as *Fick's law*.

$$J(\mathbf{r},t)=-\frac{1}{3\mu_t}\cdot\nabla\Phi(\mathbf{r},t)=-D(r)\nabla\Phi(\mathbf{r},t) \tag{5.37}$$

In Eq. (5.37), $\mu_t = \mu - g\mu_s$ is *the transport coefficient*. The parameter g represents the *anisotropy factor* and is defined in terms of elementary solid angles before and after scattering as

$$g=\langle\Omega\cdot\Omega'\rangle \tag{5.38}$$

The anisotropy factor has the meaning of the average cosine of the scattering angle, with the phase function defined in Eq. (5.6) as the weighting function. The parameter g accounts for the fact that, for large particles, the scattering pattern is not isotropic. In Eq. (5.37), $D(r)$ is the diffusion coefficient of photons in the medium and is the only parameter that contains information about the scattering medium. At this point, by combining Eqs. (5.34 and 5.37), we obtain the *diffusion equation,* which is the main result of this section and is extensively used in modeling light-tissue interactions.

$$\frac{1}{v}\frac{\partial \Phi}{\partial t} - \nabla\left[D(r)\nabla\Phi(\mathbf{r},t)\right] + \mu_a(\mathbf{r})\Phi(\mathbf{r},t) = S(\mathbf{r},t) \tag{5.39}$$

In Eq. (5.39), μ_a is the absorption coefficient and $S(\mathbf{r},t)$ is the photon source considered to be isotropic. Note that the diffusion equation breaks down for short times of evolution, that is, for $t < l_t/c$. In other words, it takes a finite amount of time for the diffusion regime to establish. Although restricted to a particular situation of multiple-light scattering, the diffusion treatment finds important applications in practice.

Using appropriate boundary conditions, Eq. (5.39) can be solved for particular geometries. Once the photon flux Φ is calculated, the current density J can be obtained using Fick's law. The current density $J(\mathbf{r},t)$ is the measurable quantity in a scattering experiment and thus deserves special attention. The total optical power measured by a detector is proportional with the integral of $J(\mathbf{r},t)$ over the area of detection. The temporal variable t represents the time of flight for a photon between the moment of emission and that of detection. When a constant group velocity is assumed, this variable can be transformed into an equivalent one—the optical path length, $s = vt$. Thus, the path-length probability density $P(s) = J(s)/\int_0^\infty J(s)\,ds$, that is, the probability that the photons traveled, within the sample an equivalent optical path length in the interval $(s, s + ds)$, can also be evaluated. This path-length distribution is crucial for situations where strong multiple scattering is present. $P(s)$ is what is physically measured in optical coherence tomography (Chapter 8).

5.3.3 Diffusive Wave Spectroscopy

The principle of dynamic light scattering (DLS), as it applies to investigating dynamic systems that weakly scatter light, has been presented in Section 5.2. The main restriction for the applicability of this technique is that the light must scatter only once before it is detected. This obviously requires the medium under investigation to be weakly scattering. However, there are numerous situations of interest where the systems scatter light strongly, and thus, the classical dynamic light scattering cannot be applied.

Diffusive wave spectroscopy (DWS) is an extension of dynamic light scattering to strongly scattering media. In DWS, just as in conventional DLS, the temporal fluctuations in the intensity of the scattered light are measured within a coherence area (speckle). These fluctuations are quantified by the intensity autocorrelation function $g^{(2)}(t)$, which can be related directly to the field autocorrelation function $g^{(1)}(t)$ through a Siegert-type relationship [Eq. (5.30)]. For a system of noninteracting particles, the diffusion coefficient of these particles is given, as in the case of DLS, by the Stokes-Einstein relationship $D = k_B T/(6\pi\eta a)$. One important characteristic time scale for the diffusion of the particles is the duration it takes on average for a particle to diffuse one wavelength of light [7]

$$\tau_0 = \frac{1}{k_0^2 D} \tag{5.40}$$

where $k_0 = 2\pi/\lambda$ is the wave number. The total correlation function is determined by summing the contributions of all possible paths, weighted by their probabilities. Figure 5.6 describes the schematic of a DWS experiment. The field autocorrelation will be an average of the intermediate scattering function [Eq. (5.24)] over all the possible angles, and for a trajectory with n scattering events, it takes the form [7]

$$g_1^n(t) = \langle \exp[-iq\Delta r(t)] \rangle_q^n \tag{5.41}$$

At short times, the scattering vector is independent of the particle's mean square displacement $<\Delta r^2(t)>$, and each scattering event can be averaged independently. In the limit of numerous scattering events, the path length followed in the medium relates to the total number of scattering events through $s = nl_s$, with l_s the mean free path. By definition, the transport mean free path l_t relates to the scattering mean free path through

$$l_t = \frac{2k_0^2 l_s}{\langle q^2 \rangle} \tag{5.42}$$

and the autocorrelation function for a trajectory with n scattering events takes the final form

$$g_1^n(t) = \exp\left[-\frac{2\tau}{\tau_0}\frac{s}{l_t}\right] \tag{5.43}$$

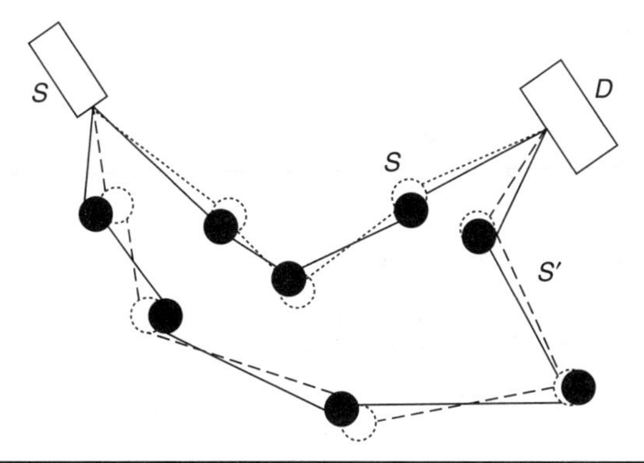

FIGURE 5.6 Multiple-light scattering.

It is obvious from Eq. (5.43) that the autocorrelation function still has negative exponential behavior, as in conventional DLS, but now it also depends on the optical path length s traveled in the medium. It is apparent that longer paths will decorrelate faster in time, whereas shorter paths will decorrelate more slowly. This fact can be easily explained by recognizing that the fluctuations of light undergoing a certain trajectory are due to a cumulative effect of the total number of scattering events. To obtain the full autocorrelation function, the contributions from all possible paths are summed and weighted by the probability density $P(s)$ [7]

$$g^{(1)}(\tau) = \int_0^{\infty} P(s) \exp\left[-\frac{2\tau}{\tau_0}\frac{s}{l_t}\right] ds \tag{5.44}$$

where $P(s)$ is the path-length probability density introduced in Section 5.3.2. For certain experimental situations, $P(s)$ can be obtained by solving the time-resolved photon diffusion equation with the appropriate boundary conditions.

In conclusion, DWS is the extension of DLS to strongly scattering media. In this case, the field autocorrelation function is no longer a negative exponential function, but involves an average over all the possible path lengths. In addition, because the light is multiple scattered, the scattering angle and wave vector do not possess the same significance as they do in a single-scattering experiment. Thus, the final angle of detection is not critical; in fact the autocorrelation function depends on the geometry of the experiment only through the boundary conditions applied to obtain the function $P(s)$. One can state that the effect of the multiple scattering is twofold: (1) it accelerates the fluctuations of the scattered light and (2) it produces an average over a larger volume size. Therefore, specific applications of DWS include an analysis of slow dynamics and large volume systems.

References

1. H. C. van de Hulst, *Light Scattering by Small Particles*, Dover Publications, Mineola, NY, 1981.
2. A. Ishimaru, *Electromagnetic Wave Propagation, Radiation, and Scattering*, Prentice Hall, Englewood Cliffs, NJ, 1991.
3. B. J. Berne and R. Pecora, *Dynamic Light Scattering: With Applications to Chemistry, Biology, and Physics*, Dover Publications, Mineola, NY, 2000).
4. A. Einstein, "Investigations on the Theory of the Brownian Movement," *Ann. Physik*. 17: 549, 1905.
5. T. G. Mason and D. A. Weitz, "Optical Measurements of Frequency-Dependent Linear Viscoelastic Moduli of Complex Fluids," *Phys. Rev. Lett.* 74: 1250–1253, 1995.
6. J. J. Duderstadt and L. J. Hamilton, *Nuclear Reactor Analysis*, Wiley, New York, 1976.
7. D. J. Pine, D. A. Weitz, P. M. Chaikin, and E. Herbolzheimer, "Diffusing Wave Spectroscopy," *Phys. Rev. Lett.* 60: 1134–1137, 1988.

CHAPTER 6

Theory of Second-Harmonic Generation

Raghu Ambekar Ramachandra Rao
Laboratory for Photonics Research of Bio/Nano Environments (PROBE)
Department of Electrical and Computer Engineering
University of Illinois at Urbana–Champaign

Kimani C. Toussaint, Jr.
Laboratory for Photonics Research of Bio/Nano Environments (PROBE)
Department of Mechanical Science and Engineering
Affiliate, Departments of Electrical and Computer Engineering, and Bioengineering
Affiliate, Beckman Institute for Advanced Science and Technology
University of Illinois at Urbana–Champaign

6.1 Introduction

Nonlinear optics deal with the interaction between optical fields in a nonlinear medium. Such interaction between fields is negligible under low optical intensities, in which case the medium is treated linearly, and the properties of the medium such as refractive index and absorption coefficient are effectively independent of intensity; optical phenomena that fall under this category are referred to as *linear optics*. However, under high intensities (~10^6 to 10^{14} W/cm^2) the medium properties become intensity-dependent, and as a result, the medium acts as a mediator for interaction between optical fields. This is the domain of nonlinear optics, and media that facilitate this interaction are referred to as *nonlinear*. Nonlinear optical processes were observed only after the advent of lasers, which could provide the required high intensities.

To further elucidate things, consider a Lorentz oscillator model. In this classic harmonic oscillator model, an electron (charge) is viewed as being bound to a much more massive atom (nucleus) by a hypothetical spring. When an oscillating electromagnetic field (light) is incident on this system, the electron experiences a force, due to the electric field, which causes it to oscillate at the frequency of the incident light. This oscillating electron in turn generates a wave at the same frequency as that of the incident light. This process is commonly referred to as *Rayleigh scattering* and can be described by Hooke's law when operating in the linear regime. However, in the nonlinear regime, the applied force is such that Hooke's law is no longer valid, and additional frequencies can be produced by the oscillating spring. Depending on the strength of the nonlinearity, many different nonlinear effects can arise. The focus of this chapter is to discuss second-harmonic generation (SHG), a specific case of a degenerate three-wave mixing process, and its application to microscopy.

6.2 Nonlinear Microscopy

Traditionally, two-photon fluorescence microscopy has been the most commonly used nonlinear microscopy technique. However, in the past decade, second-harmonic generation microscopy has gained popularity in imaging-specific biological structures due to the following advantages: (1) reduced phototoxicity, (2) inherent (endogenous) contrast, (3) preserved phase information, and (4) specificity to molecular structure organization. The first three points can be understood by invoking the use of the Jablonski diagram as shown in Fig. 6.1, and comparing it with the cases of one- and two-photon fluorescence. In linear (one-photon) fluorescence, a fluorophore is driven into an excited energy (electronic) state on absorption of a photon. Subsequently, some of this energy is lost via vibrational relaxation and then reradiated as a photon of lower energy as the fluorophore returns to its ground state (shown in Fig. 6.1*a*). Alternatively, as shown in Fig. 6.1*c*, this process could be carried out if one were to use two photons, each with energy that was half of that of the incoming photon discussed in the previous case and with the assistance of an intermediate "virtual" state. Again, the transition from electronic ground state to excited state involves molecular absorption, resulting in the emission of a lower-energy fluorescence photon. It is this absorption of energy by the system that destroys the coherence between the emitted fluorescence photon and the original excitation photons. For many incoming photons, this process could lead to potential photo damage of the fluorophore. SHG operates by a different mechanism. As shown in Fig. 6.1c, at first glance the SHG process looks similar to the two-photon fluorescence (TPF) case. However, here only virtual transitions are employed. Two lower-energy photons are annihilated while a single higher-energy photon (with twice the wavelength) is simultaneously created in a single quantum-mechanical event. Thus,

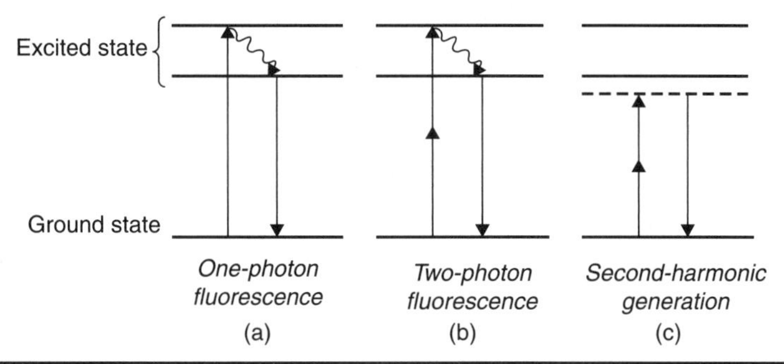

FIGURE 6.1 Jablonski diagram of linear and nonlinear processes showing (*a*) one-photon fluorescence, (*b*) two-photon fluorescence, and (*c*) second-harmonic generation.

no molecular absorption takes place, rather only scattering. As a result, photo damage is greatly reduced, and coherence is maintained between the incoming photons and the emitted SHG photon.

Nonlinear microscopy typically utilizes one of these multiphoton processes; for example, those based on m-photon excitation or absorption, or harmonic generation where m represents the number of photons that participate in the specific quantum-mechanical process. The primary advantage is that the signal generated from an mth-order nonlinear process depends on I^m, where I is the incident intensity. The requirement for high intensity for a nonlinear process means that when the incident beam is strongly focused (e.g., by a high-numerical aperture microscope objective lens) onto the sample, the nonlinear effects are primarily restricted to the focal region. This results in lower phototoxicity. Furthermore, due to excitation typically at near-infrared wavelengths, the resulting penetration depth is increased compared to linear fluorescence microscopy techniques. In fact, penetration depths as large as 0.5 mm have been obtained [1]. The next few sections present an abbreviated derivation of SHG beginning with Maxwell's equations and discuss the effect that energy and momentum conservation has on this process.

6.3 Theory of Second-Harmonic Generation (Electromagnetics Picture)

To understand the origin of SHG using (classic) electromagnetic optics, we consider Maxwell's equations for a source-free medium as given by

$$\nabla \times E = -\frac{\partial B}{\partial t} \tag{6.1}$$

$$\nabla \times H = \frac{\partial D}{\partial t} \tag{6.2}$$

$$\nabla \cdot D = 0 \tag{6.3}$$

$$\nabla \cdot B = 0 \tag{6.4}$$

where E is the electric field, H is the magnetic field, D is the electric flux density, and B is the magnetic flux density. Note that all of these fields are functions of position and time, and note that SI units will be used throughout this chapter.

The constitutive relations that define the material properties are

$$D = \varepsilon E = \varepsilon_0 E + P \tag{6.5}$$

$$B = \mu_0 H \tag{6.6}$$

where ε_0 is the free-space (electric) permittivity, μ_0 is the free-space (magnetic) permeability, and P is the polarization density, that is, the sum of the induced dipole moments. In linear optics, the polarization density P is directly proportional to the electric field E, and is given by

$$P = \varepsilon_0 \chi^{(1)} E \tag{6.7}$$

where $\chi^{(1)}$ is the linear susceptibility of the material. It is related to the relative permittivity ε_r as

$$\chi^{(1)} = \varepsilon_r - 1 \tag{6.8}$$

In nonlinear optics, P can be expressed as a power series in E.

$$P = \varepsilon_0 [\chi^{(1)} E + \chi^{(2)} E^2 + \chi^{(3)} E^3 + \ldots] \tag{6.9}$$

where $\chi^{(2)}$ and $\chi^{(3)}$ represent the second- and third-order susceptibilities. The first term varies linearly with the electric field and describes common processes that occur in nature, such as the linear absorption, reflection, and scattering of light. The second term varies quadratically with the electric field and describes second-order nonlinear processes such as SHG and sum and difference frequency generation. Finally, the third term describes third-order processes such as stimulated Raman scattering and third-harmonic generation.

For simplicity, Eq. 6.9 can be written as

$$P = P^{(1)} + P^{(2)} + P^{(3)} + P^{(4)} + \ldots \tag{6.10}$$

or

$$P = P^{(1)} + P^{(NL)} \tag{6.11}$$

where $P^{(NL)} = P^{(2)} + P^{(3)} + P^{(4)} + \ldots$

6.3.1 Nonlinear Wave Equation

Here we derive the wave equation for the propagation of light in a nonlinear medium. We begin by taking the curl on both sides of Eq. 6.1 and obtain

$$\nabla \times \nabla \times E = -\nabla \times \frac{\partial B}{\partial t} \tag{6.12}$$

This can be rewritten as

$$\nabla(\nabla \cdot E) - \nabla^2 E = -\mu_0 \frac{\partial \nabla \times H}{\partial t} \tag{6.13}$$

Using Eqs. (6.3) and (6.5), we get

$$\nabla \cdot \varepsilon E = 0 \Rightarrow \nabla \cdot E = 0 \tag{6.14}$$

Using Eqs. (6.2), (6.5), (6.11), and (6.14), Eq. (6.13) becomes

$$-\nabla^2 E = -\mu_0 \varepsilon_0 \frac{\partial^2 E}{\partial t^2} - \mu_0 \frac{\partial^2 P^{(1)}}{\partial t^2} - \mu_0 \frac{\partial^2 P^{(NL)}}{\partial t^2} \tag{6.15}$$

Next, we note that the velocity of light in a vacuum c is

$$c = \frac{1}{\sqrt{\mu_0 \varepsilon_0}} \tag{6.16}$$

and the relative permittivity is

$$\varepsilon_r = n^2 \tag{6.17}$$

where n is the refractive index of the material.

Using Eqs. (6.7), (6.8), and (6.17), Eq. (6.15) becomes

$$\nabla^2 E - \frac{n^2}{c^2} \frac{\partial^2 E}{\partial t^2} = \mu_0 \frac{\partial^2 P^{(NL)}}{\partial t^2} \tag{6.18}$$

This is the wave equation used to describe nonlinear processes.

6.3.2 Second-Order Nonlinear Polarization

Let us take a closer look at second-order nonlinearity. We can express E in terms of its complex amplitude $\tilde{E}$ such that $E(t) = \tilde{E}e^{i\omega t} + \tilde{E}e^{-i\omega t}$. For a second-order nonlinearity, $P^{(2)} = \varepsilon_0 \chi^{(2)} E^2 = \varepsilon_0 \chi^{(2)} [\tilde{E}e^{i\omega t} + \tilde{E}^* e^{-i\omega t}]^2 = \varepsilon_0 \lfloor \chi^{(2)} \tilde{E}^2 e^{i2\omega t} + \chi^{(2)} \tilde{E}^{2*} e^{-i2\omega t} + 2\chi^{(2)} |\tilde{E}|^2 \rfloor$The last term leads to the

---

* $d_{\text{eff}} = \Sigma_{ijk} d_{ijk} \cos\theta_i \cos\theta_j \cos\theta_k$, where θ_i, θ_j, and θ_k are the angles that the electric field makes with the principal axis of the crystal and d_{ijk} is the contracted notation of the χ tensor when Kleinman symmetry is valid and is given by $d_{ijk} = (1/2)\varepsilon_0 \chi_{ijk}$. The reader can refer to [2] and [3] for details on symmetry conditions and contracted notation for d_{ijk}.

generation of a static DC field in the nonlinear medium, while the first and second terms create polarization densities at $+2\omega$ and -2ω; this leads to the generation of SHG. Thus, $\tilde{P}^{(2)}(2\omega) = \varepsilon_0 \chi^{(2)} \tilde{E}^2$ or

$$\tilde{P}^{(2)}(2\omega) = 2d_{\text{eff}}\tilde{E}^2 \tag{6.19}$$

where d_{eff} is the effective value of the second-order susceptibility.*

6.3.3 Determination of Second-Harmonic Generation Intensity

To determine an expression for the SHG intensity, we recognize that the complex amplitude of the incident field can be written as

$$\tilde{E} = A_1 e^{-ik_1 z} \tag{6.20}$$

Substituting it into Eq. (6.19), we get

$$\tilde{P}^{(2)}(\omega_3) = 2d_{\text{eff}} A_1^2 e^{-i2k_1 z} \tag{6.21}$$

Let us write the generated electric field for SHG as

$$E = E_3 = A_3 e^{-i(k_3 z - \omega_3 t)} + A_3{}^* \, e^{i(k_3 z - \omega_3 t)} \tag{6.22}$$

and express the nonlinear polarization to be in the form

$$P_{NL} = \tilde{P} e^{i\omega_3 t} + \tilde{P}^* \, e^{-i\omega_3 t} \tag{6.23}$$

For SHG,

$$P_{NL} = \tilde{P}^{(2)}(\omega_3) e^{i\omega_3 t} + \tilde{P}^{(2)*}(\omega_3) e^{-i\omega_3 t} \tag{6.24}$$

where $\omega_3 = 2\omega$. Substituting Eqs. (6.21), (6.22), and (6.24) into the nonlinear wave equation, neglecting the complex conjugate part, and then simplifying, Eq. (6.8) becomes

$$\left(\frac{\partial^2 A_3}{\partial z^2} + 2ik_3 \frac{\partial A_3}{\partial z} - A_3 k_3^2 + A_3 \frac{n_3^2 \omega_3^2}{c^2} \right) e^{-i(k_3 z - \omega_3 t)} = -2d_{\text{eff}} A_1^2 e^{-i2k_1 z} e^{i\omega_3 t} \omega_3^{2*}$$

---

*
$$\frac{\partial^2 E_3}{\partial z^2} = \frac{\partial}{\partial z}\left(ikA_3 e^{-i(k_3 z - \omega_3 t)} + e^{-i(k_3 z - \omega_3 t)} \frac{\partial A_3}{\partial z} \right) \text{ and}$$

$$\frac{\partial^2 E_3}{\partial z^2} = \left(\frac{\partial^2 A_3}{\partial z^2} + 2ik \frac{\partial A_3}{\partial z} - A_3 k_3{}^2 \right) e^{-i(k_3 z - \omega_3 t)}.$$

The two terms cancel since

$$k_3 = \frac{n_3 \omega_3}{c} \tag{6.25}$$

Assuming that the spectrum of the source is a narrow band, the slowly varying amplitude approximation can be applied; that is,

$$\frac{\partial^2 A_3}{\partial z^2} << \frac{\partial A_3}{\partial z} \tag{6.26}$$

This leads to

$$\frac{\partial A_3}{\partial z} = \frac{-d_{\text{eff}} A_1^2 \omega_3^2}{2ik_3} e^{-i(2k_1 - k_3)z} \tag{6.27}$$

Integrating over the length L of the nonlinear medium, we get

$$A_3 = \frac{-d_{\text{eff}} A_1^2 \omega_3^2}{2ik_3} \int_0^L e^{-i(2k_1 - k_3)z}\, dz \tag{6.28}$$

Solving, we get

$$A_3 = \frac{d_{eff} A_1^2 \omega_3^2 L}{2ik_3} e^{-i\Delta kL/2} \operatorname{sinc}\left(\frac{\Delta kL}{2}\right) \tag{6.29}$$

where $\Delta k = 2k_1 - k_2$ is called the *phase mismatch factor.*

The intensity is given by

$$I = \frac{|A_3|^2}{2\eta} \tag{6.30}$$

where η is the material optical impedance, given by

$$\eta = \sqrt{\frac{\mu_r \mu_0}{\varepsilon_r \varepsilon_0}} \tag{6.31}$$

For a nonmagnetic material, $\mu_r = 1$, and using Eq. (6.17), it follows that

$$\eta \propto n^{-1} \tag{6.32}$$

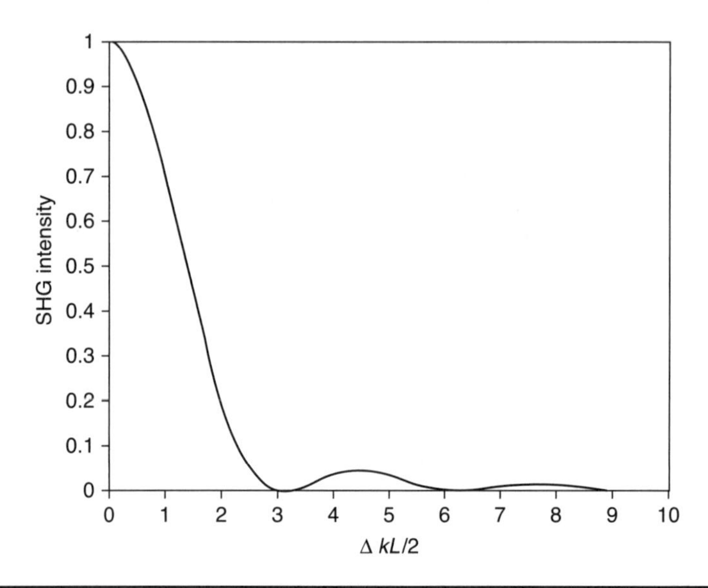

FIGURE 6.2 SHG intensity versus the phase mismatch factor.

Using Eqs. (6.31) and (6.32), Eq. (6.30) becomes,

$$I = \frac{d^2{}_{\text{eff}} c^2 I_1^2 \omega_3^2 L^2}{2n_1^2 n_3} Sinc^2\left(\frac{\Delta kL}{2}\right) \tag{6.33}$$

or,

$$I \propto sinc^2\left(\frac{\Delta kL}{2}\right) \tag{6.34}$$

Figure 6.2 is a plot of the normalized SHG intensity versus the phase mismatch factor Δk.

As we can see, the SHG intensity peaks at $\Delta kL/2 = 0$; that is, $\Delta k = 2k_1 - k_3 = 0$. Using Eq. (6.25), we have $2\omega_1 n_1/c - \omega_3 n_3/c = 0$. We note that $\omega_3 = 2\omega_1$, and thus

$$n_3 = n_1 \tag{6.35}$$

Since many materials are naturally dispersive (that is, the refractive index changes with the wavelength), this condition is difficult to achieve for isotropic samples. Hence, a common practice is to use birefringence in anisotropic materials. Here the polarization of the incident wave and angle of the anisotropic crystal is chosen such that $n_3 = n_1$.* This condition of $\Delta k = 0$ is called *perfect phase matching*.

6.3.4 Nonperfect Phase Matching

Now let us consider the case where a material is not adjusted for perfect phase matching, the so-called nonperfect phase matching; that is, $\Delta k \neq 0$. As discussed before, the SHG intensity has a sinc^2 dependence on the phase mismatch factor Δk. We observe that SHG is generated until the argument of the sinc function $\Delta kL/2 = \pi$. This defines the coherence length of SHG generation, and L_c is given by

$$L_c = \frac{2\pi}{|\Delta k|} = \frac{\lambda}{2|n_1 - n_3|}^{\dagger} \tag{6.36}$$

It is preferred that the value of Δk be small so that the coherence length is large. Larger coherence lengths allow longer interaction lengths L, which in turn allow higher SHG intensities according to Eq. 6.33. However, in practice, natural materials may be arranged in such a way that Δk may not be small. One such example is the biological tissues where SHG intensities have been observed from as small as a few hundred nanometers. The SHG intensities generated are much less than those of materials for which $\Delta k = 0$ or Δk is comparatively smaller, because of the smaller coherence lengths (therefore smaller interaction lengths). According to Eq. (6.33), the SHG intensity is also sensitive to the sample property d_{eff}. Crystalline samples such as collagen Type I have large values of d_{eff} and thus produce a stronger SHG signal compared to collagen Type III which is less crystalline.

6.3.5 Noncentrosymmetry

Yet another condition for the generation of SHG is noncentrosymmetry, which describes materials having no inversion symmetry. Inorganic materials such as beta-barium borate (BBO), lithium niobate ($LiNbO_3$), and potassium dihydrogen phosphate (KDP) possess this trait, but even biological structures such as collagen, microtubulin, and myosin display noncentrosymmetry [4]. The requirement for noncentrosymmetry can be understood by considering Eq. 6.9 [3],

$$\tilde{P}^{(2)}(2\omega) = \chi^{(2)}\tilde{E}^2 \tag{6.37}$$

---

* For a uniaxial crystal, $n_x = n_y = n_o$ and $n_z = n_e$. x, y, z represent the crystallographic axes of the crystal, and o and e represent the ordinary and extraordinary axes, respectively. If the light is incident in the x direction at an angle of θ with respect to the z axis, the angle of the crystal is adjusted such that $n_3(2\omega) = n_1(\omega)$. Specifically, $1/n_3^2 = \sin\theta/n_e^2 + \cos\theta/n_o^2$. Refer to [2] for further details.

† This coherence length is different from that defined for the laser. Coherence length of the laser is given by $l_c = c\tau_c$, where τ_c is the coherence time given by $\tau_c = 1/\Delta\nu$, and $\Delta\nu$ is the spectral width of the laser.

and comparing it to the expression for the induced polarization in a centrosymmetric medium (when the applied electric field phase is shifted by 180°) given by

$$-\tilde{P}^{(2)}(2\omega) = \chi^{(2)}(-\tilde{E})^2 = \chi^{(2)}\tilde{E}^2 \tag{6.38}$$

Comparing the two equations, one sees that they can only be equal when $\chi^{(2)} = 0$. Hence the medium needs to be noncentrosymmetric in order to produce SHG. It can also be understood qualitatively in terms of the dipole emission, which is discussed in Section 6.4.

6.3.6 Quasi-Phase Matching

Under the nonperfect phase-matching condition, $\Delta k \neq 0$. As we know, this limits the SHG intensity that can be generated. However, there is a way of increasing the SHG intensity by fabricating the material such that its second-order susceptibility is spatially modulated, according to the following equation.

$$d(z) = d\exp(-iQz) \tag{6.39}$$

The phase-matching condition in this case becomes

$$\Delta k = 2k_1 - k_3 - Q \tag{6.40}$$

and perfect phase matching is obtained when $Q = \Delta k$. However, it is difficult to fabricate a structure that has periodic variations in its second-order coefficient d. This type of periodic fabrication is called *poling*. With quasi-phase matching (QPM), the sign of d is alternated by a period equal to the coherence length. This occurs naturally in some biological structures, as will be discussed in Section 6.4.4.

6.3.7 Phase-Matching Bandwidth

So far, we have assumed that the incident beam is strictly monochromatic. However, for polychromatic light, dispersion in a nonlinear medium affects phase matching. Consider the case of second-harmonic generation from a nonlinear medium which exhibits first-order dispersion (and negligible higher-order dispersion). Here, fundamental and second harmonic waves become separated upon propagation in the medium due to the differences in their respective group velocities, v_1 and v_3. This walk-off effect, as it is known, limits the generation of SHG to a walk-off length $L_g = \tau/2|1/v_3 - 1/v_1|$, where τ is the pulse-width of the laser [2]. Hence, when using broadband sources, group velocity is an important parameter and phase matching requires that the group indices

N_1 and N_3, at the fundamental and second-harmonic frequencies, respectively, are matched. The phase-matching bandwidth is thus given by [2]

$$\Delta\nu = \frac{c_o}{2L|N_1 - N_3|} \tag{6.41}$$

The group velocity phase matching can be achieved either by using a thin crystal so that the walk-off effect is small or designing the crystal such that $N_1 \sim N_3$. In many situations, L_c is smaller than L_g, and hence the phase velocity mismatch usually dominates.

6.4 Theory on Second-Harmonic Generation (Dipole Picture)

As discussed before, second-harmonic generation (SHG) is a second-order nonlinear process wherein a noncentrosymmetric material (frequency) up-converts the incident light of frequency ω to the light of frequency 2ω.

Here we briefly discuss the mechanism of generation for SHG from a molecular dipole picture. A detailed explanation can be found in [5]. When light of frequency ω is incident on a centrosymmetric molecule, the associated electron cloud begins to symmetrically oscillate at that frequency. In turn, a wave is emitted at the same frequency in a dipole-like emission pattern, commonly referred to as Rayleigh scattering as shown in Fig. 6.3*a*. However, when light of the same

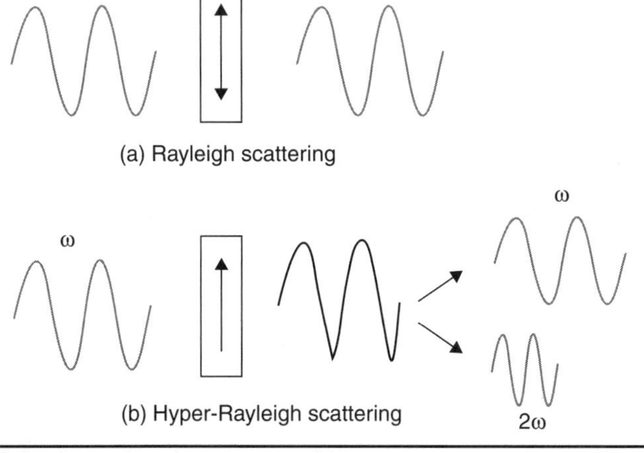

FIGURE 6.3 Scattering phenomenon of (*a*) a centrosymmetric and (*b*) a noncentrosymmetric molecule.

(a) Dipole-like HRS emission from a single molecule

(b) Directional-forward SHG emission from multiple HRS scatterers

Figure 6.4 (*a*) HRS emission from a single noncentrosymmetric molecule. (*b*) SHG emission from multiple HRS scatterers.

frequency is incident on a noncentrosymmetric molecule, the oscillation becomes asymmetric; this results in the generation of additional frequency components at 0 (DC) and 2ω (see Fig. 6.3*b*). This wave at 2ω is referred to as *hyper-Rayleigh scattered (HRS)*.

For a single noncentrosymmetric molecule, the HRS emission is dipole-like as shown in Fig. 6.4*a*. However, when there is a regular arrangement of such molecules (scatterers) in a material, then constructive interference between the generated waves occurs; this leads to a significant amount of signal generated predominantly in the forward direction. This signal is referred to as SHG. Thus, we can think of SHG as a macroscopic equivalent of HRS with an ordered arrangement of scatterers. This is an important principle to keep in mind since it explains why not all noncentrosymmetric structures (proteins such as DNA) emit SHG. DNA may emit HRS; however, it is not ordered at the scale of the wavelength of light, and thus does not produce measurable SHG. On the other hand, structures such as collagen and myosin have regularly arranged molecules that coherently build up the HRS to produce SHG.

6.4.1 Directionality

It is also important to note that the emitted dipole pattern for SHG is extremely directional; that is, SHG is emitted predominantly in the forward direction. This is due to the momentum conservation that follows directly from the phase-matching condition, $\Delta k = 2k_1 - k_3 = 0$, as shown pictorially in Fig. 6.5. However, there are two scenarios when SHG can be directed in the backward direction, and these are discussed below.

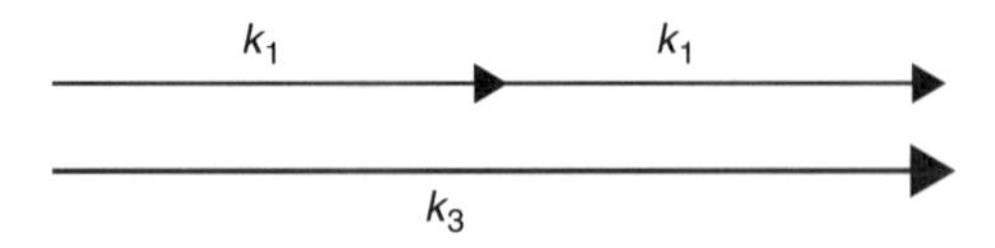

Figure 6.5 Predominant forward propagation of SHG due to phase matching.

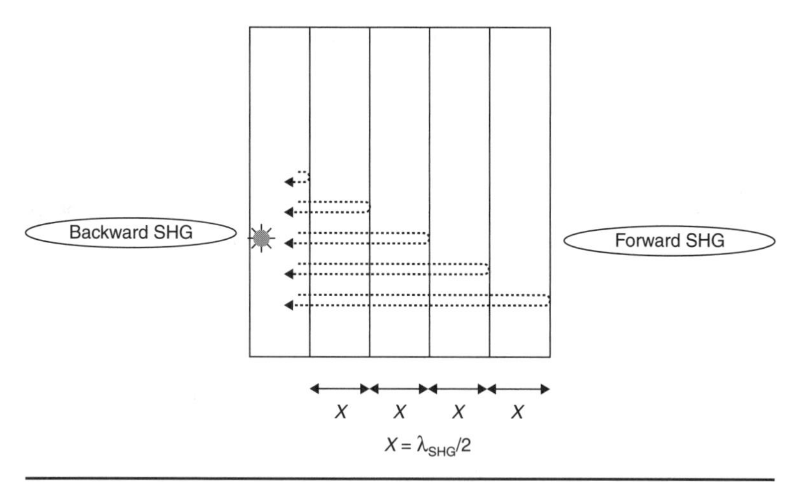

FIGURE 6.6A Coherent backward SHG generation through constructive interference of scatterers spaced $\lambda_{SHG}/2$. Both emissions are directional.

6.4.2 Backward Second-Harmonic Generation

Backward SHG can be generated when the axial scatterers are spaced by $\lambda_{SHG}/2$. This leads to constructive interference in the backward direction as shown in Fig. 6.6*a*. Here, the emission is directional in both the forward and backward directions, as indicated by the small emission cone. Moreover, the emission is coherent since phase is preserved through interference. Backward SHG may also be generated through multiple scattering of the forward-generated SHG in the backward direction as shown in Fig. 6.6*b*. In this case, due to the potentially multiple scattering events, phase information is lost and

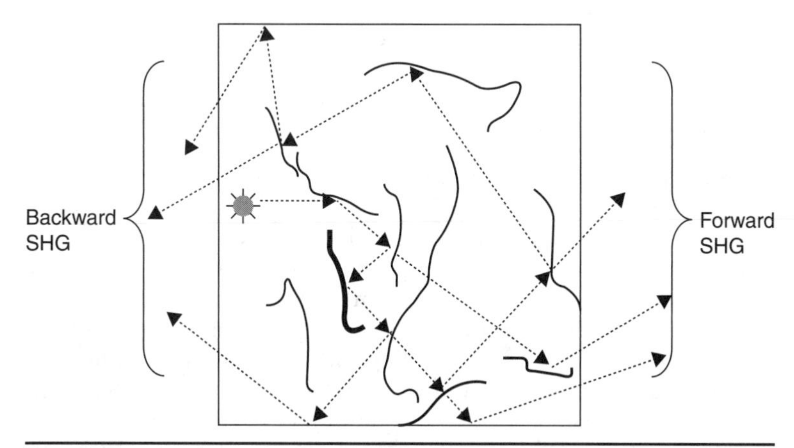

FIGURE 6.6B Incoherent backward SHG generation from multiple scattering of forward SHG. Both emissions are no longer directional.

emission is no longer directional. This is frequently observed in turbid media.

6.4.3 Effect of Focusing

As we have just discussed, SHG emission is typically in the forward direction along the optical (propagation) axis. However, usually in microscopy high-numerical-aperture objective lenses are employed to create high-spatial-resolution SHG images. For a focused Gaussian beam, there is an inherent Gouy phase shift when compared with a collimated beam. This results in a retardation in phase of the incident beam, which requires the SHG to propagate off-axis due to momentum conservation. Hence, this must be taken into account in practice if one is interested in efficient SHG collection. Details on this are provided in [5].

6.4.4 Phase Matching in Biological Tissues

As discussed before, for perfect phase matching $\Delta k = 0$. For nonlinear crystals, this is achieved by adjusting the angle of the optical axis of the crystal relative to the incident wave polarization and propagation direction. However, in noncentrosymmetric biological tissues, such as collagen fibers, we have the case of nonperfect phase matching. Thus, the thickness of a fiber should be on the order of the coherence length of SHG. This ensures that there is a sufficient number of dipoles participating in the generation of SHG to provide a measurable signal. Therefore, two conditions are absolutely necessary for efficient generation of SHG. The material must possess a sufficient density of noncentrosymmetric structures, and it must be ordered at the molecular level to generate an efficient SHG signal. Moreover, the generation of SHG in biological tissues is somewhat "relaxed." This is because the arrangement of collagen fibers along the axial direction is similar to that of quasi-phase matching—not exactly quasi-phase matching, since the collagen fibers are not poled.

6.5 Experimental Configuration

The schematic of the experimental setup for the SHG microscope is shown in Fig. 6.7. It is basically a two-photon fluorescence microscope that is modified to include both the transmission and reflection collection geometries. The beam is obtained from a wavelength-tunable (690 to 1020 nm) Ti:Sapphire laser that produces 100 femtosecond-duration, linearly polarized pulses at a 80-MHz repetition rate. The Ti:Sapphire laser is well-suited for nonlinear microscopy techniques. A high repetition rate of 80-MHz ensures that there is a large enough number of pulses incident on the sample for high-speed scanning. Pulse widths significantly <100 fs can lead to multiphoton ionization. The choice of wavelength

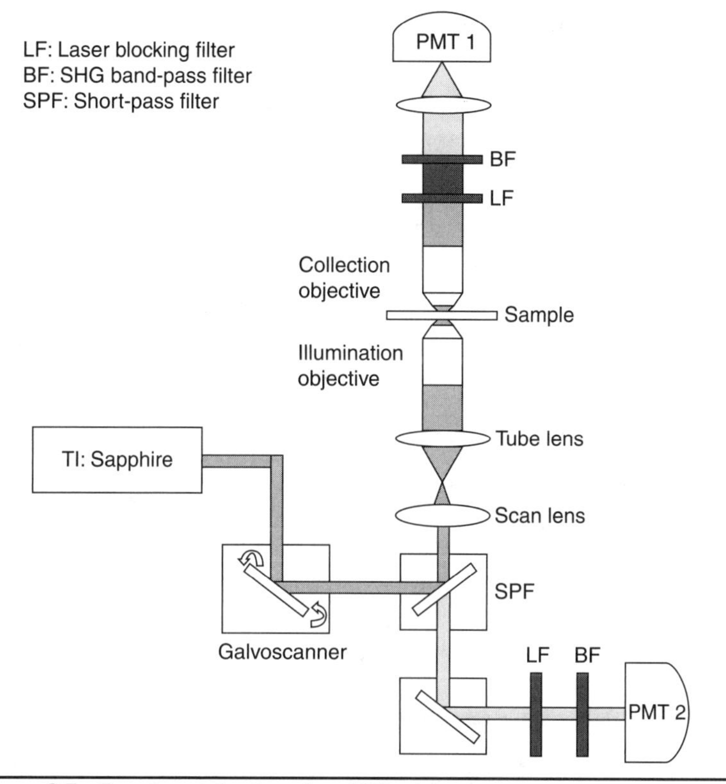

FIGURE 6.7 Experimental setup of an SHG microscope.

depends on the biological tissue under investigation. Next, the beam is spatially filtered and collimated before sending it to the galvanometer-driven *xy* scanner, which generates a raster-scan pattern. The beam is then reflected by a short-pass 680-nm dichroic beam splitter. After passing through a combination of relay lenses (scan and tube lenses), it is focused onto the sample using a high-numerical-aperture (high-NA) oil-immersion illumination objective. For thick biological tissues, a water-immersion objective is preferable to avoid aberrations due to refractive-index mismatch. The transmitted signal from the sample is collected by a second objective. For efficient SHG collection, it is better that the collection objective have an NA equal to or greater than the illumination objective. This is a direct consequence of the Gouy phase discussed earlier. The SHG signal is collected through spectral filtering using a laser-blocking filter followed by a band-pass filter that transmits only the SHG signal. The SHG intensity is recorded with a photomultiplier, and image stacks are acquired from various depths (z) into the sample. The power settings should be optimized for image contrast and avoiding unwanted multiphoton effects such as optical breakdown.

6.6 Practical Considerations

6.6.1 Distinguishing SHG

There are several ways of distinguishing SHG from other signals. However, a simple way is to look at its spectrum. The spectral width of SHG is ~70% of the illumination bandwidth which is significantly narrower compared to fluorescence. Moreover, SHG is polarized. A simple polar plot obtained from rotation of a polarizer either before or after the sample provides information on the degree of polarization of the emitted signal.

6.6.2 Resolution and Penetration Depth

The resolution of SHG is similar to that of TPF. The transverse resolution is given by [6]

$$w_t = \frac{0.7\lambda_{em}}{NA} \tag{6.42}$$

while the axial resolution is [6]

$$w_z = \frac{2.3\lambda_{em} n}{NA^2} \tag{6.43}$$

where n is the refractive index of the immersion fluid, and λ_{em} is the emission wavelength.

For example, an objective of NA = 1.4 and λ_{em} = 400 nm (for excitation at 800 nm) gives a transverse spot size of 200 nm and axial resolution of 700 nm.

The penetration depth depends on the tissue under observation. However, as discussed previously, SHG (and two-photon fluorescence) microscopy employs photons in the near-IR, where biological tissue absorption is low. This translates to a deeper penetration depth compared to linear microscopy techniques. Moreover, only a small region at the focus is excited; this allows optical histology to be performed. Depths as large as 500 μm for sclera in the eye and 400 μm in brain tissue have been imaged using SHG microscopy [1, 7].

6.6.3 Power Limitations

Once the SHG is generated at the focus, there is an unavoidable absorption of the SHG photons due to Rayleigh scattering. Moreover, there can be photothermal effects at high intensities, for example, the generation of singlet oxygen which is highly reactive

Advantages	Disadvantages
No staining	Expensive hardware requirements
Deeper penetration	Signal strength could be weak for some
Low phototoxicity	Sensitive to specific biological structures
Intrinsic sectioning capability (3D images)	
Highly sensitive to structural information	
Phase carries "extra" information	

Table 6.1 Advantages and disadvantages of SHG microscopy compared to other common microscopy techniques.

and damages the tissue. Under extremely high intensities, there can be multiphoton ionization of the tissue. Hence, it is best to keep the power as low as possible as long as there is enough SHG signal to create good image contrast. Some tissues such as tendon have been shown to withstand powers as high as 40 to 50 mW without visible damage [8].

6.6.4 Advantages and Disadvantages of SHG

The relative advantages and disadvantages of SHG microscopy compared to other microscopy techniques are summarized in Table 6.1.

6.7 Conclusion

Nonlinear microscopy has gained momentum in the past few years, partly because of the increasing availability of optical sources and detectors and because of the increasing applicability of its use in the biomedical sciences. Second-harmonic generation microscopy, in particular, has become an attractive alternative to two-photon fluorescence microscopy for imaging biological structures that exhibit noncentrosymmetry. Comparatively, SHG microscopy is less invasive due to the endogenous nature of the contrast signal generated and the fact that it is based on scattering and not molecular absorption. Although the signal yield is typically lower than fluorescence, SHG from collagen-based systems is appreciable—allowing for 3D imaging to be carried out with the same depth and spatial-resolution advantage as two-photon fluorescence. Moreover, the fact that phase information is preserved in SHG microscopy extends the possibilities for carrying out quantitative measurements when using this modality. Extracting quantitative information from SHG images is discussed in a subsequent chapter (Chap. 12) in this book.

Acknowledgments

This work was supported in part by the University of Illinois at Urbana–Champaign (UIUC) research startup funds and in part by funds from the National Science Foundation (NSF) (DBI 0839113).

References

1. M. Han, G. Giese, and J. F. Bille, "Second Harmonic Generation Imaging of Collagen Fibrils in Cornea and Sclera," *Optics Express* 13: 5791–5797, 2005.
2. B. Saleh and M. Teich, *Fundamentals of Photonics*, Wiley Interscience, Hoboken, NJ, 2007.
3. R. W. Boyd, *Nonlinear Optics*, Academic Press, San Diego, 2003.
4. R. LaComb, O. Nadiarnykh, S. S. Townsend, and P. J. Campagnola, "Phase Matching Considerations in Second-Harmonic Generation from Tissues: Effects on Emission Directionality, Conversion Efficiency, and Observed Morphology," *Opt. Commun.* 281: 1823–1832, 2008.
5. B. R. Masters and P. So, *Handbook of Biomedical Nonlinear Optical Microscopy*, Oxford University Press, New York, 2008.
6. A. Diaspro, *Confocal and Two-Photon Microscopy*, Wiley-Liss, New York, 2002.
7. D. A. Dombeck, K. A. Kasischke, H. D. Vishwasrao, M. Ingelsson, B. T. Hyman, and W. W. Webb, "Uniform Polarity Microtubule Assemblies Imaged in Native Brain Tissue by Second-Harmonic Generation Microscopy," *Proc. Natl. Acad. Sci. USA*, 100: 7081–7086, 2003.
8. W. Drexler and J. G. Fujimoto, *Optical Coherence Tomography: Technology and Applications*, Springer, Berlin, Heidelberg, New York, 2008.

CHAPTER 7

Vision Restoration in the Nanobiophotonic Era

S. Sayegh

The Eye Center, Anterior Segment and Vitreoretinal Surgery
Champaign, IL

7.1 Introduction

If one were asked to name one nanobiophotonic device, the eye would first come to mind. Vision is the most precious of senses. Given a choice, people are willing to give up any other sense rather than give up sight. The definition of blindness varies, and the prevalence and incidence of blindness vary accordingly. Vision, its loss and modern medical and surgical techniques for its restoration, will be presented in this chapter. A guide to different conditions leading to blindness or visual impairment is given in Fig. 7.1.

The presentation is targeted to biomedical researchers and practitioners as well as anterior segment and retina surgeons. It is not intended to be erudite or encyclopedic but rather to provide a clear and practical methodology on how to approach the different forms of blindness from a therapeutic point of view and shed some light on the developing use of nanophotonic strategies and tools. While there is an enormous amount of information to be gleaned from the study of a variety of animal eyes, this presentation will be restricted to the human eye.

In medicine in general and in surgery in particular, the role of accurate and precise diagnostics is key. While information gleaned from the patient is of great importance (Sir William Osler's "Listen to the patient, he is telling you the diagnosis"), actual examination and

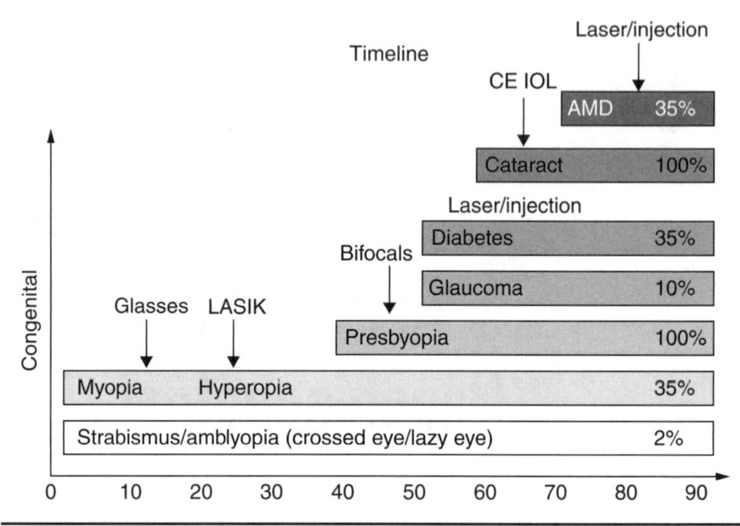

Figure 7.1 A timeline of eye conditions with horizontal axis indicating age in years with an estimate of the population affected and progressive severity. (The EYE Center, Champaign, IL)

diagnostic instruments and methods are now paramount. Imaging, in many modalities, is a fundamental tool in ophthalmic medicine and surgery, despite the naive belief that because the "eye is transparent, you can see everything." Some of the imaging modalities will be reviewed and the potential role of nanobiophotonics illustrated.

7.2 Light at Multiple Levels

Light can be understood at a variety of levels. These are discussed elsewhere in this volume. Here we make a brief mention of these levels in the context of the eye, vision, and vision restoration. While the quantum theory of light and light-matter interaction may be the "ultimate" level of understanding, as is often the case in physics, much simpler and intuitively appealing models can be used very effectively depending on the particular range of applications.

7.2.1 Ray Optics, Wave Optics, and Quantum Optics

An understanding of some level of image formation can be achieved and implemented on the basis of the simple observations that light propagates in a straight line and that certain objects are opaque. A pinhole can be fashioned out of cardboard (Fig. 7.2) and used to form an image either in a box (camera oscura) or on your retina by looking through the pinhole (if you usually wear glasses, take them off and try the pinhole instead; if you do not usually wear glasses, try some friend's glasses on to blur your vision and then improve it by looking

Figure 7.2 Pinhole used for diagnosis of refractive errors. (The EYE Center, Champaign, IL)

through the pinhole). Clinically a pinhole can be used effectively to help the clinician quickly decide whether the refractive state of the eye is related to the decrease or loss of vision. Therapeutically it can be used for improving vision in cases of high aberrations by constricting pupils (pinhole!) pharmacologically, for example, with pilocarpine drops.

Pinhole imaging has limitations such as restricting the field of view and the diffraction limit. For a larger field of view and for the understanding of imaging at larger apertures as well as the design of different diagnostic instruments, ray optics (paraxial or not) and sometimes wave optics become necessary. Theoretically ray optics can be understood on the basis of Fermat's principle, that light (mostly) travels via a path of minimum time, and of knowledge of optical properties of different materials. In practice and computationally, for example, in software for optical design (ZEMAX, CODE V, Optica), laws of refraction are used to help in ray-tracing applications. This "geometric" optics approach is considered an excellent approximation as long as the apertures and distances characteristic of the geometry at hand are significantly larger than the wavelength of light λ, say 100 λ. Given that the order of magnitude of wavelengths for visible light is around half a micron, this limit puts us out of the range of nanophotonics. We are back in range when the interactions of light occur at the wavelength and subwavelength level and photons' energy interact with the energy states of the molecules at the heart of the very structure and the function of eye and vision. Indeed, the truism that vision occurs in the visible range is a reflection of this fact.

7.3 Vision

7.3.1 Visual Acuity

One common quantitative measure of vision is the visual acuity that can be characterized by the minimum angle of resolution. The normal value is 1 minute of arc. Testing letters for normal vision are designed so that their distinguishing features are separated by 1 minute of arc. For example, a horizontal line of the letter E and the space separating it from another horizontal line are separated by 1 minute of arc. The size of a normal testing E is 5 minutes of arc. In order to test better or worse visual acuities, corresponding letter sizes are chosen according to a linear scale. The actual size of the E depends on the testing distance. For a variety of reasons, including the importance of minimizing accommodative effort, the testing distance is standardized at 6 m, or 20 ft. The common use of 6/6 or 20/20 vision for "normal vision," results from this choice. That notation, the Snellen visual acuity, is easier to understand for a value other than 20/20 and is best illustrated with an example. In order to see what an individual with normal visual acuity can see at 60 ft, a person with 20/60 visual acuity needs to be at 20 ft in order to see equally well. The angle of resolution is thus three times worse. The Snellen acuity can be read as a fraction or percentage. Coincidentally the fraction 20/20 = 1 corresponding to normal vision is the same value as the minimum angle of resolution, 1 minute of arc. For abnormal (or supernormal) vision, this correspondence is maintained.

7.3.2 Definition(s) of Blindness (Legally Blind and Legally Drive)

In the United States, legal blindness is defined by a visual acuity of 20/200 or worse or a field of vision restricted to 20 degrees. The fractional representation of blindness is thus 0.1; in other words, a blind person sees 10 times worse than normal. This also holds approximately for blindness defined as a severe restriction in the field of vision. Another practical measure is what it legally takes to be able to drive. Most states have a requirement of 20/40 and sufficient field of view for unrestricted driving. The World Health Organization definition of blindness corresponds to 20/400. According to this definition, the number of people that are bilaterally blind in the world is estimated at 45 million, with 314 million being visually impaired. The most common cause of blindness worldwide is cataracts (Fig. 7.3), an opacification of the lens that occurs most commonly with age. It is perhaps not surprising that the prevalence of certain conditions, including blindness, increases as we successfully extend our life span.

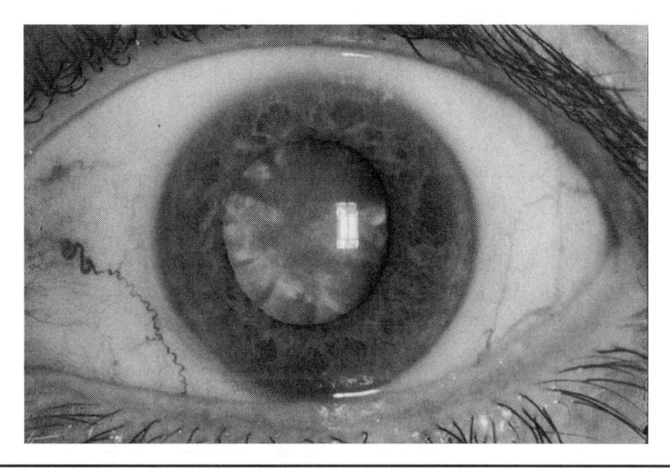

FIGURE 7.3 The lens as seen through the dilated pupil has developed a cataract. (The EYE Center, Champaign, IL)

7.3.3 Causes of Blindness: What It Takes to See . . . and Not

There are five functional requirements for vision.

1. Transparency of the media
2. Proper bending of the rays (focusing) in order to image on
3. Healthy detector (retina) and
4. Healthy conduction (optic nerve) to carry the signal to
5. Processing center (brain)

These are schematically illustrated in Fig. 7.4.

If any of these requirements fail, decreased vision, possibly to the point of blindness, ensues. Perhaps not surprisingly (Murphy's law)

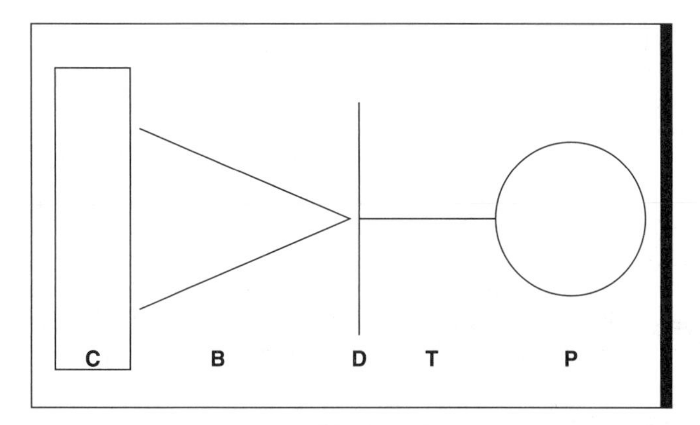

FIGURE 7.4 Schema representing requirements for vision: Clear media, Bending, Detector, Transmission, Processing. (The EYE Center, Champaign, IL)

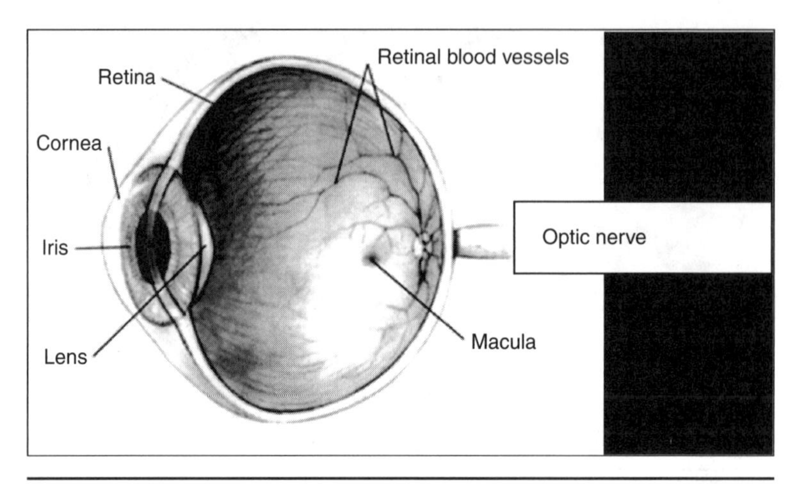

Figure 7.5 Basic anatomy of the eye. National Institutes of Health (NIH).

each of these requirements can fail, often through multiple mechanisms. A selection of the most important clinical mechanisms will be presented following a review of the anatomy of the eye.

7.3.4 Brief Introduction to the Anatomy of the Eye

There are five obvious anatomical structures that are illustrated in Fig. 7.5: cornea, lens, retina, optic nerve, and brain. These structures bear a natural correspondence to the functional requirements of vision.

7.4. The Size of Things

7.4.1 The Eye

Macroscopic

The average size of the eye from front to back is about 24 mm (~1 in.). This exact measurement from the front of the cornea to the retinal surface is of great importance to understand how vision may fail and how to restore it. For example, in myopic patients for whom this measurement may be significantly larger, resulting in failure to focus and severely decreased vision, vision may be restored via straightforward optical means, for example glasses, contact lenses, intraocular lenses, or surgical alteration of the curvature of the cornea. The accurate measurement of this length is paramount in computing the power of lens implants at the time of cataract surgery or for other types of intraocular implants.

Microscopic or Nanoscopic

The measurement needs to be well defined. For example, are we measuring from the front or from the middle of the cornea? This can account for a significant difference in surgical outcome. For example, a 1-mm error can result in a 2- to 3-diopter error on the implantable lens computation (an average lens implant is 20 diopters). Recall that the diopter is a measure of power and that a lens of D diopters will bend parallel light to a focus at $1/D$ meters. The thickness of the cornea is 1/2 mm, and blur can be perceived at 1/4 of a diopter or less. More than 1 diopter of refractive error is usually not compatible with the ability to drive, though there is no one-to-one corrsespondence between refractive error and visual acuity, due to additional factors that contribute to the final resolution at the level of the retina. Measurements at the micron level are therefore most helpful. Some of the individual structures of the eye are examined further at all three levels in what follows.

7.4.2 Cornea

Macroscopic

The cornea diameter is about 1 cm. Its size and curvature makes it into a very efficient lens with a fixed power of about 40 diopters. Normally, the curvature is maximal in the center and decreases toward the periphery, minimizing spherical aberration. In a myopic eye that undergoes refractive surgery (example: LASIK [laser *in situ* keratomileusis]) this curvature gradient is altered and possibly reversed with possible consequences on spherical aberrations that need to be minimized with an appropriate ablation profile.

Microscopic or Nanoscopic

The average thickness of a cornea is slightly above 500 microns. The cornea is itself made up of five layers, including a five-cell layer thick epithelium in front and a single-cell thick endothelium in the back. The epithelial cells are protective and play a key role in certain refractive procedures. If removed, and though this may be painful as in the case of a corneal scratch, they will grow back in a few days. The endothelial cells, on the other hand, will not regenerate *in vivo*. Their density decreases with age as well as trauma and certain pathological conditions. Their presence and appropriate density play a crucial role in maintaining corneal transparency (and therefore preventing blindness) by pumping water out of the cornea against a gradient. An epithelial cell is approximately 40 microns wide and 4 to 5 microns thick, and a healthy endothelial cell is 20 microns wide and 5 microns in height. Special microscopic procedures are available to visualize and count endothelial cells. When endothelial function fails, corneal transplantation is possible. There is a trend toward selective transplantation of endothelial tissue, while preserving the remaining corneal architecture in lieu of whole corneal transplantation which used to be the norm until recently. Both diagnostic and therapeutic modalities at the nanometer scale may be feasible for earlier characterization and intervention in this condition.

7.4.3 Example: Lens

Macroscopic

The anatomical structure known as the lens is approximately 1 cm in diameter and of variable thickness depending on age. It plays a dual functional role: that of an optical lens supplementing the power provided by the cornea with approximate power of 20 diopters and that of a structure capable of both movement and self-curvature alteration, generating a variable power allowing for focusing at different distances, for example, for driving or reading. This so-called accommodative ability of the lens diminishes with time and is therefore the site of one of the most common "pathologies" observed clinically: presbyopia. While accommodation starts declining in the teens, it becomes clinically significant when it reaches a point at around age 40, when reading becomes difficult. The theory of accommodation and presbyopia that is still accepted today is that of Hermann von Helmholtz. The accommodative effort mechanism can be identified using ultrasound biomicroscopy (UBM). The most common "cure" of presbyopia is reading glasses or the addition of a bifocal segment to existing glasses. Many alternatives have been proposed but are not universally accepted.

The lens is also the site of cataract development. A sequence of biochemical changes results in progressive opacification of the lens. This is a cataract (Fig. 7.3). The time when this opacification interferes with vision is variable and to some degree subjective, but will always happen if the individual lives long enough. As mentioned, the cataractous lens can be removed by a variety of techniques and the lens replaced by a new lens of comparable power and synthetic material. The calculation involved in obtaining the power of the lens is based on geometrical optics and sometimes statistical parameter adjustments. Perhaps surprisingly, after decades and millions of eyes implanted, this calculation is not always as accurate as desired, in particular, in situations where the patient has had previous surgeries, for example LASIK. To place this comment in perspective, however, one should note that the vast majority of lens implants in virgin eyes result in a defocus necessitating a correction on the order of a diopter or less, resulting in very good vision even without glasses and an order of magnitude improvement compared to cataract extraction *without* lens implantation.

Microscopic or Nanoscopic

The lens is enclosed in a bag. The anterior portion of the bag is about 10 microns and the posterior aspect measures about 5 microns. The modern surgical removal of a cataract preserves most of this bag and starts with the creation of a circular opening in the anterior capsule that is approximately 5 mm in diameter. This surgical maneuver, known as *capsulotomy*, is currently performed manually with the assistance of a bent needle or special forceps. It is a very delicate maneuver, and its failure can result in serious complications. A femtosecond laser can assist in the performance of well-defined capsulotomies.

The development of a cataract is governed by degenerative processes in lens proteins. Arresting or reversing this process is a key therapeutic goal that would address more than half of blindness worldwide. Probing the molecular dynamics of lens proteins and targeting their key reactions will require some tools similar to the ones described in this volume.

7.4.4 Example: Retina

Macroscopic

The retina covers the back of the eye up to the ora serata (Fig. 7.6). Its area is roughly 5 cm × 5 cm. The diameter of the fovea is approximately 1.5 mm.

Microscopic or Nanoscopic

The retina has a layered structure. In the retina proper, the photoreceptors are the last layer to be reached by light after traversal of the ganglion cell layers and intermediate cell layers. Rhodopsin is a key molecule for phototransduction. The photoreceptor types are cones for high resolution that occur centrally (in and around fovea) and rods for high-sensitivity scotopic (night) vision that have a higher density

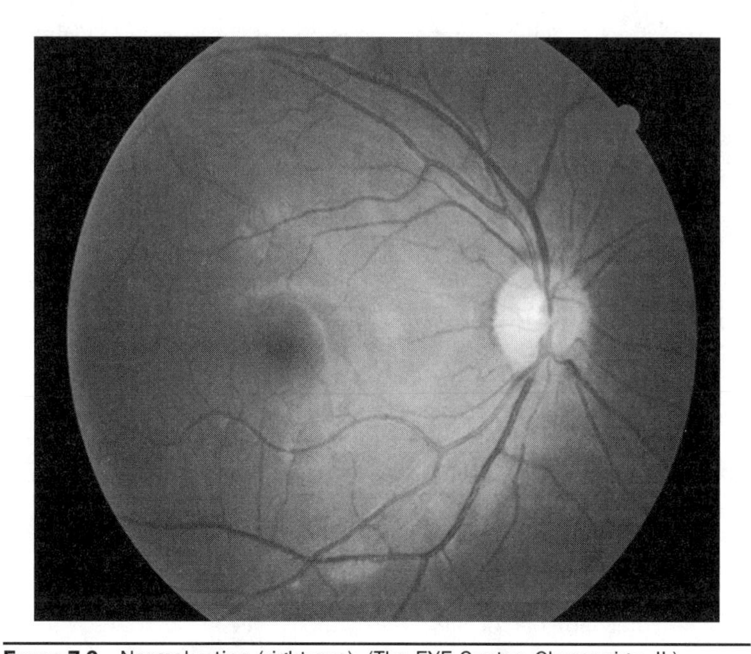

Figure 7.6 Normal retina (right eye). (The EYE Center, Champaign, IL)

in the periphery. There are approximately 5 million cones and 100 million rods in each eye. The diameter of a cone in the fovea is around 1 micron. Cones in the fovea subtend less than 1 minute of arc.

7.4.5 Optic Nerve

Macroscopic

Ganglion cells bend to become parallel to the inner aspect of the retina, forming the nerve fiber layer (NFL), and gather at the optic disk, becoming the axons of the optic nerve. These fibers continue their journey through the brain, splitting at the chiasm and ending in the lateral geniculate nucleus of the thalamus. They connect to neurons that continue the journey to the occipital cortex at the back of the brain.

Microscopic or Nanoscopic

The number of optic nerve fibers is about 1.5 million. To preserve transparency within the retina, the axons are not myelinated until they exit the retina. Myelination ensures an increase in the speed of propagation of the neural signal. A defect in myelination of the optic nerve in conditions such as multiple sclerosis can be associated with severe visual impairment.

7.5 Failure of Visual Requirements Resulting in Blindness

The five functional requirements of vision—transparency, focusing of light, a physical detector (retina), transmission mechanisms (most notably the optic nerve), and an ultimate detector/interpreter (brain, occipital cortex)—correlate highly with corresponding anatomical sites. For example, it should be obvious that the cornea and lens are largely transparent to the visible spectrum (and beyond that spectrum; this is important for certain therapeutic interventions, for example, infrared laser intervention).

7.5.1 Failure of Transparency

Transparency of the cornea can fail. This may be due to a variety of causes including trauma, infection, or dystrophies. A common mechanism is the failure of the endothelial cells and their resulting inability to pump water out of the cornea, resulting in a higher concentration of water in the cornea and subsequent opacification.

Transparency of the lens can fail. This is what constitutes a cataract. It can be genetic, secondary to trauma, or most commonly age related (Fig. 7.3). As mentioned, cataracts are the most common cause of blindness worldwide, and it is important to have efficient and cost-effective mechanisms for their removal or reversal.

Transparency of intermediate "spaces" can fail. The space between the cornea and lens (anterior chamber) can be filled with blood, for example from trauma, and result in interruption of the light signal to the retina. Similarly the vitreous gel, in the space cavity between the lens and retina, can opacify due to a number of reasons including bleeding, for example, in diabetic patients. The removal of such opacities can often restore vision.

7.5.2 Failure of Bending or Focusing

The ability of the eye to focus on an image at the retina is determined by the corneal power and lens power in harmony with the size of the eye. For example, myopia (nearsightedness) can result if the corneal power is high compared to an average value of approximately 43 diopters, unless the length of the eye is correspondingly short. Similarly, if the eye is long with a normal corneal power, myopia results. Hyperopia (farsightedness) can be understood in similar terms. The power of the cornea is determined by its index of refraction relative to that of air and water as well as the curvature of its front and back surfaces.

The lens also plays a critical role, providing about one-third of the power and the ability to accommodate. For example, if a lens is removed (say secondary to trauma or cataract surgery with no artificial lens replacement), the missing power results in a very significant shift toward hyperopia. Accomodation is also eliminated.

7.5.3 Failure of Detection

The retina can fail through a number of mechanisms. The "simplest" one is perhaps detachment of the retina. This can occur from trauma or "spontaneously" and invariably results in blindness unless reattachment intervention occurs. Another mechanism is structural and functional change due to diabetes. One common occurrence in longstanding diabetes is diabetic retinal edema or swelling of the retina where fluid builds up from leaky capillaries. Figure 7.7 illustrates a

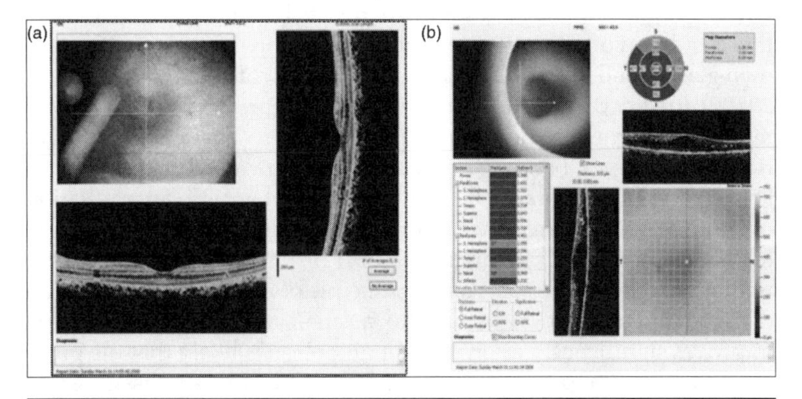

FIGURE 7.7 Optical coherence tomography (OCT) of the retina of (*a*) a normal eye and (*b*) an eye with diabetic edema (fluid-filled). (The EYE Center, Champaign, IL)

view of an optical coherence tomography of a healthy retina as well as a retina with diabetic retinal edema. Degenerative changes in the retina can also occur and result in blindness with the most common degenerative condition being age-related macular degeneration (see Fig. 7.1).

7.5.4 Failure of Conduction

Conduction can fail from trauma to the optic nerve, tumor, stroke, a degenerative condition, or more commonly glaucoma—a condition classically understood as elevated eye pressure and accelerating neuronal death in the optic nerve. As pointed out, demyelination in multiple sclerosis can cause visual impairment, due to perturbation of the conduction mechanism.

7.5.5 Failure of Processing

Signal processing of the detected signal starts at the level of the retina and continues in the lateral geniculate nucleus (LGN), to end in the visual cortex, in the occipital lobe of the brain. Injury to the visual cortex, such as that due to a stroke, can result in profound visual impairment. Cortical "brain" implants have been considered for visual restoration, but one argument for eye implants versus brain implants is the need to preserve the signal processing at early and intermediate stages.

7.6 Diagnostic Tools

If you can't see it, you can't fix it. (Stanley Chang)

We will examine a few tools for the diagnosis of eye conditions. Some are very basic but useful, and some are quite sophisticated.

The most important diagnostic test is that of *vision* since function is the ultimate goal. Testing vision with appropriate-sized letters or symbols at an appropriate distance is the first step in any "eye" evaluation. Vision should be tested, when possible and when safe, with the best-known correction for the patient, for example, with their own glasses and one eye at a time. If vision is worse than normal (say 20/20), further diagnostic testing using the algorithm outlined earlier may be performed. One can run through the five functional categories (transparency, bending or focusing, detector, conduction, higher-level processing, or brain). It is often easier to start with an evaluation of bending or focusing. A very simple test is that of a pinhole. Image formation in a camera (camara oscura) or on the retina is possible without a lens, on the basis of the principle of rectilinear propagation of light and restriction of rays to a small aperture. If the vision improves significantly upon use of a pinhole, it is likely that the problem is a refractive problem, that is, a bending or focusing issue. More detailed diagnostic testing can then be performed to determine the

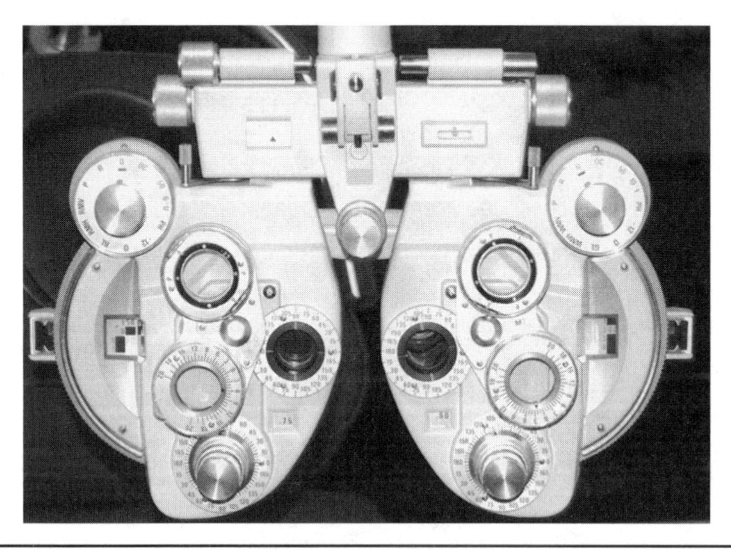

FIGURE 7.8 A phoropter: A collection of lenses allowing implementation of algorithms for determination of refractive errors. (The EYE Center, Champaign, IL)

refractive error, and possibly the higher-order aberrations. Traditionally, the refactive error is classified as "sphere"-representing defocus, cylinder (astigmatism), and axis of astigmatism. These can also be determined subjectively with the help of a phoropter (Fig. 7.8), a collection of lenses arranged in such a way as to facilitate the implementation of a forced-choice algorithm, with responses from the patient, that usually converges to the best needed correction. This correction can then be provided as a pair of glasses, contact lenses, or a prescription for a corneal or intraocular surgical procedure.

Another vision "subjective" test (i.e., depending on the patient's response) is field testing. The patient is presented with lights of decreasing intensity, according to a predetermined or adaptive algorithm, in central and peripheral vision and presses a button when the light is perceived. A map of the visual field is developed on the basis of these reposnses. Figure 7.9 shows such a map for the right eye with a dark circle to the right corresponding to the (normally occurring) blind spot. Figure 7.10 illustrates bilateral fields for a patient who has lost half of the field in each eye following a stroke to the contralateral side of the visual cortex.

One can then evaluate transparency. Once again a very quick test exists, that of the red reflex. This is similar to the "red eye" seen on photographs. For example, the lack of a red reflex may signify a cataract or even an intraocular tumor. One can then use a "biomicroscope" or "slit lamp" to examine the cornea, anterior chamber, and lens. The angle of the iris and cornea cannot be observed directly because of

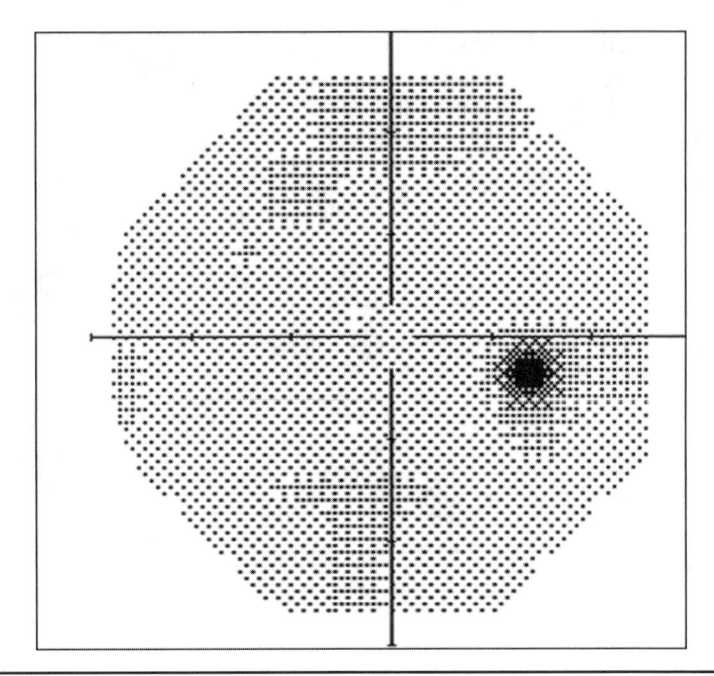

Figure 7.9 Normal field with (normal) blind spot (right eye). (The EYE Center, Champaign, IL)

total internal reflection, so a special contact lens with reflecting mirrors is used to examine that part of the anatomy. This test is called *gonioscopy* (viewing the angle, see the website gonioscopy.org).

While one could examine the anterior part of the eye with little or no magnification, it was only in the mid-19th century that an effective method of examining the eye *in vivo* behind the lens and taking a look

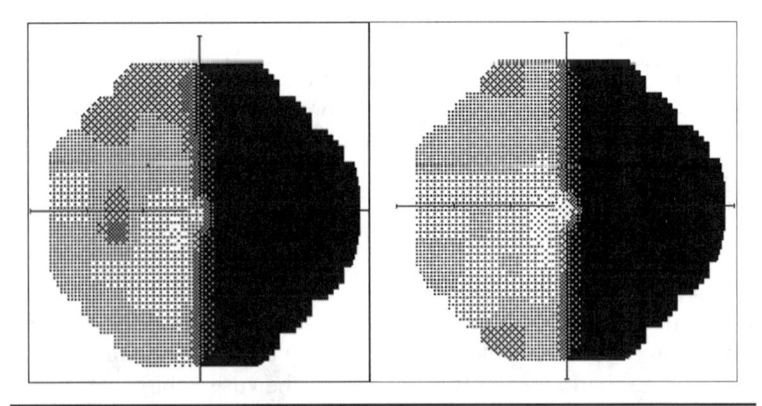

Figure 7.10 Bilateral (left and right eye) loss of half of field of vision secondary to a stroke. (The EYE Center, Champaign, IL)

at the retina became possible. One early attempt to examine the retina of a living animal was by Jean de Mery who, in 1704, reported to the French Royal Academy the immersion of a cat in water and observation of the retina.

It was in 1850, more than two centuries following the availability of telescopes and microscopes, that von Helmholtz invented the ophthalmoscope, an instrument enabling the examination of the retina *in vivo* (without drowning) and wrote to his father that it only took his knowledge of high school physics to get it done! The impact of von Helmholtz's invention is enormous. With the exception of de Mery, no one ever before had seen a live retina. Such a powerful tool caused an explosion in our understanding and classification of retinal and optic nerve conditions including some of the most common and potentially blinding, potentially curable eye conditions.

Looking at the brain beyond the eye and optic disc had to wait for X-rays and computed tomography and magnetic resonance imaging, as well as other anatomical and functional imaging modalities developed in the 20th century. Ultrasonography (US) is extremely useful in a number of contexts but specially when transparency fails (for example, from vitreous hemorrhage or dense cataract) and one needs to evaluate that the retina is attached in order for specific therapeutic surgical modalities to be considered. US is based on the generating of an echo at interfaces of impedance mismatch and time-of-flight distance computation based on a known speed of propagation of the ultrasound signal.

A number of improvements on the ophthalmoscope were introduced over the years. However, it took almost another century and a half to come to the next truly significant breakthrough, optical coherence tomography (OCT). This modality can be used to examine the individual layers of the retina. Ramon y Cajal was awarded the Nobel Prize in Physiology or Medicine along with Camillo Golgi. In addition to championing and establishing the theory of individual neurons, he developed exquisite techniques to establish the multilayer, multicellular structure of the retina. A century later, much of this structure can be revealed *in vivo*. Figure 7.7 shows a normal retinal OCT and an OCT for retinal edema in a severe diabetic. The OCT can guide the diagnosis and treatment of retinal disease and is now used routinely in retinal clinics. It can also be used in the diagnosis of anterior segment conditions of the eye.

7.7 Therapeutic Tools

7.7.1 Therapeutic Tools (Optical)

Optical therapeutic tools are usually intended for the correction of the refractive error (bending or focusing problems). Glasses and contact lenses can correct myopia, hyperopia, presbyopia, and astigmatism. A number of variations and alternative methods have been described such as adjustable glasses, glasses for the correction of higher-order aberrations, and orthokeratology (contact lenses worn at night and removed in daytime for improved vision in daytime without lenses), but these are not widely available, though they may hold some promise in the future.

7.7.2 Therapeutic Tools (Medical)

The whole armamentarium of medical treatments is available and relevant to eye conditions. For example, infections and inflammations can be treated with anti-infective and anti-inflammatory pharmacological agents. High blood pressure and diabetes, which can have a devastating effect on eye condition and vision, are treated with the standard antihypertensives and diabetic medications, including for example, beta blockers and insulin. In addition, topical drops are used among other things, such as anti-infectives (antibiotics, antivirals) and anti-inflammatories and for intraocular pressure-lowering purposes in glaucoma and other conditions. Periocular (in the eye) and intraocular or intravitreal (in the eye) injections are also used for certain conditions.

The Prostaglandin—A Story

This story relates to the pressure-lowering effect of a class of intraocular pressure-lowering drugs. Targets for pressure lowering are either receptors and sites that act as sources or sinks for aqueous humor. Source production can be reduced, and agents are known to act by blocking sources. Alternatively sinks' action can be enhanced. There are two sink systems draining the aqueous in the eye. One of them, believed to be the least significant, turned out to give rise to one of the most significant pressure-reduction drugs ever known, the prostaglandin analogs, a true paradigm shift in the treatment of glaucoma. This clinical example is cited to illustrate some of the limits of understanding and the role of prevailing dogma, and the possibility that more sophisticated nanobiophotonic imaging techniques may play a role in identifying relevant pathways resulting in more rapid and effective paradigm shifts in the future.

The Lucentis or Avastin Story

Age-related macular degeneration (ARMD) has two forms. The dry form is more common but less aggressive whereas the wet form can be devastating. Until recently, there was no treatment that stabilized or improved the vision. Vascular endothelial growth factor (VEGF) was identified as being associated with the wet form of ARMD. It was targeted and blocked by a couple of agents. A common denominator between macular degeneration and cancer is the abnormal vessels that develop and feed and worsen the condition. This is believed to be stimulated at least in part by VEGF. Drug companies have wisely designed more than one anti-VEGF medication to help shut-off the offending abnormal blood vessels and slow down cancers on the one hand, and treat macular degeneration on the other. Avastin is a drug that was originally designed and is still used to treat cancer. It was found to be very effective for treating the wet form of macular degeneration too. While the (FDA approved, very costly Lucentis) drug that was originally designed for the treatment of wet macular degeneration by the same company is still available, the vast majority of retinal specialists administer AVASTIN.

7.7.3 Therapeutic Tools (Surgical)

Restoring Transparency

As mentioned, the most common cause of blindness in the world is cataract. The only known cure for cataracts to this day is surgical cataract removal. Essentially, one needs to clear the axis of the opacity. Historically, the following summarized evolution of the procedure is presented:

1. *Couching* (from French *coucher:* "to lay down") is a procedure that has been practiced for thousands of years and has long been superseded by other methods in most areas of the world with access to modern surgical facilities and skilled surgeons. The procedure is performed by introducing a needle at the edge of the cornea and pushing the cataractous lens inside the eye. This clears the visual axis from the opacity. The complication rate of this procedure is not well known, but there are reasons to believe that it is high compared to that of modern techniques.
2. A procedure where the cataractous lens is brought out of the eye through an incision that is large enough to express the whole lens, and so requires a large incision with the need for sutures.
3. A procedure where the lens is broken inside the eye. This can be done manually at a low cost (for example, trisecting the lens and bringing each piece out in turn) or by emulsifying the lens with high-frequency ultrasound and aspirating the emulsified product. This latter so-called *phacoemulsification* process is the one most practiced in the United States today. Breaking the lens inside the eye before removing it has the distinct advantage of a very small, most often self-sealing, incision not necessitating sutures.

Removing the lens with no replacement was the standard procedure until the second half of the 20th century when intraocular lenses became available. Leaving the eye in a state of aphakia, that is, with no lens, necessitates the wearing of powerful glasses in most patients.

Another form of failure of transparency discussed earlier results from vitreous hemorrhage. The removal of vitreous hemorrhage from the vitreous cavity is done using the main surgical intervention performed by vitreoretinal surgeons: a vitrectomy. This was introduced in the early 1970s by Jean Marie Parel, one of the most innovative minds in ophthalmic technology, and the late Robert Machemer, a giant innovator in vitreoretinal surgery. A vitrectomy or trans pars plana vitrectomy is an operation that is usually performed through the creation of three incisions in the pars plana, an area that is approximately 3.5 or 4 mm beyond the cornea but not far enough posteriorly as to get into the retina. This allows access to the vitreous cavity with minimal risk of perturbing the retina proper. One of the three ports is used to introduce fluid and maintain controlled pressure in the eye; the other two ports are used for two other instruments, typically a fiber-optic light pipe and a vitrector or "cutter," whose function is to cut the vitreous and any opacities, including blood, and aspirate it outside of the eye. Until recently, a vitrectomy was performed with 20-gauge (G) instruments (0.902 mm), and this typically required a significant amount of time and dissection of tissues and suturing of all three incisions and conjunctival tissue at the end of the case. In recent years, 23-G (0.635-mm) and 25-G (0.508-mm) instruments were introduced. The technique used is that of a trocar or cannula system through which instruments can be pulled in and out (Figs. 7.11 and 7.12).

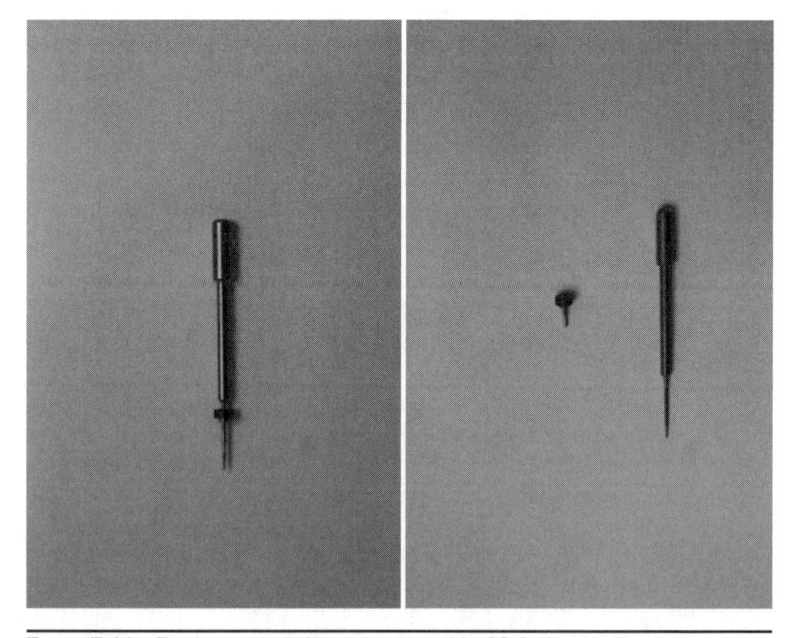

Figure 7.11 Trocar or cannula system used in 23-G vitreoretinal surgery. (The EYE Center, Champaign, IL)

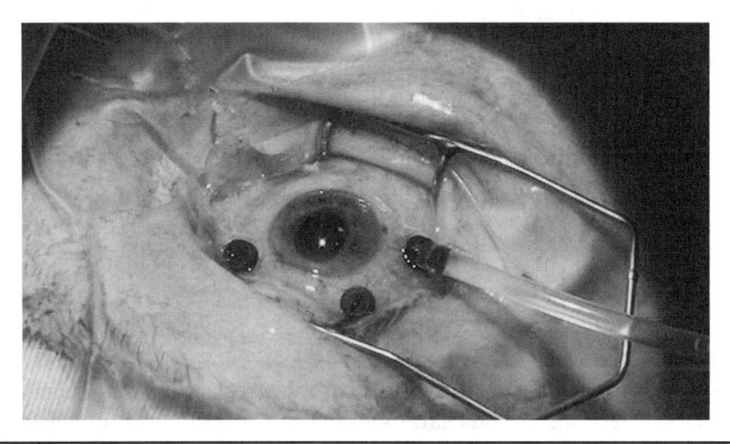

FIGURE 7.12 23-G vitrectomy setting with cannulas in place and infusion in line. (The EYE Center, Champaign, IL)

No initial cuts are needed in conjunctival tissues as the trocars are inserted transconjunctivally and no suturing is needed at the end of the vast majority of cases. In addition to being a technique to remove opacities, such as blood and dense floaters, the vitrectomy is a needed step in other more complex surgeries such as retinal detachment surgery.

Restoring Proper Bending or Focusing

This is done in a number of contexts. After cataract removal, an intraocular lens can be implanted to replace the original lens (Fig. 7.13). In modern cataract surgery, the bag in which the lens resides is largely

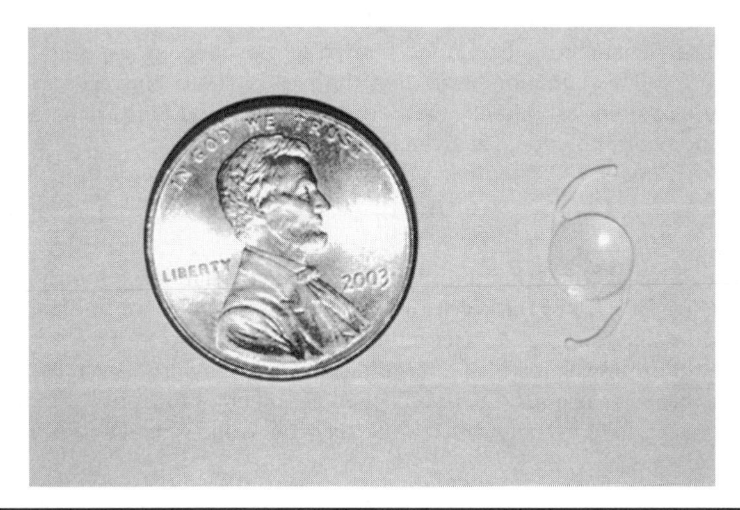

FIGURE 7.13 Intraocular lens in comparison to a penny. (The EYE Center, Champaign, IL)

preserved except for a circular opening (about 5 mm in diameter) in the front that is used to remove the cataractous lens and insert the new artificial lens. The power of the artificial lens to compensate both for the existing lens and for the correction of the patient (nearsightedness, farsightedness, etc.) can be calculated from simple anatomical measurements, mainly the axial length (length of the individual eye) and corneal power. A simple example of a formula allowing such calculation is the Sanders, Retzlaff and Kraft (SRK) formula

$$\text{Power} = A - 0.9K - 2.5L$$

where A is considered a characteristic constant, K is the power of the cornea, and L is the axial length of the eye. Dimensional analysis and other considerations can demonstrate that A is not a constant. Improved formulas are known today, though a firm general approach to the problem of intraocular lens calculation, especially in challenging cases, is nascent.

Modern artificial lenses are made of foldable acrylic that facilitates the insertion of a lens through a small incision (under 2 to 3 mm). There is currently a slight trend toward lenses that go beyond providing correction for defocus for distance vision. For example, some lenses will correct a significant degree of astigmatism and some will provide acceptable vision for both distance and near vision. These approaches, however, often come at the cost of increased aberrations and decreased contrast sensitivity.

LASIK, PRK, and All That

The surgical modality that deals with the correction of refractive error (bending or focusing) is the use of the "excimer" laser. Perhaps surprisingly, one of the basic principles of refractive surgery goes back to the "lensmaker's equation" known to Cavalieri in the mid-17th century! This equation states that the power P of a thin lens in air made of spherical refractive surfaces of radii R_1 and R_2 from material of index of refraction n is given by

$$P = (n-1)\left(\frac{1}{R_1} - \frac{1}{R_2}\right) \tag{7.1}$$

The surgical correction of myopia proceeds by flattening the curvature (radius R_1) of the cornea to yield a lesser curvature (radius R_2) and less bending, this focuses the images on the retina. This can be thought of as the removal of a thin lens whose power is given by the lensmaker's formula in Eq.(7.1). The power reduction P can be expressed in terms of removed tissue diameter S and central thickness t:

$$P = 8(n-1)\frac{t}{S^2}$$

where $n = 1.377$ is the index of refraction of the cornea.

The "excimer" laser typically used in refractive surgery is a pulsed 193-nm Xe/F UV laser and is thus not strictly speaking an excited dimer (excimer) as the two molecules are different. It breaks carbon-carbon bonds in the cornea with essentially no heat generation in neighboring tissue. The nominal values of removal per pulse for clinical ablations are on the order of 200 to 300 nm. This allows for a smooth photorefractive keratectomy (PRK) if operated at the surface and a laser *in situ* keratomileusis (LASIK) if operated after a transverse flap has been created with a mechanical keratome or a femtosecond Nd-glass laser. *Mileusis* comes from the Greek word for "sculpting," and the lens of required power is sculpted out of the cornea, this results in the required correction. Femtoseond lasers as a tool for direct intrastromal ablation without the creation of a flap were studied more than a decade ago and are now being implemented.

Restoring Detector Integrity (Retina)

Retinal Detachment Repair Retinal detachment surgery is naturally related to the understanding of the pathophysiology of this condition. Traction from the vitreous gel on the retina gives rise to a retinal tear, and fluid passing through that tear eventually detaches the retina. The modern repair of retinal detachments is more often performed starting with a vitrectomy as described above. This can be performed using a 20-, 23-, or 25-G approach and functionally relieves traction of the vitreous gel off the retina. Fluid behind the retina can be drained, and SF_6 or C_3F_8 gases can be used for tamponade. This allows the natural pumping mechanism to operate and remove residual fluid. Laser (Argon green 514 nm or frequency-doubled Nd-YAG at 532 nm) can be used to attach the retina around the tear and avoid redetachment once the gas dissolves.

Subretinal and Epiretinal Prostheses for Artificial Vision The possibility of a retinal prosthesis to accomplish "artificial vision" offers significant hope for restoration of vision in a number of diseases that affect the retina. The most common blinding disease affecting the retina is ARMD, and the most common hereditary retinal condition is retinitis pigmentosa (RP) affecting about 1 in 3000. Although ARMD is much more common, late stage RP leads to profound forms of blindness and makes even rudimentary artificial vision quite attractive. The goal with the current state of knowledge and technology is not to restore 20/20 vision but rather to provide sufficient vision for mobility. A very encouraging historic precedent is that of cochlear implants to reverse neural deafness.

A number of research groups worldwide are pursuing the goal of an artificial retina. These can be classified as the epiretinal implant or the subretinal implant approach. Each of these approaches has distinct advantages, and it is not clear at this juncture if either or any will prevail.

The diameter of a foveal photoreceptor is about 2.5 microns, and the photoreceptors are tightly packed. The optical resolution of the eye is fairly well matched to the resolution on the basis of photoreceptor density and corresponds to a vision on the order or slightly better than 1 minute of arc (20/20 vision). The resolution sought after in current implants is significantly more modest.

Following Argus I, Argus II is a project for an epiretinal device developed by Second Sight, Inc., led by Marc Humayun at the University of Southern California and supported by the Department of Energy (Fig. 7.14).

Argus Panoptes is a mythical Greek figure who had hundreds of eyes all over his body and was of superhuman strength. Some of his eyes were always open; the rest would open with sunrise and close with sunset. Hera, Zeus's wife, appointed Argus as guardian of Io, a nymph and one of Zeus's lovers. Zeus had transformed Io into a cow to deceive Hera, but Hera demanded the animal as a gift and set Argus Panoptes as its guard.

Argus I implants had 4×4 electrode arrays. Argus II now in clinical trial has an 8×8 array. There is reasonable evidence at this point that these implants are stable and can provide rudimentary vision consistent with mobility in a handful of patients. Retinal prostheses of this type may not be appropriate for the treatment of blindness

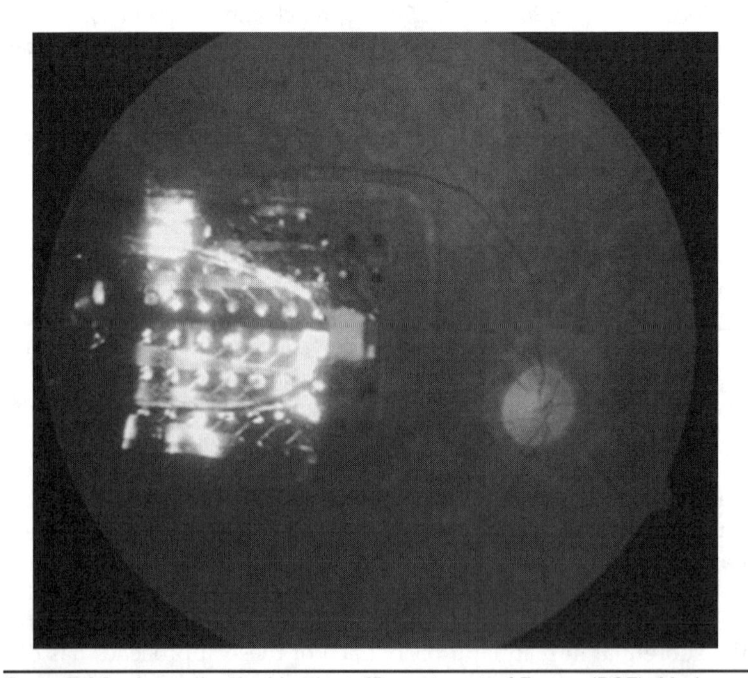

Figure 7.14 Argus II retinal implant. [Department of Energy (DOE), Mark Humayun]

secondary to diabetes or glaucoma since the optic nerve or anterior aspect of the retina may be damaged.

Zeus sent Hermes to rescue his lover, Hermes enchanted Argus by the sweetness of his music, lulled him to sleep, and then slayed him. Following his death, Hera rewarded Argus by gracing the tail of her sacred bird, the peacock, with the hundreds of Argus's eyes.

With the advent and growth of nanobiophotonics, we are positioned to witness and participate in a rapid evolution in the sophistication of future retinal implants. Perhaps our future patients will marvel at the wonders of the peacock's tail.

Restoring Conduction in the Optic Nerve and Restoring Processing (Brain)

Clinically, a success story that can be reported is that of reversal of amblyopia and perhaps some forms of glaucoma in very early childhood. Children who develop strabismus (crossed-eyes) or have had a high-refractive error or other reason for depriving the eye, and mainly the brain, of a meaningful signal, often will develop a "lazy eye" that later on will have poor vision, sometimes to the degree of legal blindness. There is a window of opportunity to correct the cause of the lazy eye and recuperate full vision. This age is believed to be around 9 years, though there may be some variability. The plasticity of the brain, at least at that juncture, is a cause of optimism as it may open an avenue to intervention in cases of neurophthalmic conditions that today are generally believed to be irreversible.

7.8 Summary

A summary of vision, its mechanisms of failure, and the state of the art in clinical and surgical diagnostic methods and therapeutic interventions were presented. Given the nature of the problem at hand and the nature of the light-matter–tissue interaction under consideration, it is no surprise that nanobiophotonics lie at the heart of many current and future modalities of diagnosis and therapy that are likely to restore vision and improve the quality of people's lives.

Suggested Reading

D. R. Hamilton, and D. R. Hardten, "Cataract Surgery in Patients with Prior Refractive Surgery," *Cur. Opin. Ophthalmol.* 14 (1): 44–53, 2003.

P. J. Rosenfeld, "Intravitreal Avastin: The Low Cost Alternative to Lucentis?" *Am J. Ophthalmol.* 142: 141–143, 2006.

S. I. Sayegh, D. Cabrera, T. Juhasz, and R. M. Kurtz, "Geometric and Biomechanical Factors in Refractive Surgery Nomograms: Implications for Femtosecond Laser Intrastromal Procedures," *ARVO 99*: 554, 1999.

S. I. Sayegh, "Customizing IOLs after LASIK: From A (Constant) to Z(ernike)," *AAO Atlanta* oral presentation abstract, 2008.

S. I. Sayegh, "Varying the A-Constant: Introduction to Customized IOL Calculation," *ASCRS Chicago,* 2008. (Best paper of session award, accessible at http://ascrs.org/Meetings/Post-Meeting-Resources/BPOS-Winners-2008.cfm.)

N. G. Basov, V. A. Danilychev, Y. Popov, and D. D. Khodkevich, "Zh. Eksp. Fiz. i Tekh. Pis'ma," *Red.* 12: 473, 1970.

C. R. Munnerlyn, S. J. Koons, and J. Marshall. "Photorefractive Keratectomy: A Technique for Laser Refractive Surgery," *J. Cataract Refract Surg.* 14 (1): 46–52, 1988.

S. I. Sayegh, D. Cabrera, and R. Kurtz, "A Hybrid Model for Excimer Laser and Femtosecond Laser Procedures Nomogram Generation," in *WRSS 1999,* abstract/oral presentation, 1999.

K. R. Sletten, K. G. Yen, S. I. Sayegh, F. Loesel, C. Eckhoff, C. Horvath, M. Meunier, T. Juhasz, and R. M. Kurtz, "An *in vivo* Model of Femtosecond Laser Intrastromal Refractive Surgery," *Ophthalmic Surgery and Lasers* 30 (9), 1999.

M. S. Humayun, J. D. Dorn, A. K. Ahuja, A. Caspi, E. Filley, G. Dagnelie, J. Salzmann, A. Santos, J. Duncan, L. Dacruz, S. Mohand-Said, D. Eliott, M. J. McMahon, and R. J. Greenberg, "Preliminary 6 Month Results from the Argus II Epiretinal Prosthesis Feasibility Study," *Conf. Proc. IEEE Eng. Med. Biol. Soc.,* 1: 4566–4568, 2009.

J. Martins, "Bioelectronic Vision: Retina Models, Evaluation Metrics & System Design" (Series on Bioengineering & Biomedical Engineering, Vol. 3, 2009).

C. Foster, D. Azar, and C. Dohlman, *Smolin and Thoft's The Cornea: Scientific Foundations and Clinical Practice,* Lippincott Williams & Wilkins, 2005.

L. Yannuzzi, "The New Retinal Atlas: Expert Consult Online and Print," Saunders, 2010.

P. Mouroulis, and J. Macdonald, *Geometrical Optics and Optical Design,* Oxford University Press, 1997.

A. Chiang and J. A. Haller, "Vitreoretinal Disease in the Coming Decade," *Curr. Opin. Ophthalmol,* 21: 197–202, 2010.

CHAPTER 8

Optical Low-Coherence Interferometric Techniques for Applications in Nanomedicine

Utkarsh Sharma and Stephen A. Boppart

Biophotonics Imaging Laboratory
Beckman Institute for Advanced Science and Technology
University of Illinois at Urbana–Champaign

8.1 Introduction

Optical interferometry has proved to be an extremely powerful tool for physicists and researchers to help gain insights into various fundamentals of basic science and nature for more than three centuries. Although the early experiments on interference by Boyle, Hooke, and Newton date back to the 17th century, the understanding of the physics behind this phenomenon continuously evolved over the next two centuries until physicists concurred on the wave-particle dual nature of photons [1–3]. However, it was the invention of the laser followed by breakthroughs in the telecommunications industry and computer technology that eventually made it feasible to exploit the fundamental principles of this phenomenon for various practical sensing applications in metrology, astronomy, and medicine.

Optical techniques play a significant role in medicine because they promise safe and low-cost solutions to many problems. Over the last one and a half decades, the field of medicine has benefited extensively from the rapid progress in the instrumentation and technology of low-coherence imaging techniques, broadly classified under the term *low-coherence interferometry* (LCI) or *optical coherence tomography* (OCT) [4–6]. The OCT technique exploits the short temporal coherence of a broadband light source and enables high-resolution, noninvasive, *in vivo* imaging of microscopic structures in scattering tissues up to depths of approximately 2 mm, depending on the tissue type. OCT has been increasingly finding applications as a potential diagnostic and imaging tool in the field of medicine [7–16]. The rapid advancements in the field of low-coherence imaging have coincided with rapid progress in the field of nanotechnology over the last decade. The progress in nanotechnology has been largely responsible for the emergence of a new field called *nanomedicine*. Although the field of nanomedicine is still at a nascent stage, it is poised to be a cutting-edge medical technology for the next decade. The low-coherence interferometry technology is proving to be one of the critical diagnostic and imaging tools for applications in nanomedicine. In this chapter, we discuss the theory and principles behind low-coherence interferometry and describe several other functional and molecular contrast–based extensions of OCT with an emphasis toward possible applications in nanomedicine.

8.1.1 Origin and Evolution of OCT Technology

The roots of the OCT technique can be traced back to the early developments of optical low-coherence reflectometry (OLCR) in the telecommunications industry during the late 1980s [17–21]. The technique was aimed at finding faults or reflection sites in miniature optical waveguides and optical fibers. There were many contributing factors such as the development of broadband semiconductor light sources, single-mode fibers, and couplers that aided in the advancement of low-coherence reflectometry techniques. Fercher et al. and Fujimoto et al. introduced this technology to ophthalmology as they used the one-dimensional optical ranging technique to measure intraocular distances in the eye in the late 1980s [22, 23]. Huang, Fujimoto et al. then extended this technique to develop the two-dimensional tomographic imaging technique known as OCT in 1991 [4]. The first *in vivo* retinal images were independently obtained by Fercher et al. and Swanson et al. in 1993 [24, 25]. Since then, the field of OCT has witnessed tremendous technological advancements which have greatly enhanced its utility for medical imaging and research. These technological advances include the development of efficient and broader bandwidth laser sources, enabling higher imaging speed and sensitivity, sophisticated integration of OCT probes with standard medical catheters and endoscopes, and development of various functional extensions such as a Doppler OCT, polarization-sensitive OCT (PS-OCT), spectroscopic OCT, magnetomotive OCT, and second-harmonic OCT [26–37].

One of the major breakthroughs in OCT technology has been the theoretical and experimental demonstration of the superior sensitivity performance of Fourier-domain OCT (FD-OCT) systems over time-domain OCT (TD-OCT) systems in 2003 [38–43]. Over the last seven years, the continuous improvements and maturation of Fourier-domain methods have resulted in the development of robust and portable FD-OCT systems and have provided a significant advantage to clinical OCT imaging as rapid image acquisition rates enabled scanning of larger tissue volumes while reducing artifacts due to patient motion [44–48].

8.1.2 Applications in Medicine

To date, the most significant clinical impact of OCT has been in the field of ophthalmology, and OCT is now established as the standard clinical imaging modality for screening and diagnosis of several retinal diseases and glaucoma [7, 8, 13, 49–52]. Currently, there are more than a dozen established and startup companies participating in the growth of the ophthalmology OCT market. Cardiology is the other field where OCT may have a significant clinical impact. Intravascular OCT has been successful at characterizing arterial plaques and visualizing interventional treatments such as stent implantations [53–61]. Intravascular OCT has already transitioned from a validation phase to clinical use. Oncology is yet another field where OCT may find applications in tumor boundary detection, image-guided surgery, and early detection of small lesions in a range of tissues [16, 62–65]. Although OCT has been demonstrated as a useful imaging technique for a wide range of biological, medical, and small animal research applications, we will not describe these in detail, as this information has been covered extensively in several books and review papers [48, 66–72].

8.1.3 Applications in Nanomedicine

Nanomedicine, a recent and rapidly progressing field with a focus toward medical applications of nanotechnology, has capitalized on the tremendous progress and breakthroughs made in the broad field of nanoscale technologies during the past decade [73–81]. Nanomedicine is a highly interdisciplinary field requiring inputs from biology, chemistry, physics, mechanical engineering, materials science, and clinical medicine. The research focus and applications of nanomedicine range from early stage disease diagnosis by detecting changes at the molecular level, novel nanotechnology platforms for targeted drug delivery and therapy at subcellular or molecular levels, and futuristic nanomachines for subcellular repairs [74, 82–87].

Low-coherence interferometry–based techniques have great potential to contribute to the field of nanomedicine. Although challenged by limited penetration depth in tissue, OCT techniques offer superior resolution for diagnostic applications as compared to ultrasound or magnetic resonance imaging (MRI). The advances in laser sources have led to ultrahigh resolution OCT capable of achieving sub-micron resolutions and the techniques like interferometric synthetic aperture microscopy (ISAM) have made it possible to perform high-resolution subcellular imaging with high transverse resolution and extended depth of focus [88–93]. These ultrahigh-resolution OCT techniques could aid in the early diagnosis of pathogenesis in tissue or monitor response to a treatment.

There are several functional extensions of LCI-based techniques that have shown great promise for current and future research in nanomedicine. Spectral-domain phase microscopy (SDPM), a functional extension of SD-OCT, has been used to study cellular dynamics by monitoring nanometer-scale motion as well as cytoplasmic flow in living cells [94–97]. SDPM can obtain subnanometer level path-length sensitivity as it detects the phase variations caused by changes in the refractive index and thickness due to cellular dynamics. Spectroscopic OCT can be used to enhance OCT image contrast from endogenous molecules as well as exogenous contrast agents (such as near-infrared dyes, plasmon-resonant gold nanoparticles, etc.) [31, 36, 98–101]. Polarization-sensitive OCT (PS-OCT) is yet another extension of OCT which can provide endogenous contrast by detecting the presence of birefringence caused by collagen in muscles or connective tissues [26, 28]. PS-OCT is capable of detecting ultrastructural changes in muscle tissue by detecting the change in the birefringence signal [102]. Magnetomotive OCT is capable of detecting the presence of magnetic iron oxide nanoparticles in tissues and may enable the possibility of tracking targeted delivery of an antibody or protein attached to a magnetic nanoparticle [33, 34]. Nonlinear interferometric vibrational imaging (NIVI), a variation of the Coherent Anti-Stokes Raman Scattering (CARS) technique with interferometric depth-resolved imaging capability, has emerged to be a promising LCI-based molecular-specific endogenous contrast imaging technique [103]. Although these techniques show great promise for applications in nanomedicine, further advances and breakthroughs in instrumentation and technology are needed before they can make a significant and wide-scale impact in the field of nanomedicine.

In the following section we will explain the basic theory of low-coherence interferometry and discuss the operational principles of a spectral-domain OCT system. The fundamentals of axial and lateral resolution, noise and signal-to-noise ratio, and sensitivity comparison of TD-OCT and FD-OCT systems will be discussed. Subsequent sections will discuss a range of functional extensions of OCT and other LCI techniques, including brief descriptions of their operational principles and some of the early research studies that are relevant to possible applications in nanomedicine. The final and fourth section will include a discussion about current trends and future prospects for the application of LCI- and OCT-based techniques in nanomedicine.

8.2 Basic Theoretical Aspects of Low-Coherence Interferometry

A standard OCT system can be described by the generalized schematic shown in Fig. 8.1. Although the operating mechanisms of time-domain OCT (TD-OCT) and Fourier-domain OCT (FD-OCT) systems differ as different broadband sources and detection and signal-processing schemes are employed, the basic principle is the same and can be explained with the help of the same schematic block diagram (Fig. 8.1). Light from the broadband source is split into two arms of the Michelson interferometer using a beam splitter or a 50/50 fiber-optic coupler. The light incident on the tissue undergoes partial backscattering due to the presence of discrete as well as a continuum of reflection sites at different depths within the tissue. At the output of

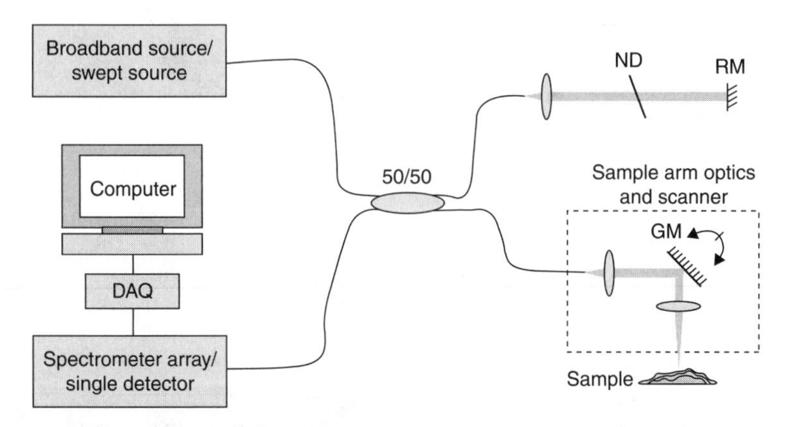

FIGURE 8.1 Schematic diagram of a generalized OCT system. Abbreviations: DAQ, data acquisition; ND, neutral density filter; 50/50, fiber-optic coupler with equal splitting ratio; RM, reference mirror; GM, galvanometer mirror.

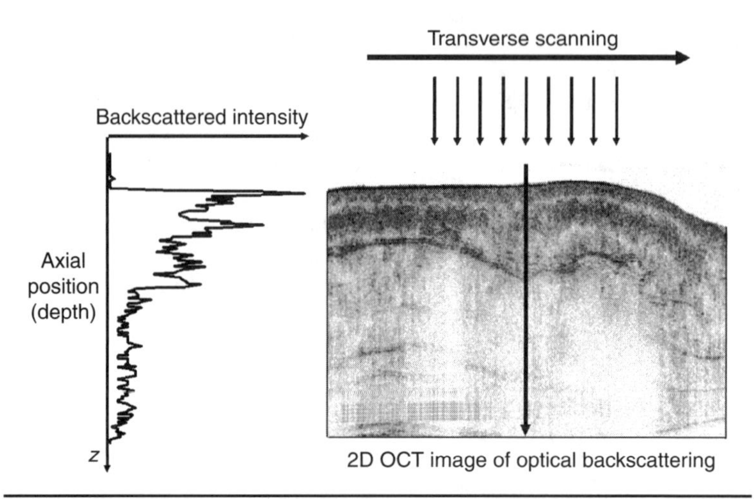

Figure 8.2 OCT image reconstruction. Cross-sectional OCT images are typically obtained by assembling a series of A-scans (axial depth profiles) as the incident beam is scanned in a lateral or transverse, direction.

the interferometer, the backscattered light from the tissue is then recombined with the light from the reference arm, and the interference signal recorded at the detection end is then used to extract axial structural information of the tissue, which can be represented as an A-scan (Fig. 8.2). The beam incident on the tissue is scanned laterally, and a series of A-scans are collected and then used to obtain the cross-sectional image of the tissue as shown in Fig. 8.2.

8.2.1 OCT Theory

If we assume a wavelength-independent splitting ratio for the coupler or beam splitter, then the broadband source light output propagating into each arm of the interferometer can be written as

$$E_{Rin} = E_{Sin} = s(k)e^{-j(\omega t - kz)} \tag{8.1}$$

where $s(k)$ is the electric field amplitude spectrum, k is the wave number, and z is the propagation distance. The reference arm typically has a variable attenuator or neutral density filter to adjust the reference light power level. The attenuated electric field reflected by the reference mirror is given as $E_R = r_R s(k)e^{-j(\omega t - 2kz_R)}$, where $R_R = |r_R|^2$ is the attenuated power reflectivity of the reference arm. The light incident on the tissue will undergo backreflection and backscattering from multiple sites due to the presence of scattering particles and refractive index variations within the tissue. Although the backscattering function of the tissue is a continuum rather than a discrete function, we define it as a series of discrete delta-functions with varying

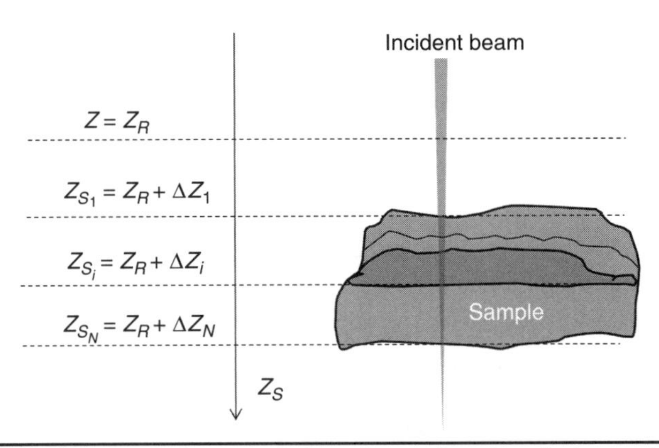

FIGURE 8.3 Representative sample based on a discrete reflector model.

reflectivity to obtain a better physical insight into this phenomenon. The sample backscattering function is therefore given as $r_S(z)=\sum_{i=1}^{N} r_{S_i}\delta(z-z_{S_i})$, where $r_{S_i}=\sqrt{R_{Si}}$ is the electric field amplitude reflectivity of the scattering site at the depth z_{S_i} (Fig. 8.3). In this model, we restrict our analysis to ballistic or least-scattered photons, and the effects of multiple scattering are ignored. The backscattered photons returning from the sample arm can be described as the convolution of the incident field and the backscattering function and can be written as

$$E_S = s(k)e^{-j(\omega t-2kz_S)} \otimes r_S(z) = s(k)\sum_{i=1}^{N} r_{S_i} e^{-j(\omega t-2kz_S)} \tag{8.2}$$

The sample and reference electric fields are recombined at the beam splitter and given by $E_{\text{det}}=(1/\sqrt{2})(E_R+E_S)$. The incident light is then converted into a photocurrent by optical detectors, which are square law intensity detection devices. The generated photocurrent is proportional to the time average of the incident electric field multiplied by its complex conjugate and is given by

$$I_{\text{Det}}(k)=\frac{\rho}{2}\langle E_R+E_S\rangle\langle E_R^*+E_S^*\rangle=\frac{\rho}{2}\left[I_R+\langle E_sE_S^*\rangle+2\operatorname{Re}\left(\langle E_RE_S^*\rangle\right)\right]=\frac{\rho S(k)}{2}\left\{R_R+\sum_{i=1}^{N}R_{S_i}\right.$$
$$\left.+\sum_{i=1}^{N}\sum_{j=i+1}^{N}\sqrt{R_{S_i}R_{S_j}}\cos\left[2k\left(z_{S_i}-z_{S_j}\right)\right]+\sum_{i=1}^{N}2\sqrt{R_RR_{S_i}}\cos\left[2k\left(z_R-z_{S_i}\right)\right]\right\} \tag{8.3}$$

where ρ is the detector responsivity (ampere/watt) and $S(k)(=|s(k)|^2)$ is the normalized spectral power density of the source. The first two terms on the right-hand side (inside the curly brackets) of the equation represent the DC component of the current. The third term corresponds to the self-interference of the sample electric field. Sample reflectivity in biological tissues is typically very small (10^{-5} to 10^{-7}) compared to reference reflectivity, and hence the signal generated due to self-interference can often be neglected. The fourth and final term in Eq. (8.3) accounts for the interference between the reference and sample electric fields and is used to extract the axial depth profile or structural information in OCT. This cross-correlation term can be further simplified by substituting $z_{S_i} = z_R + \Delta z_i$ (Fig. 8.3):

$$I_{AC}(k) = \frac{\rho S(k)}{2} \sum_{i=1}^{N} 2\sqrt{R_R R_{Si}} \cos(2k\Delta z_i) \tag{8.4}$$

The tissue can also be modeled as a source of continuous rather than discrete backscattering sites for a more realistic model.

$$I_{AC}(k, \Delta z) = \rho S(k) \sqrt{R_R R_S(\Delta z)} \cos(2k\Delta z) \tag{8.5}$$

The goal of various OCT signal-processing techniques is to extract the depth-dependent reflectivity function ($R_S(\Delta z)$) of the sample in order to obtain its axial structural profile information.

8.2.2 Spectral-Domain OCT

The working principle for both TD-OCT and FD-OCT systems can be explained by Eq. (8.5), despite differences in source design, detection schemes, and signal-processing schemes. However, in this chapter, we will cover the analysis of spectral-domain OCT (SD-OCT) systems only. In an SD-OCT system, the light output from the interferometer is directed to a spectrometer as shown in Fig. 8.4. Various spectral slices of the combined broadband output from the reference and sample arms are spatially encoded using a collimator, diffraction

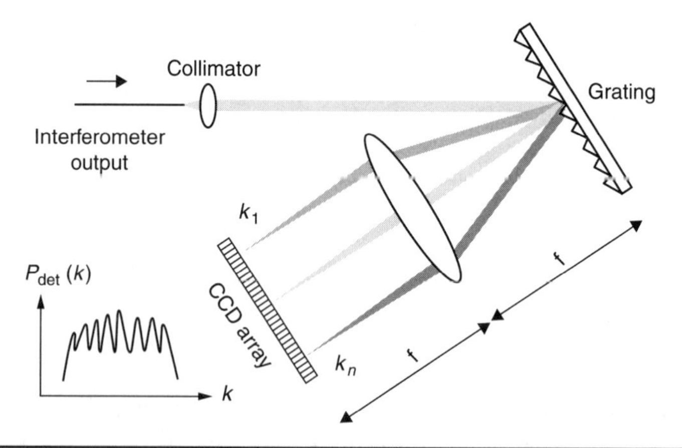

Figure 8.4 Schematic of spectrometer-based detection in an SD-OCT system. A cone of collimated beams of different wavelengths emerges from the grating plane, and each spectral slice is mapped to an individual pixel in the linear charge-couple device (CCD) array. The resulting spectrum shown is then inverse-Fourier-transformed to provide depth-dependent scattering information.

grating, and linear detector array. Resampling of the data obtained from the linear detector array is performed in order to correct for the nonlinear spatial mapping of wave numbers. After resampling and subtraction of the DC background, the depth profile structural information can be obtained by performing the inverse Fourier transform operation:

$$I_{AC}(\Delta z) = \mathrm{IFT}\,[I_{AC}(k)] = \mathrm{IFT}\left\{\frac{\rho S(k)}{2}\sqrt{R_R}\left[\int_0^{\infty}\sqrt{R_S(\Delta z)}(e^{-j2k\Delta z} + e^{j2k\Delta z})\,d\Delta z\right]\right\} \tag{8.6}$$

If we assume the depth-dependent reflectivity in the sample to be a symmetric function such that $R'_S(\Delta z) = R_S(\Delta z) + R_S(-\Delta z)$, then Eq. (8.6) can be rewritten as

$$\begin{aligned} I_{AC}(\Delta z) &= \mathrm{IFT}\left\{\frac{\rho S(k)}{2}\sqrt{R_R}\cdot \mathrm{FT}\left[\sqrt{R'_S(\Delta z)}\right]\right\} \\ &= \frac{\rho\sqrt{R_R}}{2}\{\mathrm{IFT}[S(k)]\}\otimes\sqrt{R'_S(\Delta z)} = \frac{\rho\sqrt{R_R}}{2}G(\Delta z)\otimes\sqrt{R'_S(\Delta z)} \end{aligned} \tag{8.7}$$

where $G(\Delta z)$ is the inverse Fourier transform of the source spectrum and is also known as the *coherence function*. The axial reflectivity profile thus obtained from Eq. (8.7) will have a have a mirror image of about $z = z_R$. However, only a one-sided reflectivity profile is sufficient to obtain axial structural information of the tissue in depth. The difference between TD-OCT and FD-OCT signal acquisition principles can be explained by comparing Eqs. (8.5) and (8.7). In FD-OCT systems, the interference signal is distributed and integrated over many spectral slices [Eq. (8.5)] and is inverse-Fourier-transformed to obtain the depth-dependent reflectivity profile of the sample. In TD-OCT systems, the interference signal of the broadband fields is integrated over time as the reference path delay is modulated in a periodic manner with constant speed to obtain the same information.

8.2.3 Axial and Lateral Resolution

Often it is useful practice to approximate the source spectral density function as a normalized Gaussian function: $S(k) = (1/\Delta k\sqrt{\pi})e^{-[(k-k_0)/\Delta k]^2}$, where k_0 is the central wave number and Δk is the half-width of the spectrum at $1/e$ of its maximum. The coherence function can then be expressed as $G(\Delta z) = e^{-(\Delta z \Delta k)^2} = e^{-\ln 2(2\Delta z/l_c)^2}$, where l_c is defined as the round-trip coherence length of the source. The round-trip coherence length is also a measure of axial resolution in OCT, and its relationship with the axial resolution of the system (in air) is defined as: $\Delta Z_{Res} = l_c = (2\ln 2/\pi)(\lambda_0^2/\Delta\lambda)$, where λ_0 is the central wavelength and $\Delta\lambda$ is the FWHM wavelength bandwidth of the source. It can be seen that the axial resolution is inversely proportional to the source bandwidth. A typical superluminescent diode (SLD) source having a center wavelength of $\lambda_0 = 1300$ nm and a FWHM spectral bandwidth of $\Delta\lambda = 50$ nm would have an axial resolution of approximately ~15 μm. It should be noted that in biological samples, the effective axial resolution of the system is modified to $\Delta Z'_{Res} = (2\ln 2/\pi n_s)(\lambda_0^2/\Delta\lambda)$, where n_s is the refractive index of the sample.

Lateral resolution ΔX of an OCT system is given by $\Delta X = [2\lambda_0/\pi(NA_{obj})]$, where NA_{obj} is the numerical aperture of the objective lens in the sample arm [104]. Note that the lateral resolution is decoupled from the axial resolution in an OCT system, in contrast to other microscopy imaging techniques. Another important parameter for the design of focusing optics is the depth of focus, which is generally given by $Z_{Range} = 2\lambda_0/\pi(NA_{obj})^2$. It is interesting to note that there is an inherent design trade-off between obtaining high lateral resolution and a longer depth of focus. Hence, sample arm optical designs are typically based on factors such as penetration depths in tissue and imaging application requirements.

8.2.4 SNR, Noise, and Sensitivity in OCT Systems

System sensitivity is one of the most important parameters used to evaluate the performance of an OCT system. There are many sources of noise that may degrade the OCT signal; they are broadly classified into three parts: (1) laser source (excess photon noise), (2) detection system and electronics (1/f noise, quantization error noise, and thermal noise), and (3) shot noise. Although an efficient system design can circumvent the relative contribution of the system noise sources such as quantization error noise and so on, a thorough signal-to-noise analysis of an OCT system is essential to obtain optimum system sensitivity [105–107].

In a TD-OCT system, shot noise and thermal mean square noise current can be expressed as $\langle \Delta i_{sh}^2 \rangle = 2q\langle I_{dc}\rangle \Delta f$ and $\langle \Delta i_{th}^2 \rangle = 4k_B T\Delta f / R_L$, respectively, where q is the electron charge, $\langle I_{dc}\rangle = \rho P_0 R_R/4$ is the averaged DC current (assuming a standard 50/50 Michelson interferometer configuration with $R_R >> R_S$ and P_0 being the laser source power), Δf is the detection bandwidth, R_L is the load resistance, k_B is the Boltzmann constant, and T is the temperature. Excess photon noise for a purely spontaneous source is given by $\langle \Delta i_{ex}^2 \rangle = (1+\alpha^2)\langle I_{dc}\rangle^2 \Delta f/\Delta\upsilon_{eff}$, where α is the degree of polarization and $\Delta\upsilon_{eff}$ is the effective line width [105]. Excess photon noise refers to the fluctuations in source light output intensity due to the beating of various spectral components having random phases within the spectral linewidth of the source [108]. The overall noise can be given by $\langle \Delta i_n^2 \rangle = \langle \Delta i_{th}^2 \rangle + \langle \Delta i_{sh}^2 \rangle + \langle \Delta i_{ex}^2 \rangle$, and the signal power is given by $\langle I_{ac}^2 \rangle = (\rho P_0 \sqrt{R_R R_S}/2)^2$. The SNR of the system is given as the ratio of signal-to-noise power and can be expressed as

$$(SNR)_{TD-OCT} = \frac{\langle I_{ac}^2 \rangle}{\left(\langle \Delta i_{th}^2 \rangle + \langle \Delta i_{sh}^2 \rangle + \langle \Delta i_{ex}^2 \rangle \right)} \tag{8.8}$$

System sensitivity is defined as the theoretical SNR of the system for a sample with a unity reflectivity and can be expressed as

$$Sensitivity_{TD-OCT} = \left| (SNR)_{TD-OCT} \right|_{R_S=1} = \left| \frac{\langle I_{ac}^2 \rangle}{\left(\langle \Delta i_{th}^2 \rangle + \langle \Delta i_{sh}^2 \rangle + \langle \Delta i_{ex}^2 \rangle \right)} \right|_{R_S=1} \tag{8.9}$$

FD-OCT systems present inherent advantages in terms of system sensitivity, and it was demonstrated experimentally and theoretically by three independent groups in 2003 [38–40]. FD-OCT systems exhibit higher SNR because the signal power is distributed in various spectral bins, thereby effectively reducing the noise power by a factor proportional to the number of spectral bins. Assuming that the parameters such as source power are the same, the SNR of FD-OCT systems can be related to the SNR of a TD-OCT system as:

$$(SNR)_{FD-OCT} = (SNR)_{SD-OCT} = (SNR)_{SS-OCT} = (SNR)_{TD-OCT} \frac{N}{2} \tag{8.10}$$

where N is the number of pixels in the linear CCD array in the SD-OCT system or the number of sampling points in a single sweep cycle for a swept-source OCT (SS-OCT) system. It should be noted that Eq. (8.10) holds true if the signal power is distributed equally to all the detector sampling points, and hence the actual theoretical improvement can be reduced by up to a factor of 2 for a typical Gaussian-shaped spectrum.

8.3 Functional Extensions of OCT and Other LCI-Based Techniques for Applications in Nanomedicine

In this section we will provide an overview of a range of LCI-based techniques for applications in cellular imaging, monitoring cellular dynamics, tracking the presence and localization of nanoparticles in tissue, detecting ultrastructural changes in muscle tissue, and molecular-specific endogenous as well as exogenous contrast imaging. The discussion related to the applications of the following techniques will be limited to the topics having relevance to current as well as future possible applications in nanomedicine.

8.3.1 Gold Nanoshells and Nanostructures as Contrast and Therapeutic Agents

Recently, there has been a growing interest in exploring the use of gold nanoshells for applications in imaging as well as the destruction of cancerous cells [109, 110]. Loo et al. used gold nanoshell bioconjugates to target and image human epidermal growth factor receptor 2 (HER 2) in live human breast carcinoma cells [109]. Shi et al. demonstrated binding of dendrimer-entrapped gold nanoshells with a human epithelial carcinoma cell line (KB cells) [110]. Gold nanoshell-based diagnostic and therapeutic techniques are one of the most promising avenues of research in nanomedicine because the gold nanostructures are biocompatible, easy to synthesize, manipulate, and function with other modalities. A detailed description about the fabrication, properties, and applications of various kinds of gold nanostructures such as nanoshells, nanorods, nanocages, and so on, will be covered in Chapter 11.

Gold nanoshells, consisting of a silica core and a gold shell, can be engineered to target cancerous cells and at the same time be designed to interact with specific wavelengths of light by varying nanoparticle parameters such as core radii and shell thickness [111–113]. Metallic nanoparticles interact very strongly with incoming radiation of a specific frequency due to a surface plasmon resonance, a collective excitation of electrons in the metal. This phenomenon of interaction of nanoparticles with incident electromagnetic (EM) waves can be well described by the Mie theory. Gold nanoshells can be engineered to exhibit absorption wavelengths in the near-infrared region. At the peak absorption or resonant wavelength, the electron cloud in the metal can absorb energy from the incident radiation and start to oscillate, thereby causing high temperatures locally. This effect can subsequently be used to kill cells harboring the gold nanoshells.

Plasmon-resonant gold nanoprobes, such as gold nanoshells, nanorods, and nanocages, have been explored as contrast and therapeutic agents in OCT imaging [99, 114–118]. Relatively larger nanoshells (diameter > 100 nm) exhibit large backscattering coefficients and can be detected due to an increased OCT signal [115]. Oldenburg et al. and Troutman et al. demonstrated that plasmon-resonant gold nanorods can act as an absorption or spectroscopic contrast agent in OCT [117, 118]. Bioconjugation of nanocages (< 40-nm diameter) with antibodies specific to breast cancer cells has been demonstrated [119]. Cang et al. have demonstrated that the nanocages can also provide enhanced contrast for spectroscopic OCT [99]. Adler et al. used NIR radiation to induce modulations in gold nanoshells and used a phase-sensitive swept-source OCT system to detect their presence [114]. OCT techniques, having demonstrated the capability of detecting gold nanoshells and other plasmon-resonant probes, will prove to be very useful tools in future nanomedicine research.

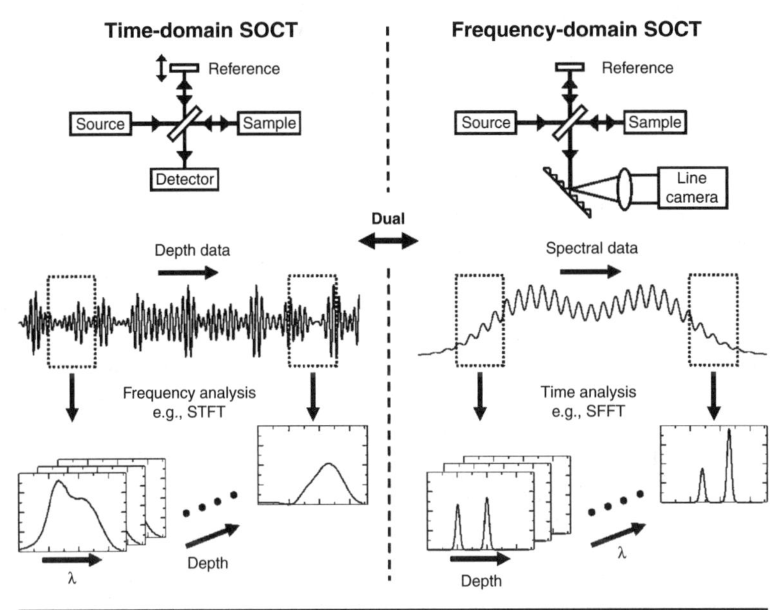

Figure 8.5 Time-frequency analysis techniques for spectroscopic OCT in time-domain (left) and Fourier-domain (right) OCT systems. (Image reprinted with permission from Oldenburg et al. (2007), *IEEE J. Sel. Top. Quant. Elect.* [100].)

8.3.2 Spectroscopic OCT

Spectroscopic OCT (SOCT) maps spatially localized spectral absorption and backscattering information of endogenous molecules as well as exogenous contrast agents in tissues by detecting and processing the interferometric OCT signal. SOCT theory and processing algorithms can be well explained by the detailed stepwise illustration in Fig. 8.5 [100]. Time-frequency analysis is commonly used to extract the SOCT signal from OCT data. For a TD-OCT system, the SOCT signal can be obtained by applying a short-time Fourier transform (STFT) to a shortened time-delay window that is computationally scanned across the temporal interferometric data. In an FD-OCT system, the short-frequency Fourier transform can be used to obtain the SOCT signal. One of the major limitations of SOCT signal processing arises from the fact that there is an inherent trade-off between spectral and depth (spatial) resolution due to the uncertainty principle. For example, if one wants to obtain better spectral resolution, then the width of the STFT time-delay window would need to be increased; this would result in decreased spatial-depth resolution.

Morgner et al. demonstrated the first *in vivo* spectroscopic OCT imaging of a *Xenopus laevis* (African frog) by using a broadband Ti:Sa laser source (650 to 1000 nm) [32]. Melanocytes were clearly distinguishable as an endogenous contrast agent in SOCT images because they exhibited higher absorption in the shorter wavelength range [32]. Yang et al. developed a technique, spectral triangulation molecular contrast OCT, in which they exploited the spectral differential absorption properties of an exogenous agent to provide the contrast [120]. They used an Food and Drug Administration (FDA)-approved dye, indocyanine green (ICG), as the contrast agent and used three spectral bands to obtain OCT data. The first spectral band was chosen to coincide with the peak absorption wavelength of the dye, and the other two bands were chosen around the first spectral band. By using a triangulation algorithm, they were able to extract differential absorptive contrast within the sample. Xu et al. also used near infrared (NIR) absorbing dyes to obtain SOCT contrast. They calculated the shift in the centroid of the SOCT response as the dye caused absorption in shorter wavelengths and were able to spatially map the presence of the NIR dye in a botanical sample [36]. Adler et al. calculated the autocorrelation bandwidth of the optical spectra and used it as a metric for contrast in imaging developing zebra fish embryos [98]. Another interesting application has been to determine sizes and densities of scatterers on the basis of the comparison of endogenous spectroscopic signals and the expected results from the Mie theory [98, 101]. As described in Section 8.1.3, plasmon-resonant gold nanorods can also act as an exogenous contrast agent for SOCT [117, 118].

8.3.3 Magnetomotive OCT

Magnetomotive optical coherence tomography (MM-OCT) is a novel extension of OCT for imaging a distribution of magnetic molecular imaging agents in biological specimens [33, 34, 121]. The MM-OCT technique was developed by Oldenburg, Boppart et al. at the University of Illinois, Urbana–Champaign. In the MM-OCT system, the magnetic field is generated using an electromagnet and *in vivo* imaging is performed while the specimen is placed within the gradient of the magnetic field. Usually, a ring-shaped, water-cooled solenoid is used and the sample is placed such that the z-axis of A-scan is along the central axis or open bore of the solenoid ring. The gradient of the magnetic field applies force on the magnetic nanoparticles (MNPs) resulting in their motion. These MNPs are often attached to the tissue matrix, and hence the force applied to these nanoparticles is coupled with viscoelastic restoring forces; this results in nanometer-scale displacements that can be detected by a conventional or phase-sensitive OCT system. The dynamic magnetomotion of the MNPs will be a function of applied magnetic forces and restoring forces from the microscopic tissue matrix, and hence these magnetomotive-based scattering signals can also be used to study the biomechanical properties of the tissues [122].

MM-OCT provides high specificity and sensitivity by exploiting the large difference between the magnetic volume susceptibilities of MNPs ($\chi \sim 1$) and biological tissue ($\chi \sim -10^{-5}$) . Super paramagnetic iron oxide MNPs, already an FDA-approved contrast agent for MRI, are promising molecular-specific imaging agents for MM-OCT owing to their versatile properties such as ease of fabrication, high biocompatibility, and ease of functionalization with a number of targeting agents including antibodies and proteins [123–125]. Target-specific functionalized MNPs also find applications in therapy involving disease and cells, specific drug delivery, and thermoablative cancer therapy [126, 127].

Oldenburg et al. used this technique to identify magnetically labeled macrophages in three-dimensional scaffolds [33]. The dynamic range of an MM-OCT signal was found to be 30 dB, and the sensitivity was found to be 220 μg per milliliter of MNPs. As a next step, they demonstrated *in vivo* imaging of *Xenopus laevis* and identified regions of higher concentration of magnetic nanoparticles [34]. Oldenburg et al. later demonstrated a phase-sensitive MM-OCT technique to obtain an increased sensitivity of up to 2-nM MNP concentration in tissue phantoms [121]. In the same study, they also monitored the diffusion of topically applied MNPs (typically 20 to 30 nm in size) in an excised rat mammary tumor tissue using MM-OCT. Figure 8.6 shows MM-OCT images of a carcinogen-induced rat mammary tumor with and without the injection of MNPs. MNP functionalization is an active area of research, and hence MM-OCT enables the

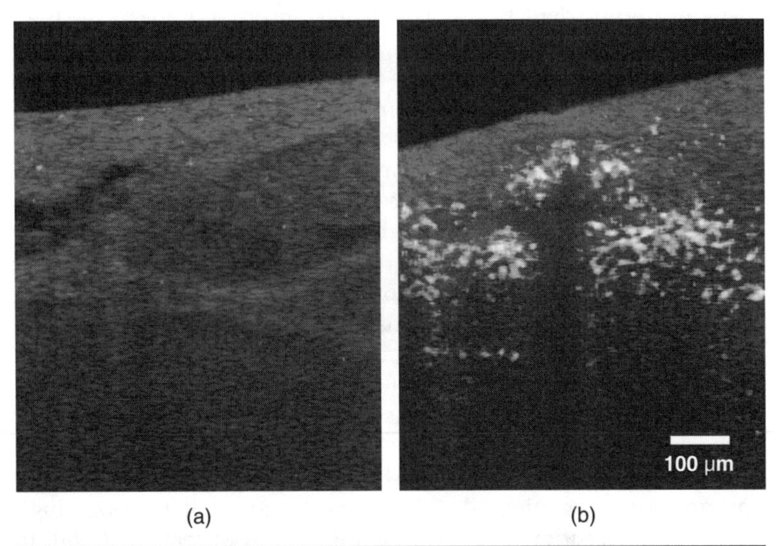

Figure 8.6 MM-OCT images of a freshly excised rat mammary tumor. (a) MM-OCT image of control tumor tissue without magnetic nanoparticles. (b) MM-OCT image of tumor after injection of magnetic nanoparticles. Bright intensity pixels correspond to the presence of nanoparticles. (Images by Oldenburg and Boppart (UIUC); color version available at group website, http://biophotonics.illinois.edu.)

possibility of molecularly sensitive OCT by using functionalized MNPs capable of targeting specific cell types. Rezaeipoor et al. recently demonstrated Fc-directed conjugation of HER-2 antibodies to MNPs, which enabled the free antibody part to target tumor-specific cell lines [128]. Readers may refer to Chapter 11 for a more detailed description of fabrication, functionalization, and therapeutic applications of MNPs.

8.3.4 Ultrahigh-Resolution OCT for Subcellular Imaging

Ultrahigh-resolution OCT has many applications in developmental, cell, and tumor biology as it can provide subcellular structural imaging capabilities. Although the research in nanomedicine is aimed toward studying the nanoscopic changes in cells, it is also essential to monitor microscopic cell behavior for a better understanding of overall biological processes. Ultrahigh-resolution OCT can be a useful diagnostic tool for nanomedicine in this regard. Although initial OCT systems exhibited modest axial resolutions (10 to 15 μm), the use of broadband Ti:Sa lasers made it possible to achieve up to 1-μm axial resolution [89, 129]. Later, Ti:Sa-pumped photonic crystal fiber–based broadband continuum sources were used to obtain even submicron-level axial resolution [91]. Since the axial and transverse resolutions are independent in OCT systems, high NA objectives are used in the sample arm to obtain a sharper focus of the incident beam. The extension of OCT that uses high NA-focusing optics in the sample arm and reconstructs *en-face* images of the sample is known as *optical coherence microscopy* (OCM) [130, 131]. Figure 8.7 shows an *in vivo* ultrahigh-resolution cross-sectional image of a *Xenopus laevis* tadpole [89]. The high-resolution imaging [1 μm (axial) × 3 μm (transverse)] can be used to study the nucleus-to-cytoplasm ratio, intracellular morphology, and processes such as mitosis. One major limitation of performing high-transverse resolution OCT imaging is the reduced depth of field. Dynamic focusing and fusion of multiple longitudinal focal plane images are the methods used to obtain enhanced depth of focus [132, 133]. Although dynamic focusing is limited to TD-OCT systems, fusion of multiple images with variable focal planes can cause artifacts due to sample movement, and increase the acquisition time by a factor of the number of focal planes.

Interferometric synthetic aperture microscopy (ISAM) is a recent and promising technique developed by Ralston et al. to obtain high transverse resolution over an extended depth of focus [92, 93]. In standard OCT processing, the scattering-field function reconstruction (i.e., the sample structural information) is best approximated for the scatterers present within the Rayleigh range of the incident beam. The scattering-field function reconstruction degrades for scatterers far away from the focal plane. ISAM is a computational

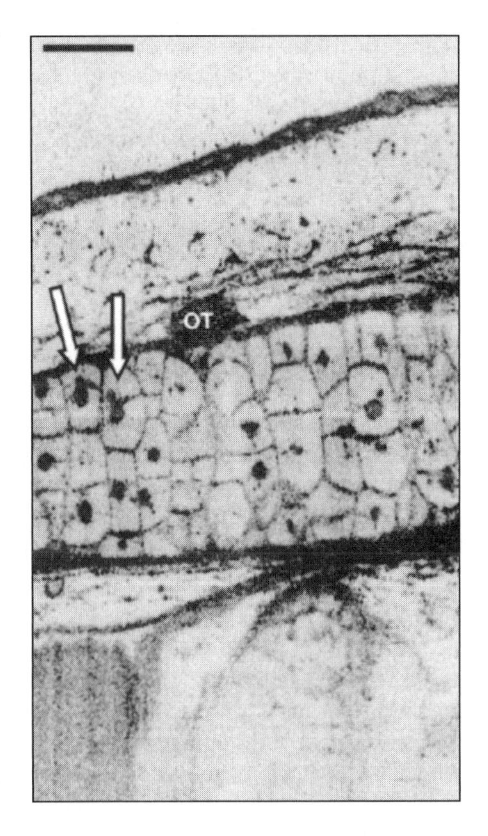

FIGURE 8.7 *In-vivo* cellular-level imaging of a *Xenopus laevis* (African frog) tadpole. Ultrahigh axial and lateral resolutions (1 μm and 3 μm, respectively) enable visualization of the olfactory tract (OT), cell membranes, nuclear and subcellular morphology. Arrows indicate mitosis occurring in two pairs of cells. Scale bar represents 100 μm. (Image reprinted with permission from Drexler et al. (1999), *Opt. Lett.* [89].)

technique that is based on the solution of the inverse-scattering problem to obtain spatially invariant resolution for scatterers in and far away from the focal zone [92, 93, 134–137]. Figure 8.8 shows volumetric reconstructions of a tissue phantom consisting of scattering particles using high-NA OCT and ISAM. ISAM has the potential to

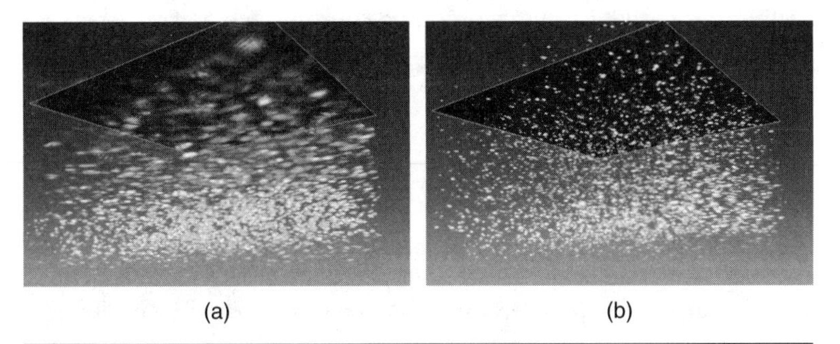

FIGURE 8.8 Three-dimensional reconstruction of a tissue phantom. (a) By standard OCT. (b) By the ISAM technique. ISAM can correct the distortion caused by out-of-focus scatterers and provide spatially invariant resolution. (Image reprinted with permission from Ralston et al. (2007), *Nat. Phy.* [92].)

significantly increase the diagnostic capabilities of OCT or OCM images by providing very high transverse resolution over an extended depth of imaging. Researchers at the University of Illinois, Urbana–Champaign, are currently pursuing a clinical study to incorporate real-time ISAM imaging for surgical guidance in breast tumor resection. A potential application of ISAM in nanomedicine could be to study early-stage tumor development. Insight into the mechanisms of tumor development and metastases would be of tremendous significance in the fields of nanomedicine and oncology. Early diagnosis and monitoring of neoplasia would also require cellular-imaging capabilities, which can be done by integrating ultrabroadband sources with high-NA OCT and ISAM techniques.

8.3.5 Phase-Sensitive LCI Techniques for Monitoring Cellular Dynamics

In vivo monitoring of cellular processes such as small-scale cell motion, small changes in cell size, cell refractive index, and cytoplasmic flow is an essential tool for nanomedicine to obtain a better understanding of these biological processes. The resolution offered by traditional OCT-imaging techniques is still inadequate to detect these nanoscale changes. Although interference data in OCT contain phase as well as amplitude information, only the latter part is utilized in standard OCT imaging. However, it is possible to obtain much higher axial displacement sensitivities if the reference arm reflector is kept fixed and relative displacements of the object in the sample arm are monitored by measuring the phase of the interferometric signal versus time. This is a very powerful technique for interferometry with a sensitivity that is limited by the stability of the path delay in the reference arm. Phase-resolved techniques can measure the effect of displacement, change in refractive index, and Doppler shift due to scatterer motion, although it is not possible to separate the relative contribution of these effects. Although quantitative phase-imaging techniques based on monochromatic sources such as Fourier phase microscopy and Hilbert phase microscopy have been successful at imaging cellular dynamics and processes, one limitation has been the lack of depth-selection capability within the sample [138–141]. Yang et al. developed a time-domain low-coherence phase-referenced interferometry technique to study effects of changes in cell volume due to immersion in hypotonic and hypertonic solutions. The study reported a system sensitivity to a small displacement of 3.6 nm and a velocity sensitivity as small as 1 nm per second [142]. Fang-Yen et al. used an improved phase-referenced heterodyne dual-beam low-coherence interferometer to perform measurements of surface motion in a nerve bundle during an action potential [143]. Spectral-domain systems exhibit higher phase stability due to the absence of any moving parts in the system.

In the last few years, many researchers have been continuing to develop a technique called *spectral-domain phase microscopy* (SDPM) to exploit the higher phase stability of spectral-domain systems and the depth-sectioning capability made possible by the use of a broadband source [94–97, 144, 145]. Although different researchers have developed slightly different system designs to implement this technique, the common-path interferometer-based OCT systems design is considered advantageous as it can mitigate the effect of phase noise due to relative vibrations in the sample and reference path delays. Choma et al. demonstrated a common-path SD-OCT and SS-OCT-based phase microscopy setup where the reference signal was derived from the reflection at the clean surface of the glass coverslip holding the sample [94]. The measured experimental sensitivity to axial displacement of a clean glass coverslip was 53 pm and 780 pm for spectral-domain and swept-source–based interferometers, respectively. The phase stability is normally defined as the standard deviation of the interference phase at the depth corresponding to the coverslip thickness. The lower displacement sensitivity of the swept-source system is attributed to the uncertainty in spectral shifts between successive sweeps. Choma et al. used this technique to monitor spontaneous beating dynamics of an embroyonic chick cardiomyocyte (Fig. 8.9). SDPM measured the changes in cell

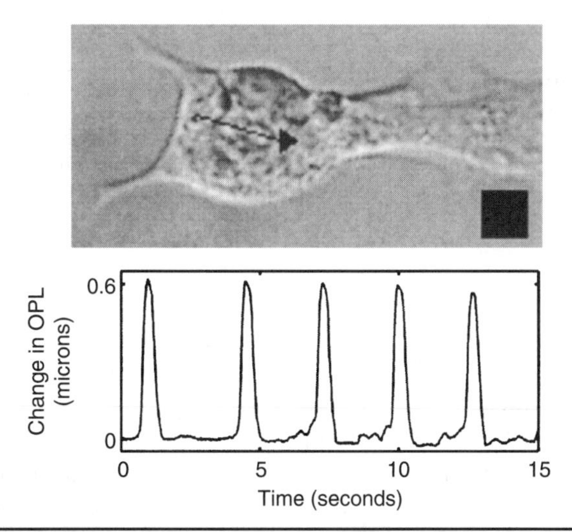

FIGURE 8.9 Monitoring contractions in a two-day-old chick embryonic cardiomyocyte using SDPM. *Top:* The tip of the arrow indicates the point of measurement, and the black box represents 10 x 10 μm². *Bottom:* SDPM was used to measure the change in thickness of the cell due to spontaneous contractions. Abbreviation: OPL, optical path length. (Image reprinted with permission from Choma et al. (2005), *Opt. Lett.* [94].)

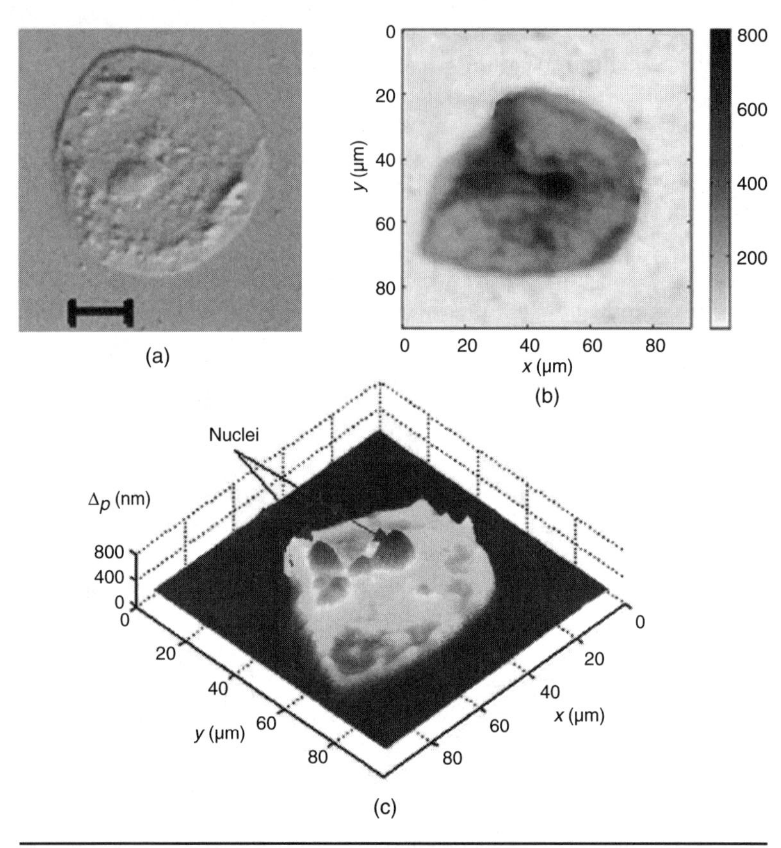

FIGURE 8.10 Images of a human epithelial cheek cell. (a) Image recorded by Normarski microscope (10x; NA, 0.3); the scale bar represents 20 μm. (b) SDPM image of the cell is plotted where the gray scale indicates the optical path length in nanometers. (c) Surface profile plot of SDPM image showing nuclei and other subcellular structures (optically thick and hence shown dark). (Image reprinted with permission from Joo et al. (2005), *Opt. Lett.* [96].)

thickness caused by the cardiomyocyte contractions. Joo et al. employed a similar technique but acquired two-dimensional *en face* images of human epithelial cheek cells by raster scanning across the sample area (Fig. 8.10) [96]. Sarunic et al. employed a phase-sensitive full-field OCT approach based on a swept source to measure the surface profile of human red blood cells [146]. Ellerbee et al. and Choma et al. applied this technique to measure cytoplasmic streaming in an *Amoeba proteus* pseudopod [95, 144]. McDowell et al. used this technique to investigate the mechanical properties of the cytoskeleton within cells. A magnetic tweezer was designed to apply force on cells and the deformation of the cells was measured using SDPM [97]. Overall, SDPM has emerged as a promising technique to image subcellular morphology and dynamics.

8.3.6 Polarization-Sensitive OCT

Polarization-sensitive optical coherence tomography (PS-OCT) maps depth and spatially resolved changes in the polarization state of light induced by anisotropic tissue properties [27, 28, 35]. By exploiting the interaction between the polarization state of light and tissue, additional structural and functional (physiological) information can be extracted [147, 148]. PS-OCT incorporates standard OCT principles, but also tracks the incident and backscattered polarization state of light. The basic experimental setup of a PS-OCT system is similar to that of a standard OCT system; however, it has several additional components such as linear polarizers, polarizer controllers, and a polarization beam splitter with an extra photodetector within the detection arm of the interferometer. Although various research groups have developed a range of sophisticated experimental designs to demonstrate PS-OCT imaging [149–152], the basic goal of these designs remains the same, to extract the information about sample birefringence. Jones matrix or Stokes vector analysis is performed to characterize the system and extract the sample properties such as birefringence, sample optical axis, and polarization biattenuance [35, 153, 154]. Hence, PS-OCT can determine if the tissue is altering the polarization state and where these alterations are occurring spatially.

Form birefringence is the main optical property present in tissues that is responsible for altering the polarization state of light. Common sources of form birefringence in tissue is the collagen and structural protein scaffolds in skin, muscle, tendons, nerve, bone, cartilage, and teeth. PS-OCT has been used to quantify the thickness of the retinal nerve fiber layer as a potential early indicator of glaucoma [155, 156]. PS-OCT has been used to assess skin morphology as well as burn depth in injured skin, since high temperatures denature proteins and alter the normal birefringence of collagen in the skin [157]. Nadkarni et al. used this technique to measure collagen and smooth muscle cell content in atherosclerotic plaques [158]. In addition to this, one of the more promising applications of PS-OCT was explored by Pasquesi et al. who investigated the birefringence-based variations in the PS-OCT signals due to ultrastructural changes in skeletal muscle [102]. The birefringence present in skeletal muscle is associated with the ultrastructure of individual sarcomeres, specifically the arrangement of A-bands corresponding to the thick myosin filaments. The experiment consisted of PS-OCT–based quantitative birefringence measurement of murine skeletal

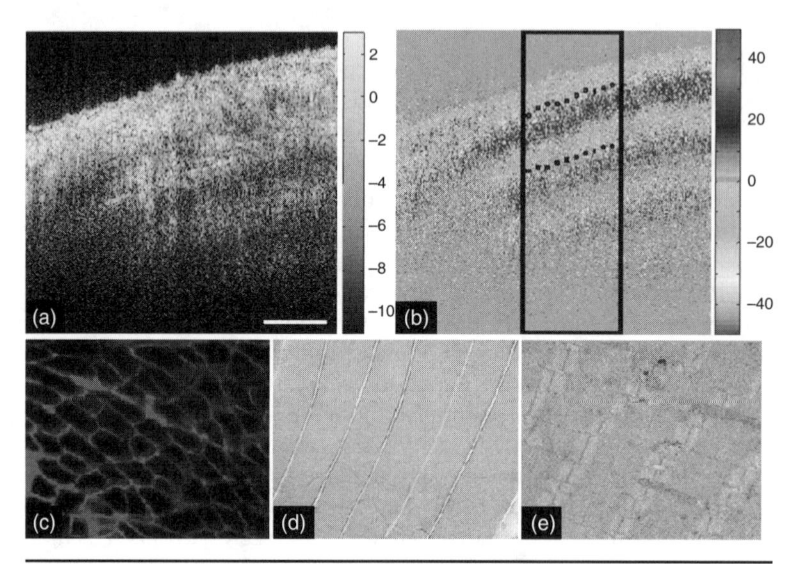

Figure 8.11 Murine skeletal muscle images. (a) Structural OCT image; scale bar represents 500 μm. (b) PS-OCT image: the spacing between the light and dark bands is a measure of tissue birefringence. (c) Evans blue fluorescence. (d) Hematoxylin and eosin-stained histological section at 20x. (e) TEM at 5000x magnification. (Image reprinted with permission from Pasquesi et al. (2006), *Opt. Exp.* [102].)

muscle in wild-type as well as genetically altered (*mdx*) mice, before and after exercise. The study suggests that there is a distinct relationship between the degree of birefringence detected using PS-OCT and the sarcomeric ultrastructure present within skeletal muscle (Fig. 8.11).

8.3.7 Molecular-Specific LCI-Based Techniques: Pump-Probe OCT and Nonlinear Interferometric Vibrational Imaging

Recently, a new set of techniques, pump-probe OCT (PP-OCT) and nonlinear interferometric vibrational imaging (NIVI), have emerged which have used low-coherence interferometry–based approaches in tandem with other molecular-specific (i.e., probing transitions in electronic or vibrational states in specific molecules) imaging techniques. The main advantages of these techniques are depth-selection capability and increased sensitivity due to interferometric detection. In a pump-probe OCT system, a pump pulse is followed by a broadband-probe pulse. The pump pulse is tuned to the peak absorption wavelength corresponding to the electronic transition in a specific molecule, resulting in a population increase of excited or intermediate states of the molecule. The population change is then monitored by a probe beam in a low-coherence interferometry

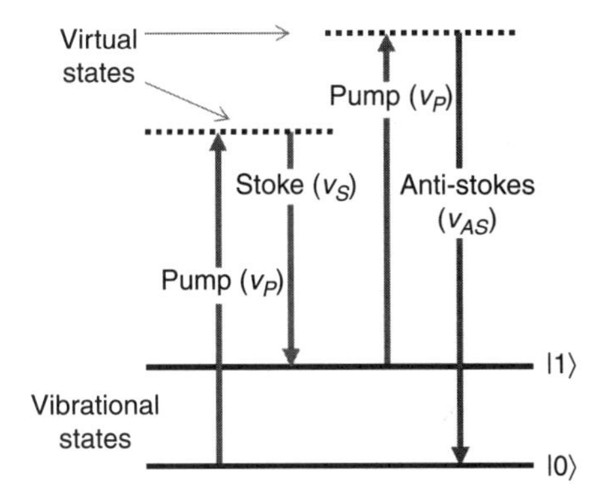

Figure 8.12 Energy-level diagrams depicting the four-wave mixing process in CARS. The anti-Stokes signal is resonant if the frequency of the incident pump and stokes photons is tuned to the molecular vibrational frequency.

setup [159–161]. These techniques have been used to detect agents such as methylene blue dye [161] and a plant protein named phytochrome A [162]. Applegate et al. used a similar technique to image hemoglobin chromophores within zebra fish. The PP-OCT signal can also differentiate between endogenous (hemoglobin) and exogenous (dyes) chromophores [159].

Coherent Anti-Stokes Raman Scattering (CARS) spectroscopy is a highly sensitive technique for molecule-specific imaging [163, 164]. CARS is sensitive to the same vibrational signatures of molecules as obtained by standard Raman spectroscopy, albeit the CARS signal is orders-of-magnitude stronger than the signal obtained from spontaneous Raman scattering. Since each molecule has a unique vibrational signature, CARS is essentially an endogenous imaging technique. It is a four-wave mixing process where three incident photons (two at pump frequency ν_P and one at Stokes frequency ν_S) mix to produce a fourth anti-Stokes photon at frequency ν_{AS} (Fig. 8.12). The anti-Strokes signal is resonant if the energy difference between pump and probe photons is equal to the energy difference between vibrational bands, that is, $\nu_P - \nu_S = \Delta\nu$. NIVI, a technique developed by Marks et al., utilizes nonlinear interferometry to detect both the amplitude and phase of the anti-Stokes signal [103]. NIVI has an inherent advantage over the amplitude-based CARS technique as it can separate resonant and nonresonant signals; it thereby improves the spectral resolution and hence the molecular specificity [165]. Bredfeldt et al. demonstrated the use of this technique to image lipids by using the specificity to the C-H vibrational transitions (2845 cm^{-1}). They imaged a layer of beef adipose tissue sandwiched between two glass slides [166]. Interferometric CARS was also used by Potma et al. to detect lipids in fibroblast cells. Benalcazar et al. used a NIVI system with higher imaging speed capability to image rat mammary tissue. Figure 8.13 shows the NIVI hyperspectral image at 2845 cm^{-1} [167]. NIVI holds great promise for future applications in nanomedicine due to its highly sensitivity endogenous molecular-specific imaging capabilities.

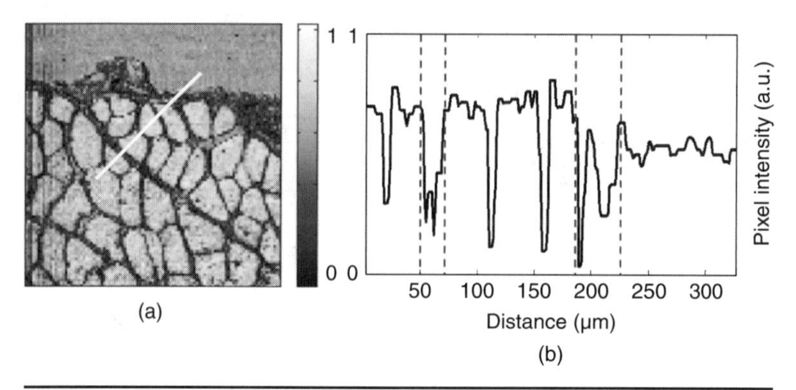

FIGURE 8.13 Nonlinear interferometric vibrational imaging. (a) NIVI image of mammary tissue at 2845 cm^{-1} (500 μm x 500 μm). Adipocytes, parenchyma, and connective tissue are shown. The gray scale bar corresponds to the intensity of the signal obtained from adipocytes (rich in CH_2 content). Nuclei can be distinguished as small black dots (low-fat content) at the periphery of the cells. (b) NIVI image intensity profile from along the white line shown in (a). Dotted lines delineate connective tissue and parenchyma from the adipocytes. (Image reprinted with permission from Benalcazar et al. (2010), *IEEE J. Sel. Top. Quant. Elect.* [167]; refer online to paper for color images.)

8.4 Conclusion

Nanomedicine is a highly interdisciplinary and rapidly expanding field of research. Despite being a relatively new field of research, it has already grown to become a multibillion dollar industry in less than a decade, and it still continues to grow at a rapid pace. It is poised to be the largest breakthrough in medicine for the next decade. Over the next 5 to 10 years, nanomedicine may be able to help answer many questions regarding our basic understanding of diseases like cancer, find novel treatments and therapeutic methods, and change the face of medicine as we know it today. Of course, the success of nanomedicine will depend on further advancements in the field of nanotechnology and an improved performance of diagnostic and imaging tools. Low-coherence interferometry–based techniques such as OCT have also progressed at a rapid pace over the last one and a half decades. Many functional extensions of OCT and other molecular-specific LCI-based techniques have already found applications in nanomedicine. LCI-based techniques are low-cost and noninvasive imaging modalities which offer superior resolution compared to standard clinical MRI and ultrasound imaging techniques.

However, it must be acknowledged that many technical challenges need to be overcome before there can be a more widespread use of LCI-based techniques in nanomedicine. One of the greatest limiting factors hindering the success of optical interferometric technique is the limited penetration depth in scattering tissues. LCI-based techniques have yet to make a successful transition to clinics in fields other than ophthalmology. Nevertheless, LCI-based technologies are exploring and creating niche applications in various fields of medicine that are willing to accept limited imaging depth for the superior imaging resolutions provided by these real-time optical imaging techniques. It is likely that the future development focus of LCI-based techniques will be centered around molecular-specific contrast techniques such as SOCT, NIVI, other nonlinear interferometric techniques, and MM-OCT, as well as many others. The synergistic improvements and advancements of molecular specific interferometry technology and nanotechnology are likely to be a significant step forward in the field of nanomedicine.

Acknowledgments

The authors would like to thank all the current and former members of the Biophotonics Imaging Laboratory at the Beckman Institute at the University of Illinois, Urbana–Champaign, for their insights and contributions to this area of research. We also wish to thank our collaborators and colleagues whose work is presented here and apologize to those whose work could not be included due to space and depth-of-coverage constraints. The work is supported in part by grants from the National Institutes of Health (Roadmap Initiative, NIBIB, R21 EB005321; NIBIB, R01 EB005221; NCI R21/R33 CA115536) and the National Science Foundation (CBET 08-52658 ARRA). Additional information and color versions of several figures can be found at http://biophotonics.illinois.edu.

References

1. M. Born and E. Wolf, *Priciples of Optics,* 7th ed., Cambridge University Press, Cambridge, 1999.
2. E. Hecht, *Optics,* 4th ed., Addison Wesley, San Francisco, 2002.
3. P. Hariharan, *Optical Interferometry,* 2nd ed., Academic Press, San Diego, 2003.
4. D. Huang, E. A. Swanson, C. P. Lin, J. S. Schuman, W. G. Stinson, W. Chang, M. R. Hee, T. Flotte, K. Gregory, C. A. Puliafito, and J. G. Fujimoto, "Optical Coherence Tomography," *Science* 254: 1178–1181, 1991.
5. G. J. Tearney, M. E. Brezinski, B. E. Bouma, S. A. Boppart, C. Pitris, J. F. Southern, and J. G. Fujimoto, "*In Vivo* Endoscopic Optical Biopsy with Optical Coherence Tomography," *Science* 276: 2037–2039, 1997.
6. J. G. Fujimoto, M. E. Brezinski, G. J. Tearney, S. A. Boppart, B. Bouma, M. R. Hee, J. F. Southern, and E. A. Swanson, "Optical Biopsy and Imaging Using Otical Coherence Tomography," *Nature Medicine* 1: 970–972, 1995.
7. M. R. Hee, C. A. Puliafito, C. Wong, J. S. Duker, E. Reichel, J. S. Schuman, E. A. Swanson, and J. G. Fujimoto, "Optical Coherence Tomography of Macular Holes," *Ophthalmology* 102: 748–756, 1995.
8. M. R. Hee, C. R. Baumal, C. A. Puliafito, J. S. Duker, E. Reichel, J. R. Wilkins, J. G. Coker, J. S. Schuman, E. A. Swanson, and J. G. Fujimoto, "Optical Coherence Tomography of Age-Related Macular Degeneration and Choroidal Neovascularization," *Ophthalmology* 103: 1260–1270, 1996.

9. M. E. Brezinski, G. J. Tearney, N. J. Weissman, S. A. Boppart, B. E. Bouma, M. R. Hee, A. E. Weyman, E. A. Swanson, J. F. Southern, and J. G. Fujimoto, "Assessing Atherosclerotic Plaque Morphology: Comparison of Optical Coherence Tomography and High Frequency Intravascular Ultrasound," *Heart* 77: 397–403, 1997.
10. G. J. Tearney, M. E. Brezinski, J. F. Southern, B. E. Bouma, S. A. Boppart, and J. G. Fujimoto, "Optical Biopsy in Human Gastrointestinal Tissue Using Optical Coherence Tomography," *American J. Gastroenterology* 92: 1800–1804, 1997.
11. G. J. Tearney, M. E. Brezinski, J. F. Southern, B. E. Bouma, S. A. Boppart, and J. G. Fujimoto, "Optical Biopsy in Human Urologic Tissue Using Optical Coherence Tomography," *J. Urology* 157: 1915–1919, 1997.
12. F. I. Feldchtein, G. V. Gelikonov, V. M. Gelikonov, R. V. Kuranov, A. M. Sergeev, N. D. Gladkova, A. V. Shakhov, N. M. Shakhova, L. B. Snopova, A. B. Terent'eva, E. V. Zagainova, Y. P. Chumakov, and I. A. Kuznetzova, "Endoscopic Applications of Optical Coherence Tomography," *Optics Express* 3: 257–270, 1998.
13. M. R. Hee, C. A. Puliafito, J. S. Duker, E. Reichel, J. G. Coker, J. R. Wilkins, J. S. Schuman, E. A. Swanson, and J. G. Fujimoto, "Topography of Diabetic Macular Edema with Optical Coherence Tomography," *Ophthalmology* 105: 360–370, 1998.
14. J. G. Fujimoto, S. A. Boppart, G. J. Tearney, B. E. Bouma, C. Pitris, and M. E. Brezinski, "High Resolution *In Vivo* Intra-Arterial Imaging with Optical Coherence Tomography," *Heart* 82: 128–133, 1999.
15. X. D. Li, S. A. Boppart, J. Van Dam, H. Mashimo, M. Mutinga, W. Drexler, M. Klein, C. Pitris, M. L. Krinsky, M. E. Brezinski, and J. G. Fujimoto, "Optical Coherence Tomography: Advanced Technology for the Endoscopic Imaging of Barrett's Esophagus," *Endoscopy* 32: 921–930, 2000.
16. S. A. Boppart, W. Luo, D. L. Marks, and K. W. Singletary, "Optical Coherence Tomography: Feasibility for Basic Research and Image-Guided Surgery of Breast Cancer," *Breast Cancer Research and Treatment* 84: 85–97, 2004.
17. R. C. Youngquist, S. Carr, and D. E. N. Davies, "Optical Coherence-Domain Reflectometry—A New Optical Evaluation Technique," *Optics Letters* 12: 158–160, 1987.
18. T. Kubota, M. Nara, and T. Yoshino, "Interferometer for Measuring Displacement and Distance," *Optics Letters* 12: 310–312, 1987.
19. K. Takada, I. Yokohama, K. Chida, and J. Noda, "New Measurement System for Fault Location in Optical Waveguide Devices Based on an Interferometric-Technique," *Applied Optics* 26: 1603–1606, 1987.
20. H. H. Gilgen, R. P. Novak, R. P. Salathe, W. Hodel, and P. Beaud, "Submillimeter Optical Reflectometry," *J. Lightwave Technology* 7: 1225–1233, 1989.
21. K. Takada, K. Yukimatsu, M. Kobayashi, and J. Noda, "Rayleigh Backscattering Measurement of Single-Mode Fibers by Low Coherence Optical Time-Domain Reflectometer with 14 μm Spatial Resolution," *Applied Physics Letters* 59: 143–145, 1991.
22. A. F. Fercher, K. Mengedoht, and W. Werner, "Eye-Length Measurement by Interferometry with Partially Coherent Light," *Optics Letters* 13: 186–188, 1988.
23. J. G. Fujimoto, S. Desilvestri, E. P. Ippen, C. A. Puliafito, R. Margolis, and A. Oseroff, "Femtosecond Optical Ranging in Biological Systems," *Optics Letters* 11: 150–152, 1986.
24. A. F. Fercher, C. K. Hitzenberger, W. Drexler, G. Kamp, and H. Sattmann, "*In-Vivo* Optical Coherence Tomography," *Amer. J. Ophthalmology* 116: 113–115, 1993.
25. E. A. Swanson, J. A. Izatt, M. R. Hee, D. Huang, C. P. Lin, J. S. Schuman, C. A. Puliafito, and J. G. Fujimoto, "*In Vivo* Retinal Imaging by Optical Coherence Tomography," *Optics Letters* 18: 1864–1866, 1993.
26. B. E. Applegate, C. H. Yang, A. M. Rollins, and J. A. Izatt, "Polarization-Resolved Second-Harmonic Generation Optical Coherence Tomography in Collagen," *Optics Letters* 29: 2252–2254, 2004.

27. J. F. de Boer, T. E. Milner, M. J. C. van Gemert, and J. S. Nelson, "Two-Dimensional Birefringence Imaging in Biological Tissue by Polarization-Sensitive Optical Coherence Tomography," *Optics Letters* 22: 934–936, 1997.
28. M. R. Hee, D. Huang, E. A. Swanson, and J. G. Fujimoto, "Polarization-Sensitive Low-Coherence Reflectometer for Birefringence Characterization and Ranging," *J. the Optical Society of Amer. B–Optical Physics* 9: 903–908, 1992.
29. J. A. Izatt, M. D. Kulkami, S. Yazdanfar, J. K. Barton, and A. J. Welch, "*In Vivo* Bidirectional Color Doppler Flow Imaging of Picoliter Blood Volumes Using Optical Coherence Tomograghy," *Optics Letters* 22: 1439–1441, 1997.
30. Y. Jiang, I. Tomov, Y. M. Wang, and Z. P. Chen, "Second-Harmonic Optical Coherence Tomography," *Optics Letters* 29: 1090–1092, 2004.
31. R. Leitgeb, M. Wojtkowski, A. Kowalczyk, C. K. Hitzenberger, M. Sticker, and A. F. Fercher, "Spectral Measurement of Absorption by Spectroscopic Frequency-Domain Optical Coherence Tomography," *Optics Letters* 25: 820–822, 2000.
32. U. Morgner, W. Drexler, F. X. Kartner, X. D. Li, C. Pitris, E. P. Ippen, and J. G. Fujimoto, "Spectroscopic Optical Coherence Tomography," *Optics Letters* 25: 111–113, 2000.
33. A. L. Oldenburg, J. R. Gunther, and S. A. Boppart, "Imaging Magnetically Labeled Cells with Magnetomotive Optical Coherence Tomography," *Optics Letters* 30: 747–749, 2005.
34. A. L. Oldenburg, F. J. J. Toublan, K. S. Suslick, A. Wei, and S. A. Boppart, "Magnetomotive Contrast for *In Vivo* Optical Coherence Tomography," *Optics Express* 13: 6597–6614, 2005.
35. C. E. Saxer, J. F. de Boer, B. H. Park, Y. H. Zhao, Z. P. Chen, and J. S. Nelson, "High-Speed Fiber-Based Polarization-Sensitive Optical Coherence Tomography of *In Vivo* Human Skin," *Optics Letters* 25: 1355–1357, 2000.
36. C. Y. Xu, J. Ye, D. L. Marks, and S. A. Boppart, "Near-Infrared Dyes as Contrast-Enhancing Agents for Spectroscopic Optical Coherence Tomography," *Optics Letters* 29: 1647–1649, 2004.
37. Y. H. Zhao, Z. P. Chen, C. Saxer, S. H. Xiang, J. F. de Boer, and J. S. Nelson, "Phase-Resolved Optical Coherence Tomography and Optical Doppler Tomography for Imaging Blood Flow in Human Skin with Fast Scanning Speed and High Velocity Sensitivity," *Optics Letters* 25: 114–116, 2000.
38. M. A. Choma, M. V. Sarunic, C. H. Yang, and J. A. Izatt, "Sensitivity Advantage of Swept Source and Fourier Domain Optical Coherence Tomography," *Optics Express* 11: 2183–2189, 2003.
39. J. F. de Boer, B. Cense, B. H. Park, M. C. Pierce, G. J. Tearney, and B. E. Bouma, "Improved Signal-to-Noise Ratio in Spectral-Domain Compared with Time-Domain Optical Coherence Tomography," *Optics Letters* 28: 2067–2069, 2003.
40. R. Leitgeb, C. K. Hitzenberger, and A. F. Fercher, "Performance of Fourier Domain vs. Time Domain Optical Coherence Tomography," *Optics Express* 11: 889–894, 2003.
41. M. Wojtkowski, T. Bajraszewski, P. Targowski, and A. Kowalczyk, "Real-Time *In Vivo* Imaging by High-Speed Spectral Optical Coherence Tomography," *Optics Letters* 28: 1745–1747, 2003.
42. S. H. Yun, G. J. Tearney, B. E. Bouma, B. H. Park, and J. F. de Boer, "High-Speed Spectral-Domain Optical Coherence Tomography at 1.3 μm Wavelength," *Optics Express* 11: 3598–3604, 2003.
43. S. H. Yun, G. J. Tearney, J. F. de Boer, N. Iftimia, and B. E. Bouma, "High-Speed Optical Frequency-Domain Imaging," *Optics Express* 11: 2953–2963, 2003.
44. D. C. Adler, Y. Chen, R. Huber, J. Schmitt, J. Connolly, and J. G. Fujimoto, "Three-Dimensional Endomicroscopy Using Optical Coherence Tomography," *Nature Photonics* 1: 709–716, 2007.
45. R. Huber, D. C. Adler, and J. G. Fujimoto, "Buffered Fourier Domain Mode Locking: Unidirectional Swept Laser Sources for Optical Coherence Tomography Imaging at 370,000 Lines/S," *Optics Letters* 31: 2975–2977, 2006.

46. R. Huber, M. Wojtkowski, and J. G. Fujimoto, "Fourier Domain Mode Locking (FDML): A New Laser Operating Regime and Applications for Optical Coherence Tomography," *Optics Express* 14: 3225–3237, 2006.
47. S. H. Yun, G. J. Tearney, B. J. Vakoc, M. Shishkov, W. Y. Oh, A. E. Desjardins, M. J. Suter, R. C. Chan, J. A. Evans, I. K. Jang, N. S. Nishioka, J. F. de Boer, and B. E. Bouma, "Comprehensive Volumetric Optical Microscopy *In Vivo*," *Nature Medicine* 12: 1429–1433, 2006.
48. A. M. Zysk, F. T. Nguyen, A. L. Oldenburg, D. L. Marks, and S. A. Boppart, "Optical Coherence Tomography: A Review of Clinical Development from Bench to Bedside," *J. Biomedical Optics* 12: 051403, 2007.
49. V. Guedes, J. S. Schuman, E. Hertzmark, G. Wollstein, A. Correnti, R. Mancini, D. Lederer, S. Voskanian, L. Velazquez, H. M. Pakter, T. Pedut-Kloizman, J. G. Fujimoto, and C. Mattox, "Optical Coherence Tomography Measurement of Macular and Nerve Fiber Layer Thickness in Normal and Glaucomatous Human Eyes," *Ophthalmology* 110: 177–189, 2003.
50. M. Hangai, Y. Jima, N. Gotoh, R. Inoue, Y. Yasuno, S. Makita, M. Yarnanari, T. Yatagai, M. Kita, and N. Yoshimura, "Three-Dimensional Imaging of Macular Holes with High-Speed Optical Coherence Tomography," *Ophthalmology* 114: 763–773, 2007.
51. T. H. Ko, J. G. Fujimoto, J. S. Duker, L. A. Paunescu, W. Drexler, C. R. Baumal, C. A. Puliafito, E. Reichel, A. H. Rogers, and J. S. Schuman, "Comparison of Ultrahigh- and Standard-Resolution Optical Coherence Tomography for Imaging Macular Hole Pathology and Repair," *Ophthalmology* 111: 2033–2043, 2004.
52. V. J. Srinivasan, M. Wojtkowski, A. J. Witkin, J. S. Duker, T. H. Ko, M. Carvalho, J. S. Schuman, A. Kowalczyk, and J. G. Fujimoto, "High-Definition and 3-Dimensional Imaging of Macular Pathologies with High-Speed Ultrahigh-Resolution Optical Coherence Tomography," *Ophthalmology* 113: 2054–2065, 2006.
53. B. E. Bouma, G. J. Tearney, H. Yabushita, M. Shishkov, C. R. Kauffman, D. D. Gauthier, B. D. MacNeill, S. L. Houser, H. T. Aretz, E. F. Halpern, and I. K. Jang, "Evaluation of Intracoronary Stenting by Intravascular Optical Coherence Tomography," *Heart* 89: 317–320, 2003.
54. A. Erglis, S. Jegere, K. Trusinskis, I. Kumsars, D. Sondore, and I. Narbute, "Stent Endothelization after Paclitaxel Eluting Stent Implantation in Left Main: A 3 Year Intravascular Ultrasound and Optical Coherence Tomography Follow-Up," *Amer. J. Cardiology* 104: 13D, 2009.
55. E. Grube, U. Gerckens, L. Buellesfeld, and P. J. Fitzgerald, "Intracoronary Imaging with Optical Coherence Tomography—A New High-Resolution Technology Providing Striking Visualization in the Coronary Artery," *Circulation* 106: 2409–2410, 2002.
56. K. Ishibashi, H. Kitabata, and T. Akasaka, "Intracoronary Optical Coherence Tomography Assessment of Spontaneous Coronary Artery Dissection," *Heart* 95: 818, 2009.
57. I. K. Jang, B. E. Bouma, D. H. Kang, S. J. Park, S. W. Park, K. B. Seung, K. B. Choi, M. Shishkov, K. Schlendorf, E. Pomerantsev, S. L. Houser, H. T. Aretz, and G. J. Tearney, "Visualization of Coronary Atherosclerotic Plaques in Patients Using Optical Coherence Tomography: Comparison with Intravascular Ultrasound," *J. Amer. College of Cardiology* 39: 604–609, 2002.
58. O. Manfrini, N. J. Miele, B. L. Sharaf, E. McNamara, and D. O. Williams, "Qualitative Results of Intracoronary Imaging during Balloon Inflation with Optical Coherence Tomography in Humans," *J. Amer. College of Cardiology* 41: 60A, 2003.
59. Y. Ozaki, M. Okumura, J. Ishii, S. Matsui, H. Naruse, S. Kato, T. Sato, Y. Nakamura, K. Inoue, S. Nakano, S. Kan, S. Hiramitsu, T. Kondo, T. F. Ismail, and H. Hishida, "Vulnerable Lesion Characteristics Assessed by Optical Coherence Tomography (OCT), Intracoronary Ultrasound (IVUS), Angioscopy and Quantitative Coronary Angiography (QCA)," *J. Amer. College of Cardiology* 47: 53B, 2006.

60. N. Rosenthal, G. Guagliumi, V. Sirbu, G. B. Zoccai, L. Fioca, G. Musumeci, A. Mathiasvili, A. Trivisonno, H. Kyono, S. Tahara, D. I. Simon, M. Costa, and H. G. Bezerra, "Comparison of Intravascular Ultrasound and Optical Coherence Tomography for the Evaluation of Stent Segment Malapposition," *J. Amer. College of Cardiology* 53: A22, 2009.
61. T. Yamaguchi, M. Terashima, T. Akasaka, T. Hayashi, K. Mizuno, T. Muramatsu, M. Nakamura, S. Nakamura, S. Saito, M. Takano, T. Takayama, J. Yoshikawa, and T. Suzuki, "Safety and Feasibility of an Intravascular Optical Coherence Tomography Image Wire System in the Clinical Setting," *Amer. J. Cardiology* 101: 562–567, 2008.
62. W. B. Armstrong, J. M. Ridgway, D. E. Vokes, S. Guo, J. Perez, R. P. Jackson, M. Gu, J. P. Su, R. L. Crumley, T. Y. Shibuya, U. Mahmood, Z. P. Chen, and B. J. F. Wong, "Optical Coherence Tomography of Laryngeal Cancer," *Laryngoscope* 116: 1107–1113, 2006.
63. P. F. Escobar, J. L. Belinson, A. White, N. M. Shakhova, F. I. Feldchtein, M. V. Kareta, and N. D. Gladkova, "Diagnostic Efficacy of Optical Coherence Tomography in the Management of Preinvasive and Invasive Cancer of Uterine Cervix and Vulva," *Inte. J. Gynecological Cancer* 14: 470–474, 2004.
64. L. P. Hariri, A. R. Tumlinson, D. G. Besselsen, U. Utzinger, E. W. Gerner, and J. K. Barton, "Endoscopic Optical Coherence Tomography and Laser-Induced Fluorescence Spectroscopy in a Murine Colon Cancer Model," *Lasers in Surgery and Medicine* 38: 305–313, 2006.
65. Y. T. Pan, T. Q. Xie, C. W. Du, S. Bastacky, S. Meyers, and M. L. Zeidel, "Enhancing Early Bladder Cancer Detection with Fluorescence-Guided Endoscopic Optical Coherence Tomography," *Optics Letters* 28: 2485–2487, 2003.
66. B. E. Bouma and G. J. Tearney, eds., *Handbook of Optical Coherence Tomography,* Marcel Dekker, New York, 2002.
67. B. E. Bouma and G. J. Tearney, "Clinical Imaging with Optical Coherence Tomography," *Academic Radiology* 9: 942–953, 2002.
68. W. Drexler and J. G. Fujimoto, eds., *Optical Coherence Tomography—Technology and Applications,* Springer-Verlag, New York, 2008.
69. A. F. Fercher, W. Drexler, C. K. Hitzenberger, and T. Lasser, "Optical Coherence Tomography—Principles and Applications," *Reports on Progress in Physics* 66: 239–303, 2003.
70. J. G. Fujimoto, C. Pitris, S. A. Boppart, and M. E. Brezinski, "Optical Coherence Tomography: An Emerging Technology for Biomedical Imaging and Optical Biopsy," *Neoplasia* 2: 9–25, 2000.
71. J. M. Schmitt, "Optical Coherence Tomography (OCT): A Review," *IEEE J. Selected Topics in Quantum Electronics* 5: 1205–1215, 1999.
72. P. H. Tomlins and R. K. Wang, "Theory, Developments and Applications of Optical Coherence Tomography," *J. Physics D-Applied Physics* 38: 2519–2535, 2005.
73. K. Bogunia-Kubik and M. Sugisaka, "From Molecular Biology to Nanotechnology and Nanomedicine," *Biosystems* 65: 123–138, 2002.
74. O. C. Farokhzad and R. Langer, "Nanomedicine: Developing Smarter Therapeutic and Diagnostic Modalities," *Advanced Drug Delivery Reviews* 58: 1456–1459, 2006.
75. K. K. Jain, "Nanomedicine: Application of Nanobiotechnology in Medical Practice," *Medical Principles and Practice* 17: 89–101, 2008.
76. G. M. Lanza, P. M. Winter, S. D. Caruthers, M. S. Hughes, T. Cyrus, J. N. Marsh, A. M. Neubauer, K. C. Partlow, and S. A. Wickline, "Nanomedicine Opportunities for Cardiovascular Disease with Perfluorocarbon Nanoparticles," *Nanomedicine* 1: 321–329, 2006.
77. K. C. P. Li, S. D. Pandit, S. Guccione, and M. D. Bednarski, "Molecular Imaging Applications in Nanomedicine," *Biomedical Microdevices* 6: 113–116, 2004.
78. Y. F. Liu and H. F. Wang, "Nanomedicine Nanotechnology Tackles Tumours," *Nature Nanotechnology* 2: 20–21, 2007.

79. Y. Y. Liu, H. Miyoshi, and M. Nakamura, "Nanomedicine for Drug Delivery and Imaging: A Promising Avenue for Cancer Therapy and Diagnosis Using Targeted Functional Nanoparticles," *Inte. J. Cancer* 120: 2527–2537, 2007.
80. S. M. Moghimi, A. C. Hunter, and J. C. Murray, "Nanomedicine: Current Status and Future Prospects," *Faseb J.* 19: 311–330, 2005.
81. V. Wagner, A. Dullaart, A. K. Bock, and A. Zweck, "The Emerging Nanomedicine Landscape," *Nature Biotechnology* 24: 1211–1217, 2006.
82. P. Gould, "Multitasking Nanoparticles Target Cancer—Nanomedicine," *Nano Today* 3: 9, 2008.
83. P. Hervella, V. Lozano, and M. Garcia-Fuentes, "Nanomedicine: New Challenges and Opportunities in Cancer Therapy," *J. Biomedical Nanotechnology* 4: 276–292, 2008.
84. D. K. Kim and J. Dobson, "Nanomedicine for Targeted Drug Delivery," *J. Materials Chemistry* 19: 6294–6307, 2009.
85. C. Shaffer, "Nanomedicine Transforms Drug Delivery," *Drug Discovery Today* 10: 1581–1582, 2005.
86. B. Sumer and J. M. Gao, "Theranostic Nanomedicine for Cancer," *Nanomedicine* 3: 137–140, 2008.
87. R. J. Zemp, "Nanomedicine Detecting Rare Cancer Cells," *Nature Nanotechnology* 4: 798–799, 2009.
88. W. Drexler, "Ultrahigh-Resolution Optical Coherence Tomography," *J. Biomedical Optics* 9: 47–74, 2004.
89. W. Drexler, U. Morgner, F. X. Kartner, C. Pitris, S. A. Boppart, X. D. Li, E. P. Ippen, and J. G. Fujimoto, "*In Vivo* Ultrahigh-Resolution Optical Coherence Tomography," *Optics Letters* 24: 1221–1223, 1999.
90. I. Hartl, X. D. Li, C. Chudoba, R. K. Ghanta, T. H. Ko, J. G. Fujimoto, J. K. Ranka, and R. S. Windeler, "Ultrahigh-Resolution Optical Coherence Tomography Using Continuum Generation in an Air-Silica Microstructure Optical Fiber," *Optics Letters* 26: 608–610, 2001.
91. B. Povazay, K. Bizheva, A. Unterhuber, B. Hermann, H. Sattmann, A. F. Fercher, W. Drexler, A. Apolonski, W. J. Wadsworth, J. C. Knight, P. S. J. Russell, M. Vetterlein, and E. Scherzer, "Submicrometer Axial Resolution Optical Coherence Tomography," *Optics Letters* 27: 1800–1802, 2002.
92. T. S. Ralston, D. L. Marks, P. S. Carney, and S. A. Boppart, "Interferometric Synthetic Aperture Microscopy," *Nature Physics* 3: 129–134, 2007.
93. T. S. Ralston, D. L. Marks, P. S. Carney, and S. A. Boppart, "Real-Time Interferometric Synthetic Aperture Microscopy," *Optics Express* 16: 2555–2569, 2008.
94. M. A. Choma, A. K. Ellerbee, C. H. Yang, T. L. Creazzo, and J. A. Izatt, "Spectral-Domain Phase Microscopy," *Optics Letters* 30: 1162–1164, 2005.
95. A. K. Ellerbee, T. L. Creazzo, and J. A. Izatt, "Investigating Nanoscale Cellular Dynamics with Cross-Sectional Spectral-Domain Phase Microscopy," *Optics Express* 15: 8115–8124, 2007.
96. C. Joo, T. Akkin, B. Cense, B. H. Park, and J. E. de Boer, "Spectral-Domain Optical Coherence Phase Microscopy for Quantitative Phase-Contrast Imaging," *Optics Letters* 30: 2131–2133, 2005.
97. E. J. McDowell, A. K. Ellerbee, M. A. Choma, B. E. Applegate, and J. A. Izatt, "Spectral-Domain Phase Microscopy for Local Measurements of Cytoskeletal Rheology in Single Cells," *J. Biomedical Optics* 12: 044008, 2007.
98. D. C. Adler, T. H. Ko, P. R. Herz, and J. G. Fujimoto, "Optical Coherence Tomography Contrast Enhancement Using Spectroscopic Analysis with Spectral Autocorrelation," *Optics Express* 12: 5487–5501, 2004.
99. H. Cang, T. Sun, Z. Y. Li, J. Y. Chen, B. J. Wiley, Y. N. Xia, and X. D. Li, "Gold Nanocages as Contrast Agents for Spectroscopic Optical Coherence Tomography," *Optics Letters* 30: 3048–3050, 2005.
100. A. L. Oldenburg, C. Y. Xu, and S. A. Boppart, "Spectroscopic Optical Coherence Tomography and Microscopy," *IEEE J. Selected Topics in Quantum Electronics* 13: 1629–1640, 2007.

101. C. Y. Xu, P. S. Carney, and S. A. Boppart, "Wavelength-Dependent Scattering in Spectroscopic Optical Coherence Tomography," *Optics Express* 13: 5450–5462, 2005.
102. J. J. Pasquesi, S. C. Schlachter, M. D. Boppart, E. Chaney, S. J. Kaufman, and S. A. Boppart, "*In Vivo* Detection of Exercise-Induced Ultrastructural Changes in Genetically Altered Murine Skeletal Muscle Using Polarization-Sensitive Optical Coherence Tomography," *Optics Express* 14: 1547–1556, 2006.
103. D. L. Marks and S. A. Boppart, "Nonlinear Interferometric Vibrational Imaging," *Physical Review Letters* 92: 123905, 2004.
104. J. A. Izatt, M. D. Kulkarni, H. W. Wang, K. Kobayashi, and M. V. Sivak, "Optical Coherence Tomography and Microscopy in Gastrointestinal Tissues," *IEEE J. Selected Topics in Quantum Electronics* 2: 1017–1028, 1996.
105. A. M. Rollins and J. A. Izatt, "Optimal Interferometer Designs for Optical Coherence Tomography," *Optics Letters* 24: 1484–1486, 1999.
106. U. Sharma, N. M. Fried, and J. U. Kang, "All-Fiber Common-Path Optical Coherence Tomography: Sensitivity Optimization and System Analysis," *IEEE J. Selected Topics in Quantum Electronics* 11: 799–805, 2005.
107. K. Takada, A. Himeno, and K. Yukimatsu, "Phase-Noise and Shot-Noise Limited Operations of Low-Coherence Optical Time-Domain Reflectometry," *Applied Physics Letters* 59: 2483–2485, 1991.
108. K. Takada, "Noise in Optical Low-Coherence Reflectometry," *IEEE J. Quantum Electronics* 34: 1098–1108, 1998.
109. C. Loo, L. Hirsch, M. H. Lee, E. Chang, J. West, N. Halas, and R. Drezek, "Gold Nanoshell Bioconjugates for Molecular Imaging in Living Cells," *Optics Letters* 30: 1012–1014, 2005.
110. X. G. Shi, S. H. Wang, S. Meshinchi, M. E. Van Antwerp, X. D. Bi, I. H. Lee, and J. R. Baker, "Dendrimer-Entrapped Gold Nanoparticles as a Platform for Cancer-Cell Targeting and Imaging," *Small* 3: 1245–1252, 2007.
111. S. J. Oldenburg, R. D. Averitt, S. L. Westcott, and N. J. Halas, "Nanoengineering of Optical Resonances," *Chemical Physics Letters* 288: 243–247, 1998.
112. S. J. Oldenburg, S. L. Westcott, R. D. Averitt, and N. J. Halas, "Surface-Enhanced Raman Scattering in the Near Infrared Using Metal Nanoshell Substrates," *J. Chemical Physics* 111: 4729–4735, 1999.
113. S. L. Westcott, S. J. Oldenburg, T. R. Lee, and N. J. Halas, "Formation and Adsorption of Clusters of Gold Nanoparticles onto Functionalized Silica Nanoparticle Surfaces," *Langmuir* 14: 5396–5401, 1998.
114. D. C. Adler, S. W. Huang, R. Huber, and J. G. Fujimoto, "Photothermal Detection of Gold Nanoparticles Using Phase-Sensitive Optical Coherence Tomography," *Optics Express* 16, 4376–4393, 2008.
115. A. Agrawal, S. Huang, A. W. H. Lin, M. H. Lee, J. K. Barton, R. A. Drezek, and T. J. Pfefer, "Quantitative Evaluation of Optical Coherence Tomography Signal Enhancement with Gold Nanoshells," *J. Biomedical Optics* 11: 041121, 2006.
116. C. S. Kim, P. Wilder-Smith, Y. C. Ahn, L. H. L. Liaw, Z. P. Chen, and Y. J. Kwon, "Enhanced Detection of Early-Stage Oral Cancer *In Vivo* by Optical Coherence Tomography Using Multimodal Delivery of Gold Nanoparticles," *J. Biomedical Optics* 14: 034008, 2009.
117. A. L. Oldenburg, M. N. Hansen, D. A. Zweifel, A. Wei, and S. A. Boppart, "Plasmon-Resonant Gold Nanorods as Low Backscattering Albedo Contrast Agents for Optical Coherence Tomography," *Optics Express* 14: 6724–6738, 2006.
118. T. S. Troutman, J. K. Barton, and M. Romanowski, "Optical Coherence Tomography with Plasmon Resonant Nanorods of Gold," *Optics Letters* 32: 1438–1440, 2007.
119. J. Chen, F. Saeki, B. J. Wiley, H. Cang, M. J. Cobb, Z. Y. Li, L. Au, H. Zhang, M. B. Kimmey, X. D. Li, and Y. Xia, "Gold Nanocages: Bioconjugation and Their Potential Use as Optical Imaging Contrast Agents," *Nano Letters* 5: 473–477, 2005.
120. C. H. Yang, L. E. L. McGuckin, J. D. Simon, M. A. Choma, B. E. Applegate, and J. A. Izatt, "Spectral Triangulation Molecular Contrast Optical Coherence

Tomography with Indocyanine Green as the Contrast Agent," *Optics Letters* 29: 2016–2018, 2004.

121. A. L. Oldenburg, V. Crecea, S. A. Rinne, and S. A. Boppart, "Phase-Resolved Magnetomotive OCT for Imaging Nanomolar Concentrations of Magnetic Nanoparticles in Tissues," *Optics Express* 16: 11525–11539, 2008.
122. X. Liang, A. L. Oldenburg, V. Crecea, E. J. Chaney, and S. A. Boppart, "Optical Micro-Scale Mapping of Dynamic Biomechanical Tissue Properties," *Optics Express* 16: 11052–11065, 2008.
123. D. Artemov, N. Mori, B. Okollie, and Z. M. Bhujwalla, "MR Molecular Imaging of the Her-2/Neu Receptor in Breast Cancer Cells Using Targeted Iron Oxide Nanoparticles," *Magnetic Resonance in Medicine* 49: 403–408, 2003.
124. D. Artemov, N. Mori, R. Ravi, and Z. M. Bhujwalla, "Magnetic Resonance Molecular Imaging of the Her-2/Neu Receptor," *Cancer Research* 63: 2723–2727, 2003.
125. M. A. Funovics, B. Kapeller, C. Hoeller, H. S. Su, R. Kunstfeld, S. Puig, and K. Macfelda, "MR Imaging of the Her2/Neu and 9.2.27 Tumor Antigens Using Immunospecific Contrast Agents," *Magnetic Resonance Imaging* 22: 843–850, 2004.
126. S. J. DeNardo, G. L. DeNardo, L. A. Miers, A. Natarajan, A. R. Foreman, C. Gruettner, G. N. Adamson, and R. Ivkov, "Development of Tumor Targeting Bioprobes (in-111-Chimeric L6 Monoclonal Antibody Nanoparticles) for Alternating Magnetic Field Cancer Therapy," *Clinical Cancer Research* 11: 7087S–7092S, 2005.
127. J. R. McCarthy and R. Weissleder, "Multifunctional Magnetic Nanoparticles for Targeted Imaging and Therapy," *Advanced Drug Delivery Reviews* 60: 1241–1251, 2008.
128. R. Rezaeipoor, R. John, S. G. Adie, E. J. Chaney, M. Marjanovic, A. L. Oldenburg, S. A. Rinne, and S. A. Boppart, "Fc-Directed Antibody Conjugation of Magnetic Nanoparticles for Enhanced Molecular Targeting," *J. Innovative Optical Health Sciences* 2: 387–396, 2009.
129. W. Drexler, U. Morgner, R. K. Ghanta, F. X. Kartner, J. S. Schuman, and J. G. Fujimoto, "Ultrahigh-Resolution Ophthalmic Optical Coherence Tomography," *Nature Medicine* 7: 502–507, 2001.
130. A. D. Aguirre, P. Hsiung, T. H. Ko, I. Hartl, and J. G. Fujimoto, "High-Resolution Optical Coherence Microscopy for High-Speed, *In Vivo* Cellular Imaging," *Optics Letters* 28: 2064–2066, 2003.
131. J. A. Izatt, M. R. Hee, G. M. Owen, E. A. Swanson, and J. G. Fujimoto, "Optical Coherence Microscopy in Scattering Media," *Optics Letters* 19: 590–592, 1994.
132. M. Pircher, E. Gotzinger, and C. K. Hitzenberger, "Dynamic Focus in Optical Coherence Tomography for Retinal Imaging," *J. Biomedical Optics* 11: 054013, 2006.
133. B. Qi, A. P. Himmer, L. M. Gordon, X. D. V. Yang, L. D. Dickensheets, and I. A. Vitkin, "Dynamic Focus Control in High-Speed Optical Coherence Tomography Based on a Microelectromechanical Mirror," *Optics Communications* 232: 123–128, 2004.
134. B. J. Davis, D. L. Marks, T. S. Ralston, P. S. Carney, and S. A. Boppart, "Interferometric Synthetic Aperture Microscopy: Computed Imaging for Scanned Coherent Microscopy," *Sensors* 8: 3903–3931, 2008.
135. D. L. Marks, T. S. Ralston, P. S. Carney, and S. A. Boppart, "Inverse Scattering for Rotationally Scanned Optical Coherence Tomography," *J. Optical Society of Amer. A—Optics Image Science and Vision* 23: 2433–2439, 2006.
136. T. S. Ralston, D. L. Marks, S. A. Boppart, and P. S. Carney, "Inverse Scattering for High-Resolution Interferometric Microscopy," *Optics Letters* 31: 3585–3587, 2006.
137. T. S. Ralston, D. L. Marks, P. S. Carney, and S. A. Boppart, "Inverse Scattering for Optical Coherence Tomography," *J. Optical Society of Amer. A—Optics Image Science and Vision* 23: 1027–1037, 2006.
138. T. Ikeda, G. Popescu, R. R. Dasari, and M. S. Feld, "Hilbert Phase Microscopy for Investigating Fast Dynamics in Transparent Systems," *Optics Letters* 30: 1165–1167, 2005.

139. N. Lue, W. Choi, G. Popescu, T. Ikeda, R. R. Dasari, K. Badizadegan, and M. S. Feld, "Quantitative Phase Imaging of Live Cells Using Fast Fourier Phase Microscopy," *Applied Optics* 46: 1836–1842, 2007.
140. G. Popescu, T. Ikeda, C. A. Best, K. Badizadegan, R. R. Dasari, and M. S. Feld, "Erythrocyte Structure and Dynamics Quantified by Hilbert Phase Microscopy," *J. Biomedical Optics* 10: 060503, 2005.
141. G. Popescu, L. P. Deflores, J. C. Vaughan, K. Badizadegan, H. Iwai, R. R. Dasari, and M. S. Feld, "Fourier Phase Microscopy for Investigation of Biological Structures and Dynamics," *Optics Letters* 29: 2503–2505, 2004.
142. C. Yang, A. Wax, M. S. Hahn, K. Badizadegan, R. R. Dasari, and M. S. Feld, "Phase-Referenced Interferometer with Subwavelength and SubHertz Sensitivity Applied to the Study of Cell Membrane Dynamics," *Optics Letters* 26: 1271–1273, 2001.
143. C. Fang-Yen, M. C. Chu, H. S. Seung, R. R. Dasari, and M. S. Feld, "Noncontact Measurement of Nerve Displacement During Action Potential with a Dual-Beam Low-Coherence Interferometer," *Optics Letters* 29: 2028–2030, 2004.
144. M. A. Choma, A. K. Ellerbee, S. Yazdanfar, and J. A. Izatt, "Doppler Flow Imaging of Cytoplasmic Streaming Using Spectral-Domain Phase Microscopy," *J. Biomedical Optics* 11: 024014, 2006.
145. S. Nezam, C. Joo, G. J. Tearney, and J. F. de Boer, "Application of Maximum Likelihood Estimator in Nano-Scale Optical Path Length Measurement Using Spectral-Domain Optical Coherence Phase Microscopy," *Optics Express* 16: 17186–17195, 2008.
146. M. V. Sarunic, S. Weinberg, and J. A. Izatt, "Full-Field Swept-Source Phase Microscopy," *Optics Letters* 31: 1462–1464, 2006.
147. M. Irving, "Birefringence Changes Associated with Isometric Contraction and Rapid Shortening Steps in Frog Skeletal-Muscle Fibers," *J. Physiology—London* 472: 127–156, 1993.
148. H. M. Jones, R. J. Baskin, and Y. Yeh, "The Molecular-Origin of Birefringence in Skeletal-Muscle—Contribution of Myosin Subfragment S-1," *Biophysical J.* 60: 1217–1228, 1991.
149. B. Cense, M. Mujat, T. C. Chen, B. H. Park, and J. F. de Boer, "Polarization-Sensitive Spectral-Domain Optical Coherence Tomography Using a Single Line Scan Camera," *Optics Express* 15: 2421–2431, 2007.
150. W. Y. Oh, B. J. Vakoc, S. H. Yun, G. J. Tearney, and B. E. Bouma, "Single-Detector Polarization-Sensitive Optical Frequency Domain Imaging Using High-Speed Intra A-Line Polarization Modulation," *Optics Letters* 33: 1330–1332, 2008.
151. W. Y. Oh, S. H. Yun, B. J. Vakoc, M. Shishkov, A. E. Desjardins, B. H. Park, J. F. de Boer, G. J. Tearney, and E. Bouma, "High-Speed Polarization Sensitive Optical Frequency Domain Imaging with Frequency Multiplexing," *Optics Express* 16: 1096–1103, 2008.
152. M. Yamanari, S. Makita, and Y. Yasuno, "Polarization-Sensitive Swept-Source Optical Coherence Tomography with Continuous Source Polarization Modulation," *Optics Express* 16: 5892–5906, 2008.
153. J. F. de Boer and T. E. Milner, "Review of Polarization Sensitive Optical Coherence Tomography and Stokes Vector Determination," *J. Biomedical Optics* 7: 359–371, 2002.
154. B. H. Park, M. C. Pierce, B. Cense, and J. F. de Boer, "Jones Matrix Analysis for a Polarization-Sensitive Optical Coherence Tomography System Using Fiber-Optic Components," *Optics Letters* 29: 2512–2514, 2004.
155. B. Cense, T. C. Chen, B. H. Park, M. C. Pierce, and J. F. de Boer, "*In Vivo* Depth-Resolved Birefringence Measurements of the Human Retinal Nerve Fiber Layer by Polarization-Sensitive Optical Coherence Tomography," *Optics Letters* 27: 1610–1612, 2002.
156. B. Cense, T. C. Chen, B. H. Park, M. C. Pierce, and J. F. de Boer, "Thickness and Birefringence of Healthy Retinal Nerve Fiber Layer Tissue Measured with Polarization-Sensitive Optical Coherence Tomography," *Investigative Ophthalmology & Visual Science* 45: 2606–2612, 2004.

157. M. C. Pierce, J. Strasswimmer, B. H. Park, B. Cense, and J. F. de Boer, "Birefringence Measurements in Human Skin Using Polarization-Sensitive Optical Coherence Tomography," *J. Biomedical Optics* 9: 287–291, 2004.
158. S. K. Nadkarni, M. C. Pierce, B. H. Park, J. F. de Boer, P. Whittaker, B. E. Bouma, J. E. Bressner, E. Halpern, S. L. Houser, and G. J. Tearney, "Measurement of Collagen and Smooth Muscle Cell Content in Atherosclerotic Plaques Using Polarization-Sensitive Optical Coherence Tomography," *J. Amer. College of Cardiology* 49: 1474–1481, 2007.
159. B. E. Applegate and J. A. Izatt, "Molecular Imaging of Endogenous and Exogenous Chromophores Using Ground State Recovery Pump-Probe Optical Coherence Tomography," *Optics Express* 14: 9142–9155, 2006.
160. B. E. Applegate, C. H. Yang, and J. A. Izatt, "Theoretical Comparison of the Sensitivity of Molecular Contrast Optical Coherence Tomography Techniques," *Optics Express* 13: 8146–8163, 2005.
161. K. D. Rao, M. A. Choma, S. Yazdanfar, A. M. Rollins, and J. A. Izatt, "Molecular Contrast in Optical Coherence Tomography by Use of a Pump-Probe Technique," *Optics Letters* 28: 340–342, 2003.
162. C. H. Yang, M. A. Choma, L. E. Lamb, J. D. Simon, and J. A. Izatt, "Protein-Based Molecular Contrast Optical Coherence Tomography with Phytochrome as the Contrast Agent," *Optics Letters* 29: 1396–1398, 2004.
163. W. M. Tolles, J. W. Nibler, J. R. McDonald, and A. B. Harvey, "Review of Theory and Application of Coherent Anti-Stokes Raman-Spectroscopy (CARS)," *Applied Spectroscopy* 31: 253–271, 1977.
164. A. Zumbusch, G. R. Holtom, and X. S. Xie, "Three-Dimensional Vibrational Imaging by Coherent Anti-Stokes Raman Scattering," *Physical Review Letters* 82: 4142–4145, 1999.
165. D. L. Marks, C. Vinegoni, J. S. Bredfeldt, and S. A. Boppart, "Interferometric Differentiation between Resonant Coherent Anti-Stokes Raman Scattering and Nonresonant Four-Wave Mixing Processes," *Applied Physics Letters* 85: 5787–5789, 2004.
166. J. S. Bredfeldt, C. Vinegoni, D. L. Marks, and S. A. Boppart, "Molecularly Sensitive Optical Coherence Tomography," *Optics Letters* 30: 495–497, 2005.
167. W. A. Benalcazar, P. D. Chowdary, Z. Jiang, D. L. Marks, E. Chaney, M. Gruebele, and S. A. Boppart, "High-Speed Nonlinear Interferometric Vibrational Imaging of Biological Tissue with Comparison to Raman Microscopy," *IEEE J. Selected Topics in Quantum Electronics* (In Press, 2010).

CHAPTER 9

Plasmonics and Metamaterials

Kin Hung Fung and Nicholas X. Fang

Department of Mechanical Science and Engineering
University of Illinois at Urbana–Champaign

9.1 Introduction

Plasmonics and metamaterials are two rapidly expanding research fields in which many novel electromagnetic wave phenomena have been realized in recent years and which promise many plausible biomedical applications. Examples of these applications include imaging and sensing of biological tissues and molecules [1–5]. Fine control and manifestation of these phenomena often require the use of fine metallic structures. For applications in the optical wavelength range, dimensions of those fine metallic structures are typically in the scale of tens to hundreds of nanometers, and electromagnetic resonances become significant due to excitation of surface plasmon. Therefore, it is our intention to combine the two topics in this chapter, following Chapter 3 about nanophotonics. This chapter explains the basic concepts required for understanding the photonic phenomena in plasmonics and metamaterials, followed by an introduction of some plausible applications based on recent discoveries, such as superresolution imaging.

Metamaterials are made of precisely fabricated constituents that are analogous to atoms and molecules in natural materials. Although artificial metamaterials and natural materials share some similarities, they are different in many aspects. Some examples of their differences are listed in Table 9.1. Although metamaterials are not homogenous in the atomic scale, they can be viewed as effectively homogenous materials possessing some effective constitutive electromagnetic parameters, such as electric

	Common Natural Materials	Metamaterials
Type of waves	Quantum or classical	Classical
Constituents	Atoms or molecules	Structured metal, dielectrics, etc.
Geometric variants	230 crystal lattices and amorphous state	Various shapes of constituents
Dopants or defect	Random	Controllable

Table 9.1 Comparison between Metamaterials and Common Natural Materials

permittivity ε_{eff} and magnetic permeability μ_{eff}. Recent advances in microfabrication technology enable the synthesis of these metamaterials with their constituents in microscale precision. This opens the possibility of engineering various electric and magnetic metamaterials and rendering their ε_{eff} and μ_{eff} driven by the small-scale physics associated with microstructure.

For simplicity, let us consider metamaterials made of periodic units, although such order is not required in the construction of metamaterials. Using natural materials (such as metals and dielectrics) as constituents, one can construct metamaterials with effective ε_{eff} and μ_{eff} that deviate a lot from that of the constituents. For example, metals and dielectrics have permeabilities $\mu \approx 1$ at high frequencies, but metamaterials consisting of only metals and dielectrics (such as metallic split rings or Ω particles [6]) can achieve $\mu_{eff} >> 1$, $\mu_{eff} < 1$, or even $\mu_{eff} < 0$ due to magnetic resonances. The ability of tuning effective ε_{eff} and μ_{eff} allows us to achieve some unconventional physical phenomena. For example, in a medium of a sparse matrix of metallic wires, the propagation modes display a surface plasmon-like dispersion with a negative permittivity down to the gigahertz range [7]. It has also been shown that a negative index of refraction can be achieved on the composites built of arrays of split copper rings [8, 9] and wires, this could lead to lenses that transfer images with a resolution far below wavelength scale [10].

Generally, we can classify materials into four classes: double-positive (DPS) medium, epsilon-negative (ENG) medium, mu-negative (MNG) medium, and double-negative (DNG) medium (see Fig. 9.1). Early research on metamaterials focuses on the phenomena of negative refraction and superlens in DNG media (i.e., materials with simultaneous negative ε and μ) because such a DNG medium has not been found in nature. Electrodynamics of a DNG medium was first theoretically predicted by Veselago in 1967 [11]. It was proposed that such a hypothetical medium would

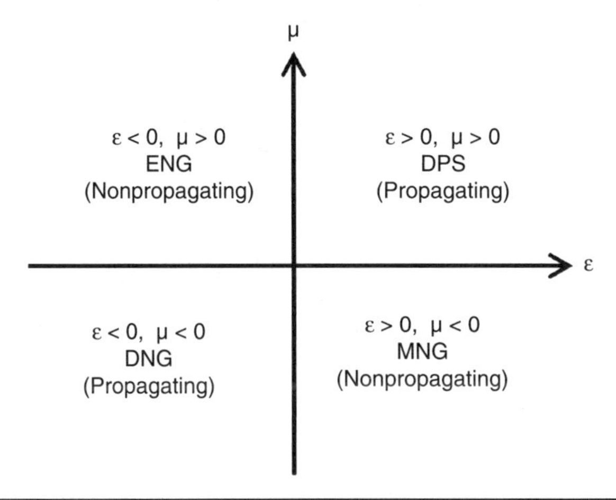

Figure 9.1 Classification of materials in the electric permittivity and magnetic permeability ($\varepsilon - \mu$) parameter space. Engineered metamaterials allow us to explore the new quadrants of electromagnetic activities that are not attainable by naturally existing materials.

support backward-propagating electromagnetic waves where energy velocity (or group velocity for a lossless medium) is opposite to its phase velocity (see Fig. 9.2). A plane wave propagating within the medium has its electric field vector **E**, associated magnetic field vector **H**, and wave vector **k** forming an orthogonal left-handed

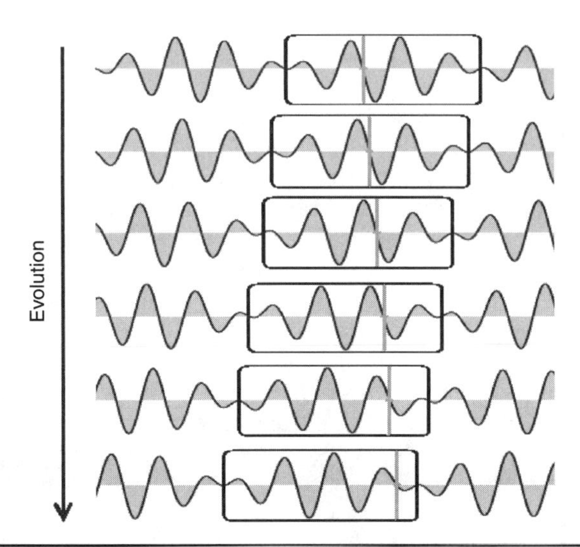

Figure 9.2 A schematic illustration of wave propagation in negative index materials, where the group and phase velocities appear in opposite directions. Boxes indicate the continuous motion of a pulse (group). Vertical lines inside boxes indicate the continuous motion of the point of fixed phase.

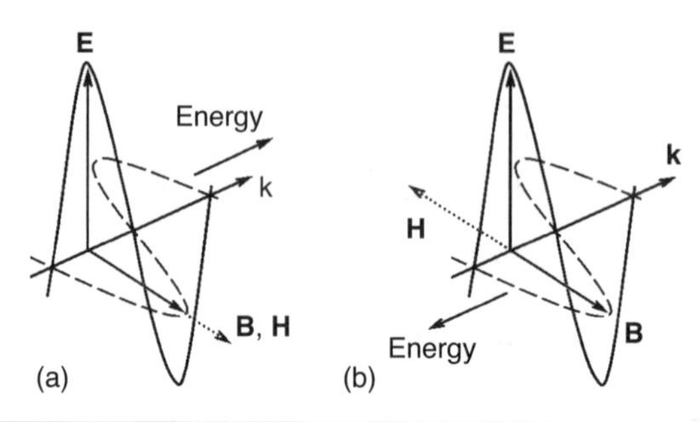

FIGURE 9.3 Relative directions of field vectors, wave vector, and energy propagation for a plane wave in (*a*) DPS and (*b*) DNG media. The flip of the relative sign between **B** and **H** fields in DNG medium changes the handedness from right hand ($\hat{\mathbf{E}} \times \hat{\mathbf{H}} = \hat{\mathbf{k}}$) to left hand ($\hat{\mathbf{E}} \times \hat{\mathbf{H}} = -\hat{\mathbf{k}}$); this makes the energy propagation direction become opposite to the wave vector.

set (i.e., $\hat{\mathbf{E}} \times \hat{\mathbf{H}} = -\hat{\mathbf{k}}$) instead of the usual right-handed set (i.e., $\hat{\mathbf{E}} \times \hat{\mathbf{H}} = \hat{\mathbf{k}}$). Therefore, such media are sometimes called left-handed media (Fig. 9.3). It should be noted that the handedness mentioned here only refers to relative orientations of **E**, **H**, and **k**. Although metamaterials can be classified into left-handed and right-handed metamaterials by chirality in geometry, the handedness mentioned here has no relation to that material geometry that commonly used in chemistry.

For an isotropic medium, one may define the refractive index n using the conventional Snell's law, $n_2/n_1 = \sin\theta_1/\sin\theta_2$, where the angles θ_1 and θ_2 are allowed to have negative values if the refracted light beam is on another side of the normal axis. When θ_1 and θ_2 have opposite signs, we call the phenomenon *negative refraction*. Negative refraction can occur at the interface between a DPS medium ($\varepsilon > 0$ and $\mu > 0$) and a DNG medium ($\varepsilon < 0$ and $\mu < 0$). Such a phenomenon is a direct consequence of the opposite sign between energy velocity and phase velocity. For electromagnetic waves, the propagation direction of energy is indicated by the Poynting vector $\mathbf{S} \equiv \mathbf{E} \times \mathbf{H}$, whereas the phase propagation direction is indicated by the wave vector **k** (see Fig. 9.4). In order to match the boundary conditions at the interface, the parallel component of **k** must be unchanged after refraction. Without knowing the direction of energy, there can be more than one (usually two) propagating solutions for the refracted beam. However, only one solution is allowed because energy cannot go back to the incident side after refraction. Therefore, the sign difference between energy and phase velocities leads to a negative refraction. Details are shown in Fig. 9.4.

The negative refraction phenomenon has been a hot topic due to the research on metamaterials. However, the more general

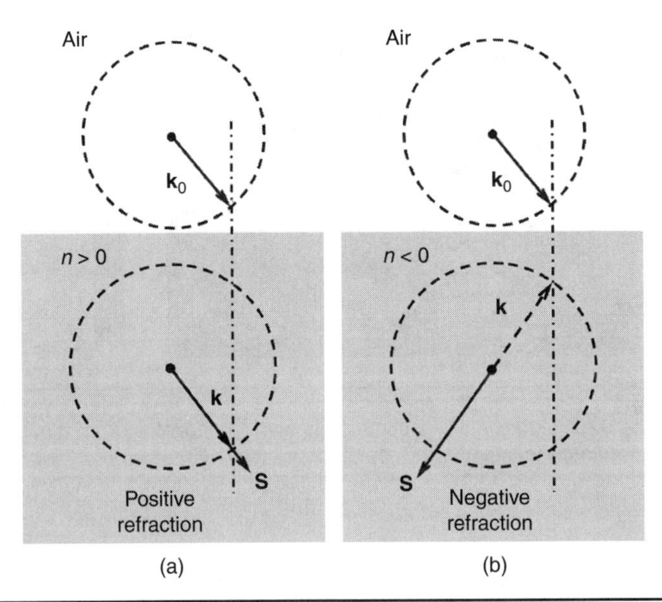

Figure 9.4 Illustrated example of (*a*) positive and (*b*) negative refraction when electromagnetic wave travels from air to other medium. Dashed circle in each medium shows all propagating solutions in the wave vector space. The vertical dashed-dotted lines indicate the conservation of the parallel wave vector after refraction. The dashed arrow in each case shows the only allowed casual solution. Solid arrows labeled by *S* show the energy propagation directions (i.e., directions of the refracted beam).

phenomenon of the DNG medium is that the wave experiences a "negative" space: this means that such a wave (propagating forward) is analogous to a wave propagating backward in a DPS medium. This leads to the "perfect lens," first proposed by Pendry [10]. When there is a flat slab of DNG medium of $\varepsilon = -1$ and $\mu = -1$, a point source from one side can be perfectly focused on the other side because the wave evolves back to its original form after passing through the slab (see Fig. 9.5). If we can make such a "perfect lens," this will be a significant breakthrough because traditional lenses are limited by the diffraction limit, which suggests that any optical image formed by a traditional lens cannot have a resolution finer than the wavelength. However, the proposal of the "perfect lens" was challenged because it is limited to near-field focusing (i.e., the high-resolution image can only be obtained at a plane extremely close to the lens) [12, 13] and the resolution is still limited by other factors, such as frequency dispersion and size of the resonant elements. In spite of these limitations, such an "imperfect lens" (later called *superlens*) can break the diffraction limit: this offers a possibility of significant improvement in imaging technology.

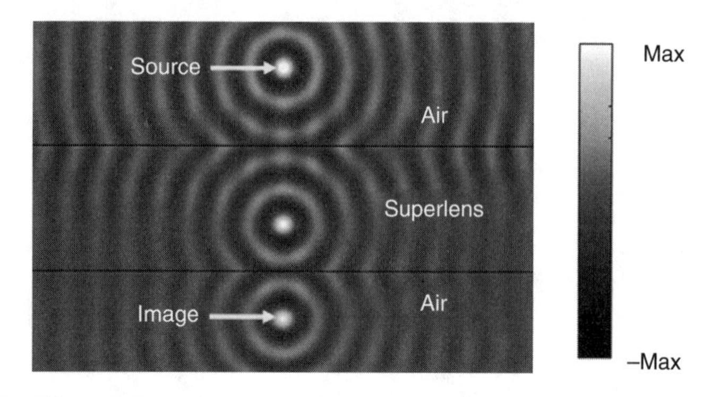

FIGURE 9.5 Imaging by a superlens with $\varepsilon \approx -1$ and $\mu \approx -1$. Intensity indicates the electric field of a continuous wave propagating from a point source through the superlens. The effect of negative space within the superlens is demonstrated. A very small material absorption is considered.

The following sections, focusing on the structural units in the nanoscale, include some basic concepts in the field of metamaterials. Since surface plasmon plays an important role in the design of metamaterials for application at optical frequencies, some basic properties of surface plasmon on metallic structures will first be described. Then, a more detailed discussion on how to design metamaterials using metallic structures is given. Two main plausible applications of metamaterials in high-resolution light imaging will be discussed. Finally, a brief outlook and conclusion will be given.

9.2 Surface Plasmon

At optical frequencies, the dominant electromagnetic response of metal is often modeled as free electrons confined within the volume of metal. These "free" electrons behave like plasma so that their quantized oscillations are called *plasmons*. Since free electrons can effectively shield electromagnetic fields, the oscillations of electrons can usually only concentrate on the surface of metal. Thus, plasmonics usually refers to the study of surface plasmons on metal. By controlling the shapes and sizes of metal, its electromagnetic response can vary significantly.

Before we discuss the surface properties, it is important to talk about the bulk properties of metal. The electromagnetic properties of a bulk metal are usually described by the dielectric function $\varepsilon(\omega)$ as a function of frequency ω. For simplicity, the dielectric function of a noble metal can be approximated as the famous Drude model.

$$\varepsilon(\omega) = 1 - \frac{\omega_p^2}{\omega^2} \tag{9.1}$$

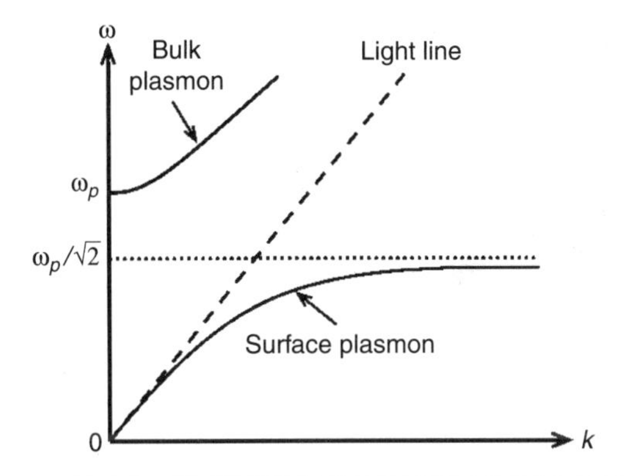

Figure 9.6 Dispersion relations of bulk plasmon and surface plasmon of a flat metal-air interface.

where ω_p is the plasma frequency. The dispersion relation for the electromagnetic wave within the metal is given by $k = \omega\sqrt{\varepsilon(\omega)}/c = \sqrt{\omega^2 - \omega_p^2}/c$, where c is the speed of light in a vacuum. Therefore, the propagating solution for bulk plasmon exists only when $\omega > \omega_p$ (see Fig. 9.6) and it is always above the light line in free space (i.e., $k < \omega/c$). In other words, the plasmon wavelength ($\lambda \equiv 2\pi/k$), is always longer than that of the free-space photon ($\lambda_0 \equiv 2\pi c/\omega$). However, when we have a thick slab of metal, the surface plasmon at the interface between metal and air (dielectric) becomes [14]

$$k_x = \frac{\omega}{c}\sqrt{\frac{\varepsilon_h \varepsilon(\omega)}{\varepsilon_h + \varepsilon(\omega)}} \tag{9.2}$$

where ε_h is the dielectric constant of air (dielectric) and k_x is the wave number of the surface plasmon. In this case, the propagating solution exists only when $\varepsilon(\omega) < -\varepsilon_h$. Using Eq. (9.1) as the material model, we can write the condition as

$$\omega < \frac{\omega_p}{\sqrt{\varepsilon_h + 1}} \tag{9.3}$$

In particular, for an air($\varepsilon_h \approx 1$)-metal interface, the surface plasmon band has an upper bound at $\omega = \omega_p/\sqrt{2}$. The schematictic dispersion relation and the surface charge-field distribution are shown in Figs. 9.6 and 9.7*a*, respectively. From Eq. (9.3), the dispersion relation for surface plasmon is always below the light line (i.e., $k_x > \omega/c$). Thus, the wavelength of such surface plasmon is shorter than that of the free-space photon. We will see that such subwavelength properties play an important role in superresolution photon imaging in the following sections.

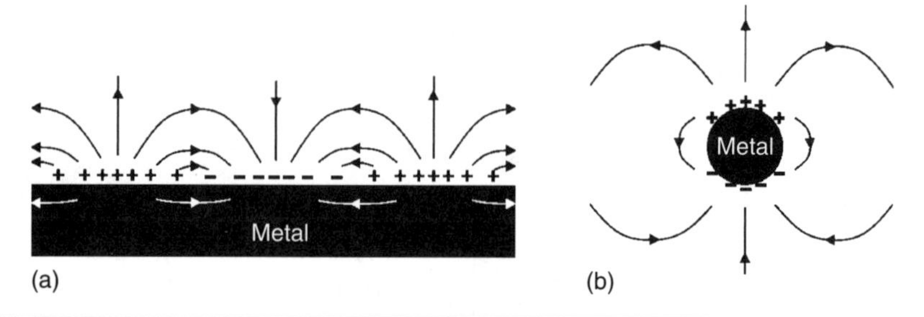

FIGURE 9.7 Schematic plot of the surface charge and field distribution of surface plasmon at (*a*) a flat metal-air interface and (*b*) a metal sphere. Panel (*b*) corresponds to a dipolar resonance.

As we have mentioned, properties of surface plasmon depend very much on the surface geometry of the metal. If we consider a finite cluster of metal, such as a metal nanoparticle, the surface plasmon frequencies will be discretized into separate values instead of a continuous spectrum as in Fig. 9.6. A simple example of this is a metal nanosphere. In the quasistatic approximation ($\lambda_0 << a$), where a is the size of the sphere, the surface plasmon (resonant) frequencies for such a metal nanosphere are given by [15]

$$\varepsilon(\omega_l) = -\frac{l+1}{l}\varepsilon_h \tag{9.4}$$

where $l = 1, 2, 3\ldots.$ Using Eq. (9.1), we get

$$\omega_l = \omega_p\sqrt{\frac{l}{(l+1)\varepsilon_h + l}} \tag{9.5}$$

At dipolar resonance ($l = 1$), the resonance condition becomes $\varepsilon(\omega) = -2\varepsilon_h$ and the plasmon frequency for a sphere embedded in air ($\varepsilon_h \approx 1$) can be approximated as $\omega_p/\sqrt{3}$ (see Fig. 9.7*b* for the charge-field distribution of a dipolar resonance). In another limit where l is very large, the plasmon frequency approaches $\omega_p/\sqrt{2}$, which is consistent with the upper bound of surface plasmon on a flat air-metal interface because the surface plasmon wavelength in this limit is too short to "see" the difference between a flat surface and a spherical surface.

For a finite metal cluster, the surface plasmon that corresponds to the lowest values of l, such as $l = 1$ or 2, is sometimes called the *localized surface plasmon* because the electromagnetic fields are localized on its finite surface, instead of propagating along a large surface. Such a localized mode can be excited by external light sources; this makes it possible to use metal nanoclusters for concentrating electromagnetic fields (or energy) from external sources.

In quasistatic approximation, the electric dipole moment for a metal sphere induced by an external incident electric field $\mathbf{E}^{inc}$ is [16]

$$\mathbf{p} = 4\pi\varepsilon_0 a^3 \left(\frac{\varepsilon(\omega) - \varepsilon_h}{\varepsilon(\omega) + 2\varepsilon_h} \right) \mathbf{E}^{inc} \tag{9.6}$$

where ε_0 is the free-space electric permittivity. We can see from Eq. (9.6) that when $\varepsilon(\omega) \to -2\varepsilon_h$ (dipolar resonance), the dipole moment diverges; this means that the local electric field at the surface of the sphere diverges too. In this case, the extinction cross section of such a metal nanosphere also has a delta-function peak with zero peak width (i.e., infinite quality factor). However, Eq. (9.6) is only a very rough formula. In the real situation, the absorption loss in the metal and the radiation loss to the environment limit the quality factor, broadening the extinction peak. Even in the quasistatic approximation, it has been shown that the quality factor is limited by approximately 80 when we take the losses into account [17].

The ability to enhance the local field at surfaces of metal nanostructures provides many plausible applications on biosensors. When molecules are attached or close to the metal surface, the surface plasmon can interact strongly with the molecules. One possible effect of the existence of molecules near metal surfaces is the shift in the extinction spectrum. This effect is usually understood as a change in the dielectric constant near the surface of metal. As we can see from Eqs. (9.2) and (9.4), the surface plasmon resonant conditions depend on the environment (ε_h); it has been proposed to use this effect for the detection of molecules (although the change in the extinction spectrum is not significant enough for a small number of molecules). Another more complicated but more sensitive phenomenon is surface-enhanced Raman scattering (SERS). Raman scattering is a process where the energy of a photon is transferred partly to mechanical vibration energy of molecules (or vice versa) so that the photon loses (or gains) energy after scattering. Since molecules can only vibrate at their own discrete energies (depending on the composition of the molecules), a frequency spectrum of Raman-scattered photons can be used as the fingerprint of a molecule. However, the Raman signal is usually very weak (many orders of magnitude smaller than the input energy), especially when the signal only comes from a single molecule. It has been, therefore, suggested to use metal nanostructures to enhance the local electric field near the molecules so that more photons can exchange energy with the molecules. The SERS enhancement factor due to field enhancement is generally approximated as $|E^{\mathrm{loc}}|^4/|E^{\mathrm{inc}}|^4$, where E^{loc} is the local electric field at the location of the molecule in the presence of metal nanostructure.

One of the most promising metal nanostructure for local field enhancement is the bowtie antenna (see Fig. 9.8). Since net charges

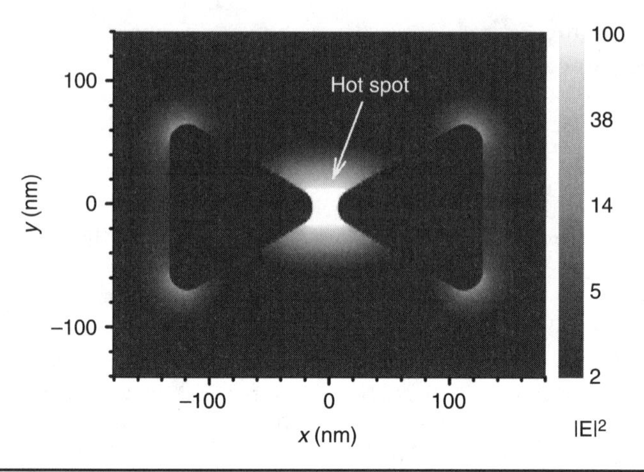

Figure 9.8 Local field enhancement by a metal nanobowtie antenna. Highest field is located near the gap between the two triangles.

tend to accumulate at the corners near the gap of the bowtie antenna, the field near the gap can be significantly enhanced. A typical enhancement factor for this kind of structures can be as high as $|E^{loc}|^4/|E^{inc}|^4 \sim 10^8$. Some numerical studies even claimed it to be $\sim 10^{12}$.

In addition to the field enhancement and resonant response in metal nanostructure, the subwavelength properties of surface plasmon can also be applied to subwavelength focusing and lithography. For example, a metal thin film can act like a double negative superlens.

9.3 Design of Metamaterials

As we have mentioned, metamaterials are nonuniform composites of some more simple materials, such as metals and dielectrics. Although the meaning of metamaterials is not restricted to using metals and dielectrics only, the material inclusions are usually chosen to possess very simple electromagnetic properties so that geometric effects from the arrangement of material inclusions can be clearly demonstrated. Therefore, we choose to focus on metamaterials that are solely composed of metals and dielectrics, this allows us to see a simpler relation between plasmonics and metamaterials.

9.3.1 Concept of Effective Medium

The most important concept in metamaterials is the concept of *effective medium*. (Without such a concept, the word *metamaterial* itself has almost no meaning, more than just a material composite with unusual electromagnetic properties.) Each metamaterial can be described by an effective medium with effective constitutive parameters (such as permittivity ε_{eff} and permeability μ_{eff}), similar to the usually constitutive parameters for conventional materials. The effective constitutive parameters should satisfy

$$\langle \mathbf{D} \rangle = \varepsilon_0 \varepsilon_{\text{eff}} \langle \mathbf{E} \rangle \tag{9.7}$$

and

$$\langle \mathbf{B} \rangle = \mu_0 \mu_{\text{eff}} \langle \mathbf{H} \rangle \tag{9.8}$$

where **E**, **B**, **D**, and **H** are, respectively, the electric field, magnetic field, displacement field, and associated magnetic field. The symbol, $\langle\ \rangle$, denotes a spatial average of the fields, and ε_0 and μ_0 are the free-space electric permittivity and magnetic permeability. It should be noted that there can be a weighting function within the averaging integral of $\langle\ \rangle$. More details can be found in [8, 18, 19]. We also note that ε_{eff} and μ_{eff} can be generalized to tensors for anisotropic metamaterials. Thanks to the current computer technology, we are able to calculate the fields for many complicated electrodynamic problems. In view of this, it looks as though as long as we know the exact solutions, we can obtain the constitutive relations like Eqs. (9.7) and (9.8). However, there are still several limitations in obtaining "good" effective parameters.

In order to ensure Eqs. (9.7) and (9.8) are valid, the spatial variations of the averaged fields have to be slow when compared to the material inhomogeneity. For example, let us consider a metamaterial that is periodic (with lattice constant α) in the x-direction. When there is a Bloch wave [$\mathbf{E}(x) = \mathbf{u}(x)e^{ik_B x}$, where $\mathbf{u}(x+a) = \mathbf{u}(x)$] propagating in the metamaterial, the Bloch wavelength $\lambda_B \equiv 2\pi/k_B$ should satisfy $\lambda_B >> a$. In addition, for an ideal effective medium, Eqs. (9.7) and (9.8) must be satisfied regardless of the excitation source. In other words, we should not be able to see any difference if the nonhomogenous metamaterial is replaced by a homogenous material of ε_{eff} and μ_{eff}. Such restricted conditions are indeed difficult to be fully satisfied. In spite of this, many current studies are based on obtaining the effective parameters from only one case of excitation, such as calculating from the transmission and reflection coefficients of a slab at normal incidence. There are some more rigorous ways to obtain effective parameters but the method of using transmission and reflection measurements is more practical in terms of experimental difficulty. Here, let us look into more details.

Using the standard S parameters (S_{11} for reflection and S_{21} for transmission), we can write the relations between incident wave ($E^{inc}e^{ik_0 x}$), transmitted wave ($E^T e^{ik_0 x}$), and reflected wave ($E^R e^{-ik_0 x}$) as $E^T = S_{21}E^{inc}$ and $E^R = S_{11}E^{inc}$. Suppose the slab of material is homogenous and has a thickness of d. By matching the boundary conditions at the two interfaces of the slab, the S parameters can be written as functions of the impedence (Z) and refractive index (n) of the material [20, 21, 22].

$$S_{11} = \frac{R_s(1 - e^{i2nk_0d})}{1 - R_s^2 e^{i2nk_0d}} \tag{9.9}$$

$$S_{21} = \frac{(1 - R_s^2)e^{ink_0d}}{1 - R_s^2 e^{i2nk_0d}} \tag{9.10}$$

where $R_s = (Z - 1)/(Z + 1)$. To determine the effective ε_{eff} $(= n/Z)$ and μ_{eff} $(= nZ)$ of the material, we have to solve the inverse problem. However, the solution is not unique mathematically [22, 23]:

$$Z = \pm\sqrt{\frac{(1 + S_{11})^2 - S_{21}^2}{(1 - S_{11})^2 - S_{21}^2}} \tag{9.11}$$

and

$$e^{ink_0d} = X \pm i\sqrt{1 - X^2} \tag{9.12}$$

where $X = [2S_{21}(1 - S_{11}^2 + S_{21}^2)]^{-1}$. Therefore, we have to reject the unphysical solutions by various restrictions on the parameters. For example, if the metamaterial is a passive medium, we require $\text{Re}(Z) > 0$ and $\text{Im}(n) > 0$. Even such requirements are, sometimes, not enough to uniquely determine the effective parameters. More details of the conditions to eliminate the unphysical solutions can be found in [22].

One important physical requirement that every effective parameter has to satisfy is the Kramers–Kronig relation, [15, 24] which is a result of the assumption of causality. It relates the real part of each effective parameter to its imaginary part [15].

$$\text{Re}(\chi(\omega)) = \frac{2}{\pi} PV \int_0^{\infty} \frac{\omega' \text{Im}(\chi(\omega'))}{\omega'^2 - \omega^2} d\omega' \tag{9.13}$$

and

$$\text{Im}(\chi(\omega)) = -\frac{2\omega}{\pi} PV \int_0^{\infty} \frac{\text{Re}(\chi(\omega'))}{\omega'^2 - \omega^2} d\omega' \tag{9.14}$$

where $\chi(\omega)$ is a response function. Here, $PV\int_0^{\infty} f(\omega')d\omega'$ means the principal-value integration of the function $f(\omega')$. For the electric response $[\langle \mathbf{P} \rangle = \varepsilon_0(\varepsilon_{\text{eff}} - 1)\langle \mathbf{E} \rangle]$, the Kramers–Kronig relation can be obtained by directly substituting $\chi(\omega) = \varepsilon_{\text{eff}}(\omega) - 1$ into Eqs. (9.13) and (9.14). A typical example of a response function satisfying the Kramers–Kronig relation is the Lorentzian function (see Fig. 9.9).

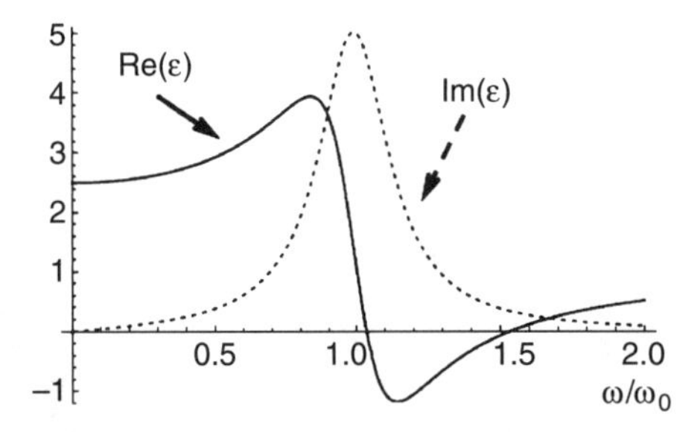

Figure 9.9 A typical dielectric function satisfying the Kramers–Kronig relation.

9.3.2 Effective Electric Permittivity (ε_{eff})

We have mentioned in the previous section that metal is an important element for constructing metamaterials because of its negative electric permittivity. However, naturally available metals have their own fixed electric permittivity functions under normal conditions. Therefore, the ability of tuning ε for specific applications is low. For example, most metals have too high plasma frequencies (in the ultraviolet range): this makes it difficult to have small negative values of ε at low frequencies. In view of this, metamaterials can provide much greater flexibility to tune the effective electric permittivity ε_{eff}.

To understand the way to reduce the plasma frequency of metal by constructing metamaterial, we consider a metamaterial that behaves as effective plasma so that its plasma frequency is given by

$$\omega_p = \sqrt{\frac{nq^2}{m\varepsilon_0}} \tag{9.15}$$

where n is the number of free charge carriers per unit volume, and q and m are the charge and the effective mass of each charge carrier, respectively. To lower the effective plasma frequency, we can reduce n, that is, reduce the volume occupied by metal. Although such a claim is not theoretically rigorous, it indeed provides us a more intuitive way to understand the change in effective plasma frequency.

One of the simplest models that can reduce the metal-occupied volume is an array of flat metal plates. Suppose we have a metamaterial consisting of alternating parallel layers of flat metal plates of thickness d_1 and dielectric plates of thickness d_2 (see Fig. 9.10). In the quasistatic limit, we can simply divide the problem into two: electric field parallel ($E_{\parallel}$) and perpendicular ($E_{\perp}$) to the metal/dielectric interfaces. In this case, we have two effective permittivities ($\varepsilon_{eff}^{\parallel}$ and $\varepsilon_{eff}^{\perp}$) for different electric-field directions. For an electric field parallel to the interfaces, we have $\varepsilon_{eff}^{\parallel} = \langle D_{\parallel}\rangle / \langle E_{\parallel}\rangle$. Using the boundary condition that $E_{\parallel}$ continuous

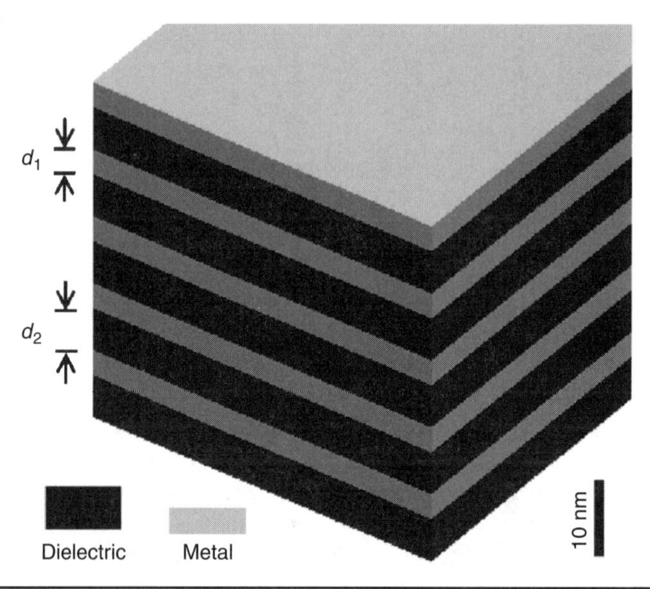

FIGURE 9.10 Schematic of a stack of parallel dielectric and metal plates.

and averaging the field simply by volume, we get $\langle E_{\parallel} \rangle = E_{\parallel,1} = E_{\parallel,2}$ and $\langle D_{\parallel} \rangle = f D_{\parallel,1} + (1-f) D_{\parallel,2} = [f\varepsilon(\omega) + (1-f)\varepsilon_h] \langle E_{\parallel} \rangle$ in the quasistatic limit, where the index 1 and 2 represent, respectively, the fields inside metal and dielectric, ε_h is the permittivity of this dielectric, and f is the filling fraction of metal. [*Note*: $f = d_1/(d_1 + d_2)$ in this example.] We thus have

$$\varepsilon_{\text{eff}}^{\parallel} = f\varepsilon(\omega) + (1-f)\varepsilon_h \tag{9.16}$$

Similarly, using the boundary condition that $D_{\perp}$ is continuous, we get

$$\frac{1}{\varepsilon_{\text{eff}}^{\perp}} = \frac{f}{\varepsilon(\omega)} + \frac{1-f}{\varepsilon_h} \tag{9.17}$$

Substituting the plasma model for metal (Eq. 9.1) into Eqs. (9.16) and (9.17), we have

$$\varepsilon_{\text{eff}}^{\parallel} = f + (1-f)\varepsilon_h - \frac{f\omega_p^2}{\omega^2} \tag{9.18}$$

and

$$\varepsilon_{\text{eff}}^{\perp} = \frac{\varepsilon_h(\omega^2 - \omega_p^2)}{\varepsilon_h f \omega^2 + (1-f)(\omega^2 - \omega_p^2)} \tag{9.19}$$

The formula looks complicated. For a simple interpretation, we now replace the dielectric with air ($\varepsilon_h = 1$), and finally, we obtain

$$\varepsilon_{\text{eff}}^{\|} = 1 - \frac{f\omega_p^2}{\omega^2} \tag{9.20}$$

and

$$\varepsilon_{\text{eff}}^{\perp} = 1 - \frac{f\omega_p^2}{\omega^2 - (1-f)\omega_p^2} \tag{9.21}$$

As expected, the reduction in the filling ratio of metal is similar to a reduction in the free-charge density; it lowers the effective plasma frequency. However, the truncation of bulk metal into pieces also induces new resonant conditions, making $\varepsilon_{\text{eff}}^{\perp}$ possess a Lorenzian-like resonant feature. Although the example shown here has a very specific geometry and material properties, it can already capture a qualitative understanding of the way to tune the electric permittivity, and the concept can also be applied to a two-dimensional (2D) array of metal wires (see Fig. 9.11) and a three-dimensional (3D) array of metal particles.

9.3.3 Effective Magnetic Permeability (μ_{eff})

Most substances are weakly diamagnetic (i.e., $0 < \mu < 1$ and μ is slightly smaller than 1). The typical value of $1 - \mu$ is smaller than 10^{-4}. Although superconductors can have $\mu = 0$, such strong diamagnetic effect so far can only be significant at very low temperature and it is limited to a static or low-frequency field. It is even more difficult to have $\mu < 0$ and $\mu >> 1$ at high frequencies, for example, optical

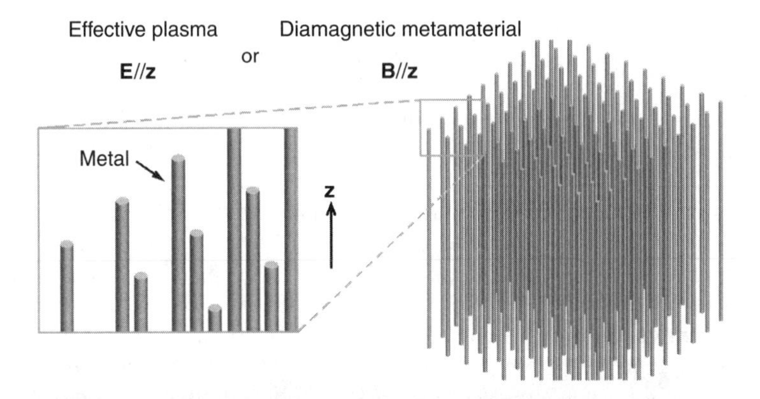

FIGURE 9.11 Two-dimensional arrays of parallel metal cylinders or wires. When the applied electric field is parallel to the *z*-axis, the metamaterial acts as an effective plasma medium with lowered plasma frequency. When the applied magnetic field is parallel to the *z*-axis, the metamaterial acts as a diamagnetic medium.

frequencies. In view of this, metamaterials provide a novel way to tune the magnetic permeability.

For tuning the value of the electric permittivity, metal is a good candidate because of its intrinsic plasma-like phenomena. Although no free magnetic charge has been found in metal, it is also very useful in tuning the value of the magnetic permeability. At normal conditions, bulk metal can be considered as possessing no magnetic response ($\mu = 1$) at high frequencies. However, when we mix metal clusters with dielectric or air, we can obtain an effective magnetic response. The simplest example is a sparse array of parallel metal cylinders with a low filling fraction f, embedded in a dielectric (see Fig. 9.11). Consider an electromagnetic wave propagating within such a metamaterial where the magnetic field is parallel to the axes of the cylinders. If we look from the outside of the metal cylinders and assume the metamaterial can only be connected to the outside medium in a way that no metal is touching the boundary of the metamaterial interface, the surface currents on the cylinders can provide an effective magnetization $\mathbf{M}'$, which can contribute to the $\langle \mathbf{H} \rangle$-field through [18]

$$\langle \mathbf{H} \rangle = \langle \mathbf{B} \rangle / \mu_0 - \mathbf{M}' \tag{9.22}$$

In this sense, the effective permeability is obtained by considering the **H**-field outside the cylinders and the **B**-field over the whole space occupied by the metamaterial [8, 19]. In the approximation that wave decays quickly inside metal (e.g., at infrared frequencies), the effective permeability in the long wavelength limit is simply given by [8, 19]

$$\mu_{\text{eff}} = 1 - f \tag{9.23}$$

where the unit magnetic permeability ($\mu_h = 1$) of the dielectric host is assumed.

The previous example demonstrated how nonmagnetic ($\mu = 1$) materials can be used to form metamaterials with $0 < \mu_{\text{eff}} < 1$. However, one of the most interesting types of metamaterials is the double-negative metamaterials, which are not included in this example. In order to include $\mu_{\text{eff}} < 0$, we have to employ metallic structures that support magnetic resonance. A typical example is the split-ring resonator (see Fig. 9.12), which consists of two concentric open rings. Such a structure can be interpreted as an LC resonator. The resonant response provides an effective magnetization $\mathbf{M}'$ that can be either the same as or opposite to the local magnetic field $\mathbf{B}_{\text{loc}}$ acting on the split-ring resonator. When $\mathbf{M}'$ is positive (negative), it reduces (increases) the average $\langle \mathbf{H} \rangle$ as well as μ_{eff}. If the resonance is strong enough, we can obtain $\mu_{\text{eff}} < 0$. However, since the sign of $\mathbf{M}'$ changes as it crosses the resonant frequency, it makes $\mu_{\text{eff}}(\omega)$ become

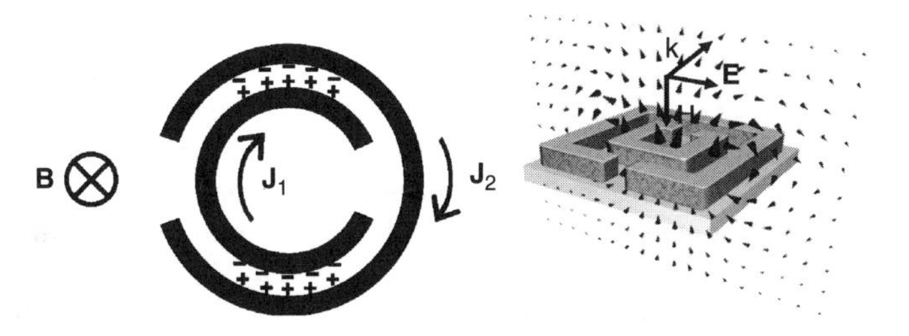

Figure 9.12 Schematic diagram of a split-ring resonator showing the current flow, charge distribution, and magnetic flux at magnetic resonance. (Reproduced with permission from Reference [44]. Copyright 2008, Materials Research Society.)

strongly dispersive [i.e., $\mu_{eff}(\omega)$ is sensitive to frequency]. A typical $\mu_{eff}(\omega)$ takes the following Lorenzian form (similar to Fig. 9.9 for dielectric function):

$$\mu_{eff} = 1 - \frac{F\omega^2}{\omega^2 - \omega_0^2 + i\omega\Gamma} \tag{9.24}$$

where F represents the strength of the resonance and ω_0 and Γ are, respectively, the resonant frequency and the peak width.

9.3.4 Double Negativity

The DNG medium is one of the most widely studied topics in the field of metamaterials because of its ability to focus light to form images in sub-wavelength resolution, that is, break the diffraction limit. Guided by our knowledge in designing single-negative metamaterials, we can design DNG metamaterials very simply, except that we have to match the resonant frequencies and take into account the coupling between the two kinds of building blocks when they are put close to each other. The first design of DNG metamaterials was demonstrated experimentally by Shelby et al. [25]. It consists of metallic split rings (for $\mu_{eff} < 0$) and wires (for $\varepsilon_{eff} < 0$). A schematic diagram of one unit cell is shown in Fig. 9.13.

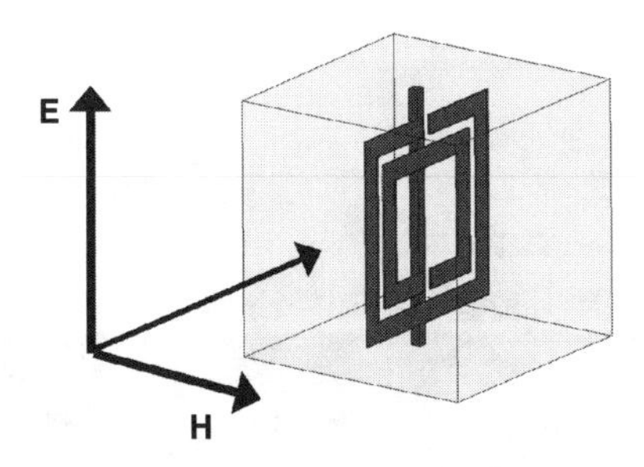

Figure 9.13 Schematic drawing of a unit cell of DNG metamaterial consisting of a metal split ring and metal wire. Arrows show the designated incident-field polarizations.

Although the building blocks (in millimeter scale) were so large that the designated working frequency was in the microwave range, this was the first time negative refraction in double-negative metamaterials was satisfactorily demonstrated experimentally. Such an experimental result suggests that metamaterials are realizable, and since then, research interests in metamaterials have grown dramatically. So far, the concept of negative refraction in a DNG medium has been widely examined at microwave domains [25–29]. It is natural to ask how to realize metamaterials at even higher frequencies. While some researchers have shown that the phonon polaritons in semiconductors offer the opportunity to engineer the negative permittivity further in the infrared frequencies [30, 31], many other researchers have been trying to reduce the scale to nanometers by simply modifying the microwave version. However, fabrication of those 3D nanostructures is an extremely challenging task. Therefore, most experiments on these nanostructures so far only focus on a single layer.

In designing 3D double-negative metamaterials to ease the fabrication processes, a very simple fishnet-like structure has been proposed recently [32]. It is fabricated by stacking layers of metal and dielectric thin films and drilling 2D arrays of identical holes on it. The size of the hole is comparable to the periodicity so that the remaining materials form a fishnet-like metamaterial (see Fig. 9.14). It has been verified experimentally that the fishnet structure supports negative refraction at an optical frequency and even visible light [33, 34].

Although the fishnet structure is a simple metamaterial in terms of its geometry and fabrication process, the effective parameters are highly anisotropic; this means that double-negative effective parameters are valid only for limited incident light directions. Therefore, it may not be possible to support large-angle negative refraction and reconstruct the evanescent fields to form a high-resolution image. Future work on improving the fishnet structure is still necessary.

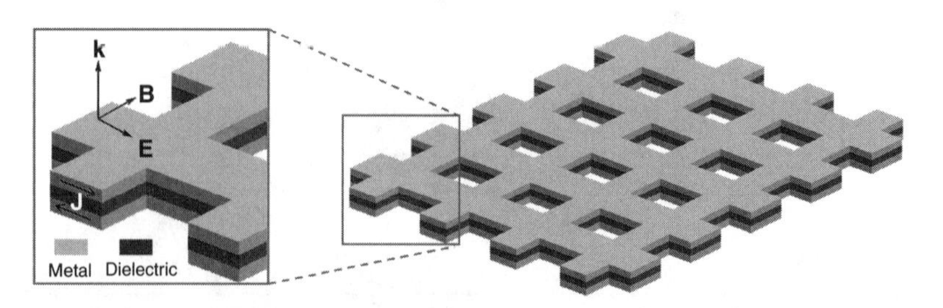

FIGURE 9.14 Schematic diagram of a metal-dielectric-metal fishnet structure. The structure is designed to support magnetic resonance.

9.4 Imaging and Lithography: Breaking the Diffraction Limit

After the first seminal prediction of the intriguing property of the "perfect lens" by Pendry in 2000, many groups launched ambitious efforts to explore the superlensing effect. Using Finite Difference Time Domain (FDTD) simulation, Ziolkowski et al. confirmed theoretically that a thin slab of dispersive DNG metamaterial can indeed focus the light spot down to 1/6 of a vacuum wavelength [35]. Pendry et al. then demonstrated the power of conformal transformation to potentially expand the design of superlenses on curved geometries [36]. The new physical constraints on the ultimate imaging resolution are central to the future application of this superlens. To this end, Feise et al. revealed that even a thin overlayer would influence the response of the DNG medium through a frequency shift of surface modes and the nonreflecting wave, which set up a lower bound of the smallest resolvable features [37]. In addition to geometry, the material properties also pose stringent requirements on the superlensing effect. This was first qualitatively explored by Shamonina et al. [38] and then quantitatively discussed by Smith et al. [39]. Experimental evidence is vital to confirm the superlens theory. Using 2D LC networks to represent a thin layer of metamaterial, microwave experiments of the in-plane propagation and lensing effect has been reported recently by Iyer et al., [40], whereas Wiltshire et al. applied the concepts to realize high-resolution magnetic resonance imaging at 21 Mhz [41]. However, due to the challenges in fabrication and material constraints, experiments at terahertz and optical frequency are still open for exploration in the long-term.

In spite of these limitations, subwavelength focusing has been shown to exist in some simple structures that do not need to employ the concept of DNG metamaterials. Two types of recently realized artificial materials that can be employed in superresolution imaging and lithography are the metal thin-film superlens and hyperlens.

9.4.1 Thin-Film Superlens

The thin-film superlens was demonstrated experimentally by Fang et al. for the first time in 2005 [42]. The fact that such a metal thin film can break the diffraction limit can be explained by the surface plasmon coupling between the two metal-air interfaces [43]. The surface plasmon offers a resonant coupling that enhances the tunneling from one side to the other side of the thin film by evanescent fields. Such an observed phenomenon is also considered as a demonstration of focusing by DNG metamaterials. It is because silver itself is an ENG medium below plasma frequency and, in the quasistatic limit, the contribution from magnetic permeability to the transfer function of the thin film

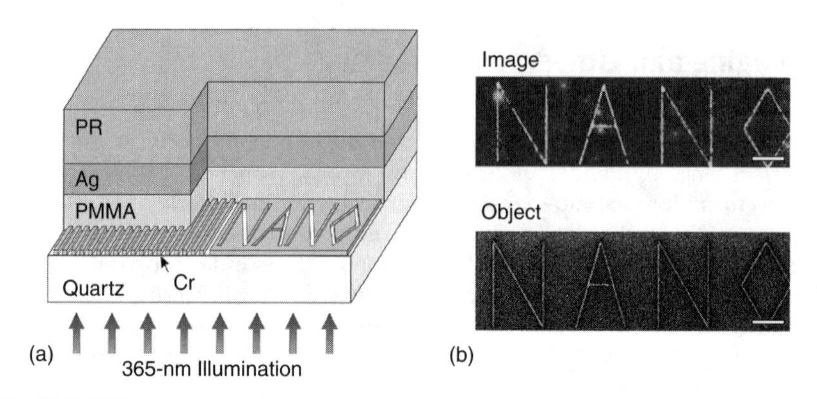

Figure 9.15 Experimental setup (*a*) and the corresponding imaging result (*b*) of the optical superlensing experiment by Fang et al. [42]. (Reprinted with permission from Reference [42]. Copyright 2005, American Association for the Advancement of Science.) Scale bar in (*b*) is 2 μm. Abbrevations: Cr, chrome; PMMA, poly(methylmethacrylate); and PR, photoresist.

can be neglected for one polarization of light, so that there is no notable difference between an image formed by ENG and DNG materials [10].

However, some important issues that limit the realistic resolution of the superlens cannot be seen from the quasistatic approximation. Those issues include material absorption loss and radiation loss due to surface roughness. The absorption loss is limited by materials. One of the lowest absorptive metals is silver, so it can be chosen for better demonstration. However, it is not easy to get an extremely smooth silver surface when the film is thin. In the experiment by Chaturvedi et al., a smooth 35-nm-thick silver thin film was used to form a focused image resolving an object of a width = $\lambda/12$, where λ is the wavelength [44]. Details of the configuration and result of the experiment are shown in Fig. 9.15.

Such a thin-film superlens could be the simplest lens that can achieve subwavelength resolution. However, the working wavelength for the lens cannot be easily tuned because, under normal conditions, the permittivity function of silver is fixed. Although one can choose different materials as a spacer next to the silver film for changing the plasmon wavelength, it is still limited to wavelengths close to ultraviolet light.

9.4.2 Hyperlens

Hyperlens was first proposed in 2006 [45, 46]. It is a curved stack of alternate layers of materials with positive and negative permittivities (similar to the stack shown in Fig. 9.10). Although each layer in the stack is curved, the imaging ability of such a hyperlens is usually

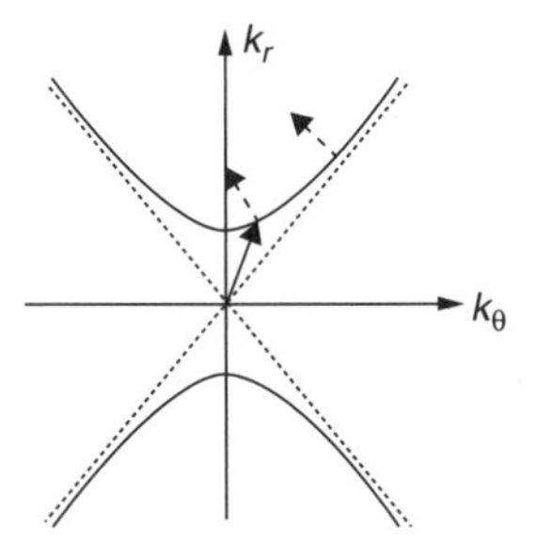

Figure 9.16 Fixed-frequency contours (solid curves) of the dispersion relation of a stack of metal and dielectric parallel layers designed for a hyperlens. Solid arrows indicate the directions of incident plane waves while dashed arrows indicate the directions of the corresponding refracted beams.

understood from the dispersion relation of the flat stack. The flat stack is considered as a metamaterial with its effective permittivity given by Eqs. (9.16) and (9.17). With such effective $\varepsilon_{\text{eff}}^{\parallel}$ and $\varepsilon_{\text{eff}}^{\perp}$, the fixed-frequency contour in the 2D reciprocal space along the stack interface is a hyperbola (see Fig. 9.16). It was shown that if one of the effective permittivity $\varepsilon_{\text{eff}}^{\parallel}$ in Eq. (9.17) is close to 0, the field distribution on each layer can be transferred point by point without distortion (provided that the material absorption can be neglected). Figure 9.17*a* is a simulated result demonstrating the point-by-point transfer of fields. As long as the bended stack has a large bending radius, such point-by-point transferring of a field pattern should still be there. With such an assumption, if the stack is bended into a large semicircle (as in Fig. 9.17*b*), closely spaced sources located at the inner surface of the hyperlens are then magnified after they propagate toward the outer surface.

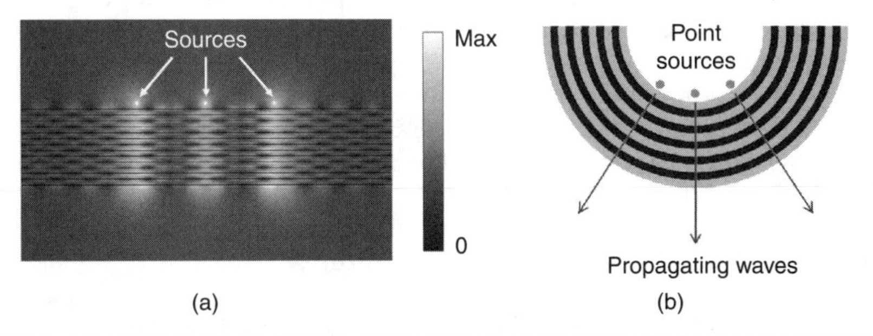

Figure 9.17 Point-by-point transfer of a field pattern by (*a*) a flat hyperlens and (*b*) a bended hyperlens. Panel (*a*) shows the simulated field pattern for three-point sources next to the hyperlens. Panel (*b*) shows the schematic plot illustrations of how a bended hyperlens converts subwavelength source signals to propagating waves.

At first glance, it looks like a hyperlens is similar to a thin-film superlens but with a more complicated structure. However, a hyperlens has an advantage over a usual superlens by its ability to convert evanescent waves to propagating waves (see Fig. 9.17*b*). This is very useful in imaging applications because we do not have to place a detector (which can strongly modify the local field) very close to the surface of the lens. Also, such a hyperlens, allowing for far-field imaging of the local field, may offer a possibility to make a real-time optical microscope with subwavelength resolution. However, one limitation that a hyperlens has still not solved is that the object that we want to image has to be almost touching the inner surface of the hyperlens.

9.5 Outlook

We have discussed in this chapter several common aspects of plasmonics and metamaterials that can play an important role in nanobiophotonics. Yet a lot more phenomena are quickly emerging in the frontier of these two exciting and novel research fields. Such examples include subwavelength waveguiding, extraordinary transparency, enhanced nonlinear phenomena, and more recently, transformation media (including the invisibility cloak). These novel concepts and techniques, also in their infancy, may become the future powerhouse in nanobiophotonics research. In particular, controlling surface plasmon and photon in the nanoscale could be essential for a set of nonlinear phenomena in biological nanostructures, such as multiphoton fluorescence, second-harmonic generation, and hyper-Raman scattering, offering unique functional imaging tools for the corresponding fundamental research.

The interaction of photons with biological matter can also lead to local heating (by converting photons to phonons) or manipulation (a well-known example is the optical tweezer for manipulation and measurement of molecules). Exploiting strong local field enhancement associated with surface plasmon resonance, might give rise to remarkable thermal effects and photonic forces. The effects may be useful in targeted therapy at the cellular scale: on the other hand, they may disrupt the physiological environment required for functional imaging. Therefore, the interaction between surface plasmon and biological matter can become a topic for important research in the near future.

9.6 Conclusion

Plasmonics and metamaterials are two new research fields based on the same sciences and technologies; this makes it difficult to define a clear distinction between the two fields. Recent research progress in the two expanding fields is unexpectedly rapid. Numerous novel phenomena have been theoretically discovered in the past decade. Thanks to the advanced computation and fabrication technologies, many of those interesting phenomena have also been experimentally verified. Some of the phenomena have been proposed for biomedical applications in the nanoscale, such as plasmonic biosensing and plasmonic/metamaterial superresolution imaging, and some experimental breakthroughs have been made recently toward true biomedical applications. It is expected that the two research fields will continue the rapid growth and make a significant impact on nanobiophotonics.

This chapter has briefly introduced some basic concepts in the fields of plasmonics and metamaterials. Many examples were explained in very ideal situations for the sake of simplicity. Other theoretical and experimental details were not able to be included due to their complexity and limited space. Readers may find more details in other related chapters of this book.

References

1. X. Zhang and Z. Liu, "Superlenss to Overcome the Diffraction Limit," *Nature Materials* 7:435–441, 2008.
2. J. N. Anker, W. P. Hall, O. Lyandres, N. C. Shah, J. Zhao, and R. P. Van Duyne, "Biosensing with Plasmonic Nanosensors," *Nature Materials* 7:442–453, 2008.
3. J. Homola, S. S. Yee, and G. Gauglitz, "Surface Plasmon Resonance Sensors: Review," *Sensors and Actuators B* 54:3–15, 1999.
4. K. Kneipp, H., Kneipp, I. Itzkan, R. R. Dasari, and M. S. Feld, "Surface-Enhanced Raman Scattering and Biophysics," *J. Physics C* 14:R597–R624, 2002.
5. A. V. Kabashin, P. Evans, S. Pastkovsky, W. Hendren, G. A.Wurtz, R. Atkinson, R. Pollard, V. A. Podolskiy, and A. V. Zayats, "Plasmonic Nanorod Metamaterials for Biosensing," *Nature Materials* 8:867–871, 2009.
6. M. M. I. Saadoun and N. Engheta, "A Reciprocal Phase Shifter Using Novel Pseudochiral or Medium," *Microwave and Optical Technology Letters* 5:184, 1992.
7. J. B. Pendry, A. J. Holden, W. J. Stewart, and I. Youngs, "Extremely Low-Frequency Plasmons in Metallic Mesostructures," *Physical Review Letters,* 76:4773, 1996.
8. J. B. Pendry, A. J. Holden, D. J. Robbins, and W. J. Stewart, "Magnetism from Conductors and Enhanced-Nonlinear Phenomena," *IEEE Transactions on Microwave Theory Techniques* 47:2075, 1999.
9. D. R. Smith, W. Padilla, D. C. Vier, S. C. Nemat-Nasser, and S. Schultz, "Composite Medium with Simultaneously Negative Permeability and Permittivity," *Physical Review Letters* 84:4184, 2000.
10. J. B. Pendry, "Negative Refraction Makes a Perfect Lens," *Physical Review Letters* 85:3966, 2000.
11. V. C. Veselago, "The Electrodynamics of Substances with Simultaneously Negative Values of ε and μ," *Soviet Physics Uspekhi* 10:509, 1968.
12. G. W. 't Hooft, "Comment on 'Negative Refraction Makes a Perfect Lens,'" *Physical Review Letters* 87:249701, 2001.
13. J. B. Pendry, "Pendry Replies," *Physical Review Letters* 87:249702, 2001.
14. H. Raether, *Surface Plasmons on Smooth and Rough Surfaces and on Gratings, Vol. 111 of Springer Tracts in Modern Physics,* Springer-Verlag, Berlin, 1988.
15. C. F. Bohren and D. R. Huffman, *Absorption and Scattering of Light by Small Particles*, Wiley, New York, 1983.

16. J. D. Jackson, *Classical Electrodynamics*, 3rd ed., Wiley, New York, 1998.
17. F. Wang and Y. R. Shen, "General Properties of Local Plasmons in Metal Nanostructures," *Physical Review Letters*, 97:206806, 2006.
18. T. C. Choy, *Effective Medium Theory, Principles and Applications*, Oxford University Press, Oxford, 1999.
19. X. H. Hu and C. T. Chan, "Diamagnetic Response of Metallic Photonic Crystals at Infrared and Visible Frequencies," *Physical Review Letters* 96:223901, 2006.
20. A. M. Nicolson and G. F. Ross, "Measurement of the Intrinsic Properties of Materials by Time-Domain Techniques," *IEEE Transactions on Instrumentation and Measurement* 19:377, 1970.
21. J. A. Kong, *Electromagnetic Wave Theory* (*EMW*), Cambridge, MA, 2000.
22. X. Chen, T. M. Grzegorczyk, B. -I. Wu, J. Pacheco, Jr., and J. A. Kong, "Robust Method to Retrieve the Constitutive Effective Parameters of Metamaterials," *Physical Review E* 70:016608, 2004.
23. D. R. Smith, S. Schultz, P. Markoš, and C. M. Soukoulis, "Determination of Effective Permittivity and Permeability of Metamaterials from Reflection and Transmission Coefficients," *Physical Review B* 65:195104, 2002.
24. J. S. Toll, "Causality and the Dispersion Relation: Logical Foundations," *Physical Review* 104:1760–1770, 1956.
25. R. A. Shelby, D. R. Smith, and S. Schultz, "Experimental Verification of a Negative Index of Refraction," *Science* 292:77–79, 2001.
26. C. G. Parazzoli, R. B. Greegor, K. Li, B. E. C. Koltenbah, and M. Tanielian, *Physical Review Letters* 90:107401, 2003.
27. R. W. Ziolkowski, "Design, Fabrication, and Testing of Double Negative Metamaterials," *IEEE Trans. Antennas and Propagation*, 51:1516, 2003.
28. A. K. Iyer, P. C. Kremer, and G. V. Eleftheriades, "Experimental and Theoretical Verification of Focusing in a Large, Periodically Loaded Transmission Line Negative Refractive Index Metamaterial," *Optics Express* 11:696, 2003.
29. P. F. Loschialpo, D. L. Smith, D. W. Forester, F. J. Rachford, and J. Schelleng, "Electromagnetic Waves Focused by a Negative-Index Planar Lens," *Physical Review E* 67:025602, 2003.
30. P. J. Hesketh, J. N. Zemel, and B. Gebhart, "Organ Pipe Radiant Modes of Periodic Micromachined Silicon Surfaces," *Nature* 324:551, 1986.
31. A. V. Shchegrov, K. Joulain, R. Carminati, and J. J. Greffet, "Near-Field Spectral Effects due to Electromagnetic Surface Excitations," *Physical Review Letters* 85:1548, 2000.
32. G. Dolling, C. Enkrich, M. Wegener, C. M. Soukoulis, and S. Linden, "Observation of Simultaneous Negative Phase and Group Velocity of Light," *Science* 312:892, 2006.
33. G. Dolling, C. Enkrich, M. Wegener, C. M. Soukoulis, and S. Linden, "A Low-Loss Negative Index Metamaterial at Telecommunication Wavelengths," *Optics Letters* 31:1800, 2006.
34. G. Dolling, M. Wegener, C. M. Soukoulis, and S. Linden, "Negative-Index Metamaterial at 780-nm Wavelength," *Optics Letters* 32:53, 2007.
35. R. W. Ziolkowski and E. Heyman, "Wave Propagation in Media Having Negative Permittivity and Permeability," *Physical Review E* 64:056625, 2001.
36. J. B. Pendry and S. A. Ramakrishna, "Near-Field Lenses in Two Dimensions," *Journal of Physics: Condensed Matter* 14:8463, 2002.
37. M. W. Feise, P. J. Bevelacqua, and J. B. Schneider, "Effects of Surface Waves on the Behavior of Perfect Lenses," *Physical Review B* 66:035113, 2002.
38. E. Shamonina, V. A. Kalinin, K. H. Ringhofer, and L. Solymar, "Imaging, Compression and Poynting Vector Streamlines for Negative Permittivity Materials," *Electronics Letters* 37:1243, 2001.
39. D. R. Smith, D. Schurig, M. Rosenbluth, S. Schultz, S. A. Ramakrishna, and J. B. Pendry, "Limitations on Subdiffraction Imaging with a Negative Refractive Index Slab," *Applied Physics Letters* 82:1506, 2003.
40. A. K. Iyer, P. C. Kremer, and G. V. Eleftheriades, "Experimental and Theoretical Verification of Focusing in a Large, Periodically Loaded Transmission Line Negative Refractive Index Metamaterial," *Optics. Express* 11:696, 2003.

41. M. C. K. Wiltshire, J. B. Pendry, I. R. Young, D. J. Larkman, D. J. Gilderdale, and J. V. Hajnal, "Microstructured Magnetic Materials for RF Flux Guides in Magnetic Resonance Imaging," *Science* 291:849, 2001.
42. N. Fang, H. Lee, C. Sun, and X. Zhang, "Sub-Diffraction-Limited Optical Imaging with a Silver Superlens," *Science* 308:534–537, 2005.
43. N. Fang, Z. Liu, T. -J. Yen, and X. Zhang, "Regenerating Evanescent Waves from a Silver Superlens," (invited paper), *Optics Express* 11:682–687, 2003.
44. P. Chaturvedi, K. -H. Hsu, S. Zhang, and N. X. Fang, "New Frontiers of Metamaterials: Design and Fabrication," *MRS Bulletin* 33:915–920, 2008.
45. A. Salandrino and N. Engheta, "Far-Field Subdiffraction Optical Microscopy Using Metamaterial Crystals: Theory and Simulations," *Physical Review B* 74:075103, 2006.
46. Z. Jacob, L. V. Alekseyev, and E. Narimanov, "Optical Hyperlens: Far-Field Imaging beyond the Diffraction Limit," *Optics Express* 14:8247, 2006.

PART III

Current Research Areas

CHAPTER 10

Infrared Spectroscopic Imaging: An Integrative Approach to Pathology

Michael J. Walsh and Rohit Bhargava

Department of Bioengineering
Micro and Nanotechnology Laboratory
Beckman Institute for Advanced Science and Technology
University of Illinois at Urbana–Champaign

10.1 Introduction

Pathology is regarded as the gold standard for disease diagnosis and is a critical component of cancer research and clinical applications. Fourier transform infrared (FT-IR) imaging represents a novel approach as an alternative to conventional pathology techniques, with a growing wealth of literature demonstrating its power. We refer throughout the chapter to other reviews and book chapters for further details on specific application areas and instrumentation.

10.1.1 Cancer Pathology

Almost 1.4 million new cancer diagnoses are made each year in the United States alone [1]. The two most prevalent cancers in the Western world are prostate cancer in men and breast cancer in women which account for ~30% of all cases. If preliminary screening results for these cancers are abnormal [2, 3], a biopsy is conducted to detect and grade or rule out cancer [4]. Histologic assessment within biopsies is the definitive diagnosis of breast and prostate cancer as well as most other cancers [5], for both initial diagnosis and therapy guidance. Current pathology has an urgent need to improve the information quality from diagnosis; this may be addressed by new technology. Current pathological cancer screening approaches are biased toward being sensitive to maximize the detection of the disease; however, this is at the cost of specificity. As a result, there are a large numbers of biopsies (e.g., > million patients per year and 1 to 23 biopsies per patient annually for just prostate and breast cancer alone [6]), which are acquired and need expert examination. This leads to problems due to economic and time constraints [7]. The availability of automated evaluation methods could be a potential and great help by reducing burdens on pathologists and pressure on the health care system. Over 80% of the biopsies are benign and contain no cancer. Current pathology requires manual examination of all samples; however, this is limited in speed and throughput. This may lead to needless delays and clinically significant anxiety in benign patients. The diversion of resources to examine all patients limits the resources available for truly suspicious cases. The removal of samples, which are diagnostically useless (for example, the tissue section contains no epithelium), from the diagnostic chain would prove economical and presents an opportunity for improvement (Fig. 10.1).

There are a number of problems associated with the current practice of pathology. The subjective nature of human evaluation leads to variation in diagnoses [8], which confounds decision making for therapy [9]. Second opinions [10] improve assessment and are cost-effective [11]; this eventually leads to lower health care costs as well as less lost wages, morbidity, or litigation [12]. Hence, objective and consistent first or second opinions are very desirable. Improvements in diagnosis are important; however, even more lacking is the availability of prognostic information. In the case of prostate cancer it is estimated that only 20% of prostate cancer patients will progress to death from the disease. Long-term studies demonstrate statistically in populations that a vast majority of lesions will not prove lethal in a man's lifetime and thus do not require treatment [13–15]. Current pathological methods do not allow for the identification of individuals who are actually at risk of death (lethal disease) from individuals who will not die of prostate cancer. Absent a definitive tool for finding nonlethal cancer, the majority of patients (~97% in the United States)

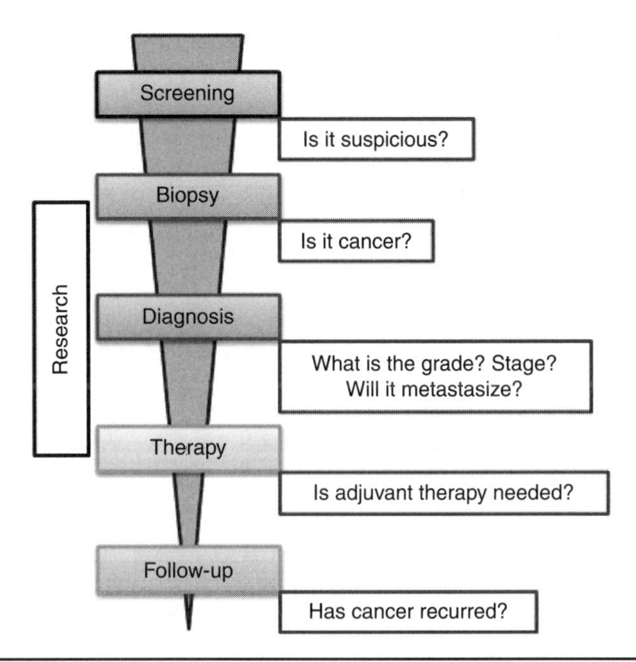

Figure 10.1 The central column shows the clinical flow for detecting to curing cancers, with the important questions for pathology highlighted in boxes. These and analogous efforts in research can be aided by infrared (IR)-spectroscopic imaging.

will choose to be treated as if they had a lethal disease. Twenty people undergo therapy for every life saved [16]. Unfortunately, most treated individuals do not truly benefit but only experience the significant downside of aggressive therapy, including personal and familial stress, financial losses [17], and medical complications. Resulting incontinence and impotence [18, 19] adversely affect quality of life in anywhere from 20% to 80% of men. Alternatively, men may choose "watchful waiting." This approach, however, does risk the advancement of prostate cancer, which may become metastatic. Nevertheless, the opportunity exists for spectroscopic technology to contribute either by itself or as part of a multimodal approach.

10.1.2 Current Practices in Pathology

To attend to the current limitations and problems of pathological approaches, high-throughput, automated and objective tools—both for clinical practice and for research—are urgently needed and represent an opportunity for spectroscopy and spectroscopic imaging. In current pathology, a biopsy sample is first fixed, with formalin for example, to prevent degradation. Next, it is embedded in a solid medium, typically paraffin wax, to allow a well-preserved two-dimensional (2D) section to be obtained. Typically, these sections are 3 to 7 μm thick and placed on a glass slide. The slide is then washed and stained in a specific protocol and prepared for viewing under a microscope. The hematoxylin and eosin (H&E) stain is commonly employed, staining protein-rich regions pink and nucleic acid-rich regions of the tissue blue. A pathologist is trained to recognize the morphology of specific cell types and their alterations that indicate disease. As over 85% of internal human cancers are epithelial in origin, morphologic patterns of epithelial cells are diagnostic of tissue health or disease. Typically, epithelial cells line three-dimensional (3D) ducts in tissue and appear as circular regions in thin sections. Patterns of distortions in organizations and shapes of these circles as well as cellular details provide evidence of cancer and characterize its severity. Recognizing disease in current pathology is thus a manual process of spatial pattern recognition. The cause of various problems in pathology is often attributed to the manual nature of examination. Variations in the appearance of tissue, with stain composition, with the skill of a histotechnologist, and under different processing conditions, further contribute to inaccurate diagnoses. The lack of a consistent presentation and preparation of artifacts has been shown to confound automated image analysis [20]. Just as it is for human diagnoses, performance is especially poor for borderline cases and mimickers of disease are problematic. Therefore, a robust means of automatically detecting epithelium and correlating its spatial patterns to disease is highly desirable but not yet clinically practical (Fig. 10.2).

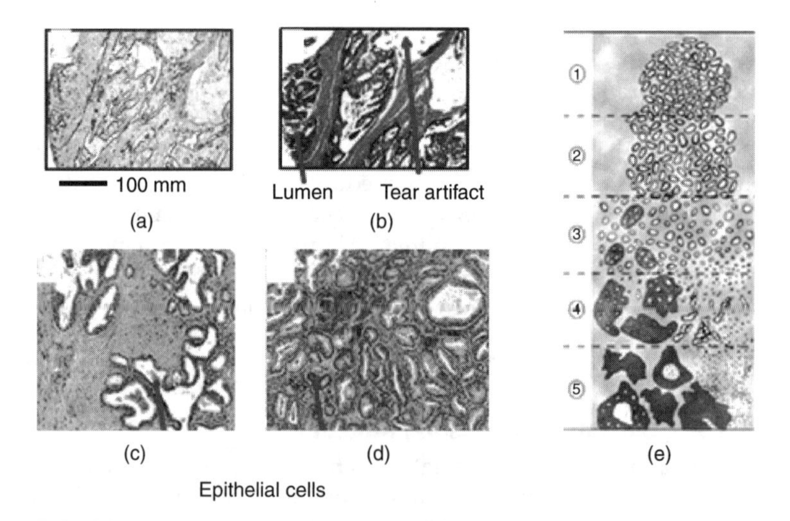

FIGURE 10.2 Prostate histology and carcinoma grading. (*a*) Unstained tissue has little contrast, requiring staining for bright field microscopy (*b*). Epithelial cells define empty regions (lumen) that are (*c*) well-differentiated in normal tissue and (*d*) poorly differentiated in cancerous tissue. (*e*) If lumen are nearly regular and well-spaced, tissue is benign. Increasing disruption of this structure leads to increasing the grade of tumors, from 1 to 5, on the Gleason scale. (Reproduced from R. Bhargava, *Anal Bioanal Chem* 389:1155–1169).

10.1.3 Molecular Pathology

Current pathology involves the visual examination of the morphology of cells and tissue architecture. Biopsies obtained are typically small; this can make the visualization and diagnosis difficult due to limited availability of material. Other confounding factors due to tissue processing can also lead to diagnostic errors. An important consideration is that the disease state of tissue may not be adequately identifiable on the basis of morphology alone. Thus pathologists are increasingly using molecular techniques to aid in characterizing tissue and for diagnosis of disease. The most common molecular technique employed by pathologists is immunohistochemical (IHC) staining which allows for both target presence and its spatial distribution. IHC staining is more important for complicated cases or those on the margins of decisions or for small tissue samples [21]. However, IHC staining also relies on the skill of the operator and the experimental conditions and methods to make an accurate diagnosis. The variance associated with IHC staining has led to it primarily being used for confirmatory purposes, rather than a primary diagnostic standard.

A number of studies have shown that a combination of spatial and biochemical information [22] may prove to be useful in pathology [23]. IHC staining requires expertise, is time consuming and expensive, and often requires multiple sections of the tissue. An alternative analysis of chemical content and changes can be achieved using label- or probe-free methods. Spectroscopy can allow for the identification of the inherent biochemistry of tissue in a fast and low-cost way. The challenge with spectroscopy is to identify and relate this spectral information to the known pathology. The combination of spectroscopy and imaging has been emerging for some time and has been termed *chemical imaging*. Among spectroscopic techniques, mid-infrared (MIR) absorption and Raman spectroscopy are chemically attractive in that they can easily harness the power of optics for imaging while providing richer molecular detail than near-infrared (NIR), visible, or ultraviolet absorption, scattering, or fluorescence spectroscopy.

10.2 FTIR Spectroscopy

10.2.1 Point Spectroscopy and Imaging

Fourier transform infrared (FT-IR) spectroscopy is a tool commonly used in the physical sciences and has numerous biochemical applications. Point FTIR spectroscopy is the tool of choice for low-cost instrumentation, fast data acquisition, and collection of high-quality spectra. Although point spectra are useful, they typically sample only a single area (typically a region of 7 to 50 μm) and their application is likely limited to cases in which other imaging technology or a pathologist requires chemical information from a single region. Recent advances in

IR imaging have made the development of FTIR imaging in a clinical setting feasible [24]. The use of FTIR imaging requires no prior knowledge to collect data or an image. The entire sample area may be imaged, and computer algorithms can be used to identify cell types or the presence of disease. This approach may be more appropriate for use in pathology as it allows for the identification of epithelium or a tumor in a section of tissue as a fully automated tool.

FT-IR spectroscopy in an imaging mode is rapidly emerging as a routine technique and is being increasingly applied for biomedical analyses. FTIR imaging can be employed to measure both chemical and structural changes without the use of any dyes or probes. FTIR provides similar imaging information in a format familiar to pathologists with the added benefit of chemical information. In conventional pathology a 2D image from stained tissue is obtained; in FTIR imaging a "data cube" is acquired without using dyes. An IR absorption spectrum, known as a *molecular fingerprint* of the material, which can subsequently be correlated to biomedical knowledge of the cell type (histology), and disease state (pathology) is obtained at every pixel. The biochemical information obtained through the IR absorbance spectra (e.g., lipid, protein, and nucleic acid levels) serves as inputs to pattern-recognition techniques. Whereas in conventional pathology stains are used, in FTIR imaging, the inherent chemical basis of the tissue is used. Unlike conventional pathology, which requires human interpretation, FTIR imaging has an added benefit and can take advantage of computer algorithms to make diagnoses automated, objective, and reproducible. To summarize, the central idea of this approach is to provide nonspectroscopists with color-coded images that are consistent with their domain-specific knowledge (Fig. 10.3).

There are two major benfits of using spectroscopic imaging. The first is that the identification of tissue structure is necessary for accurate results. Early reports of exceptional successes were generally recognized to be effects arising from histologic bias and chance due to limited samples [25]. Deriving an average spectrum of the normal and diseased samples using point spectroscopy can show apparent differences, however, this can simply arise from chemical differences of the different cell types within tissue. Hence, it may be beneficial to first identify the cell types present and then examine spectral changes in the context of cell distribution. The state of the art is summarized in recent edited books and examples of capabilities in many different tissues are enumerated (Fig. 10.4) [26, 27].

10.2.2 FTIR and Molecular Pathology

Current clinical methods typically stain tissue and then examine the morphological changes and architecture, which may both be measured by spectroscopic imaging. Further markers can be measured to aid in diagnosis or to give extra information about the disease. Prostate-Specific Antigen (PSA) is commonly tested for its presence in blood

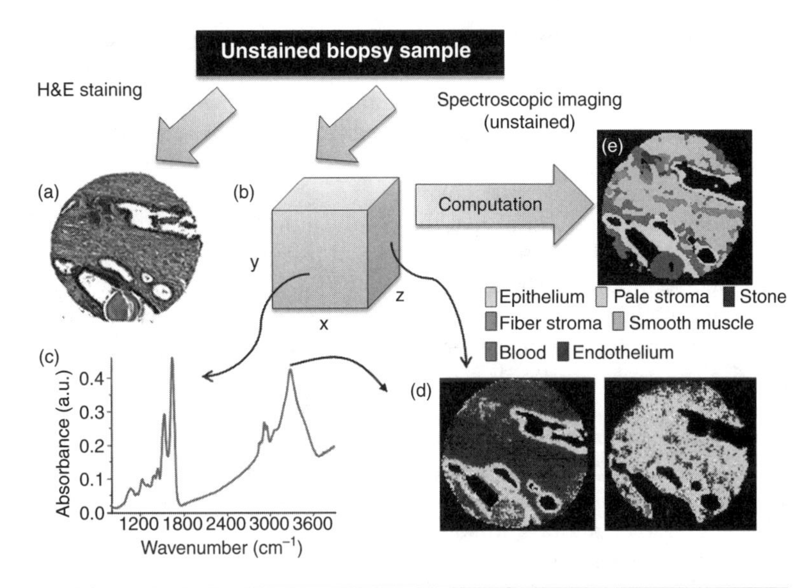

Figure 10.3 (*a*) Conventional imaging in pathology requires dyes and a human to recognize cells. In chemical imaging data (*b*), both a spectrum at any pixel (*c*) and the spatial distribution of any spectral feature can be seen, for example, in (*d*) nucleic acids (left, at ~1080 cm^{-1}), and collagen specific (right, at ~1245 cm^{-1}). Computational tools then convert data to knowledge used in pathology (*e*).

samples which can be an indicator of epithelial proliferation and leaky blood vessels in prostate cancer [28]. Histopathalogic staining of tissue for angiogenesis (factor VIII [29]) and increased proliferative capacity (ki-67 [30]) are often used to give more information in cancer samples. The treatment of cancer may also be determined by the presence of

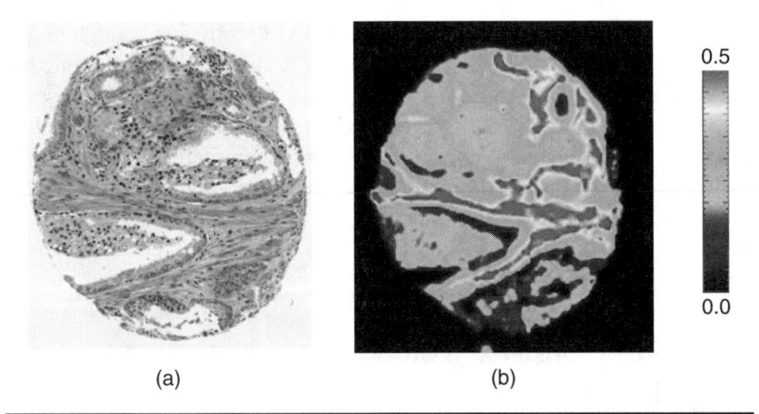

Figure 10.4 (*a*) Hematoxylin and eosin (H&E) stain of prostate cancer tissue biopsies with (*b*) corresponding IR images by protein distribution (1650 cm^{-1}).

hormonal receptors; for example, in breast cancer, treatment is determined by the presence of estrogen or progesterone receptors. Expression of HER-2/neu is also stained for as it is indicative of a more aggressive cancer [31]. Although the expression of these markers has not been studied using FTIR imaging, it has the potential for the detection of these molecular changes.

Spectroscopic imaging may be contrasted with molecular imaging [32]. Although immunohistochemical imaging techniques allow for exact monitoring of specific epitopes, dyes and reagents are required, and the functional state of cells and tissues is limited to known pathways that involve the probe target. Spectroscopic imaging, however, requires no probes but uses computation to extract information from the tissue's inherent biochemistry. But the information from spectroscopic imaging cannot be as specific as that with molecular probes. Therefore, although spectroscopic imaging may not allow for the level of detail obtainable using IHC, the flexibility and general applicability of spectrosopic imaging makes it a potentially desirable approach which may make it suitable for many applications [33]. There are some small examples of where the information content from IHC and FTIR can be related. For example, receptor status was imaged very early on, with IR spectra being correlated with steroid receptor status in breast tissue [34]. FTIR spectroscopy has also been used to monitor changes in protein conformation of the basement membrane of tumor blood vessels. The triple helix and β-turn content of the protein secondary structure of the basement membrane, primarily from collagen IV, was shown to alter during the process of angiogenesis [35].

A number of technological limitations have hindered the transfer of FTIR spectroscopy and imaging to routine clinical use. The major issues in the early 2000s were the speed of data acquisition, poor spatial resolution, and high cost of equipment. Another hindrance was lack of sufficient analytic tools for spectroscopic analysis to handle the resultant large, complex data sets. IR spectroscopy gives rise to highly complex vibrational spectra made up of many different chemical species; conventional univariate analysis is typically insufficient for diagnosis of different cell types or disease status, especially in samples with a large degree of variance which would be expected in clinical samples. Early experiments on IR spectroscopy also traditionally suffered from small sample sizes; this hindered its acceptance as a diagnostic tool. However, great strides have been made during the last decade in improving instrumentation, data collection, computational analysis, and the implementation of large-scale studies.

10.2.3 Comparison with Other (Spectral) Imaging Techniques

Raman spectroscopy offers a complementary chemical imaging technique to infrared spectroscopy. Rather than detect the fundamental vibrations of chemical bonds, it measures the inelastic scattering of incident photons. The shorter wavelengths used in Raman spectroscopy allow for images with a higher spatial resolution than FTIR [36]. Raman spectroscopy also allows for the measurement of aqueous samples as it is not affected by water [37]. In addition, Raman spectroscopy can be operated in a confocal mode [38], which allows spectra to be obtained at different depths in tissue. However the signal from Raman spectroscopy is typically much lower than that from IR spectroscopy, and thus Raman spectroscopy is limited by slow acquisition times [39]. There exists a growing literature of the potential use of Raman spectroscopy in cancer diagnosis and for understanding cell biology with a number of detailed reviews [36, 37, 39–41].

10.2.4 Progress in Instrumentation

The key advances in instrumentation and data processing have been discussed in detail previously [33, 42]. Advances in IR detector technology have led to an increased interest in the potential use of IR imaging for clinical applications. IR mapping or imaging may be performed using either single-element (referred to as *IR mapping*) or multiple-element detectors (referred to as *IR imaging*). Single-element detectors are used for IR mapping which require pixels being acquired in a sequential fashion to create a chemical map [33]. One approach taken to improve the speed of IR mapping using single-point detectors has been the use of synchrotrons as a source of highly bright and collimated IR [43, 44]. This allows for mapping in faster time than when benchtop IR sources are used; however, this approach is impractical for routine pathology due to the limited availability and the high cost of synchrotron facilities.

The declassification from defense programs of multipoint detectors has led to the development of IR imaging, which allows for significantly faster acquisition times [24]. Instead of the IR being focused to a single point at the desired spatial resolution, the IR is illuminated over the entire area of interest [33, 45]. In addition to the increase in speed of spectral acquistion, multipoint detectors allow for increases in the spatial resolution of images. Single-point detectors struggle to acquire spectra at the diffraction limit of IR due to poor signal-to-noise ratio (SNR) due to the use of apertures reducing the throughput of IR and diffraction artifacts. Multipoint detectors can allow for images at spatial resolutions of up to 5.5 μm while retaining good signal-to-noise ratio [33]. The implementation of synchrotron radiation with focal plane array (FPA) detectors has further improved achievable spatial resolution. The new beamline at the Synchrotron Radiation Center in Madison, Wisconsin, can allow for pixel sizes as small as 0.54 micrometer × 0.54 micrometer being achieved (Fig. 10.5).

Although investigations of the use of IR-based technologies have generally relied on transmission or transflection sampling measurements, other sampling approaches are being investigated that may

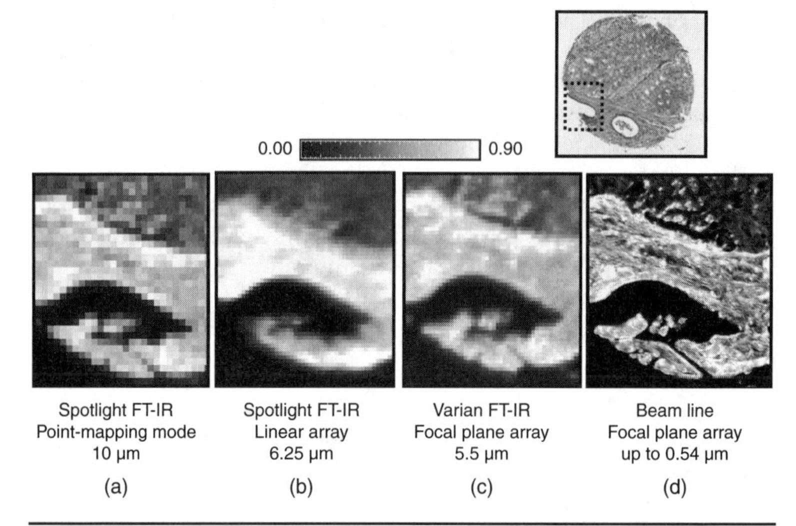

Figure 10.5 Comparison of IR images from a prostate cancer tissue biopsy using: (*a*) Spotlight FTIR in point mapping mode (10 micron spatial resolution), (*b*) Spotlight FTIR using a linear array detector (6.25 micron spatial resolution), (*c*) Varian FTIR with a focal plane array detector (5.5 micron spatial resolution), and (*d*) Using the synchrotron beamline at the Synchrotron Radiation Center (SRC) at the University of Wisconsin–Madison (0.54 micron spatial resolution). (Reproduced from Walsh et al. AIP Conference Proceedings Volume 1214).

benefit pathology [46]. Attenuated Total Reflection (ATR) FTIR is a complementary addition to FTIR in which instead of the sample directly interacting with the propagating beam, it interacts with the evanescent electric field generated from a Solid Immersion Lens (SIL), typically a crystal made of germanium, diamond, or zinc selenide. ATR-FTIR has three benefits over conventional FT-IR for pathological imaging: (1) The problem of water absorption in transmission FTIR is overcome by allowing for IR imaging of fresh or frozen biopsies. (2) Subwavelength IR imaging can be achieved up to ~ 1.25 μm spatial resolution [47]. (3) Very minimal sample preparation is required, with no need for sectioning, fixation, or staining. One limitation of ATR imaging over conventional FT-IR imaging is the requirement for intimate contact of the tissue with the SIL. This may damage the sample and potentially cause spatial changes to the image due to physical contact. Other studies are investigating the applicability of IR spectroscopy for *in vivo* tissue analysis using fiber optics [48].

Spatial resolution is a potentially important factor in the use of FTIR for pathology as IR cannot achieve the spatial resolution of visible light. Photothermal microspectroscopy (PTMS), which instead of detecting IR, detects the thermal waves subsequently released after IR irradiation [49–51], may potentially allow for improvements in spatial resolution. This has been applied to cell cycle [52], corneal [53], and intestinal stem-cell identification [54–56].

10.3 Applications of FTIR in Biomedical Research

The development of IR spectroscopy for pathology can be generally divided into two areas, basic science and clinical science. The aim of using IR spectroscopy for basic science is to understand the fundamental processes involved in pathology which can lead to new insights into how diseases begin and progress. The aim of IR spectroscopy for clinical science is primarily the development of IR methodologies for disease diagnosis or prognosis.

10.3.1 2D Cell Culture

In vitro cell culture assays represent an important model system for understanding the fundamental changes that occur in the human body. This involves the growth of either primary cell lines (isolated directly from a tumor) or a secondary cell line (immortalized cell line). With cell culture representing a model system for investigations into the behavior of cells, it has been widely utilized in IR spectroscopy studies aimed at investigating cancer pathology. The bulk of FTIR research conducted in cell culture has focused on the use of point spectroscopy due to the homogenous biochemical nature of cell cultures. Although the work in cell culture is not directly comparable to clinical pathology, it is critical that we understand the chemical changes associated with fundemental cell processes to guide the work in whole tissue. FTIR spectroscopy in point spectroscopy mode has been used to monitor the chemical changes associated with key cellular events such as proliferation [57], apoptosis and necrosis [58–61], mitosis [62, 63], growth stage [64], and cell growth and DNA synthesis [65]. It is critical to understand the spectral changes in tissue to ensure that the changes observed are real and due to differences in cell type or disease state rather than cellular events. One of the most studied cell biology systems has been the chemical changes observed with the different phases of the cell cycle [52, 58, 66, 67]. Identification of the chemical differences in the cell cycle is important, as different phases of the cell cycle can have very different cellular responses both *in vitro* and *in vivo* [68]. IR spectroscopy has also been shown to be a potentially quantitative tool for cell biology, capable of providing estimates of chemical composition comparable to more conventional biochemical approaches [69]. It has also been demonstrated that the type of fixation process can alter cellular biochemistry. This demonstrates the importance of consistency in pathology fixation; changes in this could potentially affect the accuracy of FTIR analysis [70].

FTIR spectroscopy of fresh or living tissue it is difficult to obtain good-quality spectra due to the strong interfering water absorbance in the biochemical region. Raman spectroscopy has typically been the tool of choice for living cells or tissue, due to the minimal interference of water on the spectra; however, recent studies have demonstrated that even with water absorbance, it is feasible with adjusted methodologies [71, 72]. Future work will need to determine whether the spectral features observed to change in cell culture are well-correlated with the spectral features which change in whole tisssue.

A number of exciting observations have been demonstrated, which shows the potential that FTIR may have for the future of pathology, past simple cell type or disease identification. FTIR has been applied to identify the very early chemical events associated with transformation prior to gross morphological changes. A study of cells that had been infected by the murine sarcoma virus demonstrated chemical changes prior to visual alterations being identified [73]. A separate study found chemical changes to occur in cells undergoing apoptosis earlier than the gold-standard analytical technique of flow cytometry [74]. This is highly promising, as it would be desirable to detect abnormality at an earlier time point than can be achieved using conventional pathology, in particular, in the case of aggressive cancers where early diagnosis is critical for successful treatment.

A number of cell biology studies have demonstrated that the future of FTIR for pathology could be very wide ranging and even guide treatement and monitoring of patients. One study demonstrated that the chemical changes of lung cancer cells due to exposure to the chemotherapeutic drug, gemcitabine, could be monitored [75]. Changes in the spectra of cervical cancer cells, HeLa, treated with the photosensitizer, Hypocrellin A (HA), along with photodynamic therapy were also shown [76]. Other studies have shown that it may be possible to determine whether cancer cells will respond to a certain class of drug [75] and/or the metastatic potential of a tumor [77]. Other exciting studies have demonstrated that FTIR can discriminate multidrug-resistant (MDR) cell types, that may allow for personalized treatment strategies based on IR spectra [78, 79].

10.3.2 3D Tissue Culture

In recent years, there has been growing interest in the use of 3D tissue engineering over cell cultures as a model system. Two dimensional cell culture is not entirely representative of real human tissue due to cells in the culture lacking the more complex surroundings or interactions found in real 3D tissue. Tissue engineering is expected to more closely resemble the changes that occur in human tissue. In particular, whereas 2D cell cultures give little information in regard to spatial information, 3D tissue engineering can allow for the spatial distribution of cells and to be monitored. Although point spectroscopy has been satisfactory for 2D cell cultures, FTIR imaging will be essential for examing 3D-engineered tissue to allow for the full understanding of tissue changes. Much focus has been placed on creating and developing realistic 3D tissue models which much more closely resemble human tissue [80]. The application of FTIR to 3D-engineered tissue has not yet been widely demonstrated; however, initial results show a great deal of promise which may allow for unprecedented insight into the initiation and progression of disease [81, 82]. To understand the more complex 3D processes, such as the invasive process of a primary tumor, 3D tissue engineering studies will be very useful. One model of tumor cell invasion involved the seeding of the lung cancer cell line CALU-1 on a type-1 collagen layer that acts as a basement membrane. Invasion of the tumor cells can be shown, in particular by monitoring DNA absorbance at 1080 cm^{-1}. Differences between invading and noninvading cells can also be observed [81]. A 3D melanoma model with engineered skin tissue has been used to chemically characterize the different layers of the skin and has also demonstrated that the invasion process of malignant melanocytes can be monitored [82]. Another role of FTIR in 3D tissue is monitoring drug diffusion, for example, through the skin [83–85]. Recent work has demonstrated that ATR-FTIR imaging can be used for monitoring the diffusion of chemicals through the skin in 3D [86] and for the diffusion of 12 liquid samples through pieces of skin tissue [87] (Fig. 10.6).

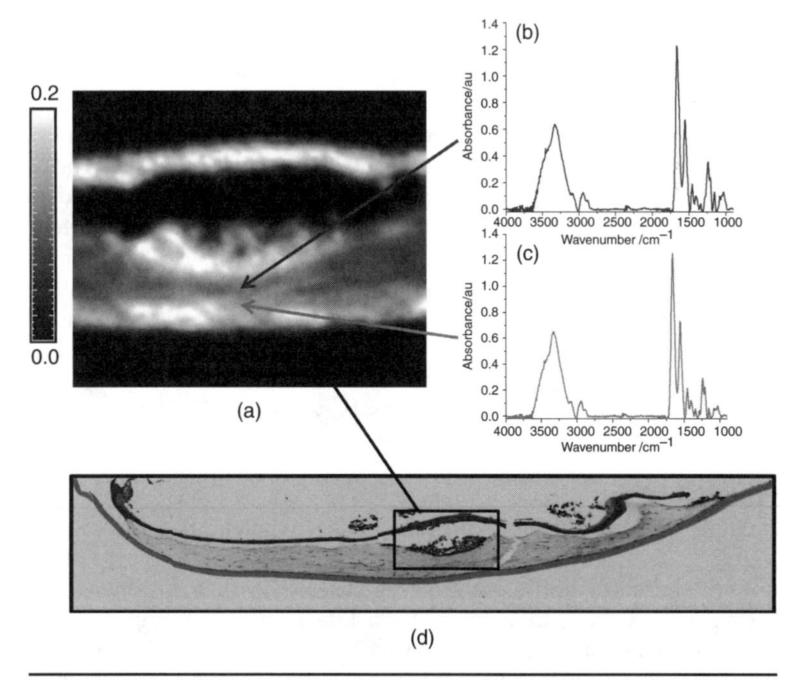

Figure 10.6 (*a*) IR image and (*d*) H&E image of an engineered skin tissue model made up of stratum corneum, epidermis, and dermis with malignant melanocytes invading the dermis. IR spectra were acquired from the dermal regions (*b*) near and (*c*) distant from the tumor growth.

10.3.3 Clinical Applications of FTIR

Breast Cancer

Breast cancer is the second most common cancer worldwide after lung cancer. Screening for breast cancer is performed by mammography, which is used to identify any abnormal regions in the breast. If an abnormal mammogram is obtained, follow-up is performed by a needle biopsy. The tissue biopsy is fixed and stained, as described previously, and diagnosed. FTIR spectroscopy in both point and imaging modes has been applied to identify chemical changes between breast cancer tumors that were identified as malignant or benign lesions, with discrimination between disease states reported [88–92]. Artificial neural networks have been shown to allow for classification of breast tissue [90]. An important observation with these studies was that there exists a strong degree of tissue microheterogenity within invasive ductal carcinomas. This required the use of high-spatial resolution IR spectroscopy to be able to adequately identify tissue changes. This demonstrates that the increased spatial resolution obtainable using IR imaging may be vital for the successful diagnosis of highly heterogeneous tissues such as in the breast. A separate study also demonstrated chemical differences between normal and cancer of breast tissue, it also showed that IR spectroscopy could detect collagen alterations between normal and cancer which correlated well with collagen changes observed using small-angle X-ray scatterings (SAXS) [93]. Accurate classification of epithelial and stromal cell types within breast tissue has also been demonstrated, which subsequently allowed for diagnosis of normal or cancerous tissue [94]. ATR-FTIR has also been used to investigate whether changes in a patient's hair can be used as a diagnostic test of breast cancer, by monitoring changes in protein structure [95]. Metastasis of breast epithelial cells to axillary lymph nodes can also be detected by IR imaging in the lymph tissue [96]. DNA extracted from breast tissue has shown that the DNA alterations between the progression from normal to primary tumor compared to the progression from primary tumor to metastatic tumor are significantly different [97]. Significant differences between the DNA of women under 30 compared to that of women over 50 was also shown [98].

Prostate Cancer

Prostate cancer is the most common male cancer and has the second-highest rate of cancer mortality among men. Prostate cancer is often an asymptomatic and slow-growing disease, with many men over the age of 50 having some form of prostate lesion. Screening for prostate cancer is commonly performed by a routine blood test for a prostate-specific antigen [99]. If prostate cancer is suspected, a biopsy is obtained by taking about 12 fine-needle biopsies. These biopsies are visually examined and are categorized by level of severity using the Gleason grading system. A number of studies have demonstrated the potential of FTIR for the examination of prostate tissue biopsies. Good sensitivity and specificity have been demonstrated between prostate biopsies by using IR spectroscopy and a three-band Gleason grading system [100–102]. The use of principal component–discriminant function analysis improved on this to give a sensitivity of 92.3% and specificity of 99.4%. A two-class criterion based on metastatic spread showed that a prediction of the clinical aggressiveness of a tumor could be made [103]. Investigations of the role of metastatic prostate cancer cells present in bone marrow have also been demonstrated [104]. The largest study to date using IR imaging for prostate cancer diagnosis demonstrated a very high level of cell type classification with a 10-class model of prostate cell types. This consisted of making a classification model on the basis of probabilistic determination and a learning algorithm. The study also demonstrated that discrimination between benign and malignant prostatic epithelium could be achieved [105]. This could lead to the development of a systems pathology approach to diagnosis, where characteristics such as the number of cells, cell ratio, cell morphology, and so on, could be quickly derived and used in a robust diagnostic tool.

The prostate is made up of two zones, the peripheral and transitional zones, with the peripheral zone being more likely to develop prostate cancer and the transitional zone being more likely to develop the nonmalignant growth, benign prostatic hypertrophy (BPH). One study using IR point spectroscopy demonstrated that the prostate cancer cells were surprisingly biochemically closer to the transitional zone cells [106]. IR spectroscopy has also been used to look at the chemical differences between normal tissue and BPH [107]. Another study used synchrotron FTIR to determine chemical differences between normal, cancer, and BPH in prostate tissue [108]. Point spectroscopy studies on DNA extracted and purified from prostate tissue has been shown to be a potentially useful tool for investigating prostate cancer. The IR spectra of prostatic DNA of healthy men over the age of 55 [77] and of histologically normal tumor-adjacent tissue [109] can have the same spectral phenotype as DNA; this demonstrates that examining extracted DNA using IR spectroscopy may serve as both an early marker of prostate cancer and more importantly allow for the identification of cancer in surrounding normal tissue. A metastatic DNA phenotype was also shown in normal tissue surrounding metastasizing prostate tumors [110]. This could have important ramifications in reducing the number of false negatives from prostate tissue biopsies. Excitingly, it was also demonstrated that a prediction of whether a tumor would metastasize could be predicted with about 90% sensitivity and specificity [77].

Colorectal Cancer

Colorectal cancer is the fourth most common cancer worldwide. Colon cancer principally arises from the abnormal proliferation of stem cells at the base of intestinal crypts. Suspected cancer is examined by an endoscopy, which allows for the visualization of the intestines for any abnormal growths. If colonoscopy is performed, a tissue biopsy can be removed, which may allow for pathological analysis. FTIR has been widely performed in point spectroscopy mode to chemically interrogate the crypts of the intestines. This work has demonstrated that FTIR may allow for the chemical differences between disease states to be identified. Good discrimination between normal, adenomatous polyps, and malignant colorectal cancer tissue was demonstrated [111]. A separate study demonstrated segregation between normal, cancer, and inflammatory bowel diseases (IBDs) could be achieved. IBD can often progress to cancer, and this study could indicate which cases would progress to cancer [112]. Crypts in the colon that were histologically normal but had abnormal proliferation were shown to be chemically distinct from normal crypts [113]. Colorectal cancer is a good example of the usefulness that spectroscopic approaches may have to pathology due to being able to chemically interrogate tissues that appear histologically normal but are actually diseased.

ATR-FTIR applied to colorectal tissue has been shown to provide good classification of normal and malignant tissue [114]. A separate study showed that DNA isolated from colon tissue could be used to discriminate between different grades of colon cancer. It was also shown that the DNA signature of patients over the age of 50 exhibited a cancer-like phenotype [115]. FTIR imaging has been applied to colorectal tissue as well and showed clear differences in histological groups with different types of cluster analysis [116, 117]. Spectral differences were demonstrated between the stem, transit-amplifying, and terminally differentiated cell types of the colon and small intestine [54]. The significant differences between the cell types of the crypts based on their differentiated states highlights the importance of imaging for robust pathological analysis, where point spectroscopy would require very careful decisions on where to take the spectra to reduce or remove spectral variance based purely on cell type.

Gastric Cancers

FTIR was applied to discriminate between normal and malignant gastric tissues. Ten spectral bands were found to be significantly increased in malignant tissue, and 88.6% diagnostic accuracy was achieved using discriminant analysis [118]. ATR-FTIR allowed for discrimination of normal and gastric tissue as well as chronic atrophic gastritis and chronic superficial gastritis [119, 120]. Endoscopic biopsies from the distal esophagus which included squamous, gastric, and Barrett's esophagous tissue samples were analyzed using ATR-FTIR coupled with partial least square fits analysis and demonstrated that discrimination could be achieved [121].

Cervical Cancer

Cervical cancer has been one of the most actively studied cancers for IR spectroscopy; it has been investigated by many groups for its potential as an alternative method to cervical cancer screening [122–124]. Cervical cancer screening represents a model system for the applicability of IR spectroscopy for disease diagnosis, due to having relatively easy access to a large number of patient samples because of the widespread implementation of cervical cancer screening programs and having well-understood premalignant stages known as cervical intraepithelial neoplasia (CIN). The successful implementation of cervical cancer screening programs in developed countries has led to this cancer dropping from the second-most common female cancer worldwide to the eleventh. Routine screening is performed by the papanicolaou (pap) smear, which consists of cells obtained by a brush from the transformation zone of the cervix. The cells are fixed and applied to a microscope slide and examined for abnormality by a cytologist. If the results are abnormal, then a follow-up cervical biopsy of the transformation zone may be obtained. The majority of the work has focused on obtaining point spectra from either exfoliated cytology smears or tissue biopsies [122, 125–136], due to its amenability to the clinical setting associated with the lower cost of equipment. A number of studies have demonstrated the value of IR imaging of cervical cancer biopsies [137–139] and the ability to reconstruct 3D IR images [140].

The application of FTIR on exfoliative cytology smears are particularly susceptible to high degrees of variance between patients due to their highly heterogenous nature. Exfoliative cervical cytology smears can consist of many different components including endo- and ectocervical cell types, mucus, red and white blood cells, fungus, and bacteria, whose confounding presence will affect derived spectra. A number of strategies to improve or remove confounding factors have been investigated [136, 141–143]. For FTIR to be applied to cervical smears, it will be important to fully understand and identify what the chemical differences between disease states mean but also the spectral changes that naturally occur in cervical smears in order to minimize the influence of confounding spectral features. A number of studies have characterized the changes in cervical tissue, such as epithelial cell maturation and differentiation of the different layers of the cervix [144, 145], differences between the squamous and columnar cell types of the cervix [146], and cyclical changes of epithelial cells throughout the menstrual cycle [147].

FTIR for cervical cancer screening has been a well-studied area of research yet has not made it any closer to the clinical setting. This may be due to the wide range of approaches, involving different instrumentation, sample preparation, spectral processing, and data analysis, which has led to little agreement over which protocols and approaches offer the greatest benefit.

Brain

Glioma is a type of brain cancer that originates from the glial cells in the brain; it is an aggressive cancer with no known cure. Glioma has been well-studied using FTIR, with the ability to distinguish and grade the cancer [148–150]. Changes in protein confirmation in benign and malignant astrocytomas have been demonstrated using reflectance

FTIR [151]. Discrimination between normal and meningiomas have been shown using FTIR coupled with Hierarchical Cluster Analysis (HCA) [152]. Brain cancer can also arise due to metastasis of a primary tumor, primarily from lung, colorectal, breast, or renal cell carcinoma. FTIR coupled with SIMCA/LDA classification can correctly distinguish the organ of origin for brain cancer metastases [153, 154].

Other brain diseases have been extensively researched using FTIR. Alzheimer's disease is caused but the misfolding and plaque-like appearance of the protein amyloid beta. Synchrotron FTIR has been used to image the structure and location of the plaque in human brain tissue [155]. The response of the brain to the drug nicotine, which may reduce the production of amyloid beta plaque, has also been monitored using FTIR [156]. Parkinson's disease, another degerentative brain disease, is characterized by degeneration of the central nervous system. FTIR was applied to determine the chemical differences between the nerve cell bodies of normal and diseased tissues [157]. FTIR has also been applied to the degenerative brain disease multiple sclerosis, with infrared maps of normal and diseased white matter showing distinct chemical changes [158]. FTIR has been applied to image the changes in scrapie-infected mouse brains [159] and to discriminate between the different transmissible spongiform encephalopathies that can cause brain diseases [160].

Skin

Skin cancer is the most commonly diagnosed cancer in the United States; however, detection is often early and treatment highly successful. There are three types of skin cancer: squamous cell carcinoma, basal cell carcinoma, and melanoma. FTIR has also been applied for the detection and characterization of skin cancer. Changes in the nucleic acid absorbance in skin tissue has been found in malignant skin [131]. Differences between melanoma, skin nevi, and normal skin have also been demonstrated [161–163]. Melanoma can be successfully distinguished from surrounding normal epidermis in skin biopsies [164]. The discrimination of the nonmelanoma skin cancers—basal cell carcinoma, squamous cell carcinoma, and Bowen's disease—have been reported using PCA and LDA [165].

FTIR has also been widely used for monitoring drug diffusion through the layers of the skin. The skin consists of an outermost layer termed the *stratum corneum,* which forms the protective layer. The ability to measure drug and skin care product diffusions and chemical effects on the stratum corneum and the deeper dermis layer has been an active research area. ATR-FTIR has been the primary tool for monitoring the diffusion and chemical changes throughout the skin for a wide range of compounds [83, 87, 166–168].

Other Cancers

One of the major aims of the application of FTIR imaging to the clinical setting is to discriminate between the different subtypes of a cancer.

For example, there are different subtypes of endometrial cancer that are important to identify to guide diagnosis and treatement [169]. The ultimate goal of the integration of FTIR imaging in the clinical setting will be to allow both for accurate grading and subtyping of cancers using both the chemical and spatial information that IR imaging can give. Future work will focues on the use of FTIR imaging to aid in the prognosis of a cancer. For example, it was demonstrated that increased DNA content in chronic lymphocytic leukemia cells can be determined; the DNA content of these cells is known to correlate with *in vivo* doubling time [170]. Differences in the DNA of mice were also shown on average 57 days prior to the appearance of tumors to have a similar structure to tumor tissue DNA [171]. Other studies have demonstrated that the detection of metastases in the brain can be imaged and the primary tissue from which the tumor originated could be identified [153, 154, 172]. Another future aim for the implementation of IR imaging in the clinical setting will be the ability to perform intraoperative diagnosis of tissue—this has been demonstrated to be feasible in cerebral glioma surgery [173]—and thyroid tissue and lymph node evaluation [96, 174]. IR spectroscopy has also been used in the detection and monitoring of a wide range of other cancers including acute lymphoblastic leukemia (ALL) [175, 176] and blood [177], brain [149, 150, 178, 179], esophagus [121], liver [180], lymph node [181–183], oral [184], ovarian [185], skin [161, 162, 164], stomach [118–120, 186], and thyroid cancer [187].

Cardiovascular Disease

Although the majority of FTIR imaging studies has been aimed toward the interrogation of tissue toward cancer diagnosis, FTIR imaging has a wide variety of biomedical applications. Cardiovascular disease is the leading cause of death in the Western world. Although tissue is not available for the diagnosis of cardiac disease, it is critical that there is a fundamental understanding of the events responsible for the initiation of heart disease. To this end, FTIR spectroscopy represents a novel approach to understanding the fundamental chemical changes that occur in heart disease. FTIR and near-IR were used to examine normal and infracted heart tissue, with increases in absorbance of collagen bands, demonstrating the deposition of collagen I following myocardial infarction [188]. The rearrangement of the extracellular matrix following myocardial infarction is also known to contribute to cardiac disfunction, and FTIR has been used to evaluate the microscopic scarring of a cardiomyopathic heart [189]. A deficiency of selenium in the diet has been linked to cardiovascular disease. FTIR of rat hearts showed, when compared to controls, that the administration of selenium leads to increases in saturated and unsaturated lipids and alterations in the protein profile [190]. Diabetes mellitus is a disease which has been strongly associated with cardiovascular disease. FTIR of rat heart and liver tissue demonstrated chemical differences with rats that had been induced to have diabetes. The ratio of CH_2 symmetric to CH_3 asymmetric and of CH_2 scissoring to CH_3 scissoring were both shown to decrease in the diabetic heart and liver [191]. Other work investigating the early stages of diabetes in rat heart demonstrated an increase in lipid content and alterations in protein profile, and increase in collagen content in the vein of the diabetic group [192].

10.4 Translation of FTIR Imaging into Routine Use in Routine Clinical Pathology

10.4.1 Clinical Considerations

The ultimate goal is for the implementation of FTIR imaging in the clinical setting. However, for FTIR imaging to be widely adopted as a tool for clinical use, there are a number of considerations that must be taken in to account. Many of the original studies have comprised a relatively small number of patients. This makes it unclear how the suggested approaches or computational analyses will be applicable to the real world. Future studies will need to comprise large numbers of patients from multiple hospitals, such that a comprehensive validation is conducted. The design of multifocal studies will be critical to determining the influence of regional factors such as diet, ethnicity, and environment, which may have a significant affect on the IR spectra of images obtained. The exact nature and significance of this has not yet been demonstrated. Other factors that may affect the robustness of IR imaging could arise from differences in sample preparation and fixation between laboratories. Even in relatively small-scale studies, large degrees of intrapatient variance can be observed [136]. Future work must determine the influence of confounding factors and intrapatient variance; this can be achieved by developing large, statistically valid multifocal studies. It may ultimately be necessary to devise algorithms that take into account different patient factors; for example, the IR spectra of DNA of prostate biopsies from young men is very different from that of older men, with the older men's DNA phenotype being closer related to the cancer DNA phenotype [77]. The use of large-scale studies will also be critical to aid in compliance of implementing IR spectroscopy in the clinical setting.

10.4.2 Spectroscopic Considerations

Although it is important to validate IR imaging as a potential approach for pathology, it will be important to identify and correct any spectroscopic issues. One important spectroscopic consideration will be to validate and standardize the IR approaches used, for example, the number of scans, spectral resolution, spatial resolution, and preprocessing techniques [24, 94, 193]. It has previously been demonstrated that the spatial resolution used is critical to differentiate between cell types [194] since at coarser resolutions some image pixels will be made up of mixed cell types [24]. The loss of spatial resolution has been shown to lead to cell-type classification errors in the prostate [24]. The implementation of FPAs has significantly improved the spatial resolution, and the implementation of ATR imaging may improve this further [46]. Other important factors that must be tightly controlled include signal-to-noise ratio and spectral resolution. Spectral resolution using a prostate cell-type classifier can still successfully classify at a resolution of 16 cm^{-1} [24]. The optimization of FTIR parameters is of paramount importance for the clinical setting, to ensure that maximal information and classification can be achieved; but to balance this with time for acquisition so that rapid diagnoses can be made. There also exists a wide variety of potential computational approaches, including both supervised and unsupervised methodologies [195].

10.4.3 Translation of FTIR Imaging to Clinic

The implementation of IR imaging to the clinical setting will be challenging; however, a number of considerations need to be taken into account which may significantly increase the speed and likelihood of uptake. To get IR spectroscopy in the clinical workplace, it will be essential that the technology, methodology, and analyses are compatible with current clinical workflow methods. In conventional pathology, when tissue biopsies are obtained, they are primarily formalin-fixed, paraffin-embedded tissue for archival material or snap-frozen tissue for intraoperative analysis [96]. FTIR imaging has been widely conducted on tissues prepared in this way. Instrumentation must be amenable to the clinical laboratory, preferably as a benchtop instrument. Cost of the infrared instrument and special setups (vibration-free tables, purging systems, liquid-nitrogen availability for detectors, etc.) will also be an important factor in the likelihood of implementation. The lack of specialized stains and expert pathologists should reduce the cost of tissue diagnosis from the expense of current approaches. FTIR imaging has the added benefit that it gives data in an intuitive format which is similar to the current analysis. Robust programs will need to identify and correct any unusual spectra or aberrations [196, 197]. The whole process will ultimately be fully automated and remove any user intervention; this will remove human subjectivity.

10.5 Discussion

FTIR imaging represents a potentially valuable tool for pathology, allowing for the rapid identification of cell types and disease states. FTIR appears to fulfill much of the criteria to improve on current methodologies of disease diagnosis. Although much research has focused on 2D and 3D cell cultures, future work must assess how realistic this is compared to real human tissue. The recent advances in detector technologies, along with reductions in costs, has made imaging a much more feasible approach for the clinical setting. Imaging not only allows for the addition of potentially valuable spatial information but will also reduce the potential problems with biased results associated with the use of point spectroscopy. The long-term goals of implementing IR spectroscopy in the clinical setting must not be just to diagnose but also to provide information about the chemical basis of the cancer with the overall aim being to personalize medical treatment.

References

1. A. Jemal, R. Siegel, E. Ward, T. Murray, J. Xu, C. Smigal, and M. J. Thun, "Cancer Statistics," *CA Cancer J. Clin.*, 56(2): 106–130, 2006.
2. S. M. Gilbert, C. B. Cavallo, H. Kahane, and F. C. Lowe, "Evidence Suggesting PSA Cutpoint of 2.5 ng/mL for Prompting Prostate Biopsy: Review of 36,316 Biopsies," *Urology*, 65(3): 549–553, 2005.
3. P. F. Pinsky, G. L. Andriole, B. S. Kramer, R. B. Hayes, P. C. Prorok, and J. K. Gohagan, "Prostate Biopsy Following a Positive Screen in the Prostate, Lung, Colorectal and Ovarian Cancer Screening Trial," *J. Urol.*, 173(3): 746–750, "Discussion" 750–751, 2005.
4. D. Grignon, P. A. Humphrey, J. R. Srigley, and M. B. Amin, *Gleason Grading of Prostate Cancer: A contemporary Approach*, Philadelphia: Lippincott Williams & Wilkins, 750–751, 2004.
5. P. A. Humphrey, *Prostate Pathology*, American Society of Clinical Pathology, Chicago, 2003.
6. S. J. Jacobsen, S. K. Katusic, E. J. Bergstralh, J. E. Oesterling, D. Ohrt, G. G. Klee, C. G. Chute, and M. M. Lieber, "Incidence of Prostate Cancer Diagnosis in the Eras before and after Serum Prostate-Specific Antigen Testing," *JAMA*, 274(18): 1445–1449, 1995.
7. R. E. Nakhleh "Patient Safety and Error Reduction in Surgical Pathology," *Archives of Pathology & Laboratory Medicine*, 132(2): 181–185, 2008.
8. A. de la Taille, A. Viellefond, N. Berger, E. Boucher, M. de Fromont, A. Fondimare, V. Molinie, D. Piron, M. Sibony, F. Staroz, M. Triller, E. Peltier, N. Thiounn, and M. A. Rubin, "Evaluation of the Interobserver Reproducibility of Gleason Grading of Prostatic Adenocarcinoma Using Tissue Microarrays," *Hum. Pathol*, 34(5): 444–449, 2003.
9. P. L. Nguyen, D. Schultz, A. A. Renshaw, R. T. Vollmer, W. R. Welch, K. Cote, and A. V. D'Amico "The Impact of Pathology Review on Treatment Recommendations for Patients with Adenocarcinoma of the Prostate," *Urol Oncol.*, 22(4): 295–299, 2004.
10. W. M. Murphy, I. Rivera-Ramirez, L. G. Luciani, and Z. Wajsman, "Second Opinion of Anatomical Pathology: A Complex Issue Not Easily Reduced to Matters of Right and Wrong," *J. Urol.*, 165(6 Pt 1): 1957–1959, 2001.
11. J. I. Epstein, P. C. Walsh, and F. Sanfilippo, "Clinical and Cost Impact of Second-Opinion Pathology. Review of Prostate Biopsies Prior to Radical Prostatectomy," *Am. J. Surg. Pathol*, 20(7): 851–857, 1996.
12. D. B. Troxel "An Insurer's Perspective on Error and Loss in Pathology," *Arch Pathol Lab Med.*, 129(10): 1234–1236, 2005.
13. P. C. Albertsen, J. A. Hanley, and J. Fine, "20-Year Outcomes Following Conservative Management of Clinically Localized Prostate Cancer," *JAMA*, 293(17): 2095–2101, 2005.
14. G. Aus, D. Robinson, J. Rosell, G. Sandblom, and E. Varenhorst, "Survival in Prostate Carcinoma—Outcomes from a Prospective, Population-Based Cohort of 8887 Men with up to 15 Years of Follow-Up: Results from Three Countries in the Population-Based National Prostate Cancer Registry of Sweden," *Cancer*, 103(5): 943–951, 2005.

15. J. E. Johansson, O. Andren, S. O. Andersson, P. W. Dickman, L. Holmberg, A. Magnuson, and H. O. Adami, "Natural History of Early, Localized Prostate Cancer," *JAMA,* 291(22): 2713–2719, 2004.
16. S. A. Tomlins, M. A. Rubin, and A. M. Chinnaiyan, "Integrative Biology of Prostate Cancer Progression," *Annu. Rev. Pathol,* 1: 243–271, 2006.
17. S. Chang, S. R. Long, L. Kutikova, L. Bowman, D. Finley, W. H. Crown, and C. L. Bennett, "Estimating the Cost of Cancer: Results on the Basis of Claims Data Analyses for Cancer Patients Diagnosed with Seven Types of Cancer During 1999 to 2000," *J. Clin. Oncol.,* 22(17): 3524–3530, 2004.
18. E. Sacco, T. Prayer-Galetti, F. Pinto, S. Fracalanza, G. Betto, F. Pagano, and W. Artibani, "Urinary Incontinence after Radical Prostatectomy: Incidence by Definition, Risk Factors and Temporal Trend in a Large Series with a Long-Term Follow-up," *BJU Int.,* 97(6): 1234–1241, 2006.
19. S. D. Kundu, K. A. Roehl, S. E. Eggener, J. A. Antenor, M. Han, and W. J. Catalona, "Potency, Continence and Complications in 3,477 Consecutive Radical Retropubic Prostatectomies," *J. Urol.,* 172(6 Pt 1): 2227–2231, 2004.
20. A. Tabesh, M. Teverovskiy, H. Y. Pang, V. P. Kumar, D. Verbel, A. Kotsianti, and O. Saidi, "Multifeature Prostate Cancer Diagnosis and Gleason Grading of Histological Images," *IEEE Trans. Med. Imaging,* 26(10): 1366–1378, 2007.
21. J. M. Harvey, G. M. Clark, C. K. Osborne, and D. C. Allred, "Estrogen Receptor Status by Immunohistochemistry Is Superior to the Ligand-Binding Assay for Predicting Response to Adjuvant Endocrine Therapy in Breast Cancer," *J. Clin. Oncol,* 17(5): 1474–1481, 1999.
22. R. L. Camp, M. Dolled-Filhart, and D. L. Rimm, "X-Tile: A New Bio-Informatics Tool for Biomarker Assessment and Outcome-Based Cut-Point Optimization," *Clinical Cancer Research,* 10(21): 7252–7259, 2004.
23. M. J. Donovan, S. Hamann, M. Clayton, F. M. Khan, M. Sapir, V. Bayer-Zubek, G. Fernandez, R. Mesa-Tejada, M. Teverovskiy, V. E. Reuter, P. T. Scardino, and C. Cordon-Cardo, "Systems Pathology Approach for the Prediction of Prostate Cancer Progression after Radical Prostatectomy," *J. Clin. Oncol,* 26(24): 3923–3929, 2008.
24. R. Bhargava, "Towards a Practical Fourier Transform Infrared Chemical Imaging Protocol for Cancer Histopathology," *Anal. Bioanal. Chem.,* 389(4): 1155–1169, 2007.
25. M. Jackson, L. P. Choo, P. H. Watson, W. C. Halliday, and H. H. Mantsch, "Beware of Connective Tissue Proteins: Assignment and Implications of Collagen Absorptions in Infrared Spectra of Human Tissues," *Biochim Biophys Acta.,* 1270(1): 1–6, 1995.
26. M. Diem, P. R. Griffiths, and J. M. Chalmers, *Vibrational Spectroscopy for Medical Diagnosis,* Wiley, New York, 2008.
27. R. Bhargava, and I. W. Levin, *Spectrochemical Analysis Using Infrared Multichannel Detectors,* Wiley-Blackwell, Oxford, UK, 2005.
28. W. J. Catalona, D. S. Smith, T. L. Ratliff, K. M. Dodds, D. E. Coplen, J. J. Yuan, J. A. Petros, and G. L. Andriole, "Measurement of Prostate-Specific Antigen in Serum as a Screening Test for Prostate Cancer," *N. Engl. J. Med.,* 324(17): 1156–1161, 1991.
29. N. Weidner, J. P. Semple, W. R. Welch, and J. Folkman, "Tumor Angiogenesis and Metastasis—Correlation in Invasive Breast Carcinoma," *N. Engl. J. Med.,* 324(1): 1–8, 1991.
30. T. Scholzen, and J. Gerdes, "The Ki-67 Protein: From the Known and the Unknown," *J. Cell Physiol,* 182(3): 311–322, 2000.
31. M. J. Piccart-Gebhart, M. Procter, B. Leyland-Jones, A. Goldhirsch, M. Untch, I. Smith, L. Gianni, J. Baselga, R. Bell, C. Jackisch, D. Cameron, M. Dowsett, C. H. Barrios, G. Steger, C. S. Huang, M. Andersson, M. Inbar, M. Lichinitser, I. Lang, U. Nitz, H. Iwata, C. Thomssen, C. Lohrisch, T. M. Suter, J. Ruschoff, T. Suto, V. Greatorex, C. Ward, C. Straehle, E. Mc Fadden, M. S. Dolci, R. D. Gelber, and H. T. S. Team, "Trastuzumab after Adjuvant Chemotherapy in HER-2-Positive Breast Cancer," *New England Journal of Medicine,* 353(16): 1659–1672, 2005.

32. P. Alivisatos, "The Use of Nanocrystals in Biological Detection," *Nature Biotechnology*, 22(1): 47–52, 2004.
33. I. W. Levin, and R. Bhargava, "Fourier Transform Infrared Vibrational Spectroscopic Imaging: Integrating Microscopy and Molecular Recognition," *Annual Review of Physical Chemistry*, 56: 429–474, 2005.
34. M. Jackson, J. R. Mansfield, B. Dolenko, R. L. Somorjai, H. H. Mantsch, and P. H. Watson, "Classification of Breast Tumors by Grade and Steroid Receptor Status Using Pattern Recognition Analysis of Infrared Spectra," *Cancer Detection and Prevention*, 23(3): 245–253, 1999.
35. K. Wehbe, R. Pinneau, M. Moenner, G. Deleris, and C. Petibois, "FT-IR Spectral Imaging of Blood Vessels Reveals Protein Secondary Structure Deviations Induced by Tumor Growth," *Analytical and Bioanalytical Chemistry*, 392(1–2): 129–135, 2008.
36. C. Matthaus, B. Bird, M. Miljkovic, T. Chernenko, M. Romeo, and M. Diem, "Chapter 10: Infrared and Raman Microscopy in Cell Biology," *Methods Cell Biology*, 89: 275–308, 2008.
37. D. I. Ellis, and R. Goodacre, "Metabolic Fingerprinting in Disease Diagnosis: Biomedical Applications of Infrared and Raman Spectroscopy," *Analyst*, 131(8): 875–885, 2006.
38. G. J. Puppels, F. F. M. Demul, C. Otto, J. Greve, M. Robertnicoud, D. J. Arndtjovin, and T. M. Jovin, "Studying Single Living Cells and Chromosomes by Confocal Raman Microspectroscopy," *Nature*, 347(6290): 301–303, 1990.
39. C. Krafft, and V. Sergo, "Biomedical Applications of Raman and Infrared Spectroscopy to Diagnose Tissues," *Spectroscopy—An International Journal*, 20(5–6): 195–218, 2006.
40. C. M. Krishna, G. D. Sockalingum, M. S. Vidyasagar, M. Manfait, D. J. Fernanades, B. M. Vadhiraja, and K. Maheedhar, "An Overview on Applications of Optical Spectroscopy in Cervical Cancers," *Cancer Research and Therapeutics*, 4(1): 26–36, 2008.
41. C. M. Krishna, J. Kurien, S. Mathew, L. Rao, K. Maheedhar, K. K. Kumar, and M. V. P. Chowdary, "Raman Spectroscopy of Breast Tissues," *Expert Review of Molecular Diagnostics*, 8(2): 149–166, 2008.
42. G. Srinivasan, and R. Bhargava, "Fourier Transform-Infrared Spectroscopic Imaging: The Emerging Evolution from a Microscopy Tool to a Cancer Imaging Modality," *Spectroscopy*, 22(7): 30–37, 2007.
43. P. Dumas, N. Jamin, J. L. Teillaud, L. M. Miller, and B. Beccard, "Imaging Capabilities of Synchrotron Infrared Microspectroscopy," *Faraday Discuss*, 126: 289–302; "Discussion" 303–311, 2004.
44. L. M. Miller, and P. Dumas, "Chemical Imaging of Biological Tissue with Synchrotron Infrared Light," *Biochim Biophys Acta.*, 1758(7): 846–857, 2006.
45. E. N. Lewis, P. J. Treado, R. C. Reeder, G. M. Story, A. E. Dowrey, C. Marcott, and I. W. Levin, "Fourier-Transform Spectroscopic Imaging Using an Infrared Focal-Plane Array Detector," *Analytical Chemistry*, 67(19): 3377–3381, 1995.
46. S. G. Kazarian, and K. L. Chan, "Applications of ATR-FTIR Spectroscopic Imaging to Biomedical Samples," *Biochim Biophys Acta.*, 1758(7): 858–867, 2006.
47. B. M. Patterson, G. J. Havrilla, C. Marcott, and G. M. Story, "Infrared Microspectroscopic Imaging Using a Large Radius Germanium Internal Reflection Element and a Focal Plane Array Detector," *Applied Spectroscopy*, 61(11): 1147–1152, 2007.
48. U. Bindig, I. Gersonde, M. Meinke, Y. Becker, and G. Muller, "Fibre-Optic IR-Spectroscopy for Biomedical Diagnostics," *Spectroscopy–An International Journal*, 17(2–3): 323–344, 2003.
49. A. Hammiche, L. Bozec, M. J. German, J. M. Chalmers, N. J. Everall, G. Poulter, M. Reading, D. B. Grandy, F. L. Martin, and H. M. Pollock, "Mid-Infrared Microspectroscopy of Difficult Samples Using Near-Field Photothermal Microspectroscopy," *Spectroscopy*, 19(2): 20–43, 2004.
50. A. Hammiche, L. Bozec, H. M. Pollock, M. German, and M. Reading, "Progress in Near-Field Photothermal Infra-Red Microspectroscopy," *Journal of Microscopy—Oxford*, 213: 129–134, 2004.

51. L. Bozec, A. Hammiche, M. J. Tobin, J. M. Chalmers, N. J. Everall, and H. M. Pollock, "Near-Field Photothermal Fourier Transform Infrared Spectroscopy Using Synchrotron Radiation," *Measurement Science & Technology*, 13(8): 1217–1222, 2002.
52. A. Hammiche, M. J. German, R. Hewitt, H. M. Pollock, and F. L. Martin, "Monitoring Cell Cycle Distributions in MCF-7 Cells Using Near-Field Photothermal Microspectroscopy," *Biophysical Journal*, 88(5): 3699–3706, 2005.
53. O. Grude, A. Hammiche, H. Pollock, A. J. Bentley, M. J. Walsh, F. L. Martin, and N. J. Fullwood, "Near-Field Photothermal Microspectroscopy for Adult Stem-Cell Identification and Characterization," *Journal of Microscopy–Oxford*, 228(3): 366–372, 2007.
54. M. J. Walsh, T. G. Fellous, A. Hammiche, W. R. Lin, N. J. Fullwood, O. Grude, F. Bahrami, J. M. Nicholson, M. Cotte, J. Susini, H. M. Pollock, M. Brittan, P. L. Martin-Hirsch, M. R. Alison, and F. L. Martin, "Fourier Transform Infrared Microspectroscopy Identifies Symmetric PO2-Modifications as a Marker of the Putative Stem Cell Region of Human Intestinal Crypts," *Stem Cells*, 26(1): 108–118, 2008.
55. M. J. Walsh, A. Hammiche, T. G. Fellous, J. M. Nicholson, M. Cotte, J. Susini, N. J. Fullwood, P. L. Martin-Hirsch, M. R. Alison, and F. L. Martin, "Tracking the Cell Hierarchy in the Human Intestine Using Biochemical Signatures Derived by Mid-Infrared Microspectroscopy," *Stem Cell Res.*, 3(1): 15-27 2009.
56. P. Heraud, and M. J. Tobin, "The Emergence of Biospectroscopy in Stem Cell Research," *Stem Cell Res.*, 2009.
57. S. Boydston-White, T. Chernenko, A. Regina, M. Miljkovic, C. Matthaus, and M. Diem, "Microspectroscopy of Single Proliferating Hela Cells," *Vibrational Spectroscopy*, 38(1–2): 169–177, 2005.
58. H. N. Holman, M. C. Martin, E. A. Blakely, K. Bjornstad, and W. R. McKinney, "IR Spectroscopic Characteristics of Cell Cycle and Cell Death Probed by Synchrotron Radiation-Based Fourier Transform IR Spectromicroscopy," *Biopolymers*, 57(6): 329–335, 2000.
59. F. Gasparri, and M. Muzio, "Monitoring of Apoptosis of HL60 Cells by Fourier-Transform Infrared Spectroscopy," *Biochemical Journal* 369: 239–248, 2003.
60. N. Jamin, L. Miller, J. Moncuit, W. H. Fridman, P. Dumas, and J. L. Teillaud, "Chemical Heterogeneity in Cell Death: Combined Synchrotron IR and Fluorescence Microscopy Studies of Single Apoptotic and Necrotic Cells," *Biopolymers*, 72(5): 366–373, 2003.
61. R. T. Yamaguchi, A. Hirano-Iwata, Y. Kimura, M. Niwano, K. I. Miyamoto, H. Isoda, and H. Miyazaki, "Real-Time Monitoring of Cell Death by Surface Infrared Spectroscopy," *Applied Physics Letters*, 91(20):203902-1–203902-3, 2007.
62. C. Matthaus, S. Boydston-White, M. Miljkovic, M. Romeo, and M. Diem, "Raman and Infrared Microspectral Imaging of Mitotic Cells," *Applied Spectroscopy*, 60(1): 1–8, 2006.
63. E. Gazi, J. Dwyer, N. P. Lockyer, J. Miyan, P. Gardner, C. A. Hart, M. D. Brown, and N. W. Clarke, "A Study of Cytokinetic and Motile Prostate Cancer Cells Using Synchrotron-Based FTIR Micro Spectroscopic Imaging," *Vibrational Spectroscopy*, 38(1–2): 193–201, 2005.
64. J. R. Mourant, Y. R. Yamada, S. Carpenter, L. R. Dominique, and J. P. Freyer, "FTIR Spectroscopy Demonstrates Biochemical Differences in Mammalian Cell Cultures at Different Growth Stages," *Biophysical J.*, 85(3): 1938–1947, 2003.
65. R. K. Sahu, S. Mordechai, and E. Manor, "Nucleic Acids Absorbance in Mid-IR and Its Effect on Diagnostic Variates During Cell Division: A Case Study with Lymphoblastic Cells," *Biopolymers*, 89(11): 993–1001, 2008.
66. S. Boydston-White, T. Gopen, S. Houser, J. Bargonetti, and M. Diem, "Infrared Spectroscopy of Human Tissue. V. Infrared Spectroscopic Studies of Myeloid Leukemia (ML-1) Cells at Different Phases of the Cell Cycle," *Biospectroscopy*, 5(4): 219–227, 1999.
67. S. Boydston-White, M. Romeo, T. Chernenko, A. Regina, M. Miljkovic, and M. Diem, "Cell-Cycle-Dependent Variations in FTIR Micro-Spectra of Single Proliferating HeLa Cells: Principal Component and Artificial Neural

Network Analysis," *Biochimica et Biophysica Acta-Biomembranes,* 1758(7): 908–914, 2006.

68. H. Y. Jiao, S. L. Allinson, M. J. Walsh, R. Hewitt, K. J. Cole, D. H. Phillips, and F. L. Martin, "Growth Kinetics in MCF-7 Cells Modulate Benzo[a]Pyrene-Induced CYP1A1 Up-Regulation," *Mutagenesis,* 22(2): 111–116, 2007.
69. J. R. Mourant, J. Dominguez, S. Carpenter, K. W. Short, T. M. Powers, R. Michalczyk, N. Kunapareddy, A. Guerra, and J. P. Freyer, "Comparison of Vibrational Spectroscopy to Biochemical and Flow Cytometry Methods for Analysis of the Basic Biochemical Composition of Mammalian Cells," *J. Biomedical Optics,* 11(6): 064024-1–064024-11, 2006.
70. E. Gazi, J. Dwyer, N. P. Lockyer, J. Miyan, P. Gardner, C. Hart, M. Brown, and N. W. Clarke, "Fixation Protocols for Subcellular Imaging by Synchrotron-Based Fourier Transform Infrared Microspectroscopy," *Biopolymers,* 77(1): 18–30, 2005.
71. M. Miljkovic, M. Romeo, C. Matthaus, and M. Diem, "Infrared Microspectroscopy of Individual Human Cervical Cancer (HeLa) Cells Suspended in Growth Medium," *Biopolymers,* 74(1–2): 172–175, 2004.
72. D. A. Moss, M. Keese, and R. Pepperkok, "IR Micro Spectroscopy of Live Cells," *Vibrational Spectroscopy,* 38(1–2): 185–191, 2005.
73. E. Bogomolny, M. Huleihel, Y. Suproun, R. K. Sahu, and S. Mordechai, "Early Spectral Changes of Cellular Malignant Transformation Using Fourier Transform Infrared Microspectroscopy," *Journal of Biomedical Optics,* 12(2): 2007.
74. K. Z. Liu, L. Jia, S. M. Kelsey, A. C. Newland, and H. H. Mantsch, "Quantitative Determination of Apoptosis on Leukemia Cells by Infrared Spectroscopy," *Apoptosis,* 6(4): 269–278, 2001.
75. J. Sule-Suso, D. Skingsley, G. D. Sockalingum, A. Kohler, G. Kegelaer, M. Manfait, and A. El Haj, "FT-IR Microspectroscopy as a Tool to Assess Lung Cancer Cells Response to Chemotherapy," *Vibrational Spectroscopy,* 38(1–2): 179–184, 2005.
76. S. Chio-Srichan, M. Refregiers, F. Jamme, S. Kascakova, V. Rouam, and P. Dumas, "Photosensitizer Effects on Cancerous Cells: A Combined Study Using Synchrotron Infrared and Fluorescence Microscopies," *Biochimica et Biophysica Acta—General Subjects,* 1780(5): 854–860, 2008.
77. D. C. Malins, P. M. Johnson, E. A. Barker, N. L. Polissar, T. M. Wheeler, and K. M. Anderson, "Cancer-Related Changes in Prostate DNA as Men Age and Early Identification of Metastasis in Primary Prostate Tumors," *Proceedings Nat. Acad. Sci. USA,* 100(9): 5401–5406, 2003.
78. C. M. Krishna, G. Kegelaer, I. Adt, S. Rubin, V. B. Kartha, M. Manfait, and G. D. Sockalingum, "Characterisation of Uterine Sarcoma Cell Lines Exhibiting MDR Phenotype by Vibrational Spectroscopy," *Biochimica et Biophysica Acta—General Subjects,* 1726(2): 160–167, 2005.
79. C. M. Krishna, G. Kegelaerl, I. Adt, S. Rubin, V. B. Kartha, M. Manfait, and G. D. Sockalingum, "Combined Fourier Transform Infrared and Raman Spectroscopic Identification Approach for Identification of Multidrug Resistance Phenotype in Cancer Cell Lines," *Biopolymers,* 82(5): 462–470, 2006.
80. K. M. Yamada, and E. Cukierman, "Modeling Tissue Morphogenesis and Cancer in 3D," *Cell,* 130(4): 601–610, 2007.
81. Y. Yang, J. Sule-Suso, G. D. Sockalingum, G. Kegelaer, M. Manfait, and A. J. El Haj "Study of Tumor Cell Invasion by Fourier Transform Infrared Microspectroscopy," *Biopolymers,* 78(6): 311–317, 2005.
82. R. Bhargava, and R. Kong, "Structural and Biochemical Characterization of Engineered Tissue Using FTIR Spectroscopic Imaging: Melanoma Progression as an Example," *Proceedings of SPIE,* 6870: 687004, 2008.
83. M. Hartmann, B. D. Hanh, H. Podhaisky, J. Wensch, J. Bodzenta, S. Wartewig, and R. H. Neubert, "A New FTIR-ATR Cell for Drug Diffusion Studies," *Analyst,* 129(10): 902–905, 2004.
84. B. D. Hanh, R. H. H. Neubert, S. Wartewig, and J. Lasch, "Penetration of Compounds through Human Stratum Corneum as Studied by Fourier Transform Infrared Photoacoustic Spectroscopy," *Journal Controlled Release,* 70(3): 393–398, 2001.

85. B. Gotter, W. Faubel, and R. H. H. Neubert, "Optical Methods for Measurements of Skin Penetration," *Skin Pharmacology and Physiology,* 21(3): 156–165, 2008.
86. M. Boncheva, F. H. Tay, and S. G. Kazarian, "Application of Attenuated Total Reflection Fourier Transform Infrared Imaging and Tape-Stripping to Investigate the Three-Dimensional Distribution of Exogenous Chemicals and the Molecular Organization in Stratum Corneum," *J. Biomedical Optics,* 13(6): 064009-01–064009-07, 2008.
87. J. M. Andanson, K. L. A. Chan, and S. G. Kazarian, "High-Throughput Spectroscopic Imaging Applied to Permeation Through the Skin," *Applied Spectroscopy,* 63(5): 512–517, 2009.
88. H. Fabian, P. Lasch, M. Boese, and W. Haensch, "Mid-IR Microspectroscopic Imaging of Breast Tumor Tissue Sections," *Biopolymers,* 67(4–5): 354–357, 2002.
89. H. Fabian, P. Lasch, M. Boese, and W. Haensch, "Infrared Microspectroscopic Imaging of Benign Breast Tumor Tissue Sections," *Journal Molecular Structure,* 661: 411–417, 2003.
90. H. Fabian, N. A. N. Thi, M. Eiden, P. Lasch, J. Schmitt, and D. Naumann, "Diagnosing Benign and Malignant Lesions in Breast Tissue Sections by Using IR-Microspectroscopy," *Biochimica et Biophysica Acta—Biomembranes,* 1758(7): 874–882, 2006.
91. Y. X. Ci, T. Y. Gao, J. Feng, and Z. Q. Guo, "Fourier Transform Infrared Spectroscopic Characterization of Human Breast Tissue: Implications for Breast Cancer Diagnosis," *Applied Spectroscopy,* 53(3): 312–315, 1999.
92. C. L. Liu, Y. Zhang, X. H. Yan, X. Y. Zhang, C. X. Li, W. T. Yang, and D. R. Shi, "Infrared Absorption of Human Breast Tissues *In Vitro,*" *J. Luminescence,* 119: 132–136, 2006.
93. G. J. Ooi, J. Fox, K. Siu, R. Lewis, K. R. Bambery, D. McNaughton, and B. R. Wood, "Fourier Transform Infrared Imaging and Small Angle X-ray Scattering as a Combined Biomolecular Approach to Diagnosis of Breast Cancer," *Medical Physics,* 35(5): 2151–2161, 2008.
94. F. N. Keith, R. K. Reddy, and R. Bhargava, "Practical Protocols for Fast Histopathology by Fourier Transform Infrared Spectroscopic Imaging," *Proceedings of SPIE,* 6853: 685306.1–685306.10, 2008.
95. D. J. Lyman, and J. Murray-Wijelath "Fourier Transform Infrared Attenuated Total Reflection Analysis of Human Hair: Comparison of Hair From Breast Cancer Patients with Hair from Healthy Subjects," *Applied Spectroscopy,* 59(1): 26–32, 2005.
96. B. Bird, M. Miljkovic, M. J. Romeo, J. Smith, N. Stone, M. W. George, and M. Diem, "Infrared Micro-Spectral Imaging: Distinction of Tissue Types in Axillary Lymph Node Histology," *BMC Clin. Pathol.,* 8: 8, 2008.
97. D. C. Malins, N. L. Polissar, and S. J. Gunselman, "Tumor Progression to the Metastatic State Involves Structural Modifications in DNA Markedly Different from Those Associated with Primary Tumor Formation," *Proceedings Natl. Acad. Sci. USA,* 93(24): 14047–14052, 1996.
98. K. M. Anderson, P. Jaruga, C. R. Ramsey, N. K. Gilman, V. M. Green, S. W. Rostad, J. T. Emerman, M. Dizdaroglu, and D. C. Malins, "Structural Alterations in Breast Stromal and Epithelial DNA: The Influence of 8,5′-Cyclo-2′-Deoxyadenosine," *Cell Cycle,* 5(11): 1240–1244, 2006.
99. W. J. Catalona, D. S. Smith, T. L. Ratliff, K. M. Dodds, D. E. Coplen, J. J. J. Yuan, J. A. Petros, and G. L. Andriole, "Measurement of Prostate-Specific Antigen in Serum as a Screening-Test for Prostate Cancer," *N. Eng. Journal Med.,* 324(17): 1156–1161, 1991.
100. E. Gazi, J. Dwyer, P. Gardner, A. Ghanbari-Siahkali, A. P. Wade, J. Miyan, N. P. Lockyer, J. C. Vickerman, N. W. Clarke, J. H. Shanks, L. J. Scott, C. A. Hart, and M. Brown, "Applications of Fourier Transform Infrared Microspectroscopy in Studies of Benign Prostate and Prostate Cancer. A Pilot Study," *Journal of Pathology,* 201(1): 99–108, 2003.
101. E. Gazi, J. Dwyer, N. Lockyer, P. Gardner, J. C. Vickerman, J. Miyan, C. A. Hart, M. Brown, J. H. Shanks, and N. Clarke, "The Combined Application of FTIR Microspectroscopy and ToF-SIMS Imaging in the Study of Prostate Cancer," *Faraday Discussions,* 126: 41–59, 2004.

102. E. Gazi, M. Baker, J. Dwyer, N. P. Lockyer, P. Gardner, J. H. Shanks, R. S. Reeve, C. A. Hart, N. W. Clarke, and M. D. Brown, "A Correlation of FTIR Spectra Derived from Prostate Cancer Biopsies with Gleason Grade and Tumour Stage," *European Urology*, 50(4): 750–761, 2006.
103. M. J. Baker, E. Gazi, M. D. Brown, J. H. Shanks, P. Gardner, and N. W. Clarke, "FTIR-Based Spectroscopic Analysis in the Identification of Clinically Aggressive Prostate Cancer," *British Journal of Cancer*, 99(11): 1859–1866, 2008.
104. E. Gazi, J. Dwyer, N. P. Lockyer, P. Gardner, J. H. Shanks, J. Roulson, C. A. Hart, N. W. Clarke, and M. D. Brown, "Biomolecular Profiling of Metastatic Prostate Cancer Cells in Bone Marrow Tissue Using FTIR Microspectroscopy: A Pilot Study," *Analytical and Bioanalytical Chemistry*, 387(5): 1621–1631, 2007.
105. D. C. Fernandez, R. Bhargava, S. M. Hewitt, and I. W. Levin, "Infrared Spectroscopic Imaging for Histopathologic Recognition," *Nature Biotechnology*, 23(4): 469–474, 2005.
106. M. J. German, A. Hammiche, N. Ragavan, M. J. Tobin, N. J. Fullwood, S. S. Matanhelia, A. C. Hindley, C. M. Nicholson, H. M. Pollock, and F. L. Martin, "Infrared Spectroscopy with Multivariate Analysis Potentially Facilitates the Segregation of Different Types of Prostate Cell. (vol 90, pg 3738, 2006)," *Biophysical Journal*, 91(2): 775–775, 2006.
107. M. J. Li, H. S. Hsu, R. C. Liang, and S. Y. Lin, "Infrared Microspectroscopic Detection of Epithelial and Stromal Growth in the Human Benign Prostatic Hyperplasia," *Ultrastructural Pathology*, 26(6): 365–370, 2002.
108. C. Paluszkiewicz, W. M. Kwiatek, A. Banas, A. Kisiel, A. Marcelli, and A. Piccinini, "SR-FTIR Spectroscopic Preliminary Findings of Non-Cancerous, Cancerous, and Hyperplastic Human Prostate Tissues," *Vibrational Spectroscopy*, 43(1): 237–242, 2007.
109. D. C. Malins, N. K. Gilman, V. M. Green, T. M. Wheeler, E. A. Barker, and K. M. Anderson, "A Cancer DNA Phenotype in Healthy Prostates, Conserved in Tumors and Adjacent Normal Cells, Implies a Relationship to Carcinogenesis," *Proceeding Natl. Acad. Sci. USA*, 102(52): 19093–19096, 2005.
110. D. C. Malins, N. K. Gilman, V. M. Green, T. M. Wheeler, E. A. Barker, M. A. Vinson, M. Sayeeduddin, K. E. Hellstrom, and K. M. Anderson, "Metastatic Cancer DNA Phenotype Identified in Normal Tissues Surrounding Metastasizing Prostate Carcinomas," *Proceedings Nat. Acad. Sci. USA*, 101(31): 11428–11431, 2004.
111. S. Argov, J. Ramesh, A. Salman, I. Sinelnikov, J. Goldstein, H. Guterman, and S. Mordechai, "Diagnostic Potential of Fourier-Transform Infrared Microspectroscopy and Advanced Computational Methods in Colon Cancer Patients," *Journal of Biomedical Optics*, 7(2): 248–254, 2002.
112. S. Argov, R. K. Sahu, E. Bernshtain, A. Salman, G. Shohat, U. Zelig, and S. Mordechai, "Inflamatory Bowel Diseases as an Intermediate Stage between Normal and Cancer: A FTIR-Microspectroscopy Approach," *Biopolymers*, 75(5): 384–392, 2004.
113. R. K. Sahu, S. Argov, E. Bernshtain, A. Salman, S. Walfisch, J. Goldstein, and S. Mordechai, "Detection of Abnormal Proliferation in Histologically 'Normal' Colonic Biopsies Using FTIR-Microspectroscopy," *Scandinavian Journal of Gastroenterology*, 39(6): 557–566, 2004.
114. M. Khanmohammadi, A. B. Garmarudi, K. Ghasemi, H. K. Jaliseh, and A. Kaviani, "Diagnosis of Colon Cancer by Attenuated Total Reflectance-Fourier Transform Infrared Microspectroscopy and Soft Independent Modeling of Class Analogy," *Med. Oncol.*, 2008.
115. V. R. Kondepati, H. M. Heise, T. Oszinda, R. Mueller, M. Keese, and J. Backhaus, "Detection of Structural Disorders in Colorectal Cancer DNA with Fourier-Transform Infrared Spectroscopy," *Vibrational Spectroscopy*, 46(2): 150–157, 2008.
116. P. Lasch, W. Haensch, E. N. Lewis, L. H. Kidder, and D. Naumann, "Characterization of Colorectal Adenocarcinoma Sections by Spatially Resolved FT-IR Microspectroscopy," *Applied Spectroscopy*, 56(1): 1–9, 2002.
117. P. Lasch, W. Haensch, D. Naumann, and M. Diem, "Imaging of Colorectal Adenocarcinoma Using FT-IR Microspectroscopy and Cluster Analysis," *Biochimica Et Biophysica Acta—Molecular Basis of Disease*, 1688(2): 176–186, 2004.

118. N. Fujioka, Y. Morimoto, T. Arai, and M. Kikuchi, "Discrimination Between Normal and Malignant Human Gastric Tissues by Fourier Transform Infrared Spectroscopy," *Cancer Detection and Prevention*, 28(1): 32–36, 2004.
119. Q. B. Li, X. J. Sun, Y. Z. Xu, L. M. Yang, Y. F. Zhang, S. F. Weng, J. S. Shi, and J. G. Wu, "Diagnosis of Gastric Inflammation and Malignancy in Endoscopic Biopsies Based on Fourier Transform Infrared Spectroscopy," *Clinical Chemistry*, 51(2): 346–350, 2005.
120. Q. B. Li, X. J. Sun, Y. Z. Xu, L. M. Yang, Y. F. Zhang, S. F. Weng, J. S. Shi, and J. G. Wu, "Use of Fourier-Transform Infrared Spectroscopy to Rapidly Diagnose Gastric Endoscopic Biopsies," *World J. Gastroenterol*, 11(25): 3842–3845, 2005.
121. T. D. Wang, G. Triadafilopoulos, J. M. Crawford, L. R. Dixon, T. Bhandari, P. Sahbaie, S. Friediand, R. Soetikno, and C. H. Contag, "Detection of Endogenous Biomolecules in Barrett's Esophagus by Fourier Transform Infrared Spectroscopy," *Proceedings Nat. Acad. Sci. USA*, 104(40): 15864–15869, 2007.
122. M. J. Walsh, M. J. German, M. Singh, H. M. Pollock, A. Hammiche, M. Kyrgiou, H. F. Stringfellow, E. Paraskevaidis, P. L. Martin-Hirsch, and F. L. Martin, "IR Microspectroscopy: Potential Applications in Cervical Cancer Screening," *Cancer Letters*, 246(1–2): 1–11, 2007.
123. C. Murali Krishna, G. D. Sockalingum, M. S. Vidyasagar, M. Manfait, D. J. Fernanades, B. M. Vadhiraja, and K. Maheedhar, "An Overview on Applications of Optical Spectroscopy in Cervical Cancers," *J. Cancer Res. Ther.*, 4(1): 26–36, 2008.
124. F. Bazant-Hegemark, K. Edey, G. R. Swingler, M. D. Read, and N. Stone, "Review: Optical Micrometer Resolution Scanning for Non-Invasive Grading of Precancer in the Human Uterine Cervix," *Technol Cancer Res. Treat.*, 7(6): 483–496, 2008.
125. P. T. Wong, R. K. Wong, T. A. Caputo, T. A. Godwin, and B. Rigas, "Infrared Spectroscopy of Exfoliated Human Cervical Cells: Evidence of Extensive Structural Changes during Carcinogenesis," *Proceeding Natl. Acad. Sci. USA*, 88(24): 10988–10992, 1991.
126. H. M. Yazdi, M. A. Bertrand, and P. T. Wong, "Detecting Structural Changes at the Molecular Level with Fourier Transform Infrared Spectroscopy. A Potential Tool for Prescreening Preinvasive Lesions of the Cervix," *Acta Cytol.*, 40(4): 664–648, 1996.
127. B. R. Wood, M. A. Quinn, F. R. Burden, and D. McNaughton, "An Investigation into FTIR Spectroscopy as a Biodiagnostic Tool for Cervical Cancer," *Biospectroscopy*, 2(3): 143–153, 1996.
128. L. Chiriboga, P. Xie, H. Yee, D. Zarou, D. Zakim, and M. Diem, "Infrared Spectroscopy of Human Tissue. IV. Detection of Dysplastic and Neoplastic Changes of Human Cervical Tissue via Infrared Microscopy," *Cell Mol. Biol. (Noisy-le-grand)*, 44(1): 219–229, 1998.
129. M. Romeo, F. Burden, M. Quinn, B. Wood, and D. McNaughton, "Infrared Microspectroscopy and Artificial Neural Networks in the Diagnosis of Cervical Cancer," *Cell Mol. Biol. (Noisy-le-grand)*, 44(1): 179–187, 1998.
130. M. A. Cohenford, and B. Rigas, "Cytologically Normal Cells from Neoplastic Cervical Samples Display Extensive Structural Abnormalities on IR Spectroscopy: Implications for Tumor Biology," *Proc. Natl. Acad. Sci. USA*, 95(26): 15327–15332, 1998.
131. S. Mordechai, R. K. Sahu, Z. Hammody, S. Mark, K. Kantarovich, H. Guterman, A. Podshyvalov, J. Goldstein, and S. Argov, "Possible Common Biomarkers from FTIR Microspectroscopy of Cervical Cancer and Melanoma," *Journal of Microscopy-Oxford*, 215: 86–91, 2004.
132. A. Podshyvalov, R. K. Sahu, S. Mark, K. Kantarovich, H. Guterman, J. Goldstein, R. Jagannathan, S. Argov, and S. Mordechai, "Distinction of Cervical Cancer Biopsies by Use of Infrared Microspectroscopy and Probabilistic Neural Networks," *Applied Optics*, 44(18): 3725–3734, 2005.
133. S. Mark, R. K. Sahu, K. Kantarovich, A. Podshyvalov, H. Guterman, J. Goldstein, R. Jagannathan, S. Argov, and S. Mordechai, "Fourier Transform Infrared Microspectroscopy as a Quantitative Diagnostic Tool for Assignment

of Premalignancy Grading in Cervical Neoplasia," *Journal of Biomedical Optics*, 9(3): 558–567, 2004.

134. C. M. Krishna, G. D. Sockalingum, B. M. Vadhiraja, K. Maheedhar, A. C. Rao, L. Rao, L. Venteo, M. Pluot, D. J. Fernandes, M. S. Vidyasagar, V. B. Kartha, and M. Manfait, "Vibrational Spectroscopy Studies of Formalin-Fixed Cervix Tissues," *Biopolymers*, 85(3): 214–221, 2007.
135. M. J. Walsh, M. N. Singh, H. M. Pollock, L. J. Cooper, M. J. German, H. F. Stringfellow, N. J. Fullwood, E. Paraskevaidis, P. L. Martin-Hirsch, and F. L. Martin, "ATR Microspectroscopy with Multivariate Analysis Segregates Grades of Exfoliative Cervical Cytology," *Biochemical and Biophysical Research Communications*, 352(1): 213–219, 2007.
136. M. J. Walsh, M. N. Singh, H. F. Stringfellow, H. M. Pollock, A. Hammiche, O. Grude, N. J. Fullwood, M. A. Pitt, P. L. Martin-Hirsch, and F. L. Martin, "FTIR Microspectroscopy Coupled with Two-Class Discrimination Segregates Markers Responsible for Inter- and Intra-Category Variance in Exfoliative Cervical Cytology," *Biomark Insights*, 3: 179–189, 2008.
137. K. R. Bambery, B. R. Wood, M. A. Quinn, and D. McNaughton, "Fourier Transform Infrared Imaging and Unsupervised Hierarchical Clustering Applied to Cervical Biopsies," *Australian Journal of Chemistry*, 57(12): 1139–1143, 2004.
138. B. R. Wood, L. Chiriboga, H. Yee, M. A. Quinn, D. McNaughton, and M. Diem, "Fourier Transform Infrared (FTIR) Spectral Mapping of the Cervical Transformation Zone, and Dysplastic Squamous Epithelium," *Gynecologic Oncology*, 93(1): 59–68, 2004.
139. W. Steller, J. Einenkel, L. C. Horn, U. D. Braumann, H. Binder, R. Salzer, and C. Krafft, "Delimitation of Squamous Cell Cervical Carcinoma Using Infrared Microspectroscopic Imaging," *Analytical and Bioanalytical Chemistry*, 384(1): 145–154, 2006.
140. B. R. Wood, K. R. Bambery, C. J. Evans, M. A. Quinn, and D. McNaughton, "A three-Dimensional Multivariate Image Processing Technique for the Analysis of FTIR Spectroscopic Images of Multiple Tissue Sections," *BMC Medical Imaging*, 6: 12, 2006.
141. B. R. Wood, M. A. Quinn, B. Tait, M. Ashdown, T. Hislop, M. Romeo, and D. McNaughton, "FTIR Microspectroscopic Study of Cell Types and Potential Confounding Variables in Screening for Cervical Malignancies," *Biospectroscopy*, 4(2): 75–91, 1998.
142. M. J. Romeo, B. R. Wood, M. A. Quinn, and D. McNaughton, "Removal of Blood Components from Cervical Smears: Implications for Cancer Diagnosis Using FTIR Spectroscopy," *Biopolymers*, 72(1): 69–76, 2003.
143. P. T. Wong, M. K. Senterman, P. Jackli, R. K. Wong, S. Salib, C. E. Campbell, R. Feigel, W. Faught, and M. Fung Kee Fung, "Detailed Account of Confounding Factors in Interpretation of FTIR Spectra of Exfoliated Cervical Cells," *Biopolymers*, 67(6): 376–386, 2002.
144. L. Chiriboga, P. Xie, H. Yee, V. Vigorita, D. Zarou, D. Zakim, and M. Diem, "Infrared Spectroscopy of Human Tissue. I. Differentiation and Maturation of Epithelial Cells in the Human Cervix," *Biospectroscopy*, 4(1): 47–53, 1998.
145. L. Chiriboga, P. Xie, V. Vigorita, D. Zarou, D. Zakim, and M. Diem, "Infrared Spectroscopy of Human Tissue. II. A Comparative Study of Spectra of Biopsies of Cervical Squamous Epithelium and of Exfoliated Cervical Cells," *Biospectroscopy*, 4(1): 55–59, 1998.
146. L. Chiriboga, P. Xie, W. Zhang, and M. Diem, "Infrared Spectroscopy of Human Tissue .3. Spectral Differences Between Squamous and Columnar Tissue and Cells from the Human Cervix," *Biospectroscopy*, 3(4): 253–257, 1997.
147. M. J. Romeo, B. R. Wood, and D. McNaughton, "Observing the Cyclical Changes in Cervical Epithelium Using Infrared Microspectroscopy," *Vibrational Spectroscopy*, 28(1): 167–175, 2002.
148. G. Steiner, S. Kuchler, A. Hermann, E. Koch, R. Salzer, G. Schackert, and M. Kirsch, "Rapid and Label-Free Classification of Human Glioma Cells by Infrared Spectroscopic Imaging," *Cytometry A.*, 73A(12): 1158–1164, 2008.

149. K. R. Bambery, E. Schultke, B. R. Wood, S. T. Rigley MacDonald, K. Ataelmannan, R. W. Griebel, B. H. Juurlink, and D. McNaughton, "A Fourier Transform Infrared Microspectroscopic Imaging Investigation into an Animal Model Exhibiting Glioblastoma Multiforme," *Biochim. Biophys. Acta.*, 1758(7): 900–907, 2006.
150. A. Beljebbar, N. Amharref, A. Leveques, S. Dukic, L. Venteo, L. Schneider, M. Pluot, and M. Manfait, "Modeling and Quantifying Biochemical Changes in C6 Tumor Gliomas by Fourier Transform Infrared Imaging," *Anal. Chem.*, 80(22): 8406–8415, 2008.
151. L. S. Lee, C. W. Chi, H. C. Liu, C. L. Cheng, M. J. Li, and S. Y. Lin, "Assessment of Protein Conformation in Human Benign and Malignant Astrocytomas by Reflectance Fourier Transform Infrared Microspectroscopy," *Oncology Research*, 10(1): 23–27, 1998.
152. K. Ali, Y. Lu, C. Christensen, T. May, C. Hyett, R. Griebel, D. Fourney, K. Meguro, L. Resch, and R. K. Sharma,"Fourier Transform Infrared Spectromicroscopy and Hierarchical Cluster Analysis of Human Meningiomas," *Int. J. Mol. Med.*, 21(3): 297–301, 2008.
153. C. Krafft, L. Shapoval, S. B. Sobottka, K. D. Geiger, G. Schackert, and R. Salzer, "Identification of Primary Tumors of Brain Metastases by SIMCA Classification of IR Spectroscopic Images," *Biochimica et Biophysica Acta—Biomembranes*, 1758(7): 883–891, 2006.
154. C. Krafft, L. Shapoval, S. B. Sobottka, G. Schackert, and R. Salzer, "Identification of Primary Tumors of Brain Metastases by Infrared Spectroscopic Imaging and Linear Discriminant Analysis," *Technology in Cancer Research & Treatment*, 5(3): 291–298, 2006.
155. L. M. Miller, Q. Wang, T. P. Telivala, R. J. Smith, A. Lanzirotti, and J. Miklossy, "Synchrotron-Based Infrared and X-Ray Imaging Shows Focalized Accumulation of Cu and Zn Co-Localized with Beta-Amyloid Deposits in Alzheimer's Disease," *Journal Structural Biology*, 155(1): 30–37, 2006.
156. P. Carmona, A. Rodriguez-Casado, I. Alvarez, E. de Miguel, and A. Toledano, "FTIR Microspectroscopic Analysis of the Effects of Certain Drugs on Oxidative Stress and Brain Protein Structure," *Biopolymers*, 89(6): 548–554, 2008.
157. M. Szczerbowska-Boruchowska, P. Dumas, M. Z. Kastyak, J. Chwiej, M. Lankosz, D. Adamek, and A. Krygowska-Wajs "Biomolecular Investigation of Human Substantia Nigra in Parkinson's Disease by Synchrotron Radiation Fourier Transform Infrared Microspectroscopy," *Archives of Biochemistry and Biophysic*, 459(2): 241–248, 2007.
158. S. M. LeVine, and D. L. Wetzel, "Chemical Analysis of Multiple Sclerosis Lesions by FT-IR Microspectroscopy," *Free Radical Biology & Medicine*, 25(1): 33–41, 1998.
159. J. Kneipp, P. Lasch, E. Baldauf, M. Beekes, and D. Naumann, "Detection of Pathological Molecular Alterations in Scrapie-Infected Hamster Brain by Fourier Transform Infrared (FT-IR) Spectroscopy," *Biochimica et Biophysica Acta*, 1501(2–3): 189–199, 2000.
160. A. Thomzig, S. Spassov, M. Friedrich, D. Naumann, and M. Beekes, "Discriminating Scrapie and Bovine Spongiform Encephalopathy Isolates by Infrared Spectroscopy of Pathological Prion Protein," *Journal of Biological Chemistry*, 279(32): 33847–33854, 2004.
161. Z. Hammody, R. K. Sahu, S. Mordechai, E. Cagnano, and S. Argov, "Characterization of Malignant Melanoma Using Vibrational Spectroscopy," *Scientific World Journal*, 5: 173–182, 2005.
162. A. Tfayli, O. Piot, A. Durlach, P. Bernard, and M. Manfait, "Discriminating Nevus and Melanoma on Paraffin-Embedded Skin Biopsies Using FTIR Microspectroscopy," *Biochim Biophys Acta*, 1724(3): 262–269, 2005.
163. E. Ly, O. Piot, R. Wolthuis, A. Durlach, P. Bernard, and M. Manfait, "Combination of FTIR Spectral Imaging and Chemometrics for Tumour Detection from Paraffin-Embedded Biopsies," *Analyst*, 133(2): 197–205, 2008.
164. Z. Hammody, S. Argov, R. K. Sahu, E. Cagnano, R. Moreh, and S. Mordechai, "Distinction of Malignant Melanoma and Epidermis Using IR Micro-Spectroscopy and Statistical Methods," *Analyst*, 133(3): 372–378, 2008.

165. E. Ly, O. Piot, A. Durlach, P. Bernard, and M. Manfait, "Differential Diagnosis of Cutaneous Carcinomas by Infrared Spectral Micro-Imaging Combined with Pattern Recognition," *Analyst*, 134(6): 1208–1214, 2009.
166. M. Boncheva, F. H. Tay, and S. G. Kazarian, "Application of Attenuated Total Reflection Fourier Transform Infrared Imaging and Tape-Stripping to Investigate the Three-Dimensional Distribution of Exogenous Chemicals and the Molecular Organization in Stratum Corneum," *J. Biomed. Opt.*, 13(6): 064009, 2008.
167. J. Tetteh, K. T. Mader, J. M. Andanson, W. J. McAuley, M. E. Lane, J. Hadgraft, S. G. Kazarian, and J. C. Mitchell,"Local Examination of Skin Diffusion Using FTIR Spectroscopic Imaging and Multivariate Target Factor Analysis," *Analytica Chimica Acta*, 642(1–2): 246–256, 2009.
168. K.L.A. Chan, and S. G. Kazarian, "Chemical Imaging of the Stratum Corneum Under Controlled Humidity with the Attenuated Total Reflection Fourier Transform Infrared Spectroscopy Method," *Journal of Biomedical Optics*, 12(4): 044010-1–044010-10, 2007.
169. J. G. Kelly, M. N. Singh, H. F. Stringfellow, M. J. Walsh, J. M. Nicholson, F. Bahrami, K. M. Ashton, M. A. Pitt, P. L. Martin-Hirsch, and F. L. Martin, "Derivation of a Subtype-Specific Biochemical Signature of Endometrial Carcinoma Using Synchrotron-Based Fourier-Transform Infrared Microspectroscopy," *Cancer Letters*, 274(2): 208–217, 2009.
170. C. P. Schultz, K. Z. Liu, J. B. Johnston, and H. H. Mantsch, "Prognosis of Chronic Lymphocytic Leukemia from Infrared Spectra of Lymphocytes," *Journal of Molecular Structure*, 408: 253–256, 1997.
171. D. C. Malins, K. M. Anderson, N. K. Gilman, V. M. Green, E. A. Barker, and K. E. Hellstrom, "Development of a Cancer DNA Phenotype Prior to Tumor Formation," *Proceedings Nat. Acad. Sci. USA*, 101(29): 10721–10725, 2004.
172. C. Krafft, M. Kirsch, C. Beleites, G. Schackert, and R. Salzer, "Methodology for Fiber-Optic Raman Mapping and FTIR Imaging of Metastases in Mouse Brains," *Analytical and Bioanalytical Chemistry*, 389(4): 1133–1142, 2007.
173. G. T. Xu, H. F. Yuan, and W. Z. Lu, "Study of Quantitative Calibration Model Suitability in Near-Infrared Spectroscopy Analysis," *Spectroscopy and Spectral Analysis*, 21(4): 459–463, 2001.
174. K. Das, C. Kendall, M. Isabelle, C. Fowler, J. Christie-Brown, and N. Stone, "FTIR of Touch Imprint Cytology: a Novel Tissue Diagnostic Technique," *J. Photochem. Photobiol. B*, 92(3): 160–164, 2008.
175. J. Ramesh, M. Huleihel, J. Mordehai, A. Moser, V. Erukhimovich, C. Levi, J. Kapelushnik, and S. Mordechai, "Preliminary Results of Evaluation of Progress in Chemotherapy for Childhood Leukemia Patients Employing Fourier-Transform Infrared Microspectroscopy and Cluster Analysis," *J. Lab. Clin. Med.*, 141(6): 385–394, 2003.
176. J. Ramesh, J. Kapelushnik, J. Mordehai, A. Moser, M. Huleihel, V. Erukhimovitch, C. Levi, and S. Mordechai, "Novel Methodology for the Follow-Up of Acute Lymphoblastic Leukemia Using FTIR Microspectroscopy," *J. Biochem. Biophys. Methods*, 51(3): 251–261, 2002.
177. M. Khanmohammadi, M. A. Ansari, A. B. Garmarudi, G. Hassanzadeh, and G. Garoosi, "Cancer Diagnosis by Discrimination between Normal and Malignant Human Blood Samples Using Attenuated Total Reflectance-Fourier Transform Infrared Spectroscopy," *Cancer Investigation*, 25(6): 397–404, 2007.
178. C. Krafft, S. B. Sobottka, K. D. Geiger, G. Schackert, and R. Salzer, "Classification of Malignant Gliomas by Infrared Spectroscopic Imaging and Linear Discriminant Analysis," *Anal. Bioanal. Chem.*, 387(5): 1669–1677, 2007.
179. G. Steiner, A. Shaw, L. P. Choo-Smith, M. H. Abuid, G. Schackert, S. Sobottka, W. Steller, R. Salzer, and H. H. Mantsch, "Distinguishing and Grading Human Gliomas by IR Spectroscopy," *Biopolymers*, 72(6): 464–471, 2003.
180. L. Chiriboga, H. Yee, and M. Diem, "Infrared Spectroscopy of Human Cells and Tissue. Part VII: FT-IR Microspectroscopy of DNase- and RNase-Treated

Normal, Cirrhotic, and Neoplastic Liver Tissue," *Applied Spectroscopy,* 54(4): 480–485, 2000.

181. P. G. L. Andrus, and R. D. Strickland, "Cancer Grading by Fourier Transform Infrared Spectroscopy," *Biospectroscopy,* 4(1): 37–46, 1998.
182. M. Isabelle, N. Stone, H. Barr, M. Vipond, N. Shepherd, and K. Rogers, "Lymph Node Pathology Using Optical Spectroscopy in Cancer Diagnostics," *Spectroscopy—An International Journal,* 22(2–3): 97–104, 2008.
183. M. J. Romeo, and M. Diem, "Infrared Spectral Imaging of Lymph Nodes: Strategies for Analysis and Artifact Reduction," *Vibrational Spectroscopy,* 38(1–2): 115–119, 2005.
184. M. J. Tobin, M. A. Chesters, J. M. Chalmers, F. J. M. Rutten, S. E. Fisher, I. M. Symonds, A. Hitchcock, R. Allibone, and S. Dias-Gunasekara, "Infrared Microscopy of Epithelial Cancer Cells in Whole Tissues and in Tissue Culture, Using Synchrotron Radiation," *Faraday Discussions,* 126: 27–39, 2004.
185. C. M. Krishna, G. D. Sockalingum, R. A. Bhat, L. Venteo, P. Kushtagi, M. Pluot, and M. Manfait, "FTIR and Raman Microspectroscopy of Normal, Benign, and Malignant Formalin-Fixed Ovarian Tissues," *Analytical and Bioanalytical Chemistry,* 387(5): 1649–1656, 2007.
186. S. C. Park, S. J. Lee, H. Namkung, H. Chung, S. H. Han, M. Y. Yoon, J. J. Park, J. H. Lee, C. H. Oh, and Y. A. Woo, "Feasibility Study for Diagnosis of Stomach Adenoma and Cancer Using IR Spectroscopy," *Vibrational Spectroscopy,* 44(2): 279–285, 2007.
187. K. Z. Liu, C. P. Schultz, E. A. Salamon, A. Man, and H. H. Mantsch, "Infrared Spectroscopic Diagnosis of Thyroid Tumors," *Journal of Molecular Structure,* 661: 397–404, 2003.
188. K. Lui, M. Jackson, M. G. Sowa, H. Ju, I. M. Dixon, and H. H. Mantsch, "Modification of the Extracellular Matrix Following Myocardial Infarction Monitored by FTIR Spectroscopy," *Biochim. Biophys. Acta,* 1315(2): 73–77, 1996.
189. K. M. Gough, D. Zelinski, R. Wiens, M. Rak, and I. M. Dixon, "Fourier Transform Infrared Evaluation of Microscopic Scarring in the Cardiomyopathic Heart: Effect of Chronic AT1 Suppression," *Anal. Biochem.,* 316(2): 232–42, 2003.
190. N. Toyran, B. Turan, and F. Severcan, "Selenium Alters the Lipid Content and Protein Profile of Rat Heart: An FTIR Microspectroscopic Study," *Arch. Biochem. Biophys.,* 458(2): 184–193, 2007.
191. F. Severcan, N. Toyran, N. Kaptan, and B. Turan, "Fourier Transform Infrared Study of the Effect of Diabetes on Rat Liver and Heart Tissues in the CH Region," *Talanta,* 53(1): 55–59, 2000.
192. N. Toyran, P. Lasch, D. Naumann, B. Turan, and F. Severcan, "Early Alterations in Myocardia and Vessels of the Diabetic Rat Heart: An FTIR Microspectroscopic Study," *Biochem. J.,* 397(3): 427–436, 2006.
193. F. N. Keith, R. Kong, A. Pryia, and R. Bhargava, "Data Processing for Tissue Histopathology Using Fourier Transform Infrared Spectral Data," *Signals, Systems and Computers, ACSSC,* 06: 71–75, 2006.
194. P. Lasch, and D. Naumann, "Spatial Resolution in Infrared Microspectroscopic Imaging of Tissues," *Biochim. Biophys. Acta,* 1758(7): 814–829, 2006.
195. L. Q. Wang, and B. Mizaikoff, "Application of Multivariate Data-Analysis Techniques to Biomedical Diagnostics Based on Mid-Infrared Spectroscopy," *Analytical and Bioanalytical Chemistry,* 391(5): 1641–1654, 2008.
196. A. Kohler, J. Sule-Suso, G. D. Sockalingum, M. Tobin, F. Bahrami, Y. Yang, J. Pijanka, P. Dumas, M. Cotte, D. G. van Pittius, G. Parkes, and H. Martens, "Estimating and Correcting Mie Scattering in Synchrotron-Based Microscopic Fourier Transform Infrared Spectra by Extended Multiplicative Signal Correction," *Appl. Spectrosc,* 62(3): 259–266, 2008.
197. M. Romeo, and M. Diem, "Correction of Dispersive Line Shape Artifact Observed in Diffuse Reflection Infrared Spectroscopy and Absorption/Reflection (Transflection) Infrared Micro-Spectroscopy," *Vibrational Spectroscopy,* 38(1–2): 129–132, 2005.

CHAPTER 11

Scattering, Absorbing, and Modulating Nanoprobes for Coherence Imaging

Renu John and Stephen A. Boppart

Biophotonics Imaging Laboratory
Beckman Institute for Advanced Science and Technology
University of Illinois at Urbana–Champaign

11.1 Introduction

Molecular imaging is defined as the characterization and observation of biological processes at a cellular or molecular level using *in vivo* methods. Tremendous developments in the field of biomedical imaging during the past two decades have resulted in the transformation of anatomical imaging to molecular-specific imaging. The developments in the field of nano- and biotechnology have created a profound impact on the biomedical imaging research community by allowing scientists to identify, follow, and quantify subcellular biological processes and pathways within a living organism. Molecular imaging is expected to play an important role in oncology, for example, by aiding the early detection of malignancies, locating metastatic disease, staging tumors, evaluating the availability of therapeutic targets, and monitoring the efficacy of treatment.

The main approaches toward imaging at a molecular level are the development of high-resolution imaging modalities with high-penetration depths and increased sensitivity and the development of molecular probes with high specificity. The optical imaging modalities have unique capabilities to distinguish cells and tissues on the basis of their optical scattering and absorption properties. Most of the optical imaging techniques rely on the absorption and scattering changes of light as it traverses biological tissue, providing optical signal changes that carry information about the tissues. Biological samples are fairly transparent in the near-infrared (NIR) window of the electromagnetic spectra (650 to 1350 nm), which is often termed the *biological window*, making this spectral region ideal for optical imaging applications. The contrast generated could be either from the endogenous signals from specific molecules or from targeted agents attached to specific molecules. The development of novel molecular contrast agents and their successes in molecular optical imaging modalities have led to the emergence of molecular optical imaging as a more versatile and capable technique for providing morphological, spatial, and functional information at the molecular level with high sensitivity and precision, compared to other imaging modalities.

This chapter will provide a brief overview of the developments and applications of various optical contrast agents in molecular imaging and address the challenges and future directions in the field of molecular optical imaging. This section gives a brief overview of various molecular imaging techniques and discusses various novel molecular contrast agents. In the next section, a variety of coherent optical imaging probes based on their nature of contrast generation will be described. Finally, we conclude with the scope and future of molecular contrast agents from the technological perspective of coherent optical imaging.

11.1.1 Molecular Imaging Techniques

The goal of biomedical imaging is to provide structural and functional information and to visualize biological processes from the nanoscale to the molecular cellular, and systems scales. Since the discovery of X-rays by Wilhelm Roentgen in 1895 [1], biomedical imaging has become an inevitable part of diagnosis and monitoring of diseases. The current field of biomedical imaging comprises a multitude of tools, like magnetic resonance imaging (MRI), X-ray computed tomography (CT), ultrasonography (US), positron emission tomography (PET), and single photon emission computed tomography (SPECT), that have developed over the last century, while optical technologies, including optical coherence tomography (OCT) and microscopy, multiphoton microscopy (MPM), and diffusion optical tomography are emerging as promising modalities for molecular imaging.

Most of the optical imaging techniques rely on the physical parameters (e.g., fluorescence, bioluminescence, absorption, reflectance, scattering, polarization, or dynamic modulation changes) as sources of contrast to provide optical information about tissues. Hence, a wide range of optical imaging modalities, like fluorescence imaging, white-light (reflectance) microscopy, confocal imaging, OCT, and related modalities, like spectroscopic (S)OCT, polarization sensitive (PS-)OCT, phase-sensitive and optical Doppler tomography, and multiphoton imaging, have been explored in recent years.

Fluorescence imaging has been a widely investigated imaging modality for biomedical imaging applications where the contrast is generated using a fluorescent molecule (e.g., dye, or proteins like green fluorescent protein (GFP) or bioluminescent agents like luciferase) [2–4]. With the extension of fluorescence to the NIR biological window, there has been a significant increase in the development of different types of near-infrared flourescence (NIRF) contrast agents with innovations like biologically activa table probes that can be activated by specific targets *in vivo* [5–8]. However, fluorescence or bioluminescence signals cannot be detected by coherence imaging modalities. The ability to record the amplitude and phase of a coherent signal using interferometric techniques [9–17] has enabled the coherence imaging modalities to retrieve the phase of the signal and extract additional valuable information from the sample. In addition, the heterodyne detection capabilities can produce images with exceptionally high signal-to-noise ratio (SNR). Offering high resolution and sensitivity at nanoscales, coherent imaging modalities enable functional imaging at the subcellular level, such as tracking biological processes, detecting flow, and sensing dynamic modulations and displacements at a nanoscale. A wide variety of molecular-specific contrast agents have been studied for coherence imaging applications. Figure 11.1 metaphorically illustrates the whole spectrum of molecular optical contrast agents on the basis of their origin of contrast and their functionality. The recent developments in nanotechnology, CCD sensors and cameras, and related technologies have added pace to the evolution of coherence imaging modalities in molecular-specific imaging.

11.1.2 Molecular Contrast Agents

Strategies to generate molecular contrast in images have evolved dramatically in the past two decades from traditional dyes to highly sophisticated molecular agents that are generally sized in the range of hundreds of microns down to a few nanometers (Fig. 11.2). They are usually made for a specific cell type, receptor, or other molecular component important in the investigated tissue, cell, process, or pathology. Chemical agents, such as dilute acetic acid, that alter the refractive index of cell nuclei and cause an increased scattering effect which is more pronounced in dysplastic regions with more nuclear density, have been in clinical use for over two decades for diagnosis

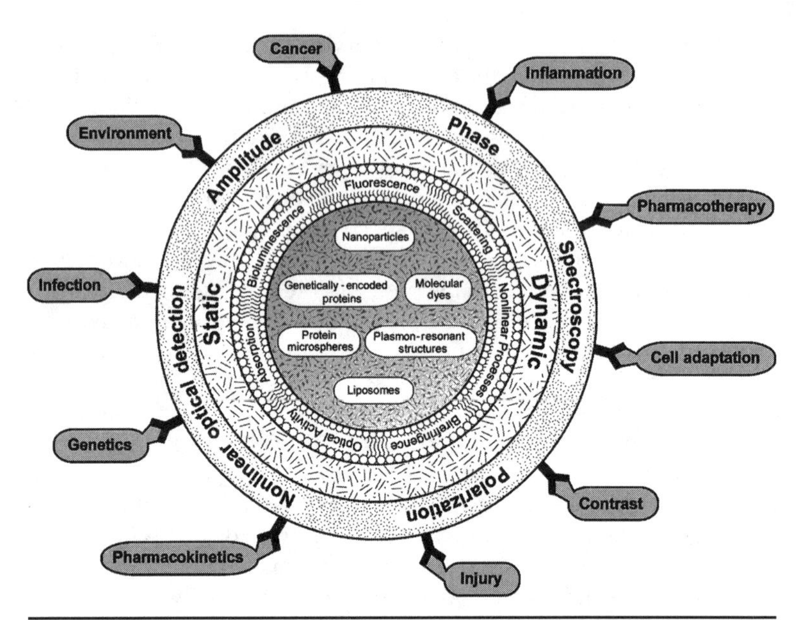

FIGURE 11.1 Metaphorical diagram of the wide spectrum of optical molecular probes, their optical signals, their optical detection parameters, and their applications. In addition to fluorescent and bioluminescent probes, a number of other probes and techniques make up optical molecular imaging, including those detected and imaged with coherent light. (Figure reprinted with permission [102].)

of cervical cancer [18]. Air-filled microbubbles, protein microspheres filled with lipids and perfluorocarbons [19], and latex microspheres [20] were the earliest reported particle-based contrast agents in imaging modalities. The recent developments in nanotechnology have contributed enormously to the field of nanomedicine, and this has resulted in a wide variety of novel agents with sizes of the order of nanometers for applications in bioimaging and biosensing.

A wide variety of nanostructures including nanoparticles, nanospheres, nanoshells, nanotubes, nanorods, nanowires, nanocages, nanostars, and quantum dots have been fabricated using materials that are optically or chemically active. Each exhibits some optical property that can be exploited by existing imaging modalities to achieve additional contrast information. Nanoparticles made of materials like gold, silver, iron oxide, gadolinium, carbon, and a wide variety of polymers have been used as optical contrast agents with different optical imaging modalities. In addition to the primary aim of generating molecular-specific contrasts, nanoprobes have been developed to be multifunctional for various applications such as site-specific delivery of drugs [21–26], nucleotides, proteins or genetic materials, and hyperthermia [25, 27, 28].

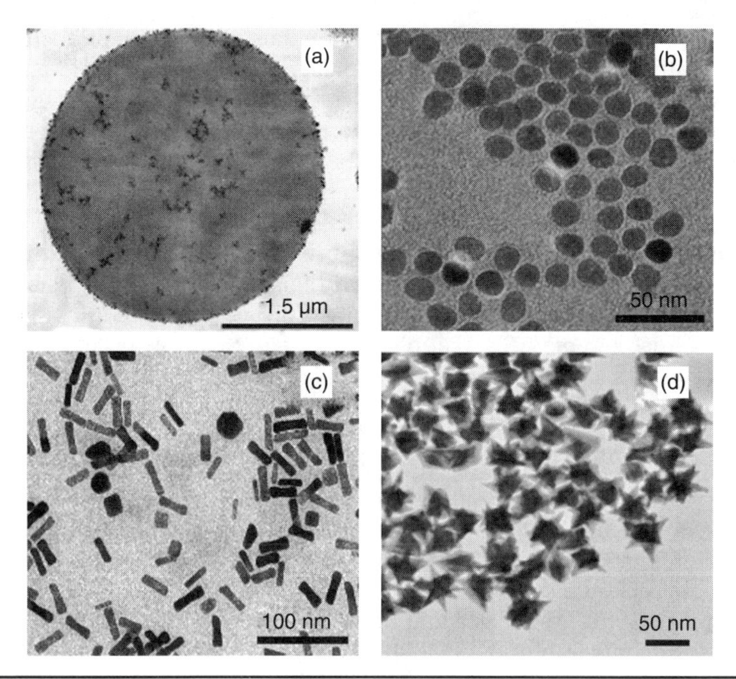

Figure 11.2 Representative molecular contrast agents. Transmission electron micrograph (TEM) images of (*a*) protein microsphere with oil and iron-oxide nanoparticles in its core, (*b*) iron-oxide nanoparticles coated with COOH terminating surfactant, (*c*) gold nanorods, and (*d*) gold nanostars. (Figures were reprinted with permission from [55, 101, 60].)

Super-paramagnetic iron-oxide magnetic nanoparticles (SPION) have a very promising role as ultrasensitive molecular-specific imaging nanoprobes. Use of iron oxide for *in vitro* diagnostic applications has been in practice for over 40 years [29]. Magnetitie (Fe_3O_4), which is a common form of iron oxide, has preferred qualities such as high magnetization values, high specific heat absorption rates (for hyperthermia), and well-known biocompatibility for *in vivo* applications. Iron-oxide nanoparticles, due to their small size and extremely large relative surface area, exhibit super-paramagnetic phenomena and find numerous applications in the field of biomedical imaging. Their unique paramagnetic properties with strong T2 and negative T2* contrast have made them attractive contrast agents in MRI [30]. The easiest method of preparation of iron-oxide nanoparticles is by the coprecipitation of ferrous and ferric ion salts in an alkaline medium in an oxygen-free environment [21, 31–34]. The size of particles depends on the ratio of the ions, types of salts, pH, and ionic strengths of the solutions. Different methods of preparation of iron-oxide nanoparticles including variations of the coprecipitation technique have been discussed in detail by Gupta and Gupta [32]. Iron-oxide nanoparticles coated with different materials like dextran or dextran derivatives [35, 36], albumin [36], polyethylene glycols (PEGs) [30, 37], and polyvinylpyrrolidone [38] have been reported to have high biocompatibility for *in vivo* applications. Numerous works have been reported over the past two decades on functionalizing iron-oxide nanoparticles with a number of targeting agents like proteins [21, 39, 40], antibodies [31, 41–43], polypeptides [25, 44, 45], and oligonucleotides [46, 47]. The ease of their fabrication with tailored surface chemistry, their biocompatibility, and their bioconjugation possibilities have made them attractive choices for use in different imaging modalities and biomedical applications.

Gold nanoparticles are nearly ideal as contrast agents for various biomedical optical imaging modalities. They come in different sizes and shapes including nanoshells [48–52], nanorods [53–56], nanocages [53, 57–59], and nanostars [60]. The ability of gold to excite surface plasmon resonances that could be tuned depending on their size, shape, coating, and environment is extremely useful in different imaging modalities such as SOCT and photothermal OCT. Gold nanoparticles have also been used extensively as molecular-specific contrast agents [61–64] in a number of imaging modalities like photoacoustic tomography (PAT) [65], MRI [66], OCT [57], and magnetomotive (MM-)OCT [54, 56, 57]. Many of these nanostructures made of gold and silver are prepared using solution-phase methods [67] where the final shape depends strongly on the crystallinity and structure of the initial seed. This seeding growth method is effective in the synthesis of gold nanorods with different aspect ratios. The gold particle aspect ratio can be controlled from 1 to 7 by simply varying the ratio of seed to metal salt in the presence of a rod-like micellar template [67]. The metal ions are reduced inside cylindrical pores of oxide or polymeric membranes. By carefully controlling the growth conditions, higher aspect ratios can be achieved. Another way is to use neutral or charged surfactants for the growth of nanostructures. In aqueous media, surfactants such as hexadecyltrimethylammonium bromide (CTAB) have been popular molecules in the synthesis of gold nanorods [68]. Another way to synthesize larger-diameter nanorods for applications requiring dominant scattering versus absorption is by electrodeposition into nanoporous membranes [69].

Following a galvanic replacement reaction, a solid-silver nanostructure can be transformed into a hollow Au/Ag-alloyed shell if refluxed with $HAuCl_4$ in water. This is possible because the standard reduction potential of $AuCl_4^+/Au$ is higher than that of Ag^+/Ag. This method is most suitable for the preparation of structures like nanoshells and nanocages. Typically, gold nanoshells are fabricated by adsorbing 1 to 2 nm-sized colloidal gold onto the surface of a silica nanosphere as a seed, then chemically reducing additional gold onto the surface [70]. Gold nanocages are synthesized from silver nanocube templates, growing gold atoms over its surface forming an ultrathin shell on a reaction with aqueous $HAuCl_4$ solution. The templates themselves gradually get etched out; this results in a hollow interior for the gold nanocages [58]. In addition to the nanostructures just described, carbon nanotubes [71–75], Gadolinium nanoparticles [76–78], various polymeric nanoparticles [79–82], and quantum dots [83–87] have been in use for different imaging modalities.

Functionalization of agents for specific targeting to molecular sites is yet another important and most desirable quality of nanoprobes. Along with the synthesis of these nanoprobes, the chemistry of bioconjugation has also contributed significantly to the development of targeted probes that can be conjugated to different biomarkers, proteins, peptides, or oligonucleotides. The development of functionalized nanoprobes for *in vivo* targeting is extremely challenging and involves multidisciplinary skills. A number of factors determine the clinical usefulness of the nanoparticles. Their physical and functional biokinetic properties, clearance profiles, biodistribution, biocompatibility, and long-term toxicity need to be carefully considered for their potential use in *in vivo* applications. Nanoparticles having sizes smaller than 50 nm show prolonged blood half-life times compared to larger ones. Larger particles are quickly uptaken by phagocytic cells. Hence, particles with longer circulation time have greater probability of reaching the target sites. Hydrophilic particles have a longer blood half-life compared to hydrophobic particles. Particles that are hydrophilic in nature with no charge on them or that have very low negative charges are known to have longer blood half-life times.

The hydrophilic nature of particles, their dispersibility in physiological medium, and their biocompatibility can all be enhanced by coating them with various natural and synthetic polymer materials [21]. The traditional approach for nanoparticle-based therapy is aimed at the passive targeting of nanoparticles to specific sites such as tumors by exploiting the enhanced permeability-and-retention (EPR) effect associated with the leaky vasculature [8]; this allows nanoparticles to readily extravasate and accumulate in the tumors. However, experiments have shown that nanoparticles most likely undergo opsonization in blood; this results in their fast uptake by the reticulo endothelial system (RES), and their final clearance from the circulation. As a result, a relatively low concentration of nanoparticles at targeting sites has become a major obstacle in using them for tumor imaging or therapy. Surface modifications, such as coating nanoparticles with polymers like PEG-thiols, have proved useful to protect the nanoparticles from being captured by the RES. Conjugation of nanoparticles to target-specific ligands is another way to direct the accumulation of nanoparticles to a specific site. Efficient ways to direct these nanoparticles to organs, tissues, tumors, and malignant cells by conjugating the nanoparticles with enzymes, proteins, antibodies, nucleotides, or peptide ligands have been reported. Various biomarkers and oncoproteins like HER-2 [31, 42, 43, 88, 89] or matrix metalloproteinases (MMPs) [90–92] that are known to be overexpressed at the sites of malignancy have been targeted with different conjugation techniques. The conjugating procedures depend on the nature of the nanoparticle coating and the conjugating ligand or proteins.

11.2 Coherence Imaging Probes

11.2.1 Scattering Probes

Optical imaging of tissues and biological specimens is possible by the direct detection of scattered light from a biological sample or by the measurement of due attenuation of incident light in the sample. This information can then be used to reconstruct the structural features of the sample. The use of air-filled or lipid-filled microbubble contrast agents [19] is a classic example for enhancing scattering, where the refractive index of the scatterer is different from that of the surrounding media. Air-filled and perfluorocarbon-filled protein microbubbles that are commercially available as ultrasound contrast agents [93, 94] have been reported in the past as scattering contrast agents in OCT [19]. Protein microspheres with different cores can be fabricated using high-frequency ultrasound. A wide variety of materials like gases, lipids, and water can be selected for the core region of these protein-shell microspheres (Fig. 11.3). The encapsulation process is mediated through sonication using high-frequency ultrasound. The protein shell is cross-linked through the reduction of disulphide bonds by the radicals formed during the sonication process. The sizes of microspheres can be controlled from hundreds of nanometers to a few tens of microns by the addition of surfactants to change the surface tension of the protein mixture.

A layer-by-layer (LBL) adhesion technique can be used to modify the surface of the protein microspheres. This technique can be used to functionalize the outer surface of the protein microsphere with targeting moieties. A wide variety of engineered microspheres with bovine serum albumen (BSA) have been fabricated using this technology. Protein microspheres consisting of a hydrophobic core composed of vegetable oil and a hydrophilic protein shell with sizes from 2 to 5 microns have been reported as scattering contrast agents in OCT (Fig. 11.4) [95, 96]. These microspheres, due to their relatively large size, would remain in the vasculature after intravenous administration and are suitable for identifying regions of altered perfusion (leaky or ruptured vessels) and for targeting specific markers that are overexpressed in regions of angiogenesis (in cancer) and atherosclerosis (in heart disease), to provide some examples. These microspheres have been encapsulated with lipids, NIR dyes, and various

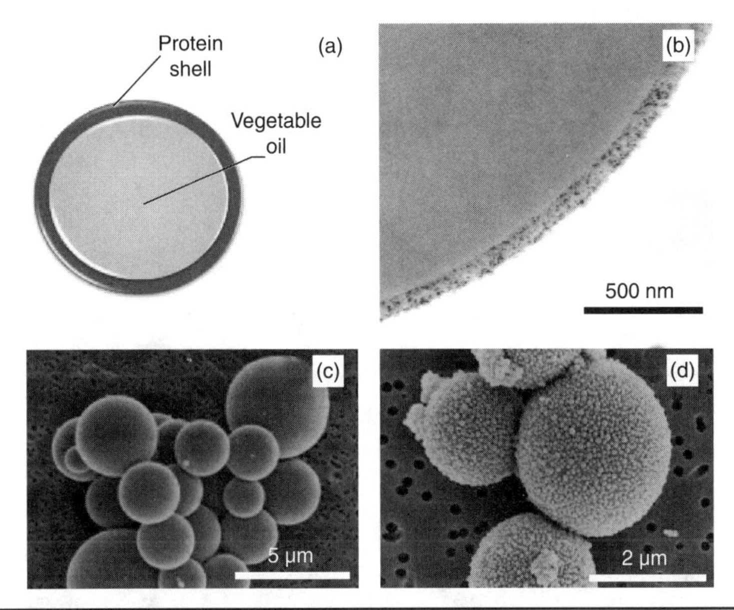

FIGURE 11.3 Protein microspheres as multifunctional multimodal contrast agents with flexible fabrication parameters allowing modifications of surface, shell, and core contents. (*a*) Schematic representation of oil-filled protein microsphere. (*b*) TEM showing magnified view of the shell. (*c*) SEM of nonconjugated protein microspheres. (*d*) Scanning electron micrograph (SEM) of RGD poly-lysine-peptide-conjugated protein microspheres.

nanoparticles including gold, carbon, melanin, and super-paramagnetic iron-oxide nanoparticles in their core and shell to effectively work as multimodal contrast agents in NIRF imaging, ultrasound, MRI, and OCT and MM-OCT [96–98]. Owing to the presence of iron-oxide nanoparticles in their core, these microspheres can be dynamically modulated using an external magnetic field; this creates a dynamic contrast in biological samples, and additionally the microspheres serve as dynamic probes.

Targeting these microspheres to alpha (v) beta (3) integrin receptors by conjugating an arginine-glycine-aspartic acid (RGD) peptide sequence to the outer surface of these microspheres has been reported [95–97]. These integrin receptors play an important role in the development of cancer or angiogenesis and are also overexpressed at sites of artherosclerosis. RGD-functionalized microspheres have been successfully tested on cell lines that overexpress these integrin receptors (Fig. 11.5). The possibility of controlling the *in vivo* rupture of microspheres to release their shell or core contents is another promising application of these agents as a means for drug administration and delivery of genetic material and proteins. These targeted protein

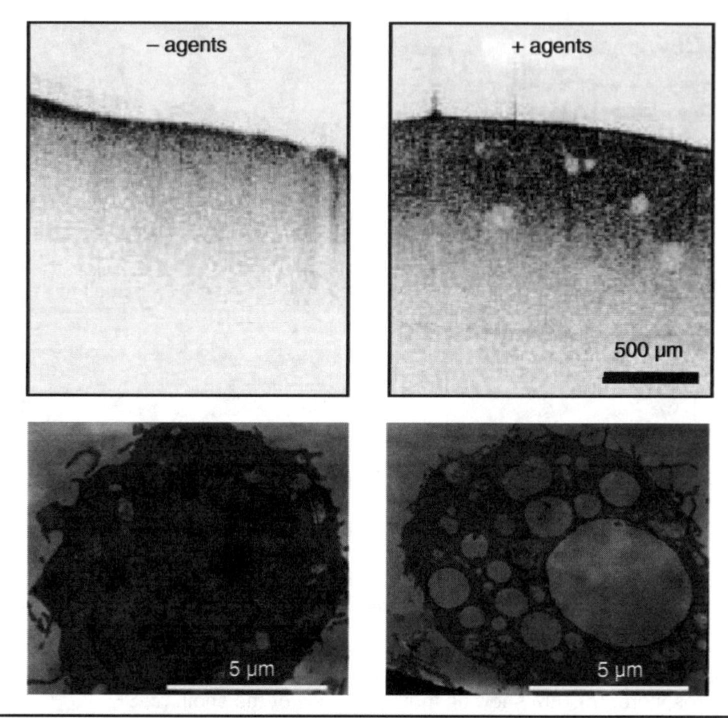

Figure 11.4 Protein microspheres as OCT contrast agents. *Top set:* OCT images of an *in vivo* mouse liver before and after tail vein injection of gold-coated protein microspheres. Post administration of the contrast agents reveals increased scattering from the liver, where Kuppfer cells had phagocytosed the scattering microspheres. The low-scattering regions are likely the liver sinusoids and vasculature. *Bottom set:* SEMs of cultured macrophages without (lower left) and with (lower right) exposure to microspheres. The large circular-appearing cavities represent cross sections of phagocytosed microspheres. (Figure reprinted with permission [95].)

microspheres would therefore serve as potential site-specific drug delivery vehicles with the capacity for relatively high payloads.

11.2.2 Dynamic Probes

All the scattering probes rely on their static scattering nature to enhance the optical signal or to provide contrast in imaging. One of the main drawbacks of scattering probes is the inherent background noise that arises from the static background scattering of the tissues. Hence, the endogenous scattering from the tissue structures can be separated from the scattering caused by the contrast agents only through prior information about the tissue structure, that is, by comparing the images before and after the administration of the contrast agents. Dynamic probes that can be modulated externally offer an

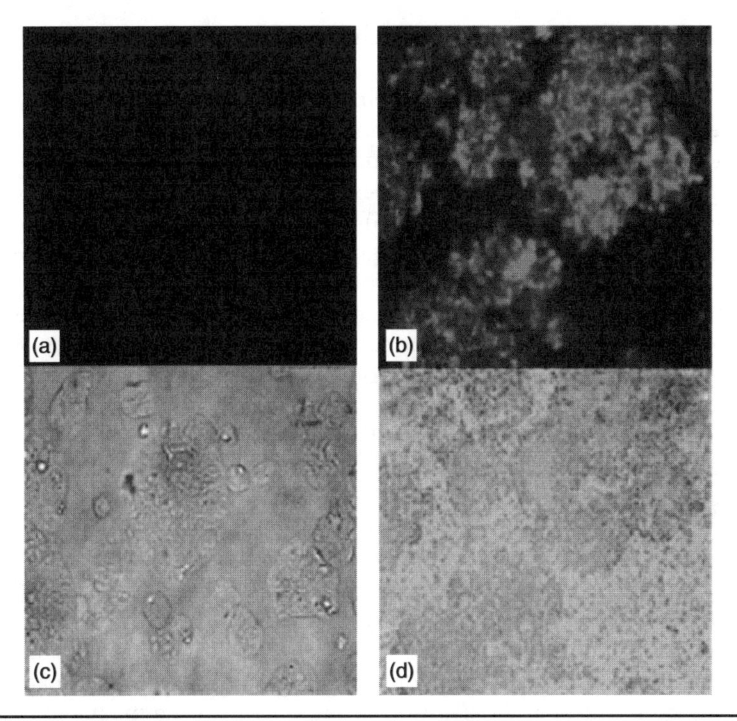

FIGURE 11.5 *In vitro* demonstration of targeted RGD-Nile red microspheres. Fluorescence microscopy of HT29 human colon cancer cells treated with (*a*) non-RGD and (*b*) RGD-functionalized microspheres. The RGD polypeptide is targeted to integrin receptors that are overexpressed on tumor cells. (*c*) and (*d*) are bright-field microscope images of cells for (*a*) and (*b*), respectively. (Figure reprinted with permission [96].)

elegant solution to this problem. Small perturbations in scattering can be created using dynamic modulation of the probes, and the unperturbed background signal can be filtered out efficiently through postacquisition data processing.

Magnetomotion of contrast agents through an externally modulating magnetic field is an excellent mechanism for achieving dynamic contrast. This can be realized by using paramagnetic particles as contrast agents. Human tissue exhibits an extremely weak magnetic susceptibility ($\chi < |10^{-5}|$). Hence the employment of magnetic probes with $\chi \sim 1$ provides a large potential dynamic range of magnetic contrast. Iron-oxide, such as magnetite, with paramagnetic properties and $\chi \sim 1$, is a good candidate with known biocompatibility after coating with polymers such as dextran. Paramagnetic iron-oxide magnetic nanoparticles (MNPs) have already been approved as contrast agents in MRI. These micron- and nano-sized MNPs can be actuated externally using a modulating magnetic field. In the presence of a high magnetic field gradient, particles with high magnetic susceptibility embedded in tissue experience a gradient force, and ferromagnetic particles rotate to align their internal magnetization along the field. The resulting magnetomotion of the MNPs and the perturbation of the surrounding cells and extracellular matrix result in a change in the scattering properties of the local tissue microenvironment under observation (Fig. 11.6) [99–101]. In an elastic medium, the particle returns to its original position and orientation after removal of the magnetic field. This permits modulation of its position by repetitively modulating the magnetic field. The resulting increase in contrast due to magnetomotion can be effectively detected by an OCT system. The magnetomotive response of the tissue phantom to different magnetic field excitations can easily be demonstrated by performing M-mode imaging at one point on a tissue phantom. A step excitation or a sweep of frequencies applied to the sample under study is used to extract information about the mechanical and viscoelastic properties of the biological tissue. Doing this, one can find the resonant mechanical frequency of the biological sample as well.

A standard OCT system can be readily modified to enable MM-OCT imaging [101]. In the MM-OCT system, the magnetic field is applied using a solenoid, and imaging is performed on the sample immediately below the solenoid bore (Fig. 11.6). Within the imaging volume of the sample, the radial components of the magnetic field are negligible and the magnetic field gradient is predominantly in the axial direction. The MNPs of size 20 to 30 nm under the effect of magnetomotion due to an axial magnetic field result in a change in amplitude and phase of the interference pattern in a spectral-domain OCT system. A sinusoidal magnetic field excitation with a frequency close

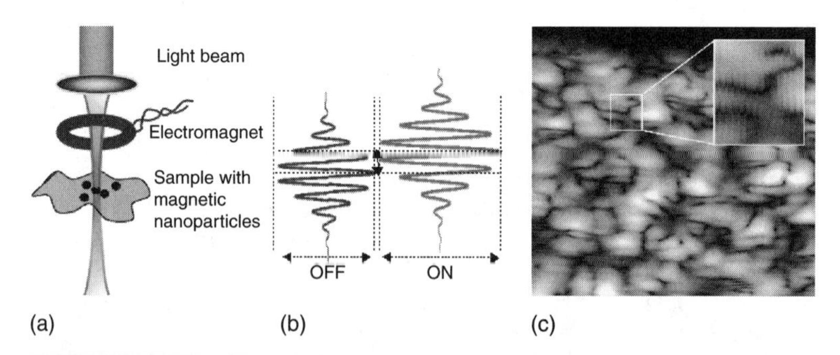

FIGURE 11.6 MM-OCT principle. (*a*) Schematic of an MM-OCT sample arm with light passing through the center of an air-core solenoid. (*b*) Amplitude and phase changes in the interference patterns with the modulating magnetic field. Arrows indicate these changes between the field is OFF and ON. (*c*) MM-OCT image (110 × 90 μm) of a tissue phantom (MNP concentration = 0.93 mg per gram). Inset is a magnified view (20 × 15 μm) showing the dynamic changes at a micron scale. (Panel (c) modified and reprinted with permission [99].)

to the mechanical resonance of the biological sample is used to perform MM-OCT imaging. The mechanical excitation of the MNPs is coupled to the B-mode OCT scanning by modulating the magnetic field several cycles during the time required to mechanically scan the imaging beam spot over one resolution point. Magnetomotive signals from a specimen under study can be isolated with the help of two B-mode MM-OCT images, one image with the modulating field off and one with the modulating field on. The system is capable of detecting MNP concentrations as low as ~2 nM within the sample, and from the phase information obtained through interferometric detection, displacements of MNPs as low as a few tens of nanometers can be measured. Iron-oxide MNPs have been topically administered to highly scattering media such as chicken skin, and the sensitivity of this technique has been demonstrated *in vitro* (Fig. 11.7) [102]. The dynamic behavior and transport of magnetic nanoprobes in biological tissues for different concentrations and temperatures can be effectively tracked using the MM-OCT technique [103]. These results will aid in understanding the actual localization and accumulation of functionalized MNPs targeted to specific sites.

Another dynamic mode of optical contrast termed *gyromagnetic imaging* [60] using gold nanostars with super-paramagnetic cores has been proposed. Gold nanostars with super-paramagnetic cores are driven by a rotating magnetic field gradient to produce periodic variations in NIR scattering intensities detected in a dark-field microscope setup. This periodic "twinkling" in response to a rotating magnetic field gradient results in a frequency-modulated signal that is converted into Fourier-domain images with a dramatic reduction in background.

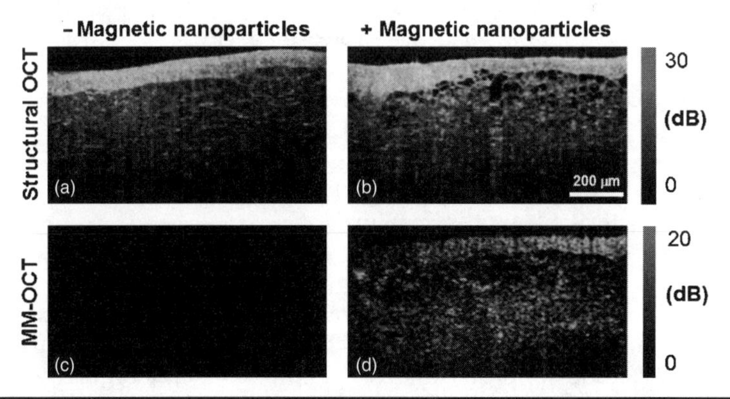

FIGURE 11.7 MM-OCT in tissue. Images were acquired from *in vitro* chicken breast tissue. Structural and MM-OCT images were acquired in tandem, without and with the administration of iron-oxide nanoparticles. (Figure reprinted with permission [102].)

Research on extracting dynamic contrast from biological tissues through biological or biochemical methods has led to the development of a new class of molecular imaging contrast agents known as *activatable molecular probes*, which has elicited a particularly high level of excitement in the molecular imaging community. These activatable molecular probes are designed to elicit a detectable change in signal on enzymatic activity or in response to specific biomolecular interactions. In many cases, these unique characteristics allow for very high signal-to-background ratios compared with conventional targeted contrast agents, and they open up the possibility of imaging intracellular targets. Some of the activatable probes have been recently developed for optical and magnetic resonance imaging platforms for the visualization of cancer biomarkers *in vivo*.

In vivo imaging of tumor-associated lysosomal protease activity in a xenograft mouse model has been demonstrated using autoquenched NIRF probes [6, 104, 105]. NIRF dye Cy5.5 molecules were bound to a long, circulating, synthetic graft copolymer consisting of poly-L-lysine protected by multiple methoxypolyethylene glycol succinate (MPEG) side chains which served as the delivery vehicle. Each polylysine backbone contained an average of 92 MPEG molecules and 11 molecules of Cy5.5; this yielded $(Cy5.5)_{11}$-PL-$MPEG_{92}$ (C-PGC). Following intravenous injection, the NIRF probe carrier accumulated in solid tumors due to its long circulation time and leakage through tumor neovasculature. In the bound state, NIR fluorescence was 15-fold lower compared with free Cy5.5 at equimolar concentrations. The released NIRF probes can be detected *in vivo* at subnanomole quantities and at depths sufficient for experimental or clinical imaging depending on the NIRF image acquisition technique [106]. *In vivo*, the polylysine backbone can be cleaved by tumor proteases and lysosomal proteases like cathepsin (Ca)- B, D, and H. Specific peptide sequences that link the fluorescent molecule and the lysine backbone can be modified to allow cleavage-protease specificity. A NIRF probe with of CaD-specific peptide sequence spacer that links Cy5 to the lysine structure has been synthesized following the same principle [107–109]. A wide variety of protease-specific probes can theoretically be possible through this technique. Recently, micro-fiber-optic catheters [110] have been developed for *in vivo* NIR optical imaging of these activatable probes in animals.

MMPs are another example of enzymes that have been engineered using the same chemistry to activate the fluorescent probes. MMPs are overexpressed in tumor cells, and they also play an important role in the progression, destabilization, and rupture of atherosclerotic lesions. Abnormal expression and regulation of MMPs in atherosclerotic plaque may ultimately lead to clinical conditions such as myocardial infarction or stroke resulting from their rupture. Successful *in vitro* demonstrations of MMP-2-activated optical contrast agents have been reported using specific peptide substrates containing motifs that are cleaved by this enzyme [92, 111, 112]. The agent was tested in animal tumor models and was optically imaged after intravenous administration of the MMP-2-sensitive probe.

The current optical imaging methods acquire the NIRF signal against a nonfluorogenic background and hence provide little spatial or anatomic information. The development of optical tomographic imaging techniques offers the prospect of providing a lower detection limit for fluorochromes and a higher spatial resolution. However, a second imaging modality will still be required to obtain the anatomical background. The main drawback in these cases is that coherence imaging technology cannot be used to detect fluorescent or bioluminescent probes that emit incoherent light.

Magnetic nanoparticles that act as MR contrast agents can provide information on the location of the probe, and can be tailored as NIRF imaging probes by conjugating cleavable Cy5.5-derivatized peptides to their surface [106]. Contrast enhancement by MR results from a core of super-paramagnetic iron-oxide, while the NIRF signal results from amino-CLIO. A magnetic nanoparticle with a surface of aminated, cross-linked dextran has been used to attach a variety of biological molecules, including the tat peptide of the HIV tat protein, transferrin, and oligonucleotides and provides a convenient platform for the attachment of Cy5.5 peptides.

Recently, another bioactivatable NIRF probe has been developed that is activated by the caspase-3 enzyme for monitoring its activity [111]. By using a peptide substrate recognized by caspase-3, caspase-3 activation in cells can be monitored. This idea of bioactivatable probes has opened up the possibility for diverse structural modification and optimization of NIRF activatable probes for *in vivo* monitoring of a variety of molecular processes.

11.2.3 Absorbing Probes

Absorption is yet another attractive contrast mechanism that can be exploited by a coherence imaging modality to enhance the imaging capabilities. Typically, all biological molecules have light-absorbing capability, and accurate measurement of differences in absorption coefficients and wavelength-dependent attenuation of light in biological tissues would reveal valuable information about the sample. Low-coherence spectroscopy is an important technique for the quantitative assessment of absorption coefficients in localized regions. Spectroscopic OCT (SOCT) combines the features of traditional structural OCT based on reflectivity of scattering coefficients with spatial spectroscopic information of the sample under study. SOCT enables the spatially resolved detection of absorption changes within the bandwidth of the OCT source [113–115]. The presence of an absorbing molecule selectively attenuates particular wavelengths depending on its inherent nature; this results in a recovered spectrum that has a shifted spectral centroid. Such a shift can be represented by a hue-saturation false-color mapping to an image-based format. The spectroscopic information can thus be used to create a spectroscopic staining of the biological tissue analogous to histological staining, which can aid in differentiating tissue structures [12, 115].

The contrast in SOCT could be endogenous from natural chemical molecules such as hemoglobin and melanin or exogenous from agents such as dyes. Structure-dependent elastic scattering also generates spectral signatures with sharp wavelength-dependent absorption profiles depending on the size, shape, and refractive index of cellular organelles as well as nanoparticle-based exogenous contrast agents. Surface plasmon resonant nanoparticles, nanospheres, nanoshells, nanorods, and nanocages constitute this class of SOCT-sensitive contrast agents. SOCT has shown the capability to distinguish between the absorption spectra of oxygenated and deoxygenated hemoglobin (Hb) in blood [116]. These results can be closely correlated with the Hb oxygen saturation (SO_2) in blood which is regarded as a current clinical health standard for monitoring patients with diseases. The traditional oximetry technique is a well-established optical method that relies on the differences in light absorption characteristics of oxy and deoxy hemoglobin at two distinct wavelengths to determine SO_2. The endogenous contrast-based SOCT, however, has limited applications in biological imaging because only a limited number of biological molecules are spectrally active in the NIR biological therapeutic window commonly used for OCT imaging. None of the major molecular components of tissue, namely, water, structural proteins without chromophores, most carbohydrates, lipids, and nucleic acids, are spectrally active in the NIR region. For this reason, anatomical and biochemical research using SOCT to date has been limited to a few types of naturally occurring NIR absorbers like melanin and oxy and deoxy Hb.

The most desirable properties for an exogenous absorbing probe are sharp, distinct, and stable spectral features in its absorption spectra within the bandwidth of the OCT source. The simplest spectroscopic absorbing probes are the NIR dyes [114, 117]. Many commercially available fluorescent NIR dyes already developed for use as fluorescent markers [117–120] and laser dyes such as Rhodamine-6G are also appropriate for absorption-based contrast applications. With their known chemical structure and nature, functional imaging by conjugating different molecules such as antibodies to the NIR dyes would be easily feasible.

Enhancement of contrast in the vascular bundles of a stalk of green celery (*Apium graveolens* var. *dulce*) using the SOCT technique has been demonstrated (Fig. 11.8) [114]. SOCT showed no difference

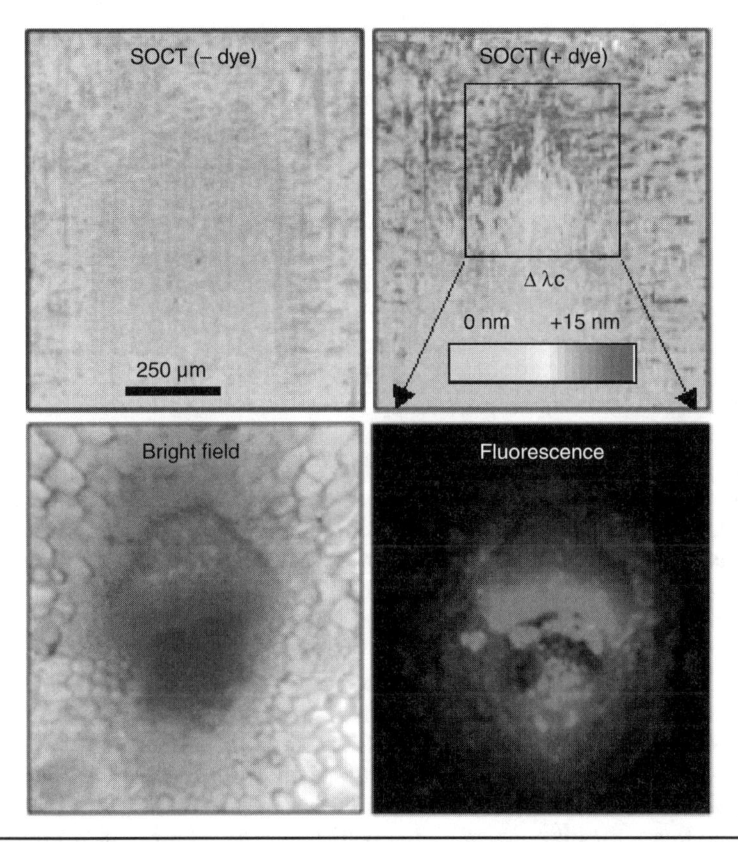

FIGURE 11.8 NIR dye contrast enhancement in SOCT. Cross-sectional SOCT images of a celery stalk without and with NIR dye present within the vascular bundle, top left and top right, respectively. Corresponding bright-field light microscopy and fluorescence microscopy show the vascular bundle and the surrounding collenchyma tissue, bottom left and bottom right, respectively. (Figure reprinted with permission [114].)

between the collenchyma tissue and the central vascular bundles in the stalk in the absence of any contrast agent. SOCT imaging was repeated after the root end of the stalk was submerged in a liquid mixture of NIR dye (ADS7460, H. W. Sands, Inc., sharp absorption at 740 nm) and Rhodamine 5G for 4 hours to allow for capillary transportation. The presence of the NIR dye caused a dramatic reduction of shorter wavelengths due to strong absorption, resulting in a red shift of the scattered SOCT spectra which was computed from the shifted spectral centroid. The SOCT image showed apparent contrast enhancement in the vascular regions containing the NIR dye, whereas the surrounding avascular collenchymas had minimal absorption changes. The results closely correlated with fluorescence images obtained from the same location.

One of the potential areas where probing spectroscopic properties of exogenous agents can be useful is in the estimation of the concentration of these molecular agents or the monitoring of their diffusion in tissues [118, 121]. Diffusion of a topically applied sensitizing dye prior to light exposure in photodynamic therapy (PDT) can be monitored using the SOCT technique. Indocyanine green (ICG), a Food and Drug Administration (FDA) approved dye commonly used in retinal fluorescence angiography, has also been a preferred choice for SOCT [122]. ICG has also been explored as a contrast-enhancing dye in a pump-probe OCT scheme because of its photobleaching property [121]. Quantum dots are another class of commercially available chromophores for pump-probe OCT as well as for nonlinear spectroscopic techniques like two-photon OCT techniques due to their large two-photon absorption cross sections [123]. A signal from Fort Orange quantum dots (Evident Technologies, Troy, NY) has also been observed using a ground-state recovery pump-probe OCT system.

The transient absorption in a biological sample by the absorbing contrast agents induced by a pump beam can be exploited to enhance image contrast using pump-probe OCT [124]. Methylene blue (MB) can be excited from its ground state (absorption peak at 650 nm) to its excited state (absorption peak at 830 nm) using optical radiation at 650 nm and has been investigated for this application [125]. By switching optical excitation on and off, the absorption property at 830 nm was modulated. However, the shorter excitation-state lifetime of MB, and hence the high pumping intensities, was a striking disadvantage. Other molecules with more stable excited molecular states were investigated to reduce the requirement in the pumping power. Phytochrome A, a naturally occurring plant protein with two distinct states of different absorption, was identified as a potential agent because one of its peak absorption states is in the commonly used OCT spectral range of the titanium:sapphire laser and has very long transition durations on the order of a second [126].

Absorption of light in the visible and NIR regions has the potential to make nanoparticle-based contrast agents ideal for both photothermal ablation therapy and coherence imaging techniques, where the photothermal modulations in the sample can simultaneously be probed for obtaining molecular contrast. When illuminated with NIR light sources, nanoparticles will act like nanoscale heat sources resulting in highly localized therapeutic agents causing heat-induced cell deaths. Temperature increase and optical path-length changes of SPIONs have been observed in response to 532-nm pulsed laser irradiation; this suggests their potential application in biomedical imaging [127]. Similarly, gold nanoshells have high-absorption cross sections in the NIR regions; this suggests their application in photoablation therapy [49, 50, 128]. Using gold nanoshells, photothermally induced cell deaths have been observed with a temperature increase of 30°C to 35°C [129].

11.2.4 Surface Plasmon Resonant Probes

Surface plasmons or surface polaritons are collective oscillations of electrons at a metal-dielectric interface when a *p*-polarized electromagnetic radiation is incident on the surface. Typically, noble metals like gold, silver, and platinum exhibit this property in the visible and near-IR regions of the electromagnetic spectra. Surface plasmons confined to metallic nanoparticle surfaces and metallic nanostructures, often referred to as *localized surface plasmons* (LSP), are excited by an electric field (light) at a particular resonant frequency. These resonant frequencies and the intensity of the surface plasmon resonance (SPR) bands highly depend on the type of metal, the size and shape of the nanostructures, and the surrounding environment of the nanoparticles. The precise properties of LSPs have prompted various researchers to exploit this concept for the development of molecular nanoprobes that are scattering or absorbing in nature. With the recent advances in nanochemistry, a wide variety of structured nanoparticles have been explored toward this goal. Among them, gold nanostructures have attained great popularity due to their exceptional properties like biocompatibility; well-established chemistry for fabrication and bioconjugation applications; and versatility in terms of size, shape, and tunability of their resonant wavelengths [130]. Plasmon-resonant nanoparticles with optical scattering or absorption in the NIR are valuable contrast agents for biophotonic imaging modalities like dark-field microscopy, SOCT, photothermal OCT, and multiphoton microscopy, as well as others.

The optical response of SPR probes can be computationally predicted by Mie theory [131]. The resonant wavelength [8, 132] can be tuned by modifying the size and shape of the particle; this provides the capability to produce multiply labeled probes for multichannel molecular imaging [55, 133–136]. Fast computation techniques for predicting the SPR spectrum from arbitrarily shaped particles are available [137]. The SPR optical extinction is dominated by scattering for larger particles and by absorption for smaller particles, nominally <80 nm.

Recently, antibody-conjugated gold nanoparticles have been demonstrated in reflectance-based optical imaging for *in vivo* applications [138]. The resonant light-scattering property of gold nanoparticles at their surface plasmon resonance has been exploited to develop them as potential molecular contrast probes for imaging biomolecular changes during carcinogenesis under reflectance-mode imaging techniques. The localization of gold bioconjugates on the epidermal growth factor receptor (EGFR) increased the reflectance properties of nasopharyngeal carcinoma (CNE2) cells and normal human lung fibroblast (NHLF) cells, and the regions of increased reflectance corresponded to regions of high EGFR expression in the cells. These gold bioconjugates were thus able to map the expression of relevant biomarkers and elicit an optical contrast for cancer cells over normal cells under confocal reflectance microscopy.

Gold nanoshells are a novel class of composite core-shell nanoparticles with an extremely tunable peak optical resonance ranging from the near-UV to the mid-IR wavelengths. They are composed of a nonconducting dielectric core (e.g., silica) and coated with an ultrathin layer of metal like gold or silver [135]. Gold nanoshells exhibit strong scattering and absorption effects due to the strong plasmon resonance of the metallic-dielectric concentric spherical configuration. These highly favorable optical and chemical properties of gold nanoshells make them extremely useful in optical molecular contrast imaging and therapeutic applications. The precise tunability of the optical resonance is achieved by varying the relative core size and shell thickness. By using current chemistries, nanoshells of a wide variety of core and shell sizes can be easily fabricated to scatter and/or absorb light with optical cross sections often several times larger than the geometric cross section. By using current fabrication protocols, it is also possible to construct gold nanoshells of varying sizes with experimental observations of gold nanoshell resonances closely matching Mie theory. Characterization of the optical behavior of gold nanoshells in tissue is important for knowing the optimal paramaters for their use as optical contrast agents. Numerous examples of nanoshell-based diagnostic and therapeutic approaches, including the development of scattering and absorbing molecular contrast agents for OCT, targeted nanoshell bioconjugates for molecular imaging, and the use of absorbing nanoshells in NIR thermal therapy of tumors, have been demonstrated in recent years.

Highly NIR scattering nanoshells with large core dimensions have been proposed as scattering contrast agents in OCT and reflectance confocal microscopy (RCM) [70, 139, 140]. The feasibility of using these targeted nanoshell bioconjugates as enhanced scattering molecular-specific contrast agents to visualize human epidermal growth factor receptor 2 (HER-2) expressions in living SKBR3 human breast carcinoma cells under dark-field microscopy has been demonstrated. Nanoshell-enhanced OCT was shown at 1310 nm in water and turbid tissue-simulating phantoms to study the effect of nanoshell concentration, core diameter, and shell thickness on OCT signal enhancement [48]. By varying the relative dimensions of the core and shell, the optical resonance of these particles can be systematically varied over a broad spectral region from the near-UV to the mid-IR. Experimental results indicated a monotonic increase in signal intensity and attenuation with increasing shell and core-size trends consistent with predicted optical properties. Threshold concentrations for a 2-dB OCT signal intensity gain were determined for several nanoshell geometries. Diffuse optical spectroscopy for the detection and estimation of concentration of gold nanoshells in tissue phantoms has also been demonstrated [52].

Gold nanoshells due to their SPR absorption are efficient agents for photothermal therapy of cancers [129, 141]. The targeted cancer cells can subsequently be selectively destroyed through photothermal therapy, paving the way for an integrated approach to image and treat cancer [139]. Selective tumor-cell destruction has been demonstrated with EGFR-targeted plasmon resonant gold nanoshells using laser-induced therapy [142]. Effective tumor ablation and dramatic contrast enhancement in OCT (Fig. 11.9) has been shown *ex vivo* in a tumor mouse model using gold nanoshells [143]. These examples suggest the growing applications and potential of gold nanoshells in *in vivo* molecular imaging and therapeutic techniques. Recently the effect of the concentration of gold nanoshells on OCT contrast in a murine *in vivo* tumor model has been studied with antibody-conjugated gold nanoshells [144, 145]. On the basis of a theoretical model of the OCT backscattered signal intensity, the concentration of gold nanoshells in the tumor tissue was produced. The results were compared with the OCT scattering contrast obtained through intratumoral delivery of nonfunctionalized gold nanoshells. It is possible to predict the effect of gold nanoshell concentration on tissue reflectance for scattering-based optical methods using Monte Carlo simulations [50]. These models have also demonstrated the importance of absorption from the nanoshells on the remitted signals, even when the optical extinction is dominated by scattering.

A new class of gold nanoshells with a super-paramagnetic iron-oxide core (SPIO-Au nanoshells) has been shown to have desirable optical and magnetic properties, enabling multimodal imaging [66, 146]. These nanoparticles can be guided to a specific site using an external guiding magnetic field, and thereby enhancing the efficacy of nanoshell-mediated photothermal therapy. With a significant absorbance in the NIR region of the electromagnetic spectrum, they would efficiently act as photothermal agents for therapy. They would also find applications as dynamic contrast agents for *in vivo* real-time MM-OCT and MRI. Gold nanoshells with gadolinium ions in their core have been demonstrated as dual-modality agents for optical and MR imaging [147]. The particles have been conjugated to anti-HER-2 antibodies to target cancer cells and have exhibited therapeutic properties based on photothermal effects. Following the protocol from Chen et al. [148], another variant of gold nanoshells was developed through surface modifications with a dextran coating [149]. These hollow gold nanoshells with a diameter of approximately 60 nm had an SPR peak in NIR and were used for photothermal therapy on

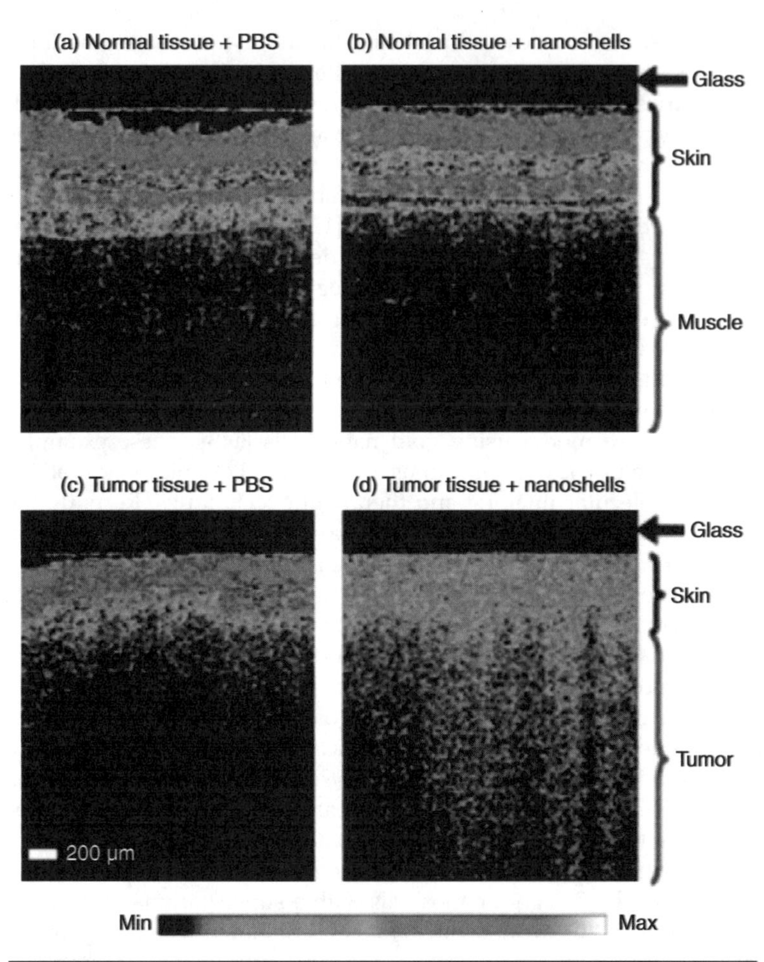

Figure 11.9 Representative OCT images from normal skin and muscle tissue areas of mice systemically injected with (*a*) phosphate buffered saline (PBS) or with (*b*) nanoshells. Representative OCT images from skin and tumors of mice systemically injected with (*c*) PBS or with (*d*) nanoshells. Analysis of all images shows a significant increase in contrast intensity after nanoshell injection in the tumors of mice treated with nanoshells while no increase in intensity is observed in the normal tissue. The glass of the probe is 200-μm thick and shows as a dark nonscattering layer. (Figure reprinted with permission [143].)

macrophage cells. Selective tumor-cell destruction has been demonstrated using EGFR-targeted gold nanoshells.

Gold nanorods (GNRs) are another class of scattering and absorbing probes for *in vivo* and *in vitro* imaging [55]. In addition to the scattering contrast, their efficient absorption in the NIR region due to surface plasmon resonances and plasmon-enhanced two-photon luminescence (TPL) make them multimodal contrast agents for biomedical imaging applications like OCT, two-photon microscopy, and photoacoustic tomography. The aspect ratio is the primary factor for determining the SPR wavelength of GNRs, with GNRs resonating at longer wavelengths for higher ratios [150]. Using the seeded surfactant method, we can produce gold nanorods typically of 10- to 20-nm width with controlled aspect ratios between ~2.5:1 and 18:1, corresponding to resonant wavelengths from 650 nm to greater than 1800 nm, respectively [67, 69]. Using well-established surface modifications and conjugation protocols, we can easily achieve cell-specific targeting using GNRs.

GNRs are excellent candidates as SOCT contrast agents because the quality of their NIR resonance is higher than most other nanostructures, including spherical gold nanoparticles [151]. The increased plasmon damping in gold nanorods explains their improved quality factors (~50-nm line width) in comparison to gold spheres [152]. Gold nanorods have been demonstrated as SOCT contrast in intralipid phantoms [115]. GNR distributions in liquid tissue phantoms have been imaged with sensitivities as low as 25 μg Au per mililiter within 2% intralipid using SOCT (Fig. 11.10). GNRs with a size of 15×45 nm and a resonance at 870 nm have been synthesized using a sulfide-arrested seeded surfactant method [153] in order to stabilize the optical response of nanorod suspensions over time periods exceeding a few months. The aspect ratio of the GNRs is appropriately adjusted for maximum absorption in the far red to create a partial spectral overlap with the short-wavelength edge of the NIR SOCT imaging band. The spectroscopic absorption profile of the GNRs is incorporated into a depth-resolved algorithm for mapping the relative GNR density within OCT images. This would improve the sensitivity of the technique further. Detection of GNRs in excised human breast tumors has been demonstrated using SOCT [54]; this shows their potential as OCT contrast agents in heterogeneous, highly scattering tissues.

The photothermal effects caused by the strong SPR absorption of GNRs can generate another level of contrast (photothermal) and may be exploited for the photothermal imaging of tissues [154]. A differential phase processing of the OCT signal can enable the noninvasive depth-resolved detection of photothermal signals from tissues. GNRs can also efficiently function as therapeutic agents by converting optical energy into heat, and thereby inflicting localized damage to tumor cells. Laser-induced heating of GNRs can disrupt cell membrane integrity and homeostasis, and result in Ca^{2+} influx and the depolymerization of the intracellular actin network [155].

Gold nanocages are another new class of promising contrast agents for conventional and spectroscopic OCT imaging. These are cubic in shape with a thin porous wall and hollow inner core. Compared to core-shell nanoparticles, gold nanocages have a large absorption

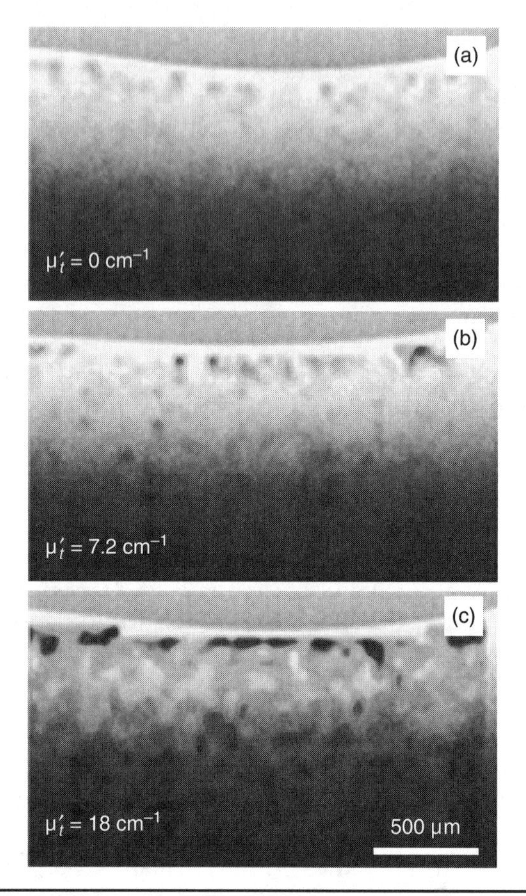

FIGURE 11.10 SOCT images of 2% intralipid (*a*) without nanorods, (*b*) with nanorods of 7.2-cm^{-1} peak attenuation, and (*c*) with nanorods of 18-cm^{-1} peak attenuation. (Figure reprinted with permission [115].)

cross section and strong optical resonance, with a peak wavelength in the optical window of biological tissue (i.e., around 800 nm), while remaining relatively small (e.g., with an edge length of 30 to 60 nm). By adjusting their size, shape, and thickness, their optical resonance peak can be easily tuned; this makes them versatile for SOCT imaging applications for a wide range of wavelengths [54]. Furthermore, gold nanocages can be easily conjugated with molecular recognition agents by using well-established gold-thiolate monolayer surface chemistry [140]. Studies have shown that gold nanocages with a 35-nm edge length exhibit very strong absorption at 800 nm; this provides excellent spectroscopic contrast enhancement. Gold nanocages were fabricated by a method based on the galvanic replacement reaction [58, 156]. The strong optical SPR peak of these nanocages can be precisely tuned from the visible to the NIR region by controlling the molar ratio between the silver template and $HAuCl_4$ [148]. Conjugation of nanocages with the tumor-specific antibody HER-2 and successful molecular-specific binding on SKBR3 cell lines has been demonstrated [58]. To demonstrate their use as OCT contrast agents, nanocages 35 nm on a side were synthesized and exhibited a resonance at 800 nm with a line width on the order of 100 nm. Nanocages at concentrations of ~1 nM provided a noticeable increase in the extinction apparent in OCT images of tissue-like phantoms (Fig. 11.11). Gold nanocages SP-resonant at 720 nm have

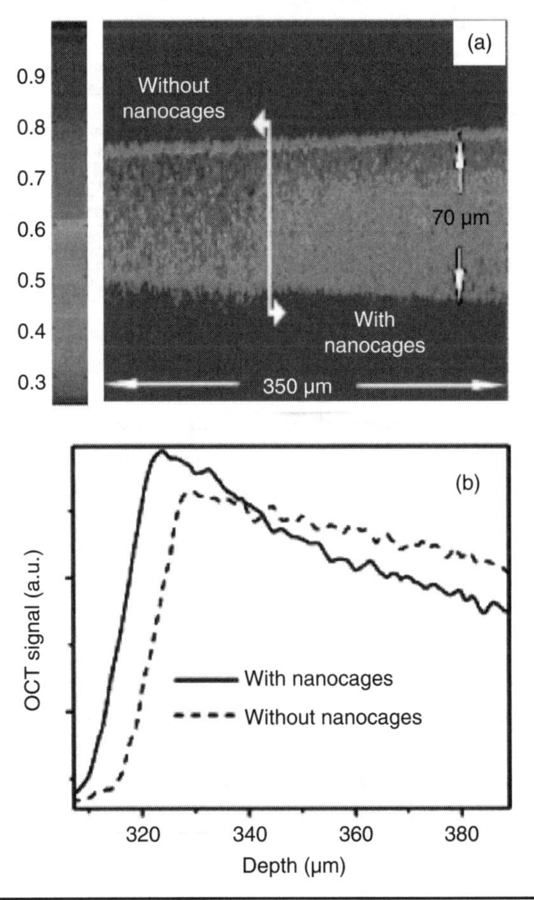

FIGURE 11.11 Scattering contrast enhancement from nanocages. (*a*) OCT image of a gelatin phantom embedded with TiO_2. The concentration of TiO_2 was controlled at 1 mg per millimiter to mimic the background scattering of soft tissues. The right portion of the phantom contains 1 nM of gold nanocages whereas the left portion did not contain any gold nanocages. (*b*) Plots of the OCT signals on a log scale as a function of depth. Note that the OCT signal recorded from the portion of phantom with gold nanocages decays faster than the portion without nanocages. (Figure reprinted with permission [58].)

shown contrast enhancement in SOCT images calculated from the red-shifted back-scattered spectra [57]. Photothermal destruction of breast cancer cells was also demonstrated [59, 157] *in vitro* using immuno-targeted Au nanocages. These reports suggest that gold nanocages could be efficient targeted contrast agents for the diagnosis and treatment of cancer.

With a large and rapidly growing variety of nanostructures with limitless material and chemical combinations, SPR-based probes can be adapted to different imaging modalities and applications. Single-walled carbon nanotubes (SWNTs) are another class of nanoprobes that have been broadly investigated as imaging agents for the evaluation of tumor targeting and localization *in vitro* and *in vivo*. SWNTs have unique physical and chemical characteristics, such as NIR intrinsic fluorescence signatures, Raman signatures, NIR absorption leading to hyperthermia, and photoacoustic signal generation, which make them ideal for noninvasive and high-sensitivity detection [71–73, 75]. Nanoparticles made up of other materials may be coupled with SPR or non-SPR probes such as SWNTs to provide multifunctional ability. Multifunctional SWNT or iron-oxide nanoparticle complexes have been demonstrated as dual magnetic and fluorescent imaging agents [71]. By encapsulation with DNA, the SWNT or iron-oxide nanoparticle complexes are individually dispersed in an aqueous solution and are introduced into a biological environment. Two-dimensional *in vitro* imaging of murine macrophage cells have been demonstrated using these nanostructures. However, significantly more investigations are needed to evaluate the preclinical uses of these probes. Many questions remain regarding short-term and long-term biocompatibility, biodistribution and pharmacokinetics, and practical usefulness in an *in vivo* environment before they can be successfully translated to clinical trials for noninvasive imaging and therapeutic applications in humans.

11.3 Conclusions and Future Perspectives

Molecular-specific optical probes are promising tools for the noninvasive *in vivo* visualization of pathological conditions at the subcellular level. These probes have also demonstrated their capabilities as therapeutic tools for targeted drug delivery and hyperthermia applications. Optical molecular imaging to date has been successful in catering to the requirements of the biological research community by providing better resolution, specificity, and sensitivity. Recent research in this field has primarily been focused on the development of molecular imaging agents and tools. Early preclinical trials have also been successful and promising. Future research investigations must now be directed at the translation of these technologies to clinical applications, having nanoparticle-based diagnostic imaging modalities enter clinical trials. In parallel, the continued evolution of more nanoparticle-based agents for imaging and therapy is anticipated over the next decade. Given the prevalence and impact of cancer on our lives, cancer detection, imaging, and therapy are expected to be the earliest applications for these molecular-specific optical probes.

Because these probes can be flexibly engineered and fabricated, imaging agents that are multimodal and multifunctional will be more attractive and are expected to find direct applications using many of the currently existing clinical imaging modalities such as MRI, ultrasound, X-ray-computed tomography, and nuclear imaging (PET and SPECT). These same probes can also find utility in image-guided procedures, contributing to their multimodal multifunctional use. Combining the potential of nanotechnology and biophotonics with our latest understanding in cancer research and molecular pathways, detailed diagnosis of diseased tissues, down to molecular levels, will likely change the state-of-the-art clinical treatment practices in years to come.

Acknowledgments

We wish to thank the members of the Biophotonics Imaging Laboratory at the Beckman Institute at the University of Illinois at Urbana–Champaign (UIUC) campus for their dedication and insight in this area of research. We also wish to thank our collaborators, including Kenneth Suslick, Department of Chemistry, UIUC, and Alex Wei, Department of Chemistry, Purdue University, for their work in fabricating novel probes. Finally, we thank our colleagues whose work is represented here for helping advance this field and apologize to those whose work was not included due to space and depth-of-coverage constraints. This research was supported in part by the National Institutes of Health (Roadmap Initiative, NIBIB, R21 EB005321; NIBIB, R01 EB005221; NIBIB R01 EB009073; NCI RC1 CA147096, to S.A.B.). Additional information can be found at http://biophotonics.illinois.edu.

References

1. G. F. Whitmore, "100 Years of X-Rays in Biological Research," *Radiation Research* **144:** 148–159, (1995).
2. S. Achilefu, R. B. Dorshow, J. E. Bugaj, and R. Rajagopalan, "Novel Receptor-Targeted Fluorescent Contrast Agents for *in vivo* Tumor Imaging," *Investigative Radiology* **35:** 479–485, (2000).
3. M. Hassan and B. A. Klaunberg, "Biomedical Applications of Fluorescence Imaging *in vivo*," *Comparative Medicine* **54:** 635–644, (2004).
4. B. Ballou, L. A. Ernst, and A. S. Waggoner, "Fluorescence Imaging of Tumors *in vivo*," *Current Medicinal Chemistry* **12:** 795–805 (2005).

5. M. Zhao, M. F. Kircher, L. Josephson, and R. Weissleder, "Differential Conjugation of Tat Peptide to Super Paramagnetic Nanoparticles and Its Effect on Cellular Uptake," *Bioconjugate Chemistry* **13:** 840–844 (2002).
6. R. Weissleder, C. H. Tung, U. Mahmood, and A. Bogdanov, "*In vivo* Imaging of Tumors with Protease-Activated Near-Infrared Fluorescent Probes," *Nature Biotechnology* **17:** 375–378 (1999).
7. R. Weissleder and V. Ntziachristos, "Shedding Light onto Live Molecular Targets," *Nature Medicine* **9:** 123–128 (2003).
8. S. Achilefu, "Lighting up Tumors with Receptor-Specific Optical Molecular Probes," *Technology in Cancer Research & Treatment* **3:** 393–409 (2004).
9. D. Huang, E. A. Swanson, C. P. Lin, J. S. Schuman, W. G. Stinson, W. Chang, M. R. Hee, T. Flotte, K. Gregory, C. A. Puliafito, and J. G. Fujimoto, "Optical Coherence Tomography," *Science* **254:** 1178–1181 (1991).
10. E. A. Swanson, J. A. Izatt, M. R. Hee, D. Huang, C. P. Lin, J. S. Schuman, C. A. Puliafito, and J. G. Fujimoto, "*In vivo* Retinal Imaging by Optical Coherence Tomography," *Optics Letters* **18:** 1864–1866 (1993).
11. M. R. Hee, J. A. Izatt, E. A. Swanson, D. Huang, J. S. Schuman, C. P. Lin, C. A. Puliafito, and J. G. Fujimoto, "Optical Coherence Tomography of the Human Retina," *Archives of Ophthalmology* **113:** 325–332 (1995).
12. U. Morgner, W. Drexler, F. X. Kartner, X. D. Li, C. Pitris, E. P. Ippen, and J. G. Fujimoto, "Spectroscopic Optical Coherence Tomography," *Optics Letters* **25:** 111–113 (2000).
13. S. A. Boppart, B. E. Bouma, C. Pitris, J. F. Southern, M. E. Brezinski, and J. G. Fujimoto, "*In vivo* Cellular Optical Coherence Tomography Imaging," *Nature Medicine* **4:** 861–865 (1998).
14. S. A. Boppart, B. E. Bouma, C. Pitris, M. E. Brezinski, and J. G. Fujimoto, "Optical Coherence Tomography of *in vivo* Developmental Cellular Dynamics," *FASEB J.* **12:** 283 (1998).
15. W. Drexler, U. Morgner, F. X. Kartner, C. Pitris, S. A. Boppart, X. D. Li, E. P. Ippen, and J. G. Fujimoto, "*In vivo* Ultrahigh-Resolution Optical Coherence Tomography," *Optics Letters* **24:** 1221–1223 (1999).
16. S. A. Boppart, "Optical Coherence Tomography: Technology and Applications for Neuroimaging," *Psychophysiology* **40:** 529–541 (2003).
17. S. A. Boppart, "Optical Coherence Tomography—Principles Applications and Advances," *Minerva Biotecnologica* **16:** 211–237 (2004).
18. T. Collier, P. Shen, B. de Pradier, K. B. Sung, R. Richards-Kortum, A. Malpica, and M. Follen, "Near Real Time Confocal Microscopy of Amelanotic Tissue: Dynamics of Aceto-Whitening Enable Nuclear Segmentation," *Optics Express* **6:** 40–48 (2000).
19. J. K. Barton, J. B. Hoying, and C. J. Sullivan, "Use of Microbubbles as an Optical Coherence Tomography Contrast Agent," *Academic Radiology* **9:** S52–S55 (2002).
20. L. C. Katz, A. Burkhalter, and W. J. Dreyer, "Fluorescent Latex Microspheres as a Retrograde Neuronal Marker for *in vivo* and *in vitro* Studies of Visual Cortex," *Nature* **310:** 498–500 (1984).
21. A. K. Gupta and S. Wells, "Surface-Modified Superparamagnetic Nanoparticles for Drug Delivery: Preparation, Characterization, and Cytotoxicity Studies," *IEEE Transactions on Nanobioscience* **3:** 66–73 (2004).
22. R. Sinha, G. J. Kim, S. M. Nie, and D. M. Shin, "Nanotechnology in Cancer Therapeutics: Bioconjugated Nanoparticles for Drug Delivery," *Molecular Cancer Therapeutics* **5:** 1909–1917 (2006).
23. B. Law, R. Weissleder, and C. H. Tung, "Peptide-Based Biomaterials for Protease-Enhanced Drug Delivery," *Biomacromolecules* **7:** 1261–1265 (2006).
24. R. Guo, R. T. Li, X. L. Li, L. Y. Zhang, X. Q. Jiang, and B. R. Liu, "Dual-Functional Alginic acid Hybrid Nanospheres for Cell Imaging and Drug Delivery," *Small* **5:** 709–717 (2009).
25. H. K. Sajja, M. P. East, H. Mao, Y. A. Wang, S. Nie, and L. Yang, "Development of Multifunctional Nanoparticles for Targeted Drug Delivery and Noninvasive Imaging of Therapeutic Effect," *Current Drug Discovery Technologies* **6:** 43–51 (2009).

26. K. M. L. Taylor-Pashow, J. Della Rocca, Z. Xie, S. Tran, and W. Lin, "Post Synthetic Modifications of Iron-Carboxylate Nanoscale Metal-Organic Frameworks for Imaging and Drug Delivery," *J. Ameri. Chemical Society* **131:** 14261–14263 (2009).
27. J. R. McCarthy and R. Weissleder, "Multifunctional Magnetic Nanoparticles for Targeted Imaging and Therapy," *Advanced Drug Delivery Reviews* **60:** 1241–1251 (2008).
28. M. E. Gindy and R. K. Prud'homme, "Multifunctional Nanoparticles for Imaging, Delivery and Targeting in Cancer Therapy," *Expert Opinion on Drug Delivery* **6:** 865–878 (2009).
29. R. K. Gilchrist, R. Medal, W. D. Shorey, R. C. Hanselman, J. C. Parrott, and C. B. Taylor, "Selective Inductive Heating of Lymph Nodes," *Annals of Surgery* **146:** 596–606 (1957).
30. H. Lee, E. Lee, D. K. Kim, N. K. Jang, Y. Y. Jeong, and S. Jon, "Antibiofouling Polymer-Coated Superparamagnetic Iron Oxide Nanoparticles as Potential Magnetic Resonance Contrast Agents for *in vivo* Cancer Imaging," *J. Ameri. Chemical Society* **128:** 7383–7389 (2006).
31. L. X. Tiefenauer, G. Kuhne, and R. Y. Andres, "Antibody Magnetite Nanoparticles *in vitro* Characterization of a Potential Tumor-Specific Contrast Agent for Magnetic Resonance Imaging," *Bioconjugate Chemistry* **4:** 347–352 (1993).
32. A. K. Gupta and M. Gupta, "Synthesis and Surface Engineering of Iron Oxide Nanoparticles for Biomedical Applications," *Biomaterials* **26:** 3995–4021 (2005).
33. A. K. Gupta and M. Gupta, "Cytotoxicity Suppression and Cellular Uptake Enhancement of Surface Modified Magnetic Nanoparticles," *Biomaterials* **26:** 1565–1573 (2005).
34. A. K. Gupta and A. S. G. Curtis, "Surface Modified Superparamagnetic Nanoparticles for Drug Delivery: Interaction Studies with Human Fibroblasts in Culture," *J. Materials Science—Materials in Medicine* **15:** 493–496 (2004).
35. C. C. Berry and A. S. G. Curtis, "Functionalisation of Magnetic Nanoparticles for Applications in Biomedicine," *J. Physics D—Applied Physics* **36:** R198–R206 (2003).
36. C. C. Berry, S. Wells, S. Charles, and A. S. G. Curtis, "Dextran and Albumin Derivatised Iron-Oxide Nanoparticles: Influence on Fibroblasts *in vitro*," *Biomaterials* **24:** 4551–4557 (2003).
37. Y. Zhang, N. Kohler, and M. Q. Zhang, "Surface Modification of Superparamagnetic Magnetite Nanoparticles and their Intracellular Uptake," *Biomaterials* **23:** 1553–1561 (2002).
38. A. J. M. D'Souza, R. L. Schowen, and E. M. Topp, "Polyvinylpyrrolidone-Drug Conjugate: Synthesis and Release Mechanism," *J. Controlled Release* **94:** 91–100 (2004).
39. A. K. Gupta and A. S. G. Curtis, "Lactoferrin and Ceruloplasmin Derivatized Superparamagnetic Iron Oxide Nanoparticles for Targeting Cell Surface Receptors," *Biomaterials* **25:** 3029–3040 (2004).
40. J. W. M. Bulte, S. C. Zhang, P. van Gelderen, V. Herynek, E. K. Jordan, I. D. Duncan, and J. A. Frank, "Neurotransplantation of Magnetically Labeled Oligodendrocyte Progenitors: Magnetic Resonance Tracking of Cell Migration and Myelination," *Proceedings of the National Academy of Sciences* USA **96:** 15256–15261 (1999).
41. X. H. Peng, X. M. Qian, H. Mao, A. Y. Wang, Z. Chen, S. M. Nie, and D. M. Shin, "Targeted Magnetic Iron Oxide Nanoparticles for Tumor Imaging and Therapy," *Int. J. Nanomedicine* **3:** 311–321 (2008).
42. F. Ibraimi, D. Kriz, M. Lu, L. O. Hansson, and K. Kriz, "Rapid One-Step Whole Blood C-Reactive Protein Magnetic Permeability Immunoassay with Monoclonal Antibody Conjugated Nanoparticles as Superparamagnetic Labels and Enhanced Sedimentation," *Analytical and Bioanalytical Chemistry* **384:** 651–657 (2006).
43. C. Gruttner, K. Muller, J. Teller, F. Westphal, A. Foreman, and R. Ivkov, "Synthesis and Antibody Conjugation of Magnetic Nanoparticles with

Improved Specific Power Absorption Rates for Alternating Magnetic Field Cancer Therapy," *J. Magnetism and Magnetic Materials* **311:** 181–186 (2007).

44. L. Josephson, C. H. Tung, A. Moore, and R. Weissleder, "High-Efficiency Intracellular Magnetic Labeling with Novel Super Paramagnetic-TAT Peptide Conjugates," *Bioconjugate Chemistry* **10:** 186–191 (1999).
45. M. Lewin, N. Carlesso, C. H. Tung, X. W. Tang, D. Cory, D. T. Scadden, and R. Weissleder, "TAT Peptide-Derivatized Magnetic Nanoparticles Allow *in vivo* Tracking and Recovery of Progenitor Cells," *Nature Biotechnology* **18:** 410–414 (2000).
46. A. Faust, S. Hermann, S. Wagner, G. Haufe, O. Schober, M. Schafers, and K. Kopka, "Molecular Imaging of Apoptosis *in vivo* with Scintigraphic and Optical Biomarkers—A Status Report," *Anti-Cancer Agents in Medicinal Chemistry* **9:** 968–985 (2009).
47. K. Wagner, A. Kautz, M. Roder, M. Schwalbe, K. Pachmann, J. H. Clement, and M. Schnabelrauch, "Synthesis of Oligonucleotide-Functionalized Magnetic Nanoparticles and Study on Their *in vitro* Cell Uptake," *Applied Organometallic Chemistry* **18:** 514–519 (2004).
48. A. Agrawal, S. Huang, A. W. H. Lin, M. H. Lee, J. K. Barton, R. A. Drezek, and T. J. Pfefer, "Quantitative Evaluation of Optical Coherence Tomography Signal Enhancement with Gold Nanoshells," *J. Biomedical Optics* **11:** 041121, (2006).
49. L. R. Hirsch, A. M. Gobin, A. R. Lowery, F. Tam, R. A. Drezek, N. J. Halas, and J. L. West, "Metal Nanoshells," *Annals of Biomedical Engineering* **34:** 15–22 (2006).
50. A. W. H. Lin, N. A. Lewinski, J. L. West, N. J. Halas, and R. A. Drezek, "Optically Tunable Nanoparticle Contrast Agents for Early Cancer Detection: Model-Based Analysis of Gold Nanoshells," *J. Biomedical Optics* **10:** (2005).
51. P. Puvanakrishnan, J. Park, P. Diagaradjane, J. A. Schwartz, C. L. Coleman, K. L. Gill-Sharp, K. L. Sang, J. D. Payne, S. Krishnan, and J. W. Tunnell, "Near-Infrared Narrow-Band Imaging of Gold/Silica Nanoshells in Tumors," *J. Biomedical Optics* **14:** 024044, (2009).
52. R. T. Zaman, P. Diagaradjane, J. C. Wang, J. Schwartz, N. Rajaram, K. L. Gill-Sharp, S. H. Cho, H. G. Rylander, J. D. Payne, S. Krishnan, and J. W. Tunnell, "*In vivo* Detection of Gold Nanoshells in Tumors Using Diffuse Optical Spectroscopy," IEEE *J. Selected Topics in Quantum Electronics* **13:** 1715–1720 (2007).
53. E. C. Cho, C. Kim, F. Zhou, C. M. Cobley, K. H. Song, J. Y. Chen, Z. Y. Li, L. H. V. Wang, and Y. N. Xia, "Measuring the Optical Absorption Cross Sections of Au-Ag Nanocages and Au Nanorods by Photoacoustic Imaging," *J. Physical Chemistry C* **113:** 9023–9028 (2009).
54. A. L. Oldenburg, M. N. Hansen, T. S. Ralston, A. Wei, and S. A. Boppart, "Imaging Gold Nanorods in Excised Human Breast Carcinoma by Spectroscopic Optical Coherence Tomography," *J. Materials Chemistry* **19:** 6407–6411 (2009).
55. A. L. Oldenburg, M. N. Hansen, D. A. Zweifel, A. Wei, and S. A. Boppart, "Plasmon-Resonant Gold Nanorods as Low Backscattering Albedo Contrast Agents for Optical Coherence Tomography," *Optics Express* **14:** 6724–6738 (2006).
56. T. S. Troutman, J. K. Barton, and M. Romanowski, "Optical Coherence Tomography with Plasmon Resonant Nanorods of Gold," *Optics Letters* **32:** 1438–1440 (2007).
57. H. Cang, T. Sun, Z. Y. Li, J. Y. Chen, B. J. Wiley, Y. N. Xia, and X. D. Li, "Gold Nanocages as Contrast Agents for Spectroscopic Optical Coherence Tomography," *Optics Letters* **30:** 3048–3050 (2005).
58. J. Chen, F. Saeki, B. J. Wiley, H. Cang, M. J. Cobb, Z. Y. Li, L. Au, H. Zhang, M. B. Kimmey, X. D. Li, and Y. Xia, "Gold Nanocages: Bioconjugation and Their Potential Use as Optical Imaging Contrast Agents," *Nano Letters* **5:** 473–477 (2005).
59. S. E. Skrabalak, J. Y. Chen, Y. G. Sun, X. M. Lu, L. Au, C. M. Cobley, and Y. N. Xia, "Gold Nanocages: Synthesis, Properties, and Applications," *Accounts of Chemical Research* **41:** 1587–1595 (2008).

60. Q. S. Wei, H. M. Song, A. P. Leonov, J. A. Hale, D. M. Oh, Q. K. Ong, K. Ritchie, and A. Wei, "Gyromagnetic Imaging: Dynamic Optical Contrast Using Gold Nanostars with Magnetic Cores," *J. Ameri. Chemical Society* **131:** 9728–9734 (2009).
61. S. Kumar, J. Aaron, and K. Sokolov, "Directional Conjugation of Antibodies to Nanoparticles for Synthesis of Multiplexed Optical Contrast Agents with both Delivery and Targeting Moieties," *Nature Protocols* **3:** 314–320 (2008).
62. K. Sokolov, J. Aaron, B. Hsu, D. Nida, A. Gillenwater, M. Follen, C. MacAulay, K. Adler-Storthz, B. Korgel, M. Descour, R. Pasqualini, W. Arap, W. Lam, and R. Richards-Kortum, "Optical Systems for *in vivo* Molecular Imaging of Cancer," *Technology in Cancer Research & Treatment* **2:** 491–504 (2003).
63. K. Sokolov, M. Follen, J. Aaron, I. Pavlova, A. Malpica, R. Lotan, and R. Richards-Kortum, "Real-Time Vital Optical Imaging of Precancer Using Anti-Epidermal Growth Factor Receptor Antibodies Conjugated to Gold Nanoparticles," *Cancer Research* **63:** 1999–2004 (2003).
64. A. Wax and K. Sokolov, "Molecular Imaging and Darkfield Microspectroscopy of Live Cells Using Gold Plasmonic Nanoparticles," *Laser & Photonics Reviews* **3:** 146–158 (2009).
65. D. L. Chamberland, A. Agarwal, N. Kotov, J. B. Fowlkes, P. L. Carson, and X. Wang, "Photoacoustic Tomography of Joints Aided by an Etanercept-Conjugated Gold Nanoparticle Contrast Agent—an *ex vivo* Preliminary Rat Study," *Nanotechnology* **19:** (2008).
66. X. J. Ji, R. P. Shao, A. M. Elliott, R. J. Stafford, E. Esparza-Coss, J. A. Bankson, G. Liang, Z. P. Luo, K. Park, J. T. Markert, and C. Li, "Bifunctional Gold Nanoshells with a Super Paramagnetic Iron-Oxide Silica Core Suitable for Both MR Imaging and Photothermal Therapy," *J. Physical Chemistry C* **111:** 6245–6251 (2007).
67. N. R. Jana, L. Gearheart, and C. J. Murphy, "Wet Chemical Synthesis of High Aspect Ratio Cylindrical Gold Nanorods," *J. Physical Chemistry B* **105:** 4065–4067 (2001).
68. B. Nikoobakht and M. A. El-Sayed, "Preparation and Growth Mechanism of Gold Nanorods (NRs) Using Seed-Mediated Growth Method," *Chemistry of Materials* **15:** 1957–1962 (2003).
69. V. M. Cepak and C. R. Martin, "Preparation and Stability of Template-Synthesized Metal Nanorod Sols in Organic Solvents," *J. Physical Chemistry B* **102:** 9985–9990 (1998).
70. C. Loo, A. Lowery, N. Halas, J. West, and R. Drezek, "Immunotargeted Nanoshells for Integrated Cancer Imaging and Therapy," *Nano Letters* **5:** 709–711 (2005).
71. J. H. Choi, F. T. Nguyen, P. W. Barone, D. A. Heller, A. E. Moll, D. Patel, S. A. Boppart, and M. S. Strano, "Multimodal Biomedical Imaging with Asymmetric Single-Walled Carbon Nanotube/Iron Oxide Nanoparticle Complexes," *Nano Letters* **7:** 861–867 (2007).
72. A. De La Zerda, C. Zavaleta, S. Keren, S. Vaithilingam, S. Bodapati, Z. Liu, J. Levi, B. R. Smith, T. J. Ma, O. Oralkan, Z. Cheng, X. Y. Chen, H. J. Dai, B. T. Khuri-Yakub, and S. S. Gambhir, "Carbon Nanotubes as Photoacoustic Molecular Imaging Agents in Living Mice," *Nature Nanotechnology* **3:** 557–562 (2008).
73. H. Hong, T. Gao, and W. B. Cai, "Molecular Imaging with Single-Walled Carbon Nanotubes," *Nano Today* **4:** 252–261 (2009).
74. Y. T. Lim, Y. W. Noh, J. N. Kwon, and B. H. Chung, "Multifunctional Perfluorocarbon Nanoemulsions for F-19–Based Magnetic Resonance and Near-Infrared Optical Imaging of Dendritic Cells," *Chemical Communications*, **45:** 6952–6954 (2009).
75. M. Pramanik, M. Swierczewska, D. Green, B. Sitharaman, and L. V. Wang, "Single-Walled Carbon Nanotubes as a Multimodal-Thermoacoustic and Photoacoustic-Contrast Agent," *J. Biomedical Optics* **14:** 034018 (2009).
76. V. Ntziachristos, A. G. Yodh, M. Schnall, and B. Chance, "Concurrent MRI and Diffuse Optical Tomography of Breast after Indocyanine Green Enhancement," *Proceedings of the National Academy of Sciences USA* **97:** 2767–2772 (2000).

77. A. Y. Louie, M. M. Huber, E. T. Ahrens, U. Rothbacher, R. Moats, R. E. Jacobs, S. E. Fraser, and T. J. Meade, "*In vivo* Visualization of Gene Expression Using Magnetic Resonance Imaging," *Nature Biotechnology* **18:** 321–325 (2000).
78. P. Caravan, J. J. Ellison, T. J. McMurry, and R. B. Lauffer, "Gadolinium(III) Chelates as MRI Contrast Agents: Structure, Dynamics, and Applications," *Chemical Reviews* **99:** 2293–2352 (1999).
79. R. Kumar, T. Y. Ohulchanskyy, I. Roy, S. K. Gupta, C. Borek, M. E. Thompson, and P. N. Prasad, "Near-Infrared Phosphorescent Polymeric Nanomicelles: Efficient Optical Probes for Tumor Imaging and Detection," *ACS Applied Materials & Interfaces* **1:** 1474–1481 (2009).
80. J. H. Kim, K. Park, H. Y. Nam, S. Lee, K. Kim, and I. C. Kwon, "Polymers for Bioimaging," *Progress in Polymer Science* **32:** 1031–1053 (2007).
81. E. Glynos, V. Koutsos, W. N. McDicken, C. M. Moran, S. D. Pye, J. A. Ross, and V. Sboros, "Nanomechanics of Biocompatible Hollow Thin-Shell Polymer Microspheres," *Langmuir* **25:** 7514–7522 (2009).
82. C. A. Boswell, P. K. Eck, C. A. S. Regino, M. Bernardo, K. J. Wong, D. E. Milenic, P. L. Choyke, and M. W. Brechbiel, "Synthesis, Characterization, and Biological Evaluation of Integrin Alpha(v) Beta(3)-Targeted PAMAM Dendrimers," *Molecular Pharmaceutics* **5:** 527–539 (2008).
83. A. M. Smith, G. Ruan, M. N. Rhyner, and S. M. Nie, "Engineering Luminescent Quantum Dots for *in vivo* Molecular and Cellular Imaging," *Annals of Biomedical Engineering* **34:** 3–14 (2006).
84. M. N. Rhyner, A. M. Smith, X. H. Gao, H. Mao, L. L. Yang, and S. M. Nie, "Quantum Dots and Multifunctional Nanoparticles: New Contrast Agents for Tumor Imaging," *Nanomedicine* **1:** 209–217 (2006).
85. W. J. Parak, T. Pellegrino, and C. Plank, "Labelling of Cells with Quantum Dots," *Nanotechnology* **16:** R9–R25 (2005).
86. N. Y. Morgan, S. English, W. Chen, V. Chernomordik, A. Russo, P. D. Smith, and A. Gandjbakhche, "Real Time *in vivo* Non-Invasive Optical Imaging Using Near-Infrared Fluorescent Quantum Dots," *Academic Radiology* **12:** 313–323 (2005).
87. S. Mazumder, R. Dey, M. K. Mitra, S. Mukherjee, and G. C. Das, "Review: Biofunctionalized Quantum Dots in Biology and Medicine," *J. Nanomaterials* **2009:** 815734 (2009).
88. J. M. C. Luk and A. A. Lindberg, "Rapid and Sensitive Detection of Salmonella (0–6,7) by Immunomagnetic Monoclonal Antibody-Based Assays," *J. Immunological Methods* **137:** 1–8 (1991).
89. I. Koh, R. Hong, R. Weissleder, and L. Josephson, "Nanoparticle-Target Interactions Parallel Antibody-Protein Interactions," *Analytical Chemistry* **81:** 3618–3622 (2009).
90. J. Q. Chen, C. H. Tung, J. R. Allport, S. Chen, R. Weissleder, and P. L. Huang, "Near-Infrared Fluorescent Imaging of Matrix Metalloproteinase Activity after Myocardial Infarction," *Circulation* **111:** 1800–1805 (2005).
91. C. Bremer, C. H. Tung, and R. Weissleder, "*In vivo* Molecular Target Assessment of Matrix Metalloproteinase Inhibition," *Nature Medicine* **7:** 743–748 (2001).
92. C. Bremer, S. Bredow, U. Mahmood, R. Weissleder, and C. H. Tung, "Optical Imaging of Matrix Metalloproteinase-2 Activity in Tumors: Feasibility Study in a Mouse Model," *Radiology* **221:** 523–529 (2001).
93. N. Dejong and L. Hoff, "Ultrasound Scattering Properties of Albunex Microspheres," *Ultrasonics* **31:** 175–181 (1993).
94. H. J. Bleeker, K. K. Shung, and J. L. Barnhart, "Ultrasonic Characterization of Albunex, a New Contrast Agent," *J. Acoustical Society of Amer.* **87:** 1792–1797 (1990).
95. T. M. Lee, A. L. Oldenburg, S. Sitafalwalla, D. L. Marks, W. Luo, F. J. J. Toublan, K. S. Suslick, and S. A. Boppart, "Engineered Microsphere Contrast Agents for Optical Coherence Tomography," *Optics Letters* **28:** 1546–1548 (2003).
96. F. J. J. Toublan, K. S. Suslick, S. A. Boppart, T. M. Lee, and A. Oldenburg, "Modification of Protein Microspheres for Biomedical Applications," *Abstracts of Papers of the Amer. Chemical Society* **225:** 324–POLY (2003).

97. F. J. J. Toublan, S. A. Boppart, and K. S. Suslick, "Tumor Targeting by Surface-Modified Protein Microspheres," *J. Amer. Chemical Society* **128:** 3472–3473 (2006).
98. M. A. McDonald, L. Jankovic, K. Shahzad, M. Burcher, and K. C. P. Li, "Acoustic Fingerprints of Dye-Labeled Protein Submicrosphere Photoacoustic Contrast Agents," *J. Biomedical Optics* **14:** 034032 (2009).
99. A. L. Oldenburg, F. J. J. Toublan, K. S. Suslick, A. Wei, and S. A. Boppart, "Magnetomotive Contrast for *in vivo* Optical Coherence Tomography," *Optics Express* **13:** 6597–6614 (2005).
100. A. L. Oldenburg, J. R. Gunther, and S. A. Boppart, "Imaging Magnetically Labeled Cells with Magnetomotive Optical Coherence Tomography," *Optics Letters* **30:** 747–749 (2005).
101. A. L. Oldenburg, V. Crecea, S. A. Rinne, and S. A. Boppart, "Phase-Resolved Magnetomotive OCT for Imaging Nanomolar Concentrations of Magnetic Nanoparticles in Tissues," *Optics Express* **16:** 11525–11539 (2008).
102. S. A. Boppart, A. L. Oldenburg, C. Y. Xu, and D. L. Marks, "Optical Probes and Techniques for Molecular Contrast Enhancement in Coherence Imaging," *J. Biomedical Optics* **10:** 041208 (2005).
103. R. John, E. J. Chaney, and S. A. Boppart, (October 06, 2009) "Dynamics of Magnetic Nanoparticle-Based Contrast Agents in Tissues Tracked Using Magnetomotive Optical Coherence Tomography," *IEEE J. Selected Topics Quantum Electronics* In press, (2010) 10.1109/JSTQE.2009.2029547.
104. U. Mahmood, C. H. Tung, A. Bogdanov, and R. Weissleder, "Near-Infrared Optical Imaging of Protease Activity for Tumor Detection," *Radiology* **213:** 866–870 (1999).
105. U. Mahmood, C. H. Tung, and R. Weissleder, "Optical Imaging of Gene Expression Using Enzyme-Specific Activatable Contrast Agents," *Radiology* **213P:** 102 (1999).
106. L. Josephson, M. F. Kircher, U. Mahmood, Y. Tang, and R. Weissleder, "Near-Infrared Fluorescent Nanoparticles as Combined MR/Optical Imaging Probes," *Bioconjugate Chemistry* **13:** 554–560 (2002).
107. C. H. Tung, S. Bredow, U. Mahmood, and R. Weissleder, "Preparation of a Cathepsin D Sensitive Near-Infrared Fluorescence Probe for Imaging," *Bioconjugate Chemistry* **10:** 892–896 (1999).
108. C. H. Tung, U. Mahmood, S. Bredow, and R. Weissleder, "*In vivo* Imaging of Proteolytic Enzyme Activity Using a Novel Molecular Reporter," *Cancer Research* **60:** 4953–4958 (2000).
109. C. H. Tung, Y. H. Lin, W. K. Moon, and R. Weissleder, "A Receptor-Targeted Near-Infrared Fluorescence Probe for *in vivo* Tumor Imaging," *Chembiochem* **3:** 784–786 (2002).
110. H. Alencar, M. A. Funovics, J. Figueiredo, H. Sawaya, R. Weissleder, and U. Mahmood, "Colonic Adenocarcinomas: Near-Infrared Microcatheter Imaging of Smart Probes for Early Detection—Study in Mice," *Radiology* **244:** 232–238 (2007).
111. C. Bremer, V. Ntziachristos, U. Mahmood, C. H. Tung, and R. Weissleder, "Advances in Optical Imaging," *Radiologe* **41:** 131–137 (2001).
112. C. B. Bremer, S. Bredow, R. Weissleder, and C. H. Tung, "*In vivo* Imaging of Matrix-Metalloproteinase Activity: A Key Enzyme for Tumor Progression and Angiogenesis," *Radiology* **221:** 517–518 (2001).
113. D. C. Adler, T. H. Ko, P. R. Herz, and J. G. Fujimoto, "Optical Coherence Tomography Contrast Enhancement Using Spectroscopic Analysis with Spectral Autocorrelation," *Optics Express* **12:** 5487–5501 (2004).
114. C. Y. Xu, J. Ye, D. L. Marks, and S. A. Boppart, "Near-Infrared Dyes as Contrast-Enhancing Agents for Spectroscopic Optical Coherence Tomography," *Optics Letters* **29:** 1647–1649 (2004).
115. A. L. Oldenburg, C. Y. Xu, and S. A. Boppart, "Spectroscopic Optical Coherence Tomography and Microscopy," IEEE *J. Selected Topics in Quantum Electronics* **13:** 1629–1640 (2007).
116. D. J. Faber, E. G. Mik, M. C. G. Aalders, and T. G. van Leeuwen, "Toward Assessment of Blood Oxygen Saturation by Spectroscopic Optical Coherence Tomography," *Optics Letters* **30:** 1015–1017 (2005).

117. C. Y. Xu, P. S. Carney, and S. A. Boppart, "Wavelength-Dependent Scattering in Spectroscopic Optical Coherence Tomography," *Optics Express* **13:** 5450–5462 (2005).
118. T. Storen, A. Royset, L. O. Svaasand, and T. Lindmo, "Functional Imaging of Dye Concentration in Tissue Phantoms by Spectroscopic Optical Coherence Tomography," *Journal of Biomedical Optics* **10:** 024037 (2005).
119. B. Hermann, K. Bizheva, A. Unterhuber, B. Povazay, H. Sattmann, L. Schmetterer, A. F. Fercher, and W. Drexler, "Precision of Extracting Absorption Profiles from Weakly Scattering Media with Spectroscopic Time-Domain Optical Coherence Tomography," *Optics Express* **12:** 1677–1688 (2004).
120. C. H. Yang, "Molecular Contrast Optical Coherence Tomography: A Review," *Photochemistry and Photobiology* **81:** 215–237 (2005).
121. Z. Yaqoob, E. McDowell, J. Wu, X. Heng, J. Fingler, and C. Yang, "Molecular Contrast Optical Coherence Tomography: a Pump-Probe Scheme Using Indocyanine Green as a Contrast Agent," *J. Biomedical Optics* **11:** 054017 (2006).
122. C. H. Yang, L. E. L. McGuckin, J. D. Simon, M. A. Choma, B. E. Applegate, and J. A. Izatt, "Spectral Triangulation Molecular Contrast Optical Coherence Tomography with Indocyanine Green as the Contrast Agent," *Optics Letters* **29:** 2016–2018 (2004).
123. D. R. Larson, W. R. Zipfel, R. M. Williams, S. W. Clark, M. P. Bruchez, F. W. Wise, and W. W. Webb, "Water-Soluble Quantum Dots for Multiphoton Fluorescence Imaging *in vivo*," *Science* **300:** 1434–1436 (2003).
124. K. D. Rao, M. A. Choma, S. Yazdanfar, A. M. Rollins, and J. A. Izatt, "Molecular Contrast in Optical Coherence Tomography by Use of a Pump-Probe Technique," *Optics Letters* **28:** 340–342 (2003).
125. M. I. Canto, "Methylene Blue Chromoendoscopy for Barett's Esophagus: Coming Soon to Your GI Unit?," *Gastrointest. Endosc.* **54:** 403–409 (2001).
126. C. H. Yang, M. A. Choma, L. E. Lamb, J. D. Simon, and J. A. Izatt, "Protein-based Molecular Contrast Optical Coherence Tomography with Phytochrome as the Contrast Agent," *Optics Letters* **29:** 1396–1398 (2004).
127. J. Kim, J. Oh, H. W. Kang, M. D. Feldman, and T. E. Milner, "Photothermal Response of Super paramagnetic Iron-Oxide Nanoparticles," *Lasers in Surgery and Medicine* **40:** 415–421 (2008).
128. S. Lal, S. E. Clare, and N. J. Halas, "Nanoshell-Enabled Photothermal Cancer Therapy: Impending Clinical Impact," *Accounts of Chemical Research* **41:** 1842–1851 (2008).
129. L. R. Hirsch, R. J. Stafford, J. A. Bankson, S. R. Sershen, B. Rivera, R. E. Price, J. D. Hazle, N. J. Halas, and J. L. West, "Nanoshell-Mediated Near-Infrared Thermal Therapy of Tumors under Magnetic Resonance Guidance," *Proceedings of the National Academy of Sciences USA* **100:** 13549–13554 (2003).
130. M. Hu, J. Y. Chen, Z. Y. Li, L. Au, G. V. Hartland, X. D. Li, M. Marquez, and Y. N. Xia, "Gold Nanostructures: Engineering Their Plasmonic Properties for Biomedical Applications," *Chemical Society Reviews* **35:** 1084–1094 (2006).
131. C. F. Bohren and D. R. Huffman, *Absorption and Scattering of Light by Small Particles*, Wiley, New York, 1998.
132. M. E. Brezinski, G. J. Tearney, S. A. Boppart, B. E. Bouma, C. Pitris, J. F. Southern, and J. G. Fujimoto, "Ultrahigh Resolution Catheter Based *in vivo* Imaging of the Rabbit Aorta with Optical Coherence Tomography," *Circulation* **96:** 3294–3294 (1997).
133. S. L. Westcott, J. B. Jackson, C. Radloff, and N. J. Halas, "Relative Contributions to the Plasmon Line Shape of Metal Nanoshells," *Physical Review B* **66:** 155431 (2002).
134. J. Yguerabide and E. E. Yguerabide, "Light-Scattering Submicroscopic Particles as Highly Fluorescent Analogs and Their Use as Tracer Labels in Clinical and Biological Applications—I. Theory," *Analytical Biochemistry* **262:** 137–156 (1998).
135. S. J. Oldenburg, R. D. Averitt, S. L. Westcott, and N. J. Halas, "Nanoengineering of Optical Resonances," *Chemical Physics Letters* **288:** 243–247 (1998).

136. R. D. Averitt, S. L. Westcott, and N. J. Halas, "Linear Optical Properties of Gold Nanoshells," *J. Optical Society of Amer. B* **16:** 1824–1832 (1999).
137. B. T. Draine and P. J. Flatau, "Discrete-Dipole Approximation for Scattering Calculations," *J. Optical Society of Amer. A* **11:** 1491–1499 (1994).
138. J. C. Y. Kah, M. C. Olivo, C. G. L. Lee, and C. J. R. Sheppard, "Molecular Contrast of EGFR Expression Using Gold Nanoparticles as a Reflectance-Based Imaging Probe," *Molecular and Cellular Probes* **22:** 14–23 (2008).
139. C. Loo, L. Hirsch, M. H. Lee, E. Chang, J. West, N. Halas, and R. Drezek, "Gold Nanoshell Bioconjugates for Molecular Imaging in Living Cells," *Optics Letters* **30:** 1012–1014 (2005).
140. C. Loo, A. Lin, L. Hirsch, M. H. Lee, J. Barton, N. Halas, J. West, and R. Drezek, "Nanoshell-Enabled Photonics-Based Imaging and Therapy of Cancer," *Technology in Cancer Research & Treatment* **3:** 33–40 (2004).
141. D. P. O'Neal, L. R. Hirsch, N. J. Halas, J. D. Payne, and J. L. West, "Photo-Thermal Tumor Ablation in Mice Using Near Infrared-Absorbing Nanoparticles," *Cancer Letters* **209:** 171–176 (2004).
142. M. P. Melancon, W. Lu, Z. Yang, R. Zhang, Z. Cheng, A. M. Elliot, J. Stafford, T. Olson, J. Z. Zhang, and C. Li, "*In vitro* and *in vivo* Targeting of Hollow Gold Nanoshells Directed at Epidermal Growth Factor Receptor for Photothermal Ablation Therapy," *Molecular Cancer Therapeutics* **7:** 1730–1739 (2008).
143. A. M. Gobin, M. H. Lee, N. J. Halas, W. D. James, R. A. Drezek, and J. L. West, "Near-Infrared Resonant Nanoshells for Combined Optical Imaging and Photothermal Cancer Therapy," *Nano Letters* **7:** 1929–1934 (2007).
144. J. C. Y. Kah, M. Olivo, T. H. Chow, K. S. Song, K. Z. Y. Koh, S. Mhaisalkar, and C. J. R. Sheppard, "Control of Optical Contrast Using Gold Nanoshells for Optical Coherence Tomography Imaging of Mouse Xenograft Tumor Model *in vivo*," *J. Biomedical Optics* **14:** 054015 (2009).
145. J. C. Y. Kah, T. H. Chow, B. K. Ng, S. G. Razul, M. Olivo, and C. J. R. Sheppard, "Concentration Dependence of Gold Nanoshells on the Enhancement of Optical Coherence Tomography Images: a Quantitative Study," *Applied Optics* **48:** D96–D108 (2009).
146. Q. H. Lu, K. L. Yao, D. Xi, Z. L. Liu, X. P. Luo, and Q. Ning, "Synthesis and Characterization of Composite Nanoparticles Comprised of Gold Shell and Magnetic Core/Cores," *J. Magnetism and Magnetic Materials* **301:** 44–49 (2006).
147. Y. T. Lim, M. Y. Cho, J. K. Kim, S. Hwangbo, and B. H. Chung, "Plasmonic Magnetic Nanostructure for Bimodal Imaging and Photonic-Based Therapy of Cancer Cells," *Chembiochem* **8:** 2204–2209 (2007).
148. J. Y. Chen, B. Wiley, Z. Y. Li, D. Campbell, F. Saeki, H. Cang, L. Au, J. Lee, X. D. Li, and Y. N. Xia, "Gold Nanocages: Engineering their Structure for Biomedical Applications," *Advanced Materials* **17:** 2255–2261 (2005).
149. Y. T. Lim, M. Y. Cho, B. S. Choi, Y. W. Noh, and B. H. Chung, "Diagnosis and Therapy of Macrophage Cells Using Dextran-Coated Near-Infrared Responsive Hollow-Type Gold Nanoparticles," *Nanotechnology* **19:** 375105 (2008).
150. S. Link, M. A. El-Sayed, and M. B. Mohamed, "Simulation of the Optical Absorption Spectra of Gold Nanorods as a Function of Their Aspect Ratio and the Effect of the Medium Dielectric Constant," *J. Physical Chemistry B* **103:** 3073–3077 (1999).
151. P. K. Jain, K. S. Lee, I. H. El-Sayed, and M. A. El-Sayed, "Calculated Absorption and Scattering Properties of Gold Nanoparticles of Different Size, Shape, and Composition: Applications in Biological Imaging and Biomedicine," *J. Physical Chemistry B* **110:** 7238–7248 (2006).
152. C. Sonnichsen, T. Franzl, T. Wilk, G. von Plessen, J. Feldmann, O. Wilson, and P. Mulvaney, "Drastic Reduction of Plasmon Damping in Gold Nanorods," *Physical Review Letters* **88:** 077402 (2002).
153. D. A. Zweifel and A. Wei, "Sulfide-Arrested Growth of Gold Nanorods," *Chemistry of Materials* **17:** 4256–4261 (2005).
154. S. A. Telenkov, D. P. Dave, S. Sethuraman, T. Akkin, and T. E. Milner, "Differential Phase Optical Coherence Probe for Depth-Resolved Detection of Photothermal Response in Tissue," *Physics in Medicine and Biology* **49:** 111–119 (2004).

155. L. Tong, Q. S. Wei, A. Wei, and J. X. Cheng, "Gold Nanorods as Contrast Agents for Biological Imaging: Optical Properties, Surface Conjugation and Photothermal Effects," *Photochemistry and Photobiology* **85:** 21–32 (2009).
156. B. Wiley, Y. G. Sun, J. Y. Chen, H. Cang, Z. Y. Li, X. D. Li, and Y. N. Xia, "Shape-Controlled Synthesis of Silver and Gold Nanostructures," *MRS Bulletin* **30:** 356–361 (2005).
157. S. E. Skrabalak, J. Chen, L. Au, X. Lu, X. Li, and Y. Xia, "Gold Nanocages for Biomedical Applications," *Advanced Materials* **19:** 3177–3184 (2007).

CHAPTER 12

Second-Harmonic Generation Imaging of Collagen-Based Systems

Monal R. Mehta

Laboratory for Photonics Research of Bio/Nano Environments (PROBE)
Department of Mechanical Science and Engineering
University of Illinois at Urbana–Champaign

Raghu Ambekar Ramachandra Rao

Laboratory for Photonics Research of Bio/Nano Environments (PROBE)
Department of Electrical and Computer Engineering
University of Illinois at Urbana–Champaign

Kimani C. Toussaint, Jr.

Laboratory for Photonics Research of Bio/Nano Environments (PROBE)
Department of Mechanical Science and Engineering
Affiliate, Departments of Electrical and Computer Engineering, and Bioengineering
Affiliate, Beckman Institute for Advanced Science and Technology
University of Illinois at Urbana–Champaign

12.1 Introduction

Second-harmonic generation (SHG) microscopy is a nonlinear microscopy technique whereby intense light, with frequency ω_0, propagating through a material exhibiting no inversion symmetry (noncentrosymmetry) is converted to light at half the wavelength of the original, with frequency $2\omega_0$ (as shown in Fig. 12.1). This process involves transitions from an electronic ground state to higher

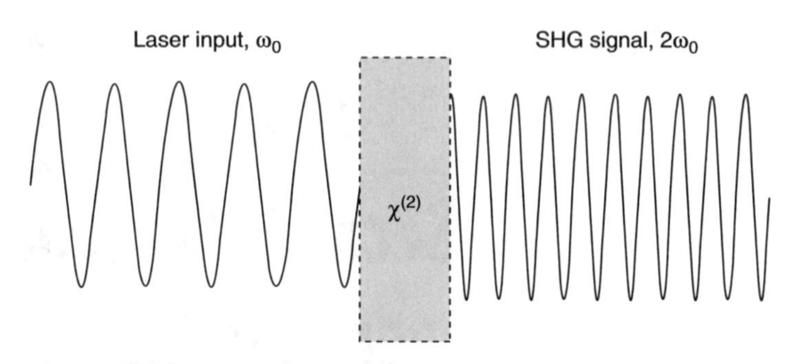

FIGURE 12.1 Schematic of the SHG process from a nonlinear medium.

"virtual" states only; no molecular absorption takes place. Therefore, the SHG process is based strictly on nonlinear scattering and does not transfer energy into the system, and thus reducing the possibility for photo damage.

Typically, SHG occurs in nonlinear optical crystals and biological systems with spatially ordered molecular arrangements. Biological structures such as fibrillar collagen, microtubules, and myosin yield strong SHG signals. This makes the technique attractive for noninvasive biological imaging since it obviates the use of exogenous contrast agents. The efficiency of SHG depends on the properties of the material ($\chi^{(2)}$), the excitation light's polarization properties, and the interaction time inside the material. It is also proportional to the square of the input laser intensity which is at its peak at the microscope focus; this provides an inherent optical sectioning capability. That is, the emitted SHG signal is confined to a focal volume of approximately a few hundred femtoliters (10^{-15} liters).

Within the past decade, SHG microscopy has found use in biomedical and basic biology applications to obtain images of highly ordered structures such as collagen fibers, microtubulin, and skeletal muscle, with high resolution and contrast [1–10]. The study of collagen fibers with SHG microscopy is particularly interesting due to the ubiquity of this protein in the human body and the relatively strong SHG signal that it produces without the need for staining. In general, collagens account for 25% of the total protein mass in mammals [11]. Collagen molecules arrange into highly organized hierarchical fibrillar structures that form fiber bundles that can be found in the cornea, bone, cartilage, and skin [11–15]. Thus, these systems are amenable to imaging, using SHG microscopy.

Aside from the impressive strides made over the years in SHG imaging, most studies in the past have relied on the qualitative assessment of obtained images for analysis [5, 6, 16]. In recent years, there has been a strong push to develop quantitative metrics for this imaging modality [2, 7–9, 17]. In principle, such a noninvasive imaging technique could complement existing methods for diagnosing structural damage to tissues, resulting from either injury or disease.

12.2 Approaches to Obtaining Quantitative Information from Second-Harmonic Generation Images

It was discussed in the previous section that historically much effort has been placed on qualitative assessment of SHG images. However, the right choice of quantitative metrics could help in identifying the markers of a specific disease or in assessing wound healing. Due to this strong diagnostic potential, researchers have begun to pursue a variety of routes to quantitative second-harmonic generation imaging.

12.2.1 The Ratio of Forward-to-Backward Second-Harmonic Generation Intensity

A straightforward way of deriving quantitative information from SHG images is to obtain the ratio of the intensity in the forward-to-backward direction (F/B ratio). Although a relatively simple metric, much information can be obtained from it. For example, the Webb group used this technique to understand the environment of collagen [7]. It was found that the ionic strength of the solution played a major role in determining the F/B ratio and was attributed to small structural changes in collagen organization. Figure 12.2 compares the difference in the forward and backward SHG for the collagen fibers belonging to rat tail tendons and immersed in different concentrations of sodium chloride (NaCl). On the other hand, Cox et al. used the F/B ratio to differentiate between types of collagens [8]. Type I collagen which is highly crystalline produces a strong F/B signal, where as Types III and IV collagen which are less crystalline produce weaker F/B signals. These findings suggest that the F/B ratio is sensitive to structural changes well below the diffraction limit.

12.2.2 Distribution of Lengths in Structures

The Campagnola group investigated the skeletal muscles of mice (in particular, gastrocnemius), human skin, and the subcutaneous layer and quantified the disease stage on the basis of the sarcomere patterns in myofibrils [10]. Regularity in the period and orientation of sarcomeres means that the muscle tissue is healthy. Any gaps, splits, or ruptures between the sarcomeres are a direct representation of the severity of the disease. The length of the sarcomeres for normal and diseased tissues from the SHG images were quantified; it was observed that diseased tissues had more hypercontracted (disintegrated) sarcomeres below 1.6 microns when compared to healthy tissues (Fig. 12.3).

12.2.3 $\chi^{(2)}$ Tensor Imaging

A very interesting approach to quantitative SHG imaging is to determine the nonlinear susceptibility tensor $\chi^{(2)}$ of a specimen. For an

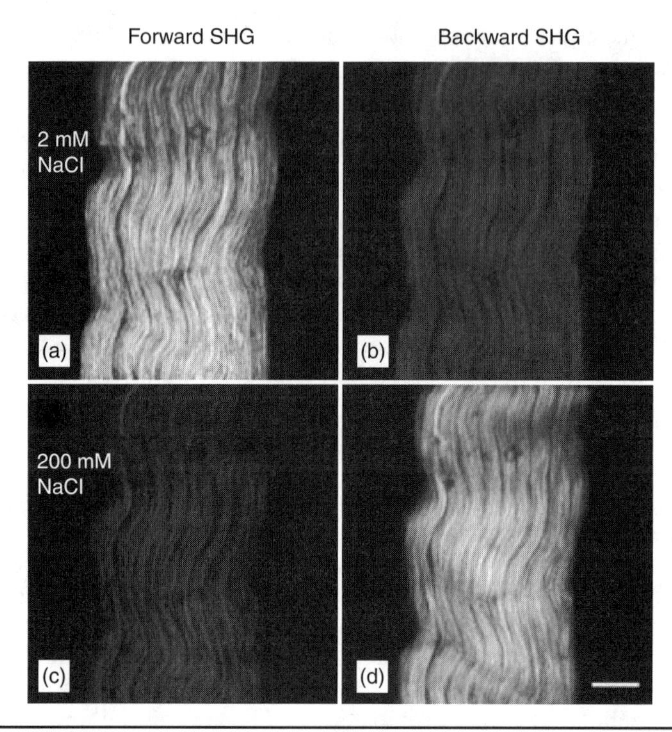

Figure 12.2 Forward and backward SHG intensities from the collagen fibers change depending on the ionic strength of the surrounding solution. (Reprinted with permission)

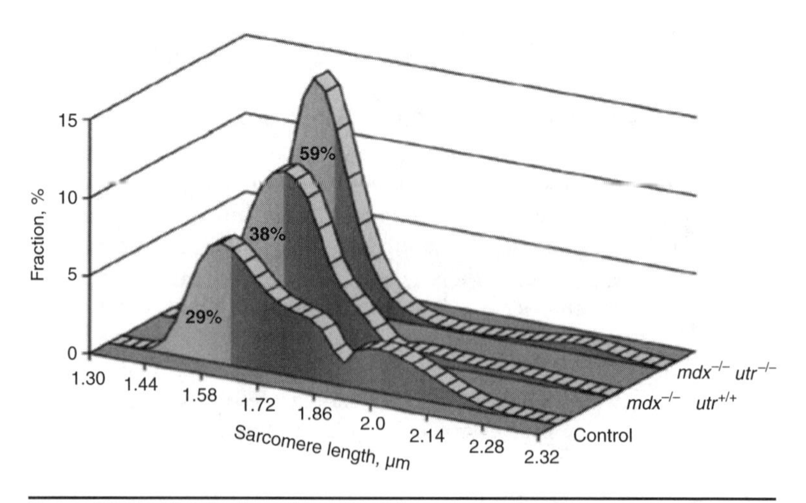

Figure 12.3 Distribution of length of the sarcomeres for normal and diseased muscle tissue in normal and diseased muscle tissue. (Reprinted with permission)

anisotropic second-order nonlinear crystal, the induced polarization density P at frequency 2ω can be written in matrix form as

$$\begin{bmatrix} P_x(2\omega) \\ P_y(2\omega) \\ P_z(2\omega) \end{bmatrix} = 2 \begin{bmatrix} d_{11} & d_{12} & d_{13} & d_{14} & d_{15} & d_{16} \\ d_{21} & d_{22} & d_{23} & d_{24} & d_{25} & d_{26} \\ d_{31} & d_{32} & d_{33} & d_{34} & d_{35} & d_{36} \end{bmatrix} \begin{bmatrix} E_x(\omega)^2 \\ E_y(\omega)^2 \\ E_z(\omega)^2 \\ 2E_z(\omega)E_y(\omega) \\ 2E_x(\omega)E_z(\omega) \\ 2E_y(\omega)E_x(\omega) \end{bmatrix} \tag{12.1}$$

where $E_x(\omega)$, $E_y(\omega)$, and $E_z(\omega)$ and $P_x(2\omega)$, $P_y(2\omega)$, and $P_z(2\omega)$ are the components of the electric-field vector and polarization density, respectively, along the principal axes of the nonlinear medium.

It is fairly established that collagen fibers exhibit cylindrical symmetry (class *C6* crystal symmetry) [18]. The *C6* matrix has only three nonzero components, d_{15}, d_{31}, and d_{33}, as depicted in Fig. 12.4. Therefore, considering light that is propagating along the x direction ($E_x(\omega) = 0$), and collagen oriented along z direction, as shown in Fig. 12.4*b*, we obtain

$$\mathrm{P}_x(2\omega) = 0 \tag{12.2}$$

$$P_y(2\omega) = 4d_{15}E_y(\omega)E_z(\omega) \tag{12.3}$$

and

$$P_z(2\omega) = 2[d_{31}E_y^2(\omega) + d_{33}E_z^2(\omega)] \tag{12.4}$$

$$\begin{bmatrix} 0 & 0 & 0 & 0 & d_{15} & 0 \\ 0 & 0 & 0 & d_{15} & 0 & 0 \\ d_{15} & d_{15} & d_{33} & 0 & 0 & 0 \end{bmatrix}$$

(a)

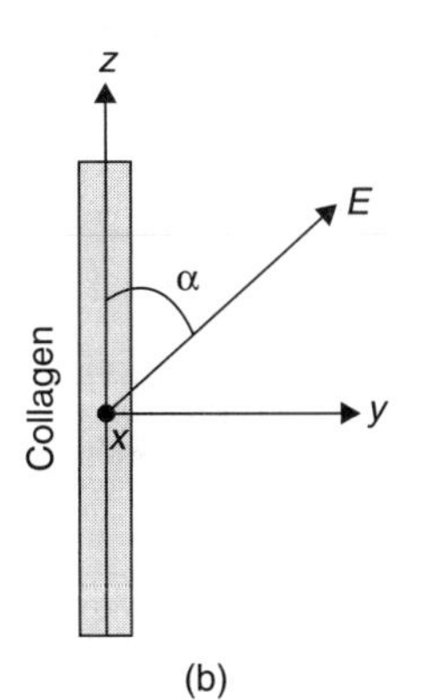

FIGURE 12.4 (*a*) Second-order nonlinear susceptibility matrix for *C6* crystal class. (*b*) Geometry showing collagen fiber with respect to laboratory coordinates *x*, *y*, *z*.

Therefore, the SHG intensity is given by

$$I(2\omega) \sim P_y^2(2\omega) + P_z^2(2\omega) \tag{12.5}$$

Using $E_y(\omega) = \sin\alpha$ and $E_Z(\omega) = \cos\alpha$, where α is the angle that the electric field makes with respect to the z-axis, we have

$$I(2\omega) \sim \sin^2 2\alpha + \left(\frac{d_{31}}{d_{15}}\sin^2\alpha + \frac{d_{33}}{d_{15}}\cos^2\alpha\right)^2 \tag{12.6}$$

Therefore, the ratios of d elements, d_{31}/d_{15} and d_{33}/d_{15}, can be calculated by obtaining a set of SHG intensities for different polarization angles α.

Tiaho et al. used this approach to obtain the orientation of myosin and collagen in tissues in amphibians and mammals [9]. Here, SHG images were obtained at different incident polarizations, and the SHG intensity is recorded for each angle as shown in Fig. 12.5*a* and *b*. It is then fitted with Eq. (12.6) to obtain the ratio of the nonlinear susceptibility elements d_{31}/d_{15} and d_{33}/d_{15}. From these ratios, they determined the orientation angles of myosin and collagen harmonophores as 62° and 49°, respectively.

Cheng et al. calculated the ratios, d_{15}/d_{31}, and d_{33}/d_{31} as a function of space and displayed them as images for human dermis and a chicken muscle-tendon junction as shown in Fig. 12.6*a* [2]. Only the ratio d_{33}/d_{31} showed better contrast since the other ratio d_{15}/d_{31} is vulnerable to fluctuations due to $\sin^2 2\alpha$ dependence. The image of the ratio d_{33}/d_{31} revealed information on the different sources of SHG generators, myosin, and collagen fibers of a chicken wing. Furthermore, a histogram of the image of the same ratio on human dermis as shown in Fig. 12.6*b* produced two distinct peaks; this suggests that the SHG signal is produced not only from Type I but also from Type III collagen, the other major constituent in dermis.

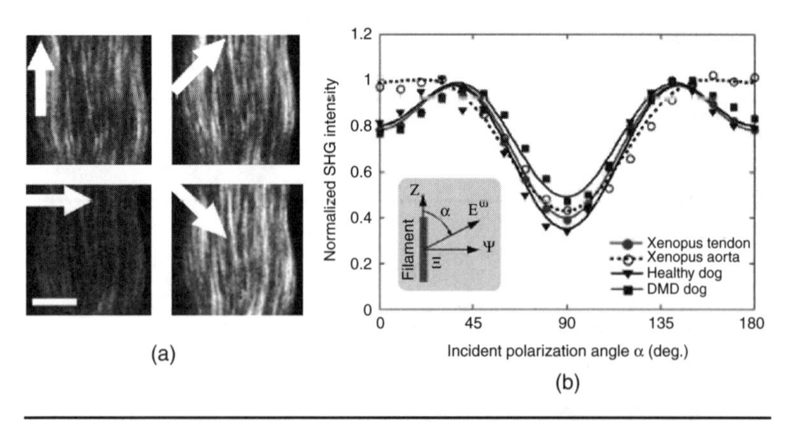

Figure 12.5 (*a*) SHG images of collagen in adult xenopus tendon for different polarization angles (indicated by arrows). (*b*) Collection of SHG intensities at different polarization angles lead to determination of the nonlinear susceptibility of the collagen fibers. (Reprinted with permission)

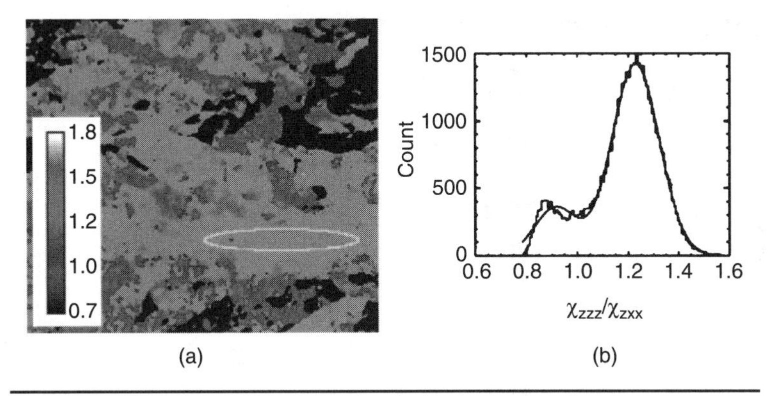

Figure 12.6 (*a*) Image of the ratio for d_{33}/d_{31} for human dermis. (*b*) Histogram showing two distinct peaks corresponding to Types I and III collagen. (Reprinted with permission)

12.2.4 Fourier-Transform Second-Harmonic Generation Microscopy

Another recent quantitative tool that has been applied to SHG images is harmonic analysis via the Fourier transform (FT). Typically, the FT has been used in image processing as a technique for noise removal [19]. However, the dependence of SHG imaging on structure, rather than chemical composition, in noncentrosymmetric biological materials allows for the FT to be used as a method of presenting information in the spatial-frequency domain that directly relates to the spatial organization of these physical structures. For this reason, Fourier-transform second-harmonic generation (FT-SHG) was introduced as a new method for the analysis of biological structures [20]. Its application in the literature has focused on the analysis of collagen fibers; however, it can be applied to any structure that emits SHG, such as muscle fibers and tubulin [21]. In this section, three parameters (preferred orientation, maximum spatial frequency, and peaks in the magnitude spectrum) will be presented which have been implemented thus far to characterize the FT of SHG images. They are derived within the context of potential use as quantitative metrics for assessing tissue morphology. Additional details can be found in [20, 22].

Preferred Orientation

The orientation of collagen fibers has been assessed using an assortment of techniques with varying degrees of complexity [17, 23–26]. FT-SHG allows for the analysis of collagen fiber orientation through image postprocessing in a simple manner that does not require complex algorithms. Specifically, the *preferred orientation* refers to the average direction in which fibers are oriented in a given plane.

To understand how the FT can be used to obtain preferred orientation, let us first consider the image in Fig. 12.7*a* and its corresponding

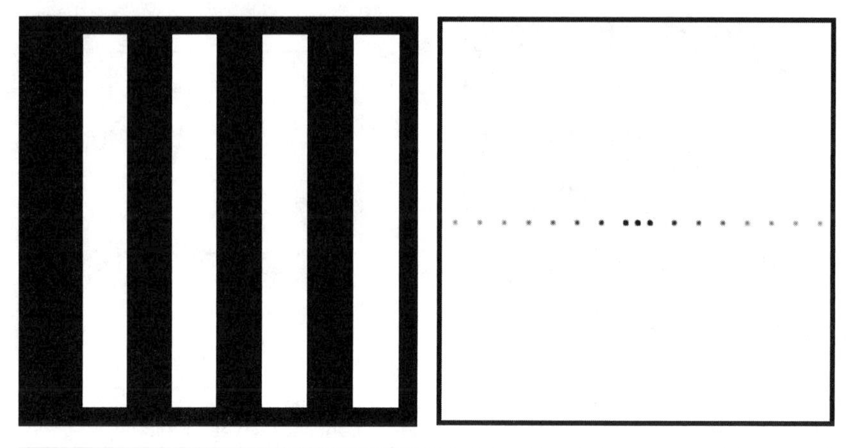

Figure 12.7 (*a*) An image with vertical stripes. (*b*) Magnitude of the 2D FT of the image on the left (where black indicates high magnitude and white indicates low magnitude). Note that the resulting peaks are aligned horizontally.

two-dimensional (2D) magnitude spectrum in Fig. 12.7*b*. It can be seen that the vertical lines in (*a*) lead to a horizontal set of peaks in (*b*). Put another way, the intensity of pixels in the image does not change in the vertical direction. Changes in intensity are only seen in the horizontal direction.

Figure 12.8 depicts what happens when the stripes are not aligned along the same direction. Here, the peaks in the magnitude spectrum of the FT align perpendicularly with the stripes. Again, this is because the intensity pattern along any given stripe is constant, while the other directions will have changes (namely at the center of the image).

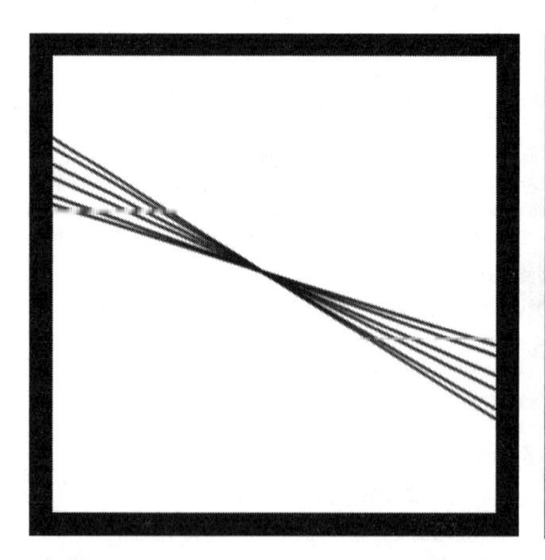

Figure 12.8 (*a*) An image with arbitrarily oriented stripes. (*b*) Magnitude of the 2D FT of the image on the left (where black indicates high magnitude and white indicates low magnitude). Note that the resulting peaks are aligned perpendicularly to the stripes.

This quality of the FT can be utilized to determine the preferred orientation of a region of interest. The process can be summarized in four steps.

1. Take the 2D FT of the region of interest.
2. Convert the image to binary using a threshold.
3. Fit the 1s in the binary image to a line.
4. Take the direction perpendicular to the line as the direction of orientation.

The 2D FT is converted to a binary image so that we focus only on the dominant (high amplitude) frequencies. One would not want noise or texture in an image to affect the calculation of preferred orientation. Furthermore, once the image is binary, it can be treated as a set of points lying on a 2D plane. This will allow a line to be fit to these points that lie on the plane of the image. The fit will not be affected by the relative amplitudes of the dominant frequencies. Figure 12.9 describes this process using an SHG image taken from [20] of porcine ear (elastic) cartilage. The preferred orientation is perpendicular to the line fit to the binary image. As fibers become less organized (or more randomly oriented), the error from fitting the line will increase. For this reason, the standard deviation of the fit is also a useful metric of the entropy in fiber orientation in a given region of interest.

Maximum Spatial Frequency

The second parameter of interest is the maximum spatial frequency (F_{high}) in the magnitude spectrum of the FT. This informs us about the minimum observable size (or feature) in real space, and is usually taken along a single direction, simplifying the 2D FT down to

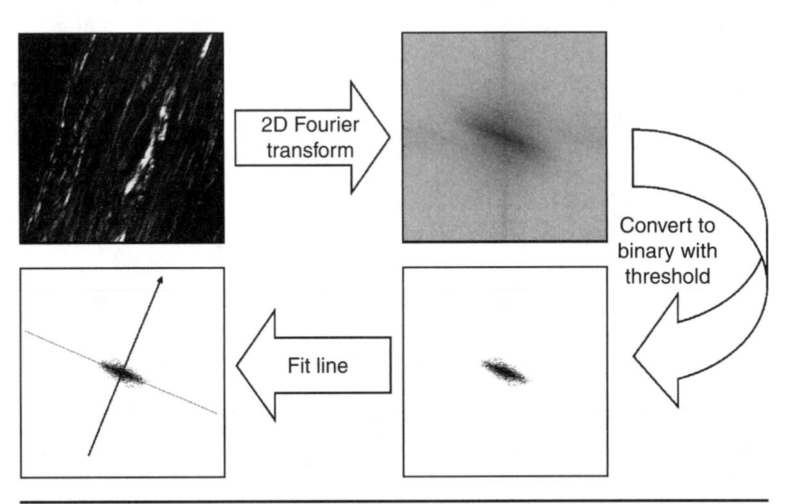

Figure 12.9 The process for obtaining the preferred orientation of fibers using FT-SHG.

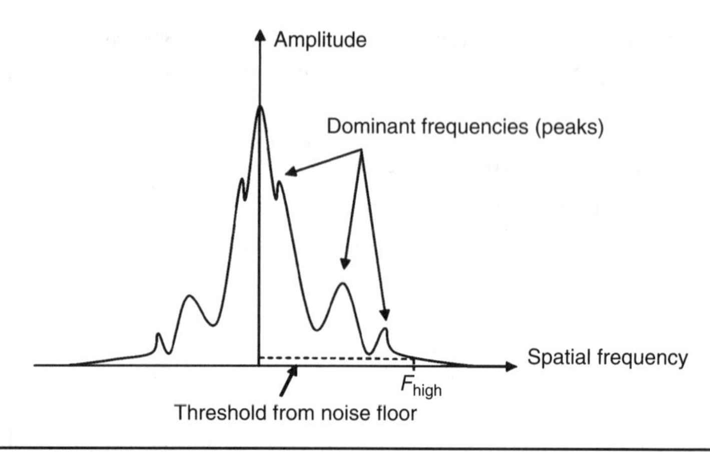

FIGURE 12.10 A visual representation of F_{high} and peaks in the magnitude spectrum of the FT.

one-dimension (1D). F_{high} represents the highest (spatial) frequency observable along a given direction, which is usually chosen to be the direction of preferred orientation. Ideally, one could simply take the highest (nonzero) spatial frequency in the magnitude spectrum. However, noise makes the situation more difficult. One method to account for noise is to take the 2D FT over a dark region of an image, one where the sample produces no SHG signal. This way the only intensity fluctuations in the region are caused by photon shot noise. The noise level is then assumed to be the maximum magnitude of the FT from this blank region of interest. A threshold may then be used as 10% above the noise level, and the highest frequency above this threshold along the chosen direction in the region of interest can be taken as the maximum spatial frequency. Figure 12.10 gives an illustration of a 1D FT and F_{high}.

F_{high} can be used to evaluate biological systems in different ways. One example can be taken from a comparison done in [20] between elastic cartilage in the ear and hyaline cartilage in the trachea. This comparison is shown in Fig. 12.11.

It can be seen in Fig. 12.11 that the distribution for F_{high} for the elastic cartilage in the ear is narrower than that obtained for the hyaline cartilage in the trachea. This shows a better consistency in organization for the fibers in the ear compared to those in the trachea and can readily be verified by inspection of the SHG images in Fig. 12.11*a* and *b*. Note that for a better comparison, F_{high} was determined along the horizontal direction for both the ear and trachea images.

Peaks in the Magnitude Spectrum

The final parameter of interest is the peaks in the magnitude spectrum of the Fourier transform, or Fourier peaks. Fourier peaks are shown in Fig. 12.12. They are simply the peaks in the magnitude spectrum for the FT taken along a chosen direction within a region of interest, above

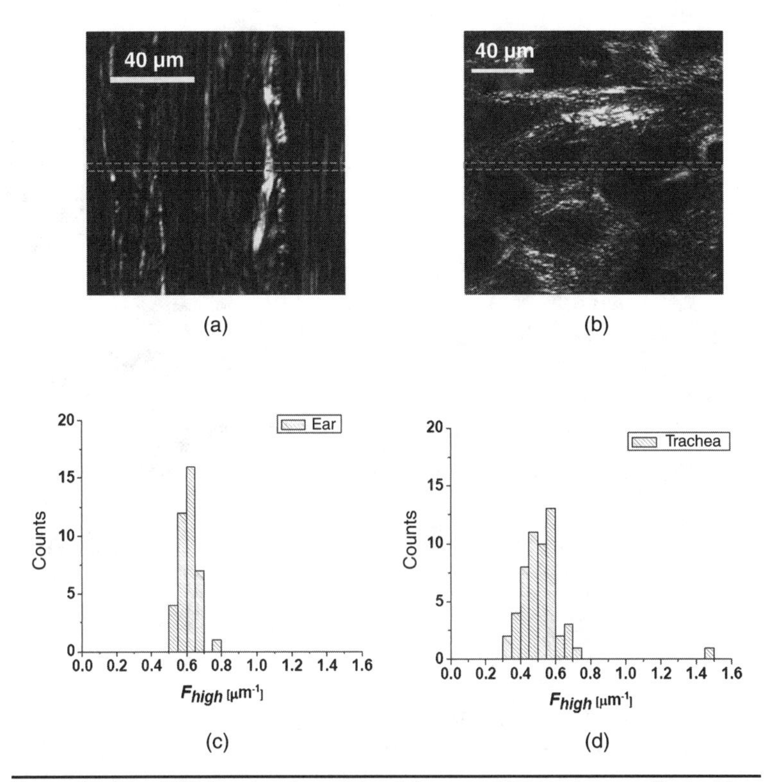

Figure 12.11 SHG image of a porcine (*a*) ear and (*b*) trachea cartilage. The region of interest for estimating F_{high} is a rectangular box with a of height 2.87 μm. A histogram of the values of F_{high} from all the regions of interest for the (*c*) ear and (*d*) trachea. (Reprinted with permission)

the noise threshold. They collectively represent the information content along a certain direction. Similar to F_{high}, Fourier peaks are typically evaluated for the direction of preferred orientation.

12.3 Quantitative Comparison of Forward and Backward Second-Harmonic Generation Images

One of the major advantages given to SHG is its diagnostic potential. Diagnostic applications of SHG are only practical if collected in reflection. However, most SHG images of biological systems are acquired in the transmitted direction [3, 27]. This is due to the fact that the signal from SHG is typically strongest in the transmitted direction. In order to consider practical applications of reflected SHG for diagnostic purposes, a study was done to compare forward and backward SHG images using FT-SHG [22]. Forward and backward SHG images of collagen fibers in a porcine tendon, ear cartilage, and sclera were analyzed. These images are shown in Fig. 12.12.

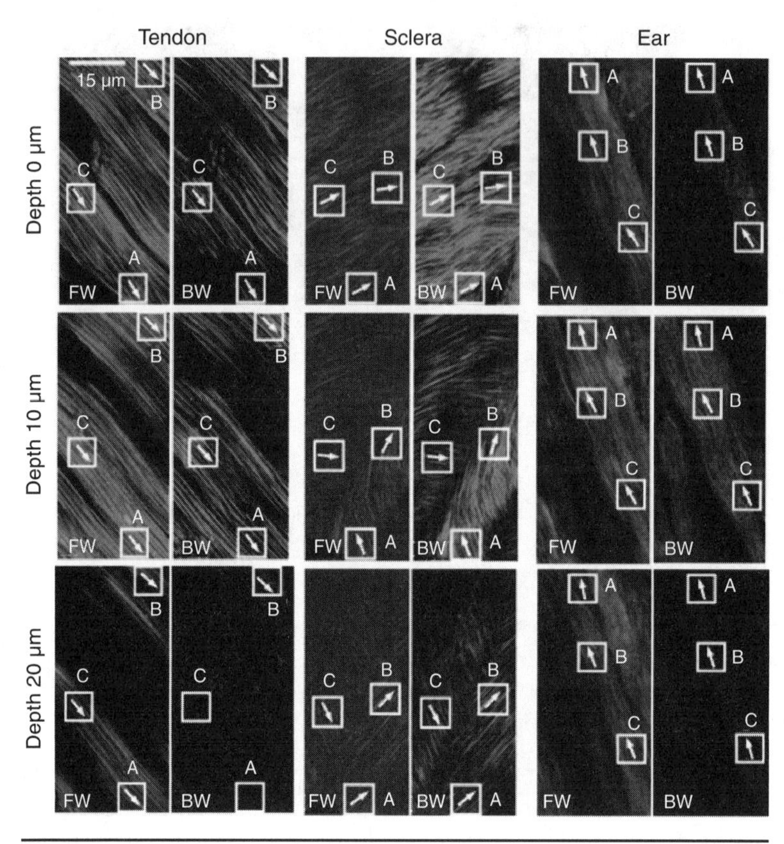

FIGURE 12.12 Forward and backward SHG images of collagen fibers in the tendon, sclera, and ear cartilage at various depths within the sample. Scale bar: 15 μm (same for all samples). The arrows represent the preferred orientation of the fibers within each region of interest A, B, and C. (Reprinted with permission)

The samples were cut to 40-μm thick, which may not be thick enough for comparison to *in vivo* imaging. However, the technique used for comparison can be applied to thicker samples. The forward and backward images in Fig. 12.12 show similarities. It should also be noted that the backward signal for the tendon and ear is lost at a depth of 20 μm.

12.3.1 Preferred Orientation

The preferred orientation was evaluated for each region of interest shown in Fig. 12.12. The difference between preferred orientation measured in the forward image and preferred orientation measured in the backward image ($\Delta\gamma$) was calculated. The results are shown in Fig. 12.13.

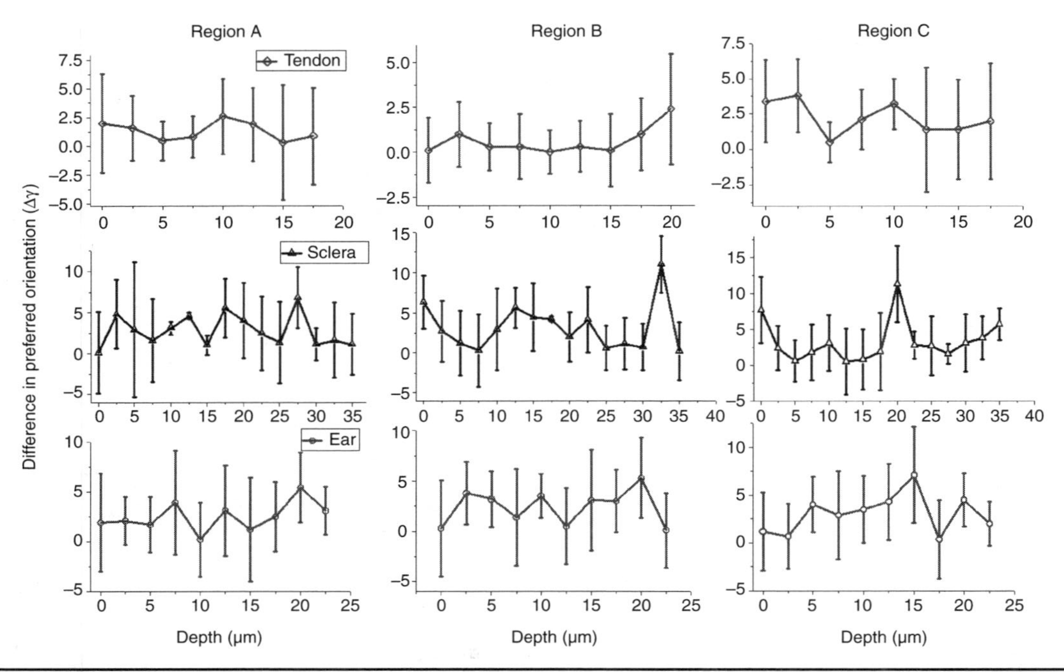

Figure 12.13 The difference in the preferred orientation, $\Delta\gamma$, of collagen fibers between forward and backward SHG versus depth for the tendon, sclera, and ear. The regions of interest are the same as those shown in Fig. 12.1. (Reprinted with permission)

We obseve that $\Delta\gamma$ for the tendon remains below 3.8° with a standard deviation of less than 4.5°; this suggests that the forward and backward images of collagen fibers give similar preferred orientations. For the sclera, $\Delta\gamma$ reaches a maximum 11°, and its standard deviation reaches 8°; this suggests the greatest randomness in fiber orientation and the least consistency hold in the backward image. Overall, $\Delta\gamma$ and its standard deviation remain relatively small; this suggests that the backward SHG images maintain essentially the same information regarding preferred orientation as the forward SHG images.

12.3.2 Peaks in Magnitude Spectrum

Aside from comparing preferred orientation, the study also looked at the Fourier peaks between forward and backward images. The results are shown in Fig. 12.14. Fourier peaks were analyzed at three different depths (0, 10, and 20 μm). The forward peaks are represented by

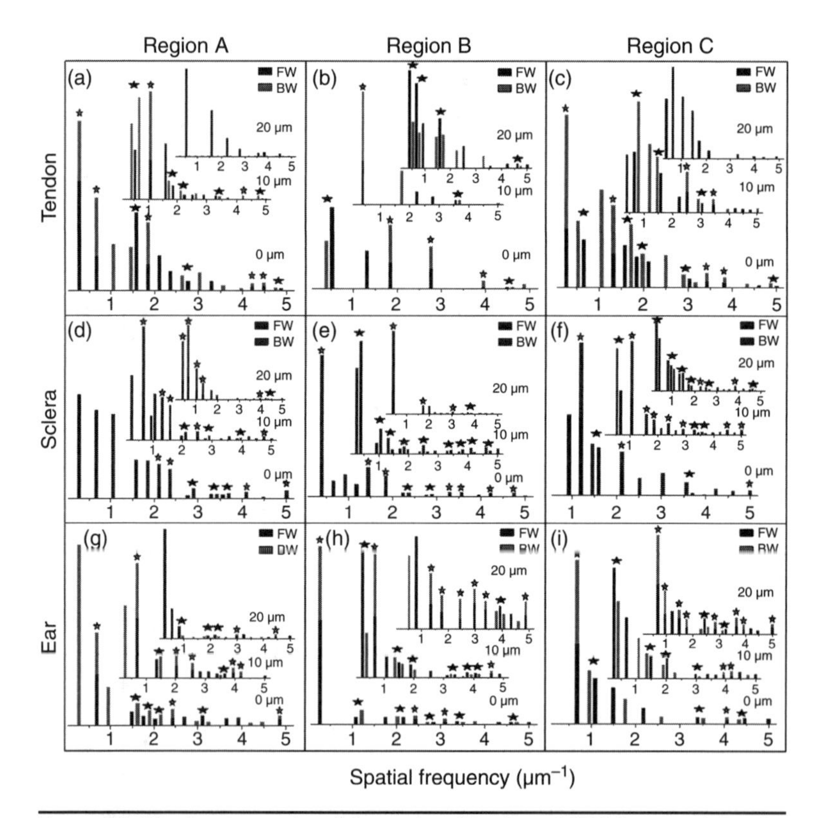

Figure 12.14 Fourier peaks of forward and backward SHG images, taken to analyze the direction of preferred orientation. Small stars show frequencies that are peaks in both forward and backward images, while wide stars show peaks that are within error for the forward and backward images. (Reprinted with permission)

black bars (the height of each bar represents the amplitude of the peak), while the backward peaks are represented by gray bars.

It can be seen that many of the Fourier peaks overlap between the forward and backward collected images; again this suggests that much of the information collected in the forward direction can be obtained from the backward direction.

12.4 Conclusion

Second-harmonic generation microscopy has the potential to be a useful tool for diagnostics due to the fact that it (1) is noninvasive, (2) offers high 3D spatial resolution, (3) does not require staining, (4) produces low phototoxicity, (5) preserves phase information, and (6) is sensitive to structural organization. Several recent techniques have taken advantage of many of these characteristics in order to derive quantitative metrics that assess the morphology of the tissue under study. An obvious next step will be to apply these techniques to live imaging and perform real-time quantitative analysis.

Acknowledgments

This work was supported in part by the University of Illinois at Urbana–Champaign (UIUC) research startup funds and by funds from the National Science Foundation (NSF) (DBI 0839113).

References

1. H. T. Chen, H. F. Wang, M. N. Slipchenko, Y. K. Jung, Y. Z. Shi, J. B. Zhu, K. K. Buhman, and J. X. Cheng, "A Multimodal Platform for Nonlinear Optical Microscopy and Microspectroscopy," *Opt. Exp.* 17: 1282–1290, 2009.
2. W. L. Chen, T. H. Li, P. J. Su, C. K. Chou, P. T. Fwu, S. J. Lin, D. Kim, P. T. C. So, and C. Y. Dong, "Second-Harmonic Generation Chi Tensor Microscopy for Tissue Imaging," *Appl. Phys. Lett.* 94: 3, 2009.
3. P. J. Campagnola and L. M. Loew, "Second-Harmonic Imaging Microscopy for Visualizing Biomolecular Arrays in Cells, Tissues and Organisms," *Nat. Biotechnol.* 21: 1356–1360, 2003.
4. M. Han, G. Giese, and J. F. Bille, "Second-Harmonic Generation Imaging of Collagen Fibrils in Cornea and Sclera," *Opt. Exp.* 13: 5791–5797, 2005.
5. C. H. Yu, S. P. Tai, C. T. Kung, I. J. Wang, H. C. Yu, H. J. Huang, W. J. Lee, Y. F. Chan, and C. K. Sun, "*In Vivo* and *Ex Vivo* Imaging of Intra-Tissue Elastic Fibers Using Third-Harmonic Generation Microscopy," *Opt. Exp.* 15: 11167–11177, 2007.
6. S. W. Teng, H. Y. Tan, J. L. Peng, H. H. Lin, K. H. Kim, W. Lo, Y. Sun, W. C. Lin, S. J. Lin, S. H. Jee, P. T. C. So, and C. Y. Dong, "Multiphoton Autofluorescence and Second-Harmonic Generation Imaging of the *Ex Vivo* Porcine Eye," *Invest. Ophthalmol. Vis. Sci.* 47: 1216–1224, 2006.
7. R. M. Williams, W. R. Zipfel, and W. W. Webb, "Interpreting Second-Harmonic Generation Images of Collagen I Fibrils," *Biophys. J.* 88: 1377–1386, 2005.
8. G. Cox, E. Kable, A. Jones, I. K. Fraser, F. Manconi, and M. D. Gorrell, "3-Dimensional Imaging of Collagen Using Second-Harmonic Generation," *J. Struct. Biol.* 141: 53–62, 2003.

9. F. Tiaho, G. Recher, and D. Rouede, "Estimation of Helical Angles of Myosin and Collagen by Second-Harmonic Generation Imaging Microscopy," *Opt. Exp.* 15: 12286–12295, 2007.
10. S. V. Plotnikov, A. M. Kenny, S. J. Walsh, B. Zubrowski, C. Joseph, V. L. Scranton, G. A. Kuchel, D. Dauser, M. S. Xu, C. C. Pilbeam, D. J. Adams, R. P. Dougherty, P. J. Campagnola, and W. A. Mohler, "Measurement of Muscle Disease by Quantitative Second-Harmonic Generation Imaging," *J. Biomed. Opt.* 13: 11, 2008.
11. B. Alberts, A. Johnson, J. Lewis, M. Raff, K. Roberts, and P. Walter, *Molecular Biology of the Cell,* 5th ed., Garland Science, New York, 2008.
12. D. R. Keene, L. Y. Sakai, and R. E. Burgeson, "Human Bone Contains Type-III Collagen, Type-VI Collagen, and Fibrillin—Type-III Collagen is Present on Specific Fibers That May Mediate Attachment of Tendons, Ligaments, and Periosteum to Calcified Bone Cortex," *J. Histochem. Cytochem.* 39: 59–69, 1991.
13. Y. Komai and T. Ushiki, "The 3-Dimensional Organization of Collagen Fibrils in the Human Cornea and Sclera," *Invest. Ophthalmol. Vis. Sci.* 32: 2244–2258, 1991.
14. R. C. Billinghurst, L. Dahlberg, M. Ionescu, A. Reiner, R. Bourne, C. Rorabeck, P. Mitchell, J. Hambor, O. Diekmann, H. Tschesche, J. Chen, H. VanWart, and A. R. Poole, "Enhanced Cleavage of Type II Collagen by Collagenases in Osteoarthritic Articular Cartilage," *J. Clin. Invest.* 99: 1534–1545, 1997.
15. L. Klein and J. Chandrarajan, "Collagen Degradation in Rat Skin But Not in Intestine during Rapid Growth—Effect on Collagen Type-I and Type-III from Skin," *Proc. Natl. Acad. Sci. USA.* 74: 1436–1439, 1977.
16. N. Morishige, A. J. Wahlert, M. C. Kenney, D. J. Brown, K. Kawamoto, T. Chikama, T. Nishida, and J. V. Jester, "Second-Harmonic Imaging Microscopy of Normal Human and Keratoconus Cornea," *Invest. Ophthalmol. Vis. Sci.* 48: 1087–1094, 2007.
17. P. Matteini, F. Ratto, F. Rossi, R. Cicchi, C. Stringari, D. Kapsokalyvas, F. S. Pavone, and R. Pini, "Photothermally-Induced Disordered Patterns of Corneal Collagen Revealed by SHG Imaging," *Opt. Exp.* 17: 4868–4878, 2009.
18. I. Freund, M. Deutsch, and A. Sprecher, "Connective-Tissue Polarity—Optical 2nd-Harmonic Microscopy, Crossed-Beam Summation, and Small-Angle Scattering in Rat-Tail Tendon," *Biophys. J.* 50: 693–712, 1986.
19. J. C. Russ, "Processing Images in Frequency Space," in *The Image Processing Handbook,* 5th ed., CRC Press, Boca Raton, FL, 335–398, 2007.
20. R. Ambekar Ramachandra Rao, M. R. Mehta, and K. C. Toussaint, "Fourier Transform Second-Harmonic Generation Imaging of Biological Tissues," *Opt. Exp.* 17: 14534–14542, 2009.
21. P. J. Campagnola, A. C. Millard, M. Terasaki, P. E. Hoppe, C. J. Malone, and W. A. Mohler, "Three-Dimensional High-Resolution Second-Harmonic Generation Imaging of Endogenous Structural Proteins in Biological Tissues," *Biophys. J.* 82: 493–508, 2002.
22. R. Ambekar Ramachandra Rao, M. R. Mehta, S. Leithem, and K. C. Toussaint, "Quantitative Analysis of Forward and Backward Second-Harmonic Images of Collagen Fibers Using Fourier Transform Second-Harmonic-Generation Microscopy," *Opt. Lett.* 34: 3779–3781, 2009.
23. C. Bayan, J. M. Levitt, E. Miller, D. Kaplan, and I. Georgakoudi, "Fully Automated, Quantitative, Noninvasive Assessment of Collagen Fiber Content and Organization in Thick Collagen Gels," *J. Appl. Phys.* 105: 102042–102042-11, 2009.
24. C. P. Fleming, C. B. M. Ripplinger, B. Webb, I. R. Efimov, and A. M. Rollins, "Quantification of Cardiac Fiber Orientation Using Optical Coherence Tomography," *J. Biomed. Opt.* 13: 030505, 2008.
25. K. M. Meek and C. Boote, "The use of X-Ray Scattering Techniques to Quantify the Orientation and Distribution of Collagen in the Corneal Stroma," *Prog. Retin. Eye Res.* 28: 369–392, 2009.
26. B. M. Palmer and R. Bizios, "Quantitative Characterization of Vascular Endothelial Cell Morphology and Orientation Using Fourier Transform Analysis," *J. Biomech. Eng. Trans. ASME* 119: 159–165, 1997.
27. A. H. Reshak, "Enhancing the Resolution of the Forward Second Harmonic Imaging Using the Two-Photon Laser Scanning Microscope," *Micron* 40: 750–755, 2009.

CHAPTER 13

Plasmonics: Toward a New Paradigm for Light Manipulation at the Nanoscale

Maxim Sukharev

Department of Applied Sciences and Mathematics
Arizona State University at the Polytechnic Campus
Mesa, Arizona

13.1 Introduction

Subwavelength optical systems composed of noble metals have long been attracting considerable attention due to their unique properties in the visible. Recent advancements in fabrication methods of such materials [1] and tremendous progress in laser technology [2] are now offering a wide variety of exciting opportunities for researchers, ranging from applications in medicine [3], biology [4], and single-atom molecule manipulation [5], to fundamental [6] and applied physics [7]. Among several fascinating applications of nanostructural materials is the long-standing question of controlling light in a subdiffraction scale aiming for optical nanodevices [8] and nanoscaled coherent sources operating in the visible region of the spectrum [9]. From the fundamental physics point it is also important to understand general behavior of electromagnetic (EM) near-fields and their interaction with atoms and molecules. From that perspective, knowledge of EM field dynamics at the subwavelength scale and the ability to predict an EM field's behavior in the presence of mesoscopic materials, such as metal tips, metallodielectric nanoparticles (NPs), and NP arrays, are needed.

The problem of EM near-field dynamics in the vicinity of nanostructural materials has been the topic of rapidly increasing interest in both experimental and theoretical works [10]. It has been realized that structures composed of metal NPs are capable of strong EM field enhancements [11] and coherent EM energy transport in space [12]. The physical origin of this enhancement is easily seen from the famous Mie's expression for the extinction cross section of a metallic sphere of a radius R in the long wavelength limit ($\lambda >> R$) [13]

$$\sigma \sim \frac{R^3}{\lambda}\frac{\mathrm{Im}(\varepsilon)}{(\mathrm{Re}(\varepsilon)+2)^2+(\mathrm{Im}(\varepsilon))^2} \tag{13.1}$$

where Re(ε) and Im(ε) are the real and imaginary parts of the metal NP's dielectric constant ε. It is obvious that for materials where the real part of the dielectric constant is negative (as is the case for the noble metals in the visible part of the spectrum, for instance) and close to −2 at some particular frequency, the spectrum experiences resonant enhancement, due to the excitation of collective motion of free electrons (referred to as *plasmon oscillations*) in the metallic region. The resulting spectral peaks are often referred to as *localized surface plasmon-polariton resonance* (LSPPR) [14]. A second NP, placed sufficiently close to the resonantly excited NP, experiences similar collective oscillations of free electrons, which in this case are induced by the excited neighbor particles, rather than directly [12]. In an array of NPs, such indirect excitations propagate via near-neighbor interactions [15], this provides a mechanism for transmission of EM energy through properly configured NP arrays.

To illustrate the sensitivity of the optical response to the NP's shape and symmetry, we perform simulations of extinction for silver NPs of four different shapes (starting from the ideal sphere and approaching the cube). It should be emphasized that throughout these simulations we keep the volume of an NP constant while changing its shape so any features appearing in the spectra are governed solely by the NP's shape only. The steady-state intensity distributions at the resonant conditions are shown in the inset of Fig. 13.1, demonstrating the well-known lightning rod effect when the EM intensity tends to be localized near sharp features of the system [16]. It is interesting to note that once the shape of an NP exhibits any sharp features, additional resonances appear in the extinction spectra.

The demonstration of the sensitivity of LSPPR to the relative arrangement of individual NPs in the array is shown in Fig. 13.2, where we present results of simulations of four different systems composed of silver ellipsoids. We again compare extinction cross

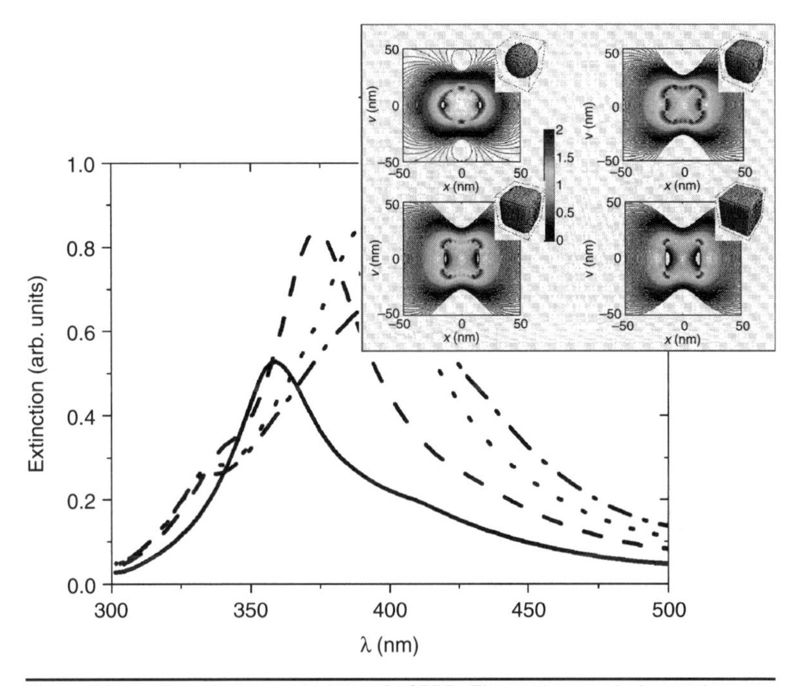

Figure 13.1 The shape sensitivity of LSPPR. The main panel shows the extinction cross section (in arbitrary units) as a function of the incident wavelength (in nm) for four silver NPs: sphere (solid line), cube (dash-dotted line), and two intermediate shapes (dashed and dotted lines). The inset demonstrates steady-state intensity distributions in a logarithmic scale in the units of enhancement at the plane bisecting an NP. Note that external laser radiation propagates perpendicular to the figure plane and is horizontally polarized.

sections of several NP arrays, illustrating how LSPPR depends not only on the shape of an individual NP (as we have seen in Fig. 13.1) but also on how we arrange the NPs in the array.

Although local EM field enhancements are relatively high for an NP with sharp corners, the regions with a high intensity are usually affected by various factors including boundary scattering and radiation losses. For example, the quality factor Q of a typical LSPPR is on the order of 5 or less.* One is encouraged to examine structures where radiation losses are suppressed. Such structures are metal films and gratings that support surface plasmon waves.

The physics of surface plasmons has long been studied [17]. When propagating at a half-space separating a bulk metal from an adjacent dielectric, radiation is kinematically forbidden; damping only involves

*The Q-factor is defined as a ratio of a line width at the half-power line and a resonant wavelength.

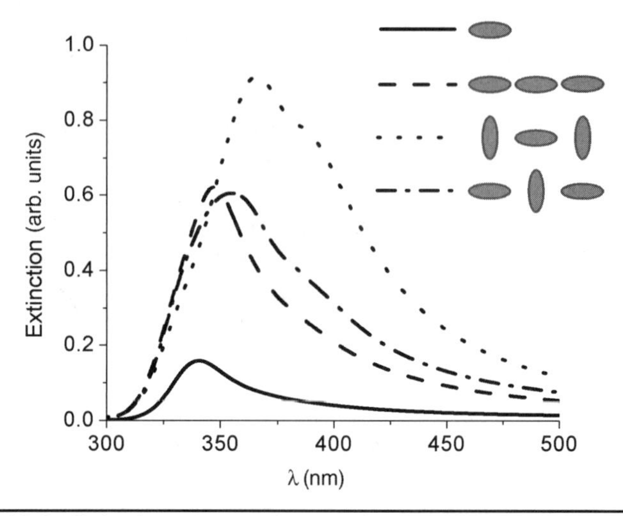

FIGURE 13.2 Extinction cross section as a function of the incident wavelength in nm for a single silver ellipsoid (solid line) and arrays of three ellipsoids with various arrangements.

scattering from inhomogeneities (e.g., a rough surface) and dissipation in the metal itself. In fact, the very existence of a surface plasmon requires that its frequency ω_p and wave vector k_p not match light propagating at any angle in the surrounding dielectric. For this reason it is common to couple to surface plasmons via an evanescent light wave and carefully control the coupling strength. In the commonly used Otto geometry [18] this is accomplished by forming a thin dielectric layer between a prism and the free surface of the metal adjacent to it. By adjustment of the thickness of the coupling layer, the reflected beam in the prism can be made to vanish at some critical angle resulting in critical coupling of light to the surface plasmon mode, which leads to high EM fields localized at the metal-dielectric interface.

In this review we discuss both localized plasmon-polaritons (LSPP) and surface plasmon-polariton (SPP) modes. We will review the computational aspects of nano-optics in Section 13.2. Next we introduce the phase-polarization control scheme that allows one to manipulate EM energy transfer along chains of closely spaced NPs in Section 13.3. The idea of classical trapping of light by plasmonic crystals is discussed in Section 13.4. Optical properties of NPs with no center of inversion symmetry are illustrated by the example of L-shaped silver particles in Section 13.5. Section 13.6 is an overview of the optimal design approach for plasmonic materials. Section 13.7 reviews both theory and recent experiments on sinusoidal silver gratings. Section 13.8 concludes this review with a brief outlook to new research areas in nano-optics.

13.2 Computational Nano-Optics

Among several numerical techniques that allow one to predict optical properties of nanostructural materials, the finite-difference time-domain (FDTD) approach is considered to be the most efficient and yet relatively simple to implement [19]. In brief the FDTD scheme is a numerical integrator of Maxwell equations in a time domain that spatially separates electric and magnetic fields (Yee's grid). This is one of the major advantages of this approach, that is, boundary conditions are automatically satisfied at all grid points. This allows a straightforward programming of complex structures.

In spatial regions occupied by material with a dielectric constant, ε depends on frequency, ω. Maxwell equations include polarization currents depending on what type of media is considered. It is usually assumed that the Drude model correctly describes optical properties of a metal with the dielectric constant given by [14]

$$\varepsilon(\omega) = \varepsilon_r - \frac{\omega_p^2}{\omega^2 - i\Gamma\omega} \tag{13.2}$$

where ε_r is the high frequency limit of the dielectric constant, ω_p is the bulk plasma frequency, and Γ is the damping constant. Maxwell equations in the metal regions read

$$\varepsilon_0\varepsilon_r \frac{\partial \vec{E}}{\partial t} = \nabla \times \vec{H} - \vec{J}$$

$$\mu_0 \frac{\partial \vec{H}}{\partial t} = -\nabla \times \vec{E} \tag{13.3}$$

with ε_0 and μ_0 being dielectric permittivity and magnetic permeability of vacuum, respectively. Density currents $\overline{J}$ satisfy a simple equation,

$$\frac{\partial \overline{J}}{\partial t} = -\Gamma \overline{J} + \varepsilon_0 \omega_p^2 \overline{E} \tag{13.4}$$

A more general description of optical properties of metals adds additional terms to Eq. (13.2) in the Lorentz form leading to the Drude–Lorentz model [20]. The Drude model is limited to a narrow spectrum, the Drude–Lorentz model correctly describes optical properties of various materials in a wider range of wavelengths [21]. Figure 13.3 shows a comparison of both models for silver with the following set of Drude parameters: $\varepsilon_r = 8.926$, $\omega_p = 1.7601 \times 10^{16}$ rad/second, and $\Gamma = 3.0841 \times 10^{14}$ rad/second. The fitting parameters for the Drude–Lorentz model can be found in [22].

For simulations of open systems, one needs to impose artificial absorbing boundaries in order to avoid reflection of outgoing EM

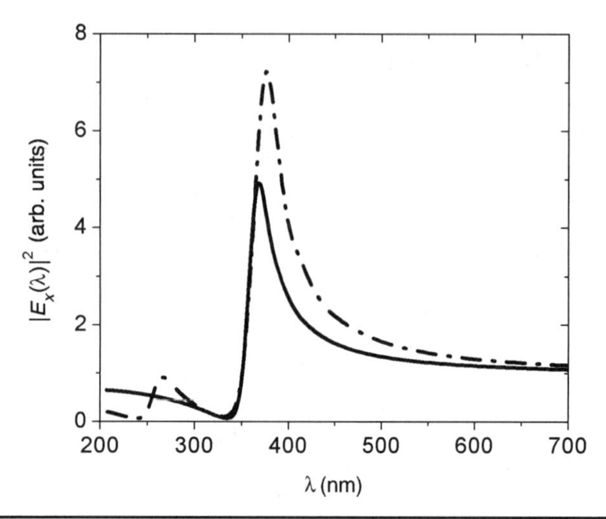

Figure 13.3 Spectral response of a silver sphere of the diameter 40 nm to the external plane wave excitation calculated using the Drude model (solid line) and the Drude-Lorentz model (dash-dotted line).

waves back to the simulation domain. Among various approaches that address this numerical issue, the perfectly matched layer (PML) technique is considered to be the most adequate [23]. It reduces the reflection coefficient of outgoing waves at the simulation region boundary to 10^{-8}. Essentially, the PML approach surrounds the simulation domain by thin layers of nonphysical material that efficiently absorbs outgoing waves incident at any angle. We implement the most efficient and least memory-intensive method, convolution perfectly matched layers (CPML) absorbing boundaries, at all six sides of the three-dimensional (3D) modeling space [24]. Through extensive numerical experimentation, we have empirically determined optimal parameters for the CPML boundaries that lead to almost no reflection of the outgoing EM waves at all incident angles.

A particular advantage of the FDTD method is its ability to obtain the optical response of the structure (assuming linear response) in the desired spectral range in a single run [19]. The system is excited with an ultrashort optical pulse constructed from Fourier components spanning the frequency range of interest. Next, Maxwell's equations are propagated in time for several hundred femtoseconds and the components of the EM field are detected at the point of interest. Fourier transforming the detected EM field on the fly yields intensities that can be easily processed into the spectral response. Since we also have access to the field components, we can evaluate the intensity enhancement relative to the incident field as well. This provides the capability for straightforward evaluation of the "coupling efficiency" between EM radiation and plasmonic structures in the spectral range of interest.

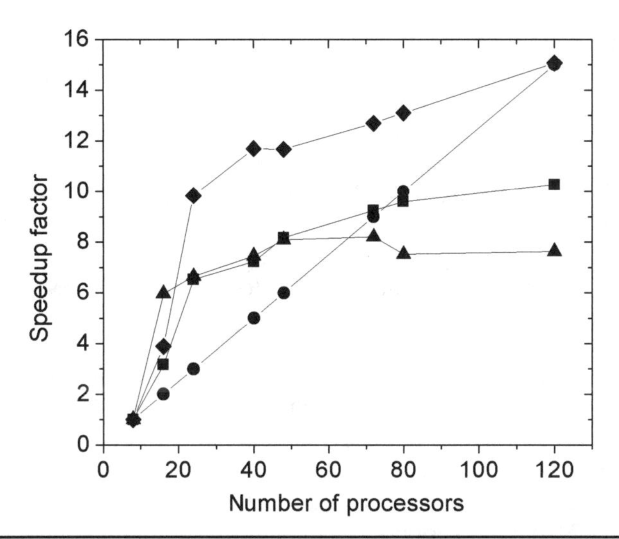

FIGURE 13.4 Speedup factor defined in (5) as a function of a number of processors for the two-dimensional (2D) FDTD simulations performed at the home-built Arizona State University (ASU) supercomputer plasmon (squares), Abe supercomputer at University of Illinois at Urbana–Champaign (UIUC) (triangles), and Hewlett-Packard (HP) cluster at Western Kentucky University (WKU) (diamonds). The ideal speedup is also shown (circles).

Another important advantage of the FDTD scheme is related to its simple and straightforward parallelization procedure. In order to partition the FDTD scheme onto a parallel grid, we divide the simulation cube into N slices (N corresponds to the number of available processors) along the z-axis and implement point-to-point Message-Passing-Interface (MPI) communication subroutines at the boundaries between the slices. We then need to perform scaling tests to identify the most efficient number of xy-planes carried by a single processor, which obviously varies from one multiprocessor computer to another depending on the network latency and memory.

In order to test the performance and efficiency of the parallel technique described above, we investigate the speedup factor $S(n)$, defined as

$$S(n) = T(8)/T(n) \tag{13.5}$$

where $T(8)$ is the algorithm execution time on a single node with 8 processors* and $T(n)$ is the execution time on n processors. Such a speedup factor for the case of 2D simulations of the optical response of a silver cylinder with a diameter of 20 nm is demonstrated in Fig. 13.4. It is interesting to note that for a small number of occupied processors the

*Here we utilized a set of nodes equipped with dual quad-core processors resulting in 8 total cores per node.

speedup factor exceeds its ideal value (according to Amdahl's law the ideal speedup in our case corresponds to the ratio $n/8$ [25]) for almost all tested clusters. The nonlinear deviation of the simulated speedup from its ideal value can be explained as follows: By increasing the number of processors involved in the simulations while keeping the problem size constant, we effectively decrease the amount of swapping that the code performs. In other words, for those points in Fig. 13.4 at which the simulated speedup is higher than its ideal value, the memory occupied by the locally propagated matrices (consisting of the EM field components) is less than the cash size of each processor.

A general case of the 3D FDTD speedup is shown in Fig. 13.5, where we compare speedup factors for an open system (with CPML layers imposed at all 6 sides of the 3D FDTD grid) with the one obtained for a periodic structure (with CPML boundaries applied only at the top and bottom sides of the FDTD cube while Bloch periodic conditions are imposed at the other four sides) utilizing a BlueGene/L cluster at Argonne National Laboratories (ANL). Here we simulate the interaction of a 35-fs long laser pulse with silver nanodisks of 40-nm diameter and of 40-nm height. The laser radiation is modeled as a plane wave that is generated above the nanoparticles and propagates along the negative z-direction. The cube size is $120 \times 120 \times 1842$ nm with a spatial step of 1.199 nm, and the total number of time iterations is 20,000. The observable interest in these studies is the vector eigenmodes of interacting nanoparticles. The measured speedup factor is nearly ideal. It degrades with an increasing number of processors

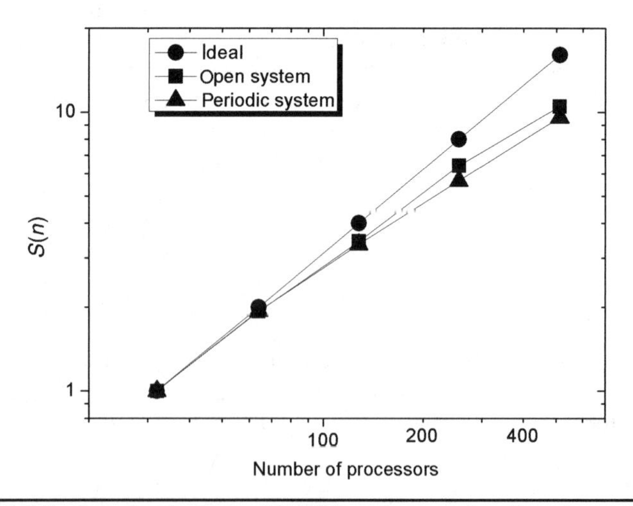

Figure 13.5 Speedup factor for 3D FDTD codes as a function of a number of processors at the double logarithmic scale for a single silver nanodisk (squares) and periodic array of nanodisks (triangles). The ideal speedup factor is shown using circles.

since in the model simulations, the grid size was fixed and hence the number of xy layers per processor decreased with an increasing number of processors; this resulted in more sending and receiving of MPI operations per iteration with consequent latency of the network.

A test of the accuracy of the FDTD scheme is presented in Fig. 13.6, which directly compares FDTD simulations with experimental data. Recent experiments in nanoplasmonics [26], which offer both exceptional precision and interesting physics, measured the interference patterns arising from slit-groove structures, as depicted in the inset of Fig. 13.6. The thick dashed contour represents the far-field detection contour in the FDTD simulations. Interference of waves directly propagating through the slit with waves originating from the groove and propagating to the slit as surface waves gives rise to the simulated interference pattern shown in the main panel. Comparison of the data with the FDTD calculations shows remarkable agreement. We note that in order to efficiently model the EM radiation scattering at large slit-groove distances, parallel simulations have to be implemented. In particular, we carried out FDTD simulations on 256 processors for slit-groove distances ranging from 4.5 to 6 μm so as to capture correctly the long propagation lengths of surface waves [27].

Another comparison between several fully vectorial methods has been recently applied to the same slit-groove scattering problem depicted in the inset of Fig. 13.6 [28]. The benchmarked methods, all

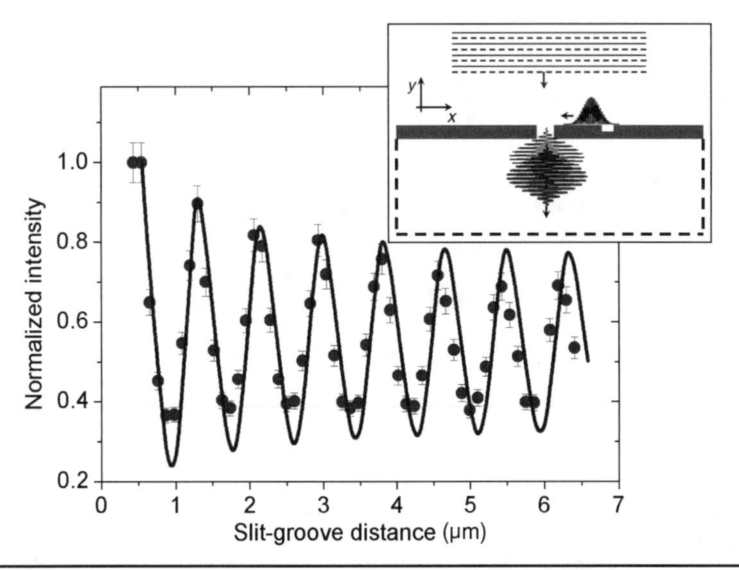

FIGURE 13.6 Direct comparison of FDTD simulations with experimental data on surface wave generation by subwavelength slit-groove structures. The circles in the main panel show the experimental intensity as a function of the slit-groove distance. The solid line presents the FDTD modeling.

of which used in-house developed software, included a broad range of fully vectorial approaches from finite-element methods, volume-integral methods, and FDTD methods to various types of modal methods based on different expansion techniques. It has been shown that all those techniques resulted in nearly perfect agreement.

13.3 Phase-Polarization Control Scheme

In this section we illustrate the possibility of guiding the EM energy into one or the other of two branches of the symmetric T-junction solely by wave interference. The inset of Fig. 13.7 depicts schematically the T-junction, showing the spatial positions of the incident source (LS) and two detectors (D_1 and D_2). The radius of the silver spheres is 25 nm and the center-to-center distance between the spherical particles is 75 nm. We use the cylindrical symmetry of the experiment to restrict attention to two (x and y) E-components and one (z) H-component (TE_z-mode). As an incident source, we implement a pointwise elliptically polarized electric field.

$$\vec{E}_{inc} = f(t)\left[\vec{e}_x \cos\frac{\xi}{2}\cos(\omega t + \phi) + \vec{e}_y \sin\frac{\xi}{2}\sin(\omega t)\right] \tag{13.6}$$

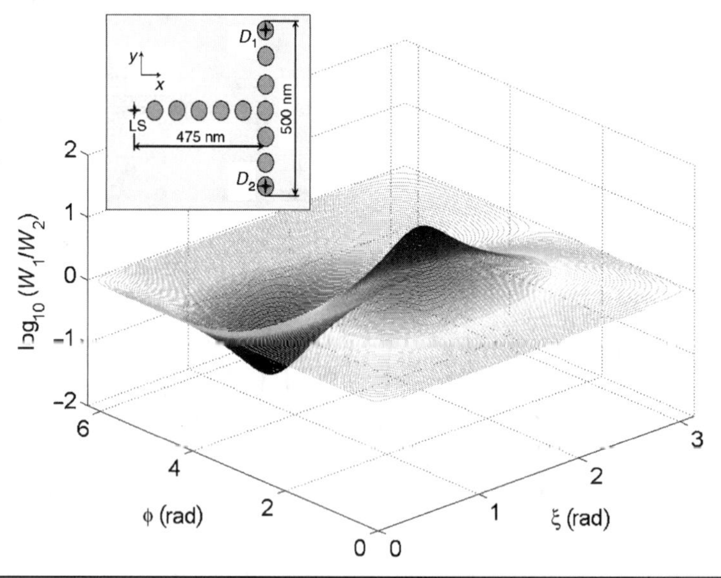

Figure 13.7 Phase-polarization control of EM energy transport through the T-structure depicted schematically in the inset. Logarithm of the ratio W_1/W_2 of the time-averaged EM energies in the upper (D_1) and lower (D_2) ends of a T-structure (see inset) as a function of the laser polarization parameters, ξ and ϕ. The incident pulse is centered at the LSPP resonant wavelength $\lambda = 300$ nm.

where $f(t)$ is the pulse envelope, $\bar{e}_{x,y}$ is a unit vector along the $x(y)$-axis, ω is the optical frequency, and ξ and ϕ specify the polarization vector (ξ reduces to ellipticity for $\phi = 0$). A pulse duration τ of 15 fs is used in the following calculations: Since the approach is based on resonance effects, control improves with τ, saturating as τ converges to the resonance lifetime. In all simulations, a 2D grid ranges from $(x, y) = -850$ nm to 850 nm. The results are converged to within 0.5% with spatial grid steps $\delta x = \delta y = 1.1$ nm and a temporal step $\delta t = \delta x/(3^{1/2}c)$, where c denotes the speed of light in a vacuum.

The main frame of Fig. 13.7 shows the ratio of the time-averaged EM energies W_1/W_2 detected at the upper (D_1) and lower (D_2) detection points as a function of the polarization parameters, ξ and ϕ, at the incident wavelength corresponding to one of the two LSPP resonances. We note here that the transmission spectrum of the system (the time-averaged EM energy at detector D_n as a function of the incident wavelength) exhibits double-peak behavior for the elliptically polarized excitation, which the widely used model of interacting Hertzian dipoles does not predict [29]. This suggests the significance of quadrupole-quadrupole interactions between NPs.

In the case of a linear polarized source, $W_1/W_2 = 1$, nonlinear polarization of LS allows efficient (by two orders of magnitude for the pulse of 15 fs) control over the branching ratio. The physics underlying the phase and polarization control can be understood by a study of the time evolution of the EM energy propagation. Essential to the finding is the excitation of a coherent superposition of transverse and longitudinal plasmon modes, whose relative phase and spatial distribution are adjusted by the polarization parameters ξ and ϕ of the incident field. The elliptical polarization enables bending about the corner with little losses. The optical phase breaks the symmetry so as to funnel the energy into one or the other of the T-junction branches. To check the sensitivity of the control results to deviation from perfect regularity, we introduce a random size and location deviation of 2 nm for each particle. The sensitivity of the results to the particle size was checked by varying the NP's radii in the 25- to 30-nm range. Simulations show that, while the details of Fig. 13.7 are not invariant, the extent of control is not affected by either deviation from perfect regularity of small-size modifications.

To quantify the effect of losses, we have computed the power transmission coefficient, defined as the ratio of the time-averaged EM energy at the center of the NP closest to the incident source NP to the time-averaged EM energy at detection points D_1 and D_2. We found a power transmission coefficient of 2.6 (that is about 50 times smaller than the controlled branching ratio shown in Fig. 13.7). Whereas the control ratio W_1/W_2 improves with increasing duration of the incident pulse, the power transmission coefficient is at most essentially independent on the pulse duration.

We found that the phase-polarization control is equally effective in larger T-structures consisting of two interacting linear chains of NPs of 900-nm length each as demonstrated in Fig. 13.8, where we plot the time-averaged EM intensity for the case of EM energy transmission to the lower detector. The results of Fig. 13.7 suggest a potential nano-scale plasmonic switch or inverter with an operational time comparable to the incident pulse duration of tens of fs [30].

Potential applications of the phase-polarization control scheme are many and varied [31]. For instance, colloidal NPs and synthetic nanomaterials (such as dented and striped multilayer nanorods) with engineered optical properties have attracted much interest in recent years [32], due to the rapid progress of fabrication techniques such as template electrochemical synthesis physical/chemical vapor deposition, and chemical lithography [33]. Controllability over the properties of such systems translates into control over their dielectric constant and therefore over the EM wave dispersion and plasmon resonances. This suggests an opportunity for coherent control strategies, with which a variety of desired targets (e.g., minimization of losses, enhancement of a given mode, sensitivity to a given species, etc.) could be potentially attained by adjusting the width and length of different layers of individual NPs and selecting different materials ranging from noble metals to insulating solids.

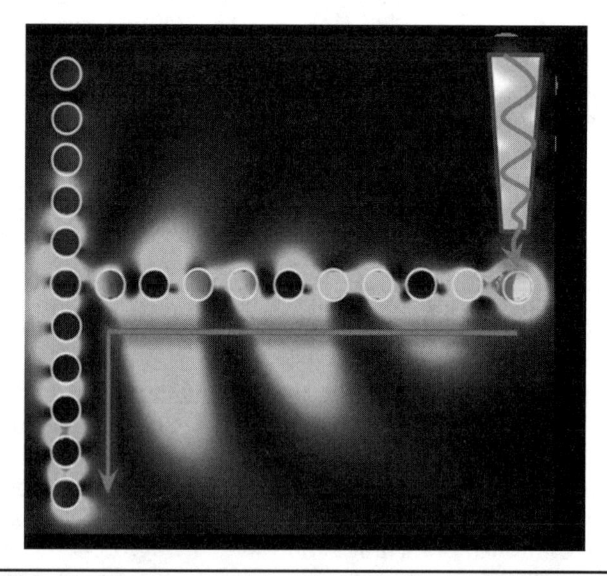

FIGURE 13.8 The time-averaged intensity distribution for the phase-polarization control scheme at the parameters of the incident source corresponding to the EM-energy transfer solely to the detector at the bottom of the T-junction.

13.4 Plasmon-Driven Light Trapping and Guidance by 1D Arrays of Nanoparticles

In this section we describe a recently proposed scheme of selective absorption of light by an infinite periodic and finite one-dimensional (1D) metal crystal [34]. The structure considered is shown in the inset of Fig. 13.9. It is periodic in the horizontal direction and is composed of silver NPs with diameters of 55 nm and center-to-center distances of 65 nm. NPs are excited by a horizontally polarized plane wave, and the transmitted EM energy is detected along the horizontal line shown as a dashed line in the inset. The transmission through the crystal is expressed in terms of the ratio of the time-averaged EM energy in the presence of the nanoconstruct and without periodic arrays, which is shown in the main frame of Fig. 13.9 for five arrays with different numbers of layers. Transmission clearly exhibits a minimum located at $\lambda = 360$ nm and becomes more pronounced when the number of layers in the nanostructure increases. The transmission reaches the value of 10^{-6} for five layers. It is important to emphasize that the calculated reflection coefficient is almost flat with a small spike centered at 360 nm. Moreover, reflection falls with the increase of the number of layers in the crystal. This results in almost complete

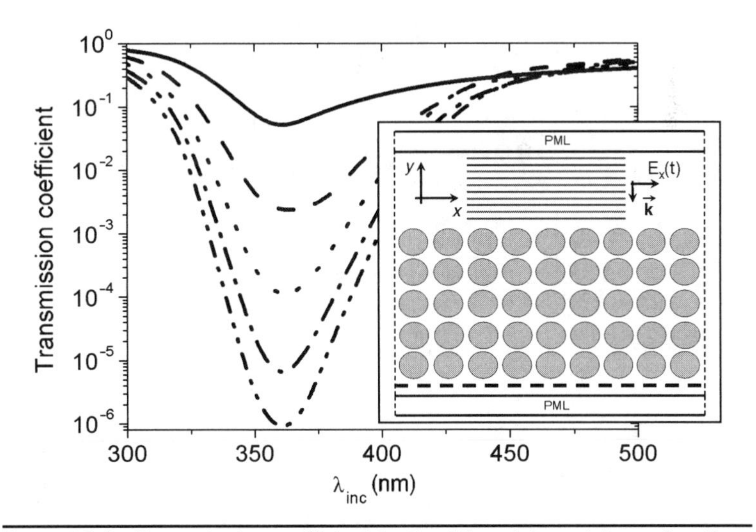

FIGURE 13.9 Transmission coefficient for the 1D metal crystal as a function of the incident wavelength (in nm). Solid line shows the transmission through the single horizontal layer of NPs, dashed line represents the transmission through the two layers, dotted line indicates the transmission for the three-layer system, dash-dotted line indicates four layers, and dash-dot-dotted line indicates five layers. The inset shows the geometry of the simulations, where two dashed vertical lines represent periodic boundaries.

absorption of light for a five-layer crystal. The origin of the minimum corresponds to the excitation of longitudinal (x-polarized) plasmons at each layer of the crystal, and hence, an incident light, which initially had the propagation k-vector along the y-axis, is now guided along the horizontal layers. We also note that the transmission exponentially decays with a number of layers in the y-dimension.

Such plasmon-assisted absorption suggests an exciting opportunity to control absorption and scattering of light by adjusting the individual LSPP resonances of each layer in NP arrays. For example, we can adjust shapes of individual NPs; this leads to a noticeable LSPP shift as shown in Section 13.1.

Another inviting possibility is to consider nanoconstructs composed of NPs with well-distinguished nonoverlapping individual LSPP modes such as silver and gold [34], so that the whole system absorbs light at more than a single wavelength as illustrated in Fig. 13.10. For material parameters used in the simulations, LSPP resonances of Ag and Au linear chains are separated by almost 120 nm resulting not only in the selective absorption of light but also in the spatial separation of different resonant wavelengths.

For example, a pulse with a frequency corresponding to the LSPP resonance for Ag, $\lambda = 340$ nm, is absorbed by the upper-right

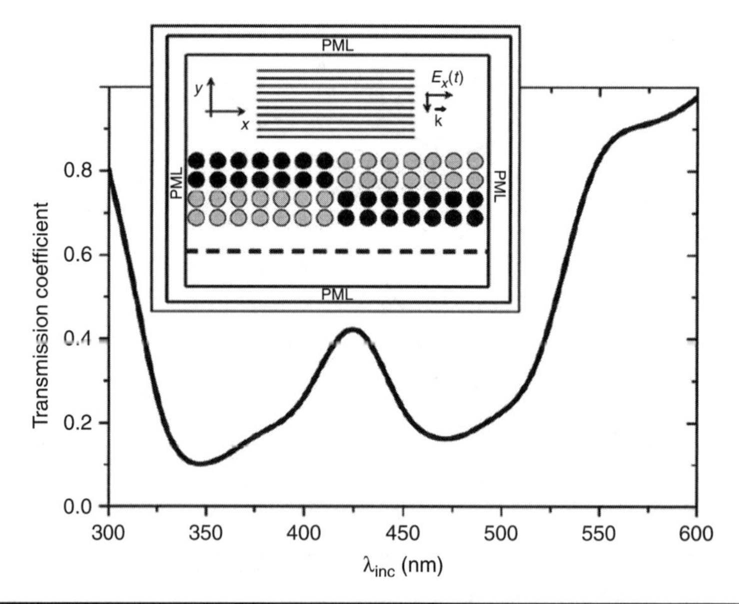

FIGURE 13.10 The transmission coefficient as a function of the incident wavelength (in nm) for the hybrid structure depicted schematically in the inset. Silver NPs are shown as gray circles and gold NPs are presented as black circles. Both silver and gold particles are 40-nm in diameter with a center-to-center distance of 50 nm.

and lower-left silver layers (see the inset in Fig. 13.10), while passing through the Au layers with almost no or a very small reflection. The transparency of different parts of the structure is due to the fact that the effective thickness of the skin layer at wavelengths we consider is of the same order of magnitude as the sizes of the NPs. Skin depth d at optical frequencies ω, much larger than a plasmon frequency ω_p, read $d \approx c/\omega_p$. This results in 17 and 22 nm for silver and gold, respectively.

The effect of wavelength-dependent optical absorption suggests an opportunity to use such nonsymmetric (with respect to the y-axis) constructs to guide the light in different directions. This is illustrated in Fig. 13.11, where we plot the time-averaged intensity distribution at the incident wavelength corresponding to the LSPP resonance for the silver part of the system. At $\lambda = 340$ nm, the upper-right part of the structure depicted in the inset of Fig. 13.10 absorbs the light, while the upper-left part composed of Au NPs is transparent. Excited at their LSPP resonant wavelength, Ag layers guide EM energy to the right at the top of the system. On the other hand, the lower-left part of the system also traps the light but now guides it to the left side. Clearly, an incident plane wave, initially symmetric, is broken by the hybrid system into two parts.

To summarize this section: it has been shown that due to the excitation of collective longitudinal plasmons in linear chains of closely spaced NPs, an incident wave can be selectively localized in the direction vertical to the light propagation, while remaining delocalized laterally. By proper design, EM excitation can thus be guided in space. A variety of extensions of the phenomena of selective light-trapping and funneling can be envisioned. Three-dimensional subwavelength crystals could be

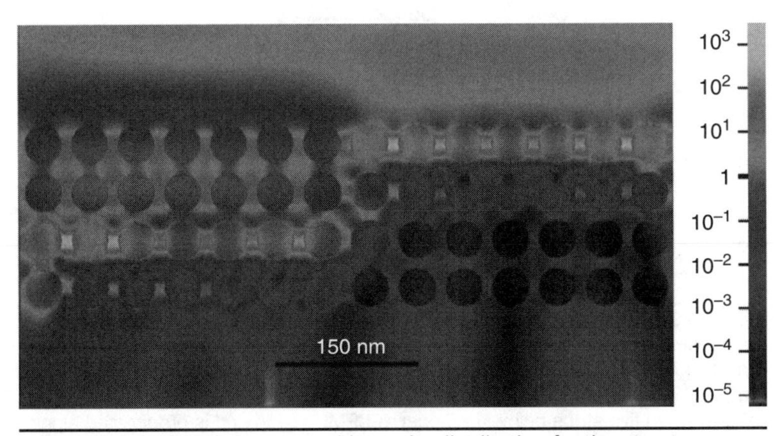

FIGURE 13.11 The time-averaged intensity distribution for the structure shown in the inset of Fig. 13.10 at the incident wavelength $\lambda = 340$ nm corresponding to the excitation of longitudinal plasmons in silver NPs.

used to guide EM energy in 3D. The interaction of white light with periodic arrays could also serve to make interesting optical sources.

13.5 Optics of Metal Nanoparticles with No Center of Inversion Symmetry

Recent experiments on silver L-shaped NPs and their arrays raise questions about the physical origin of the multiband scattering spectrum and its dependence on the external field polarization [35]. In this section we present a systematic study of the optical response of a single L-shaped metal NP and scrutinize the behavior of EM near-fields at the resonant frequencies in order to explain the observed plasmon resonances and their physical nature. The geometry considered is depicted in Fig. 13.12 with dimensions corresponding to experimental data.

First, we compare the FDTD simulations described with measurements of scattering and extinction spectra of a single silver *L*-particle

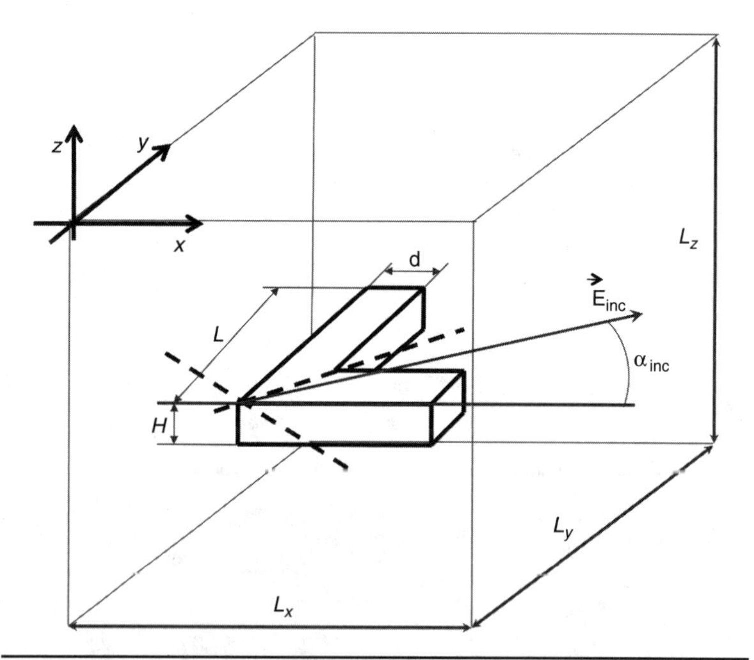

Figure 13.12 A schematical setup of FDTD simulations for the L-shaped silver NP. $L_{x,y,z}$ denotes the size of the simulation cube. *L*, *d*, and *H* represent the arm length, thickness, and the height of the L-particle, respectively; here $L = 145$ nm, $d = 60$ nm, and $H = 30$ nm. The particle is excited by an EM plane wave that is generated five steps below the upper *xy*-CPML region and is propagated along the *z*-direction. The parameter α_{inc} defines the relative orientation of the particle with respect to the incident EM field. Two dashed lines represent the two axes of symmetry of the particle.

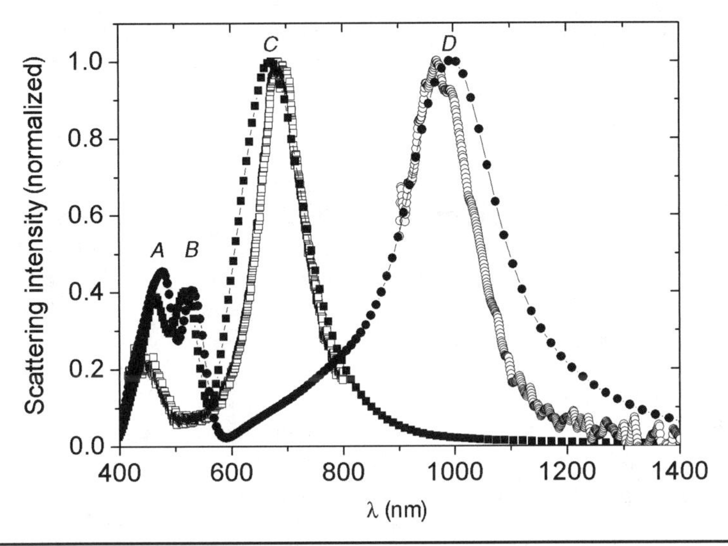

FIGURE 13.13 Normalized experimental (open squares and circles) and theoretical (solid squares and circles) scattering intensity as a function of the incident wavelength. The squares correspond to $\alpha_{inc} = \pi/4$. The circles correspond to $\alpha_{inc} = 3\pi/4$. *A*, *B*, *C*, and *D* indicate the four resonances that are discussed in the text.

at different polarizations of the incident field. Figure 13.13 presents both the experimental data and the numerical simulations. The latter involve 10 independent FDTD runs, in which we calculate the scattering intensity along the incident field polarization for an *L*-particle in vacuum, taking $\alpha_{inc} = \pi/4$ and $3\pi/4$ (see Fig. 13.12 for details). In order to estimate the substrate effects, we introduce an effective refractive index n_{eff}. Random deviations of the particle dimensions are accounted for by performing several simulations and averaging the resulting spectra over the ensemble of different-sized particles. By comparing experimental and simulated resonant wavelengths for the blue ($\alpha_{inc} = \pi/4$) and red ($\alpha_{inc} = 3\pi/4$) bands, we found that $n_{eff} = 1.3$ best matches the experimental observations.

Clearly, the optical response of the *L*-particle is very sensitive to the incident field polarization and exhibits four resonant bands. We note that if the incident field polarization is not along one of two axes of symmetry ($\alpha_{inc} \neq \pi/4$ or $3\pi/4$), the spectrum exhibits all the resonant features, with the amplitudes depending on α_{inc}. The physical origin of the blue (C) and red (D) bands can be understood through extension of the simple case of the eigenmodes of a simple metallic wire. Collective oscillations of conductive electrons along the wire axis leads to a low-energy resonance, whereas oscillations perpendicular to the wire axis give rise to a higher-energy band.

For the case of a more complex structure, such as the *L*-structure considered here, the concept of a functional shape provides an analogous picture [36]. The shape functionals determine the polarizability tensor of the particle and hence the response of the confined electrons to an external field. Let us consider a charged particle in a 2D L-shaped box driven into oscillations along the x-axis (corresponding to an L-particle exposed to an x-polarized external EM field). The resonant wavelength of these oscillations is proportional to the average oscillation length $<x>$ of the particle inside the box and depends on the orientation of the particle with respect to the x-axis. The dependence of $<x>$ on the orientation exhibits two extrema, at $\alpha_{inc} = \pi/4$ and $3\pi/4$, corresponding to the blue (C) and red (D) bands. It is important to note that the two low-energy bands become degenerated in the limit of a parallelepiped shape.

The origin of the resonance labeled B in Fig. 13.13 is due to the sharp corners of the L-particle in FDTD simulations. It is well known that sharp features of metallic structures tend to accumulate surface charges and hence lead to strong EM field enhancements (the lightning rod effect [11]). Once sharp features in the particle are rounded, the B band completely disappears from the spectrum.

A particularly interesting result is the observation of the high-energy band, labeled A in Fig. 13.13. Its origin differs markedly from that of the features discussed above. Whereas the EM fields corresponding to resonances B, C, and D are mostly localized on the surface of the particle and can be readily explained in terms of surface plasmon modes, the EM field distributions in the case of the A resonance reveal complex volume oscillations. Figure 13.14 shows the xy-distribution of the steady-state EM intensity at the plane bisecting the particle for all four bands. It is evident that the EM energy is localized in the interior of the particle at the energy of band A. By contrast, similar plots at the energies of the remaining bands emphasize the surface character of the response. Figure 13.14*a* also shows remarkably large fields inside the particle, reaching almost a three-order of magnitude enhancements throughout the entire volume. Additional simulations, in which the height H of the particle was varied illustrate the strong sensitivity of the A resonance to the volume of the particle, whereas the lower-energy bands are essentially invariant to the volume. We also analyzed the density currents for all resonances and found that surface currents dominate the B, C, and D bands, whereas the volume current in the particle plays a dominant role in the high-energy band A. Although the behavior of the density currents at the energy of band A is similar to that of multipole oscillations, they differ qualitatively from the latter case in that (by contrast to the multipole modes [37]) they are not localized on the surface.

By numerically bending a metallic wire and calculating the scattering spectra as a function of the curvature of such arcuate particle, we verified that it is the breakdown of the inversion symmetry that gives rise to the volume plasmon modes, which are dipole-forbidden for the symmetric particles. Another interesting observation, which

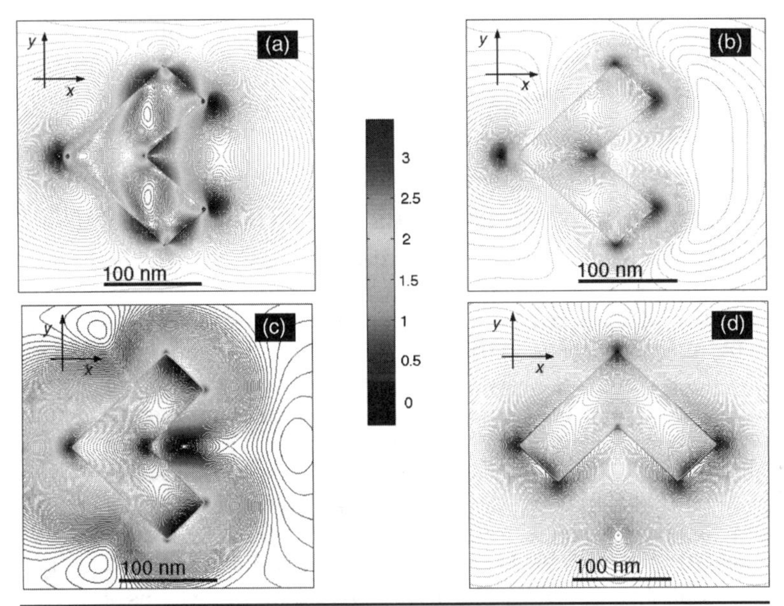

Figure 13.14 The steady-state EM intensity distribution of the plasmon eigenmodes in the *L*-particle at the plane bisecting the particle on a logarithmic scale. (*a*) The *A* resonance, (*b*) the *B* resonance, (*c*) the blue band, and (*d*) to the red band. The incident EM field is polarized along the *x*-axis with the propagation *k*-vector is perpendicular to the plane of the figure.

distinguishes the high-energy band from the lower-energy ones, is the number of oscillations per resonance lifetime—the resonance quality factor *Q*. We found that this ratio is about 3/4 for the *C* and *D* resonances, depending on the incident polarization, but noticeably larger (ca 8) for the volume plasmon band. We note that a similar high-energy band has been observed in recent experiments on crescent-shaped nanoparticles [38].

A natural extension of this work is to consider optics of the arrays of L-shaped particles. This task has been successfully accomplished in our recent work, where we applied both a computational approach and experimental measurements [39]. The experiments used polarization optical methods to describe the birefringence properties of 2D arrays of L-shaped silver nanoparticles. Among several intriguing features, we observed a relative phase delay of 30° between the two overlapping dipole resonant wavelengths, corresponding to the *C* and *D* bands in Fig. 13.13. Although the FDTD model predicted a larger effect of up to 105°, this might be due to the statistical variation of NP shapes in the experimental arrays. The arrays were fabricated by electron beam lithography, and the size of the particles was 145 and 155 nm in nominal total-edge length, 63 nm in arm width, and 30 nm in height. We note that the observed phase delay is significantly larger than the one for the conventional birefringence materials with the same thickness. This study suggests the possible application of 2D NP arrays or single particles as wavelength-tunable, extremely thin birefringence materials.

13.6 Genetic Algorithm Design of Advanced Plasmonic Materials for Nano-Optics

It has been recognized that the optical response of metallodielectric NPs and their arrays depends sensitively on shape and size, along with the NP's relative arrangement in the system [30]. The ability to experimentally control structural parameters calls for the development of a systematic design tool, which can be used on the one hand, as a general approach to construct metallodielectric nanostructures with predefined properties, and on the other, to gain some insight into SPP dynamics and propagation of light at the nanoscale.

In this section we consider the general problem of the interaction of EM fields with a nanoscale construct consisting of NPs, the optical response of which is determined by a set of adjustable parameters (material, shapes, sizes, etc.). The objective is defined in terms of specific properties, such as predefined EM field distributions in the vicinity of the structure or in terms of a desired propagation pathway of EM energy.

Figure 13.15 schematically depicts the optimization procedure. Structural and/or incident field parameters are used as inputs. Next,

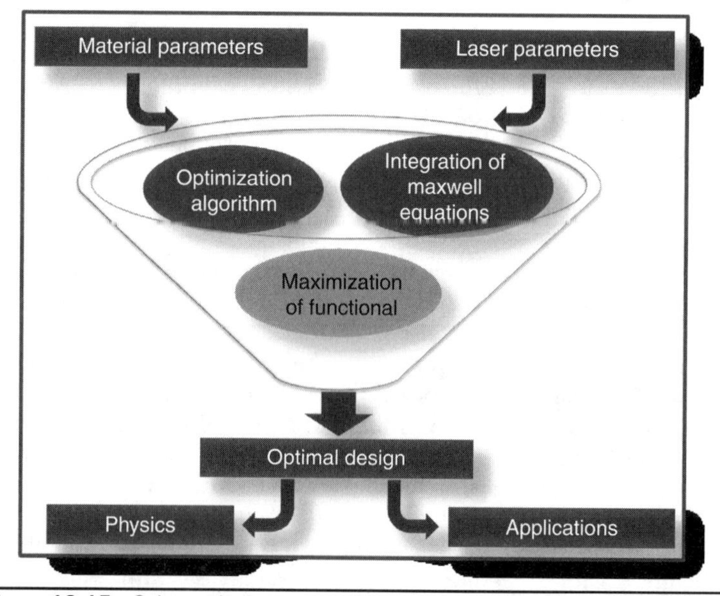

Figure 13.15 Schematical setup of the optimal design approach in nano-optics.

equipped with highly parallelized FDTD simulations, an optimization algorithm iteratively evaluates optical properties of the construct, adjusting input parameters in order to achieve a desired outcome, such as a particular EM field distribution. The output produces an optimal set of parameters, which is then analyzed in order to understand the physics behind an optimally designed system.

The choice of the optimization algorithm is somewhat arbitrary. However, it has to be emphasized that the optimization problems we encounter deal with a large parameter space, which would require fast convergence of the algorithm. In that perspective, utilization of evolutionary algorithms, such as the widely used genetic algorithm (GA), is more attractive than standard gradient methods [40]. Evolutionary algorithms belong to the class of search algorithms inspired by biological evolution by means of natural selection, these algorithms perform robustly on problems characterized by ill-behaved search spaces, where standard techniques usually fail. The stochastic nature of the GA makes it a numerically stable optimizer. However, as in most stochastic approaches, a large number of evaluations are necessary in order to find the global extremum. Incorporated with fast parallel FDTD codes, GAs are stable, robust, and very efficient optimizers [30, 31].

The choice of objectives is obviously spacious. It ranges from the determination of optimal shapes, sizes, and material of individual NPs to the design of optical nanodevices that can guide the light along predefined pathways. We consider a general nanoconstruct via which transmission of EM energy or the scattering of light depends on a set of adjustable parameters, including both field parameters (laser pulse duration, wavelength, polarization, pulse shape, etc.) and material parameters (size, shape, and relative arrangement of individual NPs). The standard GA procedure involves the following steps [40]:

1. Start by generating a set (population) of possible solutions, usually by choosing random values for all model parameters.
2. Evaluate the fitness function of each member of the current population.
3. Select pairs of solutions (parents) from the current population, with the probability of a given solution being selected made proportional to that solution's fitness.
4. Breed the two solutions selected in step 3, and produce two new solutions (the "offspring," or in some textbooks, "a new generation").
5. Repeat steps 3 and 4 until the number of offspring produced equals the number of individuals in the current population.
6. Use the new population of offspring to replace the old population.
7. Repeat steps 2 through 6 until a (problem-dependent) termination criterion is satisfied.

Each individual is encoded by one additional numerical parameter, the number of significant digits, referred to as the *number of genes*.

We are motivated in part by recent studies in which it was demonstrated that single L-shaped silver NPs and arrays thereof exhibited strong birefringence as discussed in Section 13.5. The interest in the development of nanoscaled materials with adjustable birefringence properties for applications in optics has been discussed in recent literature [39]. The combination of the GA with FDTD in full 3D, presents a numerical challenge since the 3D FDTD calculations that are repeatedly performed within the GA are numerically costly. Here an individual is evaluated by partitioning the FDTD grid onto 256 processors and implementing point-to-point communication MPI subroutines as discussed in Section 13.2. The average execution time of the GA iterations for 150 generations is close to 5 days on the BlueGene/L cluster at ANL [41].

We consider a periodic array of X-shaped silver NPs, as depicted schematically in Fig. 13.16, which is excited by an x-polarized laser

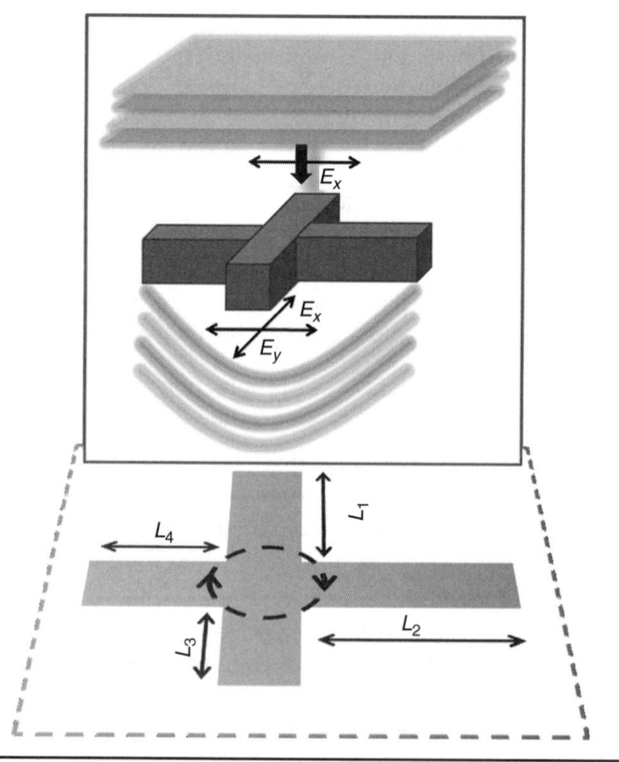

FIGURE 13.16 A schematic setup of a 3D birefringent material. A 2D array of X-shaped silver NPs is excited with a linearly polarized incident field at the normal incidence. Dashed lines represent periodic boundaries. Optimization parameters are a length of each arm of the particle, $L_{1,2,3,4}$, the angle positioning each NP in the array ϕ, and the incident wavelength λ.

field propagating along the z-axis. The target of the optimization is to enhance the birefringence of the periodic array, that is, to maximize the ratio of the time-averaged EM energy in the field component perpendicular to the incident field polarization, $W_y = E_y^2$, to the corresponding energy in the field component parallel to the incidence polarization, $W_x = E_x^2$. Clearly, the limit of $W_y/W_x \rightarrow 0$ corresponds to poor or no birefringence, whereas the opposite case of $W_y/W_x \rightarrow 1$ corresponds to equal magnitudes of the incident polarization and the field component that is purely induced by the array.

Figure 13.17 presents the evolution of the ratio W_y/W_x with respect to a number of GA iterations, and the inset shows the optimal unit cell of the array after 149 generations. For the given incident pulse duration, the optimized ratio exceeds 0.5 and so results in a strong depolarization of the EM field by the array of X-shaped NPs. Since the enhanced birefringence relies mainly on resonant excitation of LSPP waves, a further increase of the incident pulse duration at the optimal wavelength leads to higher W_y/W_x. We also note that the optimal incident wavelength, λ = 554.3 nm, obtained by the GA runs corresponds to the global maximum of W_y/W_x. An optimally designed array exhibits complex dynamics of EM near-fields. Simulations of EM eigenmodes of the array (not shown) reveal the basic concept of strong birefringence; namely, at the resonant wavelength, an incident EM plane wave excites LSPP waves that significantly absorb the

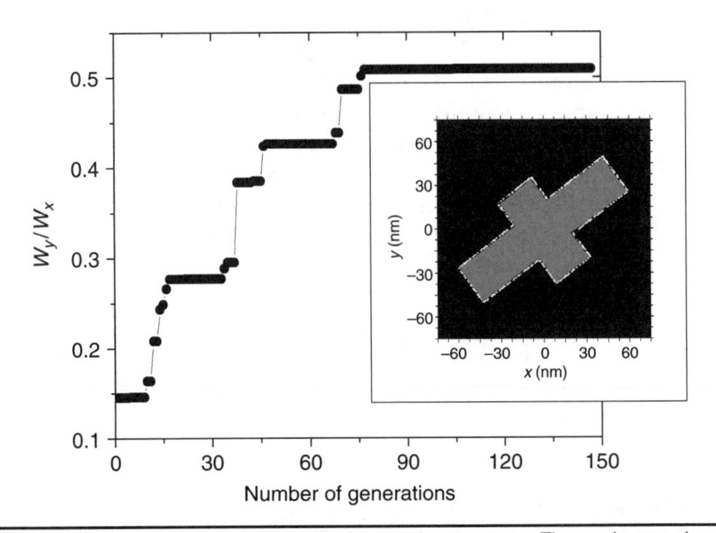

FIGURE 13.17 The GA-designed 3D plasmonic structure. The main panel shows the evolution of the fitness function, W_y/W_x, defined in the text with respect to a number of generations. The optimal unit cell in the xy-plane bisecting the array is depicted in the inset, where gray color corresponds to spatial regions occupied by metal and dark color represents the free space.

x-component of the light at the array's plane in a similar manner to the plasmonic crystal as discussed in Section 13.4, while the y-component is strongly induced by the combination of longitudinal and transverse LSPP modes confined within an NP's volume. It is then scattered in a direction lateral to the array's plane due to the low in-plane EM coupling between neighbor NPs in the array. Another interesting observation is a high relative phase between E_x and E_y in the far-field region. The phase along with the ratio W_y/W_x defines an actual polarization of the EM field that passed through the array. We also note that for the optimal parameters, the phase approaches 211° at the distance of 656 nm on the output side of the array and is almost independent on x and y. The Stockes parameters defining the polarization state of the resulting EM field are (1,0.24, −0.83, −0.50) corresponding to the left-elliptically polarized light [42].

13.7 Perfect Coupling of Light to Plasmonic Diffraction Gratings

Although the local fields associated with small structures such as arrays of NPs can be quite high, they are in practice limited by various effects. For intermediate sizes, finite mean-free-path effects enhance the losses. More generally, radiation losses limit the attainable fields. This suggests examining structures where radiation is suppressed or entirely eliminated. Such systems are thin metal films and metal gratings. When the metal film is sandwiched between two dielectric layers (so-called Otto geometry [18]) one can show that the system supports two SPP modes, which are characterized by their symmetry with respect to the midplane of the metal [43]. The electric field for the mode having the lower frequency is symmetric with respect to the film, that is, it points into and out of the film on opposing surfaces at a given instant. The high-frequency mode is antisymmetric. Since the field amplitude is largest within the metal for the symmetric mode, it leads to the largest dissipation. The mode is also soft in the sense that its frequency for a given wavelength falls rapidly as the film is thinned. This mode can be referred to as a *short-range mode* since it decays much faster than the antisymmetric mode. On the other hand, the antisymmetric mode, which has a node at the film midpoint, has a damping which falls as the film is thinned; in addition, the frequency associated with a given wavelength approaches a limiting value on thinning the film. Hence this mode is called a *long-range plasmon* mode referring to its very low damping in comparison with the short-range mode mentioned above.

In modeling, the overall behavior of the incoming beam in the symmetric dielectric environment, one can regard the system as a stack of dielectric layers and apply the Fresnel conditions at all interfaces, while simultaneously accounting for the propagation between interfaces. For a featureless, flat, metal film and fixed excitation frequency, the only method available is varying the thickness of the dielectric (coupling) layer between the coupling prism and the metal film. The total reflection coefficient can be calculated from the generalized Fresnel coefficients, accounting for the multiple reflections from the layers involved. The excitation of the SPP modes in the film is indicated by minima in the total reflection. The minimum corresponding to the long-range mode can then be tuned to 0 by adjusting the thickness of the dielectric layer. This is the critical coupling layer thickness and corresponds to perfect coupling or impedance matching. The results for the Q-factor of long-range SPP mode are shown in Fig. 13.18. Note that Q significantly increases as the silver layer thins and reaches four orders of magnitude at a thickness of 5 nm; this implies a corresponding buildup of the stored energy. The electric field, which scales as the square root of the energy density, is similarly enhanced.

For applications, it is important to consider the coupling of EM waves normal to a thin silver film that forms an oscillatory grating embedded between two otherwise uniform, semi-infinite half-spaces. Two grating structures considered are presented in Fig. 13.19, in the top one of which the midpoint of the Ag film remains fixed whereas the thickness varies sinusoidally and in the bottom one the midpoint oscillates sinusoidally whereas the film thicknesses remains fixed. Provided that the gratings do not significantly depart from flat films, we expect resonant wavelengths of the long- and short-range modes to be approximately given by the flat

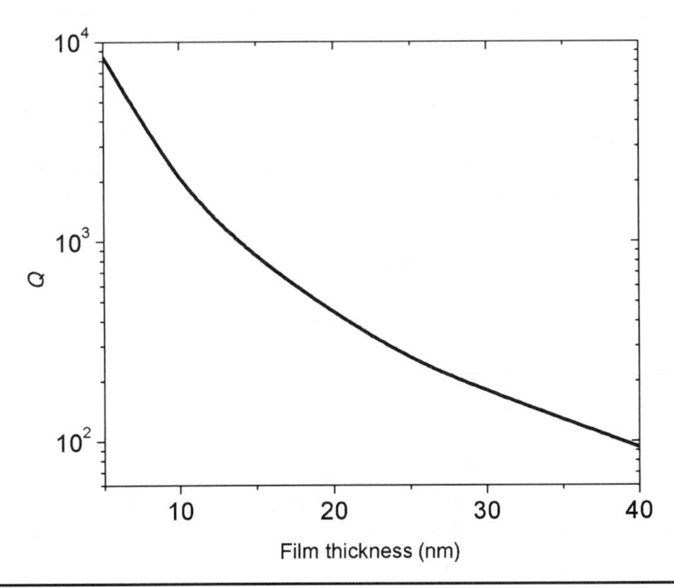

Figure 13.18 The Q-factor of the long-range SPP mode for the silver film as a function of the film thickness t in nm.

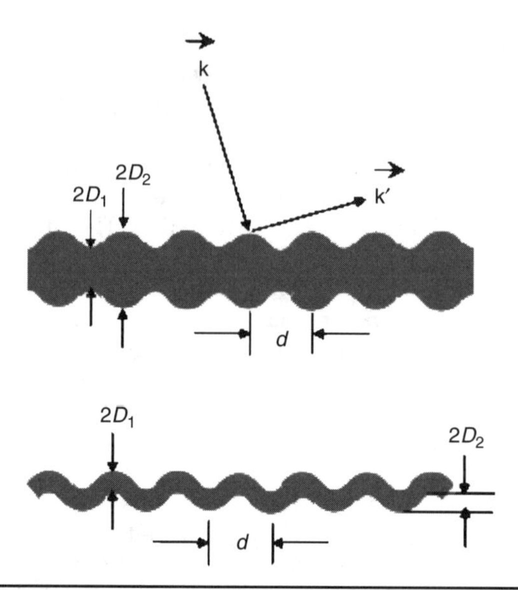

FIGURE 13.19 The setup of the out-of-phase (top) and the in-phase (bottom) sinusoidal gratings of the period *d* with parameters D_1 and D_2 defining the thickness and amplitude of the gratings.

film modes, which are governed by the film thickness as discussed above. For the in-phase gratings, the thickness is given by $2D_1$, and we can then use the parameter $2D_2$ to control the coupling [44]. For the out-of-phase sinusoids the resonant wavelength would, in the first approximation, be governed by the average thickness, $2(D_1 + D_2)$; the difference, $2(D_1 - D_2)$, can then be adjusted to control the coupling in order to find structural parameters that result in impedance-matching conditions. At such conditions incident EM radiation is efficiently transformed into propagating surface plasmons, this leads to strong field enhancements on the surface of the gratings. Moreover, it is possible to control which mode is excited by sending incident plane waves from both sides of the grating simultaneously and adjusting their relative phase delays (we refer to this scheme as *double-ended excitation*). Owing to the antisymmetric nature of the long-range mode we use the phase delay of 0 or π depending on the symmetry of the gratings (see Fig. 13.19 for details). This allows us to excite solely the desired mode.

Fig. 13.20*a* presents intensity enhancement factors averaged over the metal surface for the out-of-phase and in-phase sinusoidal gratings under the double-ended excitation conditions (here we consider only long-range modes). We note that high-EM fields, significantly enhanced at the impedance-matching conditions, are not localized near specific spatial regions, as in the case of tip-enhanced optical probes [21], for instance, but are rather distributed over the surface of

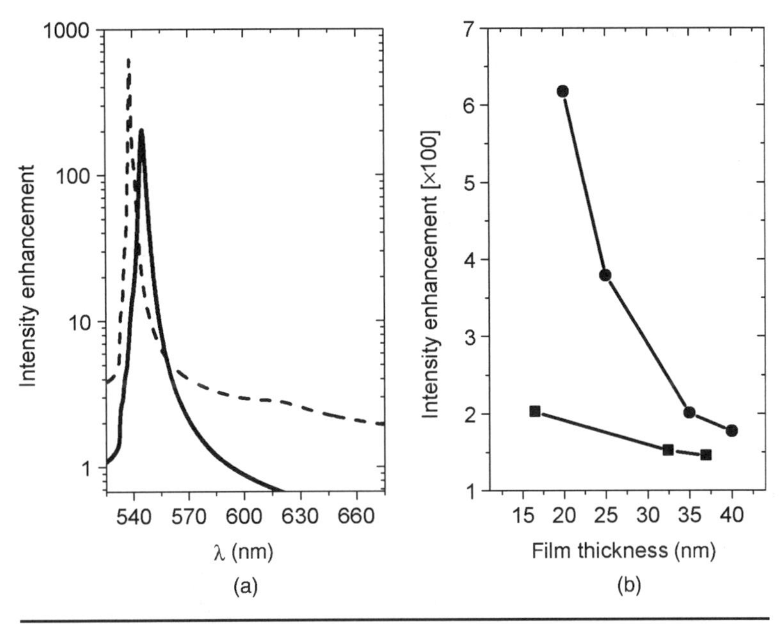

Figure 13.20 (*a*) Surface-averaged intensity enhancement as a function of the incident wavelength for out-of-phase (solid line) and in-phase (dashed line) silver gratings at the impedance matching conditions for the long-range mode (see Fig. 13.19 for details). (*b*) Surface-averaged intensity enhancement as a function of the silver film thickness at the impedance-matching conditions for long-range modes for the out-of-phase (squares) and in-phase (circles) sinusoidal gratings.

the grating. Following the results for the flat films, we simulated the enhancement for several grating thicknesses as shown in Fig. 13.20*b*. The results confirm our general conclusions that with thinner films, intensity enhancements are much higher as follows from the high-quality factors *Q*.

In addition to the discussed numerical simulations, we have also fabricated line gratings from periodically etched fused silica on which a thin silver film is deposited that is in turn covered with a silica-matched fluid [45]. This dielectrically symmetric geometry supports independent long- and short-range SPP. The associated plasmonic band structure has been probed. By extrapolating the angular dependence of the position long- and short-range SPP lines, a small band gap (associated with the second Brillouin zone point) is determined, which arises from an asymmetry of the grating amplitude. The experimental results are in good agreement with FDTD simulations.

We recall that the large intensity enhancements for our grating structures were calculated on a per-unit-area basis. This property is particularly important when considering sensor applications involving fluorescence-based spectroscopy or Surface-Enhanced Raman

Spectroscopy (SERS) where we assume we have no control over where the target molecule will actually attach. It is often assumed that the enhancement of SERS involves the square of the intensity; while taking no position on this issue, we note that the surface averaged square of the intensity is also greatly enhanced. Although various plasmon resonant nanostructures can have "hot spots" with rather large intensity enhancements, exploiting these requires that the target molecules "find" these hot spots; this in turn, requires effective mixing or long exposure times. If receptors are involved, we would want them attached only on the hot spots; if they are widely distributed, they will compete with the hot spots for target molecules and thereby deplete the latter.

13.8 Conclusion

The research areas of nano-optics in general and plasmonics in particular have benefitted from recent progress in fabrication techniques such as electron-beam lithography [46]. In addition, applications of lasers have been extended to subwavelength scales. Combined together, they have resulted in an enormous growth of applications of nanoscaled materials in various fields. As stated in previous sections, owing to the coherent nature of SPP resonance and its great sensitivity to a wide range of parameters we now have the tools to manipulate EM radiation far beyond the diffraction limit. We have the ability to focus light to tiny local spots which may be used to probe individual atoms and molecules [47]. By using polarization dependence of SPP waves, a specifically designed plasmonic material has the capability to efficiently guide EM radiation along predefined pathways [30]. Due to material dispersion of metals, we may also use plasmonic systems to shape laser pulses aiming to coherent control [48]. Moreover, a new field, namely, molecular nanopolaritonics [49], studying the molecular influence on field propagation, is used as a tool for developing molecular switches [50]. The latter utilizes nonadiabatic alignment of a molecule on a semiconductor surface under a tip of a scanning tunneling microscope. Recent developments in experimental techniques capable of measuring the optical response of current-carrying molecular junctions [51] have led to theoretical formulations suitable for simultaneous descriptions of both transport and optical properties of molecular devices [52].

All these and other intriguing applications, however, mainly utilize the linear optical response of SPP resonance. Most of the research has been centered at linear optics of plasmonic structures [10] with several attempts to go beyond linear regime and consider nonlinear properties of materials including noble metal NPs aiming for a wide variety of nonlinear phenomena to appear [53–57]. A newly emerging research field of nonlinear plasmonics takes two obvious routes, namely, combining linear plasmonic systems with highly nonlinear media [58] and taking into account strongly inhomogeneous EM fields associated with surface plasmons that lead to spatial dependence of conductive electron density in metals and hence result in nonlinear phenomena such as second-harmonic generation [59]. The first route deals with nonlinear media at the macroscopic level where even a single atom or molecule located near the plasmonic structure at resonant conditions may be considered as a nonlinear optical system due to high plasmon near-fields [60]. For example, it is well known that a two-level atom exposed to strong resonant EM radiation has an average dipole moment that may be written as a sum over odd powers of the electric field amplitude [61]. Such a dipole coupled to plasmon fields via polarization currents in Maxwell equations results in a nonlinear optical system and obviously is able to lead to new venues of research now including nonlinear optics. It is thus important to develop an approach capable of capturing both the size effects at the nanoscale and the time dynamics of EM fields. Generally speaking, such an approach has to take into account Maxwell equations capturing EM fields and the quantum dynamics of atoms or molecules in the vicinity of plasmonic materials. The latter can be accomplished by employing Bloch formalism [62]. The resulting system of coupled Maxwell–Bloch equations will contain the most accurate description of nonlinear phenomena driven by surface plasmon waves.

References

1. E. Hutter and J. H. Fendler, *Adv. Mat.* 16:1685, 2004.
2. R. E. Slusher, *Rev. Mod. Phys.* 71:00S471, 1999.
3. A. H. Haes and R. P. van Duyne, *Expert Rev. Mol. Diagn.* 4:527, 2004.
4. P. Alivisatos, *Nat. Biotech.* 22:47, 2004.
5. R. P. Van Duyne, *Science* 306:985, 2005.
6. L. Novotny and B. Hecht, *Principles of Nano-Optics*, Cambridge University Press, Cambridge, England, 2006.
7. S. A. Maier and H. A. Atwater, *J. Appl. Phys.* 98:011101, 2005.
8. E. Ozbay, *Science* 311:189, 2006.
9. W. A. Murray and W. L. Barnes, *Adv. Mat.* 19:3771, 2007.
10. S. A. Maier, *Plasmonics: Fundamentals and Applications*, Springer, 2007.
11. P. K. Aravind, A. Nitzan, and H. Metiu, *Surf. Sci.* 110:189, 1981.
12. M. Quinten, A. Leitner, J. R. Krenn, and F. R. Aussenegg, *Opt. Lett.* 23:1331, 1998.
13. M. Born and W. Wolf, *Principles of Optics: Electromagnetic Theory of Propagation, Interference, and Diffraction of Light*, Pergamon Press, 1965.
14. U. Kreibig and M. Vollmer, *Optical Properties of Metal Clusters*, Springer, New York, 1995.
15. S. A. Maier, P. G. Kik, H. A. Atwater, S. Meltzer, E. Harel, B. E. Koel, and A. G. Requicha, *Nature Mat.* 2:229, 2003.
16. J. I. Gersten and A. Nitzan, *J. Chem. Phys.* 73:3023, 1980.
17. H. Raether, *Surface Plasmons on Smooth and Rough Surfaces and on Gratings*, Springer, Berlin, 1988.
18. A. Otto, *Z. Phys.* 216:398, 1968.
19. A. Taflove and S. C. Hagness, *Computational Electrodynamics: The Finite-Difference Time-Domain Method*, 3rd ed., Artech House, Boston, 2005.

20. W. H. P. Pernice, F. P. Payne, and D. F. G. Gallagher, *Opt. Quant. Electron.* 38:843, 2007.
21. M. Sukharev and T. Seideman, *J. Phys. Chem. A* 113:7508, 2009.
22. A. D. Rakić, A. B. Djurišić, J. M. Elazar, and M. L. Majewski, *Appl. Opt.* 37:271, 1998.
23. J. P. Berenger, "Perfectly Matched Layer (PML) for Computational Electromagnetics," ed. C.A. Balanis, Lecture #8 in Synthesis Lectures on Computational Electromagnetics, Morgan and Claypool, San Rafel, California, 2007.
24. J. A. Roden and S. D. Gedney, *Microw. Opt. Techn. Let.* 27:334, 2000.
25. W. Gropp, E. Lusk, and T. Sterling, *Beowulf Cluster Computing with Linux*, MIT Press, Boston, 2003.
26. G. Gay, O. Alloschery, B. Viaris de Lesegno, C. O'Dwyer, J. Weiner, and H. J. Lezec, *Nature Phys* 2: 262, 2006.
27. G. Gay, O. Alloschery, J. Weiner, H. J. Lezec, C. O'Dwyer, M. Sukharev, and T. Seideman, *Phys. Rev. E* 75:016612, 2007.
28. M. Besbes, J. P. Hugonin, P. Lalanne, S. van Haver, O. T. A. Jansse, A. M. Nugrowati, M. Xu, S. F. Pereira, H. P. Urbach, A. S. van de Nes, P. Bienstman, G. Granet, S. Helfert, M. Sukharev, T. Seideman, F. I. Baida, B. Guizal, and D. Van Labeke, *J. European Opt. Soc.* 2:07022, 2007.
29. M. L. Brongersma, J. W. Hartman, and H. A. Atwater, *Phys. Rev. B* 62:R16356, 2000.
30. M. Sukharev and T. Seideman, *Nano Lett.*, 6:715, 2006.
31. M. Sukharev and T. Seideman, *J. Chem. Phys.*, 124:144707, 2006.
32. "Nanoscale Materials Special Issue," *Acc. Chem. Res.* 32:387, 1999.
33. S. A. Maier and H. A. Atwater, *J. Appl. Phys.* 98:011101, 2005.
34. M. Sukharev and T. Seideman, *J. Chem. Phys.* 126:204702, 2007.
35. J. Sung, E. M. Hicks, R. P. van Duyne, and K. G. Spears, *J. Phys. Chem. C* 111: 10368, 2007.
36. M. Sukharev, J. Sung, K. G. Spears, and T. Seideman, *Phy. Rev. B* 76:184302, 2007.
37. R. Fuchs, *Phys. Rev. B* 11:1732, 1975.
38. H. Rochholz, N. Bocchio, and M. Kreiter, *New J. Phys.* 9:53, 2007.
39. J. Sung, M. Sukharev, E. M. Hicks, R. P. van Duyne, T, Seideman, and K. G. Spears, *J. Phys. Chem. C* 112:3252, 2008.
40. R. L. Haupt and S. E. Haupt, *Practical Genetic Algorithms*, John Wiley, New York, 2004.
41. J. Yelk, M. Sukharev, and T. Seideman, *J. Chem. Phys.* 129:064706, 2008.
42. H. C. van de Hulst, *Light Scattering by Small Particles*, Dover, New York, 1981.
43. D. Sarid, *Phys. Rev. Lett.* 47, 1927, 1981.
44. M. Sukharev, P. R. Sievert, T. Seideman, and J. B. Ketterson, *J. Chem. Phys.* 131:034708, 2009.
45. W. Mu, D. B. Buchkolz, M. Sukharev, J. Jang, R. P. H. Chang, and J. B. Ketterson, *Opt. Lett.* submitted 2009.
46. C. R. K. Marrian, E. A. Dobisz, and J. A. Dagata, *J. Vac. Sci. Technol B* 10:2877, 1992.
47. A. Cerezo,T. J. Godfrey, and G. D. W. Smith, *Rev. Sci. Instrum.* 59:862, 1998.
48. M. Aeschlimann, M. Bauer, D. Bayer, T. Brixner, F. J. García de Abajo, W. Pfeiffer, M. Rohmer, C. Spindler, and F. Steeb, *Nature* 446:301, 2007.
49. K. Lopata and D. Neuhauser, *J. Chem. Phys.* 130:104707, 2009.
50. M. G. Reuter, M. Sukharev, and T. Seideman. *Phys. Rev. Lett.* 101:208303, 2008.
51. S. W. Wu, G. V. Nazin, and W. Ho, *Phys. Rev. B* 77:205430, 2008.
52. M. Sukharev and M. Galperin, *Phys. Rev. B* (submitted, 2009), arXiv:0911.2499.
53. N. -C. Panoiu and R. M. Osgood, Jr., *Nano Lett.* 4:2427, 2004.
54. R. V. Hanglund Jr., in *Photon-Based Nanoscience and Nanobiotechnology*, ed. J. J. Dubowski and S. Tanev, Springer, New York, 2006.
55. S. Palomba and L. Novotny, *Phys. Rev. Lett.* 101:056802, 2008.
56. N. Kroo, S. Varro, G. Farkas, P. Dombi, D. Oszetzky, A. Nagy, and A. Czitrovszky, *J. Modern Opt.* 55:3203, 2008.
57. G. A. Wurtz and A.V. Zayats, *Laser Phot. Rev.* 2:125, 2008.

58. A. V. Krasavin, K. F. MacDonald, A. S. Schwanecke, and N. I. Zheludev, *Appl. Phys. Lett.* 89:031118, 2006.
59. Y. Zeng, W. Hoyer, J. Liu, S. W. Koch, and J. V. Moloney, *Phys. Rev. B* 79:235109, 2009.
60. D. Neuhausera and K. Lopata, *J. Chem. Phys.* 127:154715, 2007.
61. L. Mandel and E. Wolf, *Optical Coherence and Quantum Optics,* Cambridge University Press, Cambridge, England, 1995.
62. G. Slavcheva, J. M. Arnold, I. Wallace, and R. W. Ziolkowski, *Phys. Rev. A* 66:063418, 2002.

CHAPTER 14

Plasmon Resonance Energy Transfer Nanospectroscopy

Logan Liu

Department of Electrical and Computer Engineering
University of Illinois at Urbana–Champaign

14.1 Introduction

Nanoparticle plasmon resonance [1, 2] is distinctive from the propagating surface plasmon resonance on metallic thin film [3]. It is spatially confined within the physical boundary of the nanoparticle; however, it can be continuously transferred to adjacent metallic nanoparticles through plasmon coupling [4, 5]. It has also been conjectured for a long time that the plasmon resonance energy can be transferred to chemical or biological molecules adsorbed on metallic nanostructures [6, 7]. This chapter reports the highest-sensitivity ever nanoscopic biomolecular absorption spectroscopy enabled by the first-time direct observation of plasmon resonance energy transfer (PRET) from a single metallic nanoparticle to a conjugated biomolecule. Due to PRET, quantized plasmon quenching dips are observed in single nanoparticle scattering spectra which correspond to the absorption spectral peaks of conjugated biomolecules. The hybrid nanoparticle–biomolecule PRET system allows near-single molecular sensitivity absorption spectroscopy with nanoscale spatial resolution for *in vivo* functional molecular imaging.

Although not directly observed before, PRET has accounted for one of the possible explanations for surface-enhanced Raman scattering [8, 9] and fluorescence [10] on single nanoparticles. Experimentally, the plasmon resonance of gold and silver nanoparticles conjugated with various biomolecules such as DNA [11], peptides [12], and biotin-streptavidin [13] has been studied by single-particle Rayleigh scattering spectroscopy [14, 15]. All these previous studies demonstrated the shift of a plasmon resonant wavelength by changing dielectric medium due to structural changes of biomolecule conjugated on the surface of single metallic nanoparticles. Since most of the previous cases have conjugated biomolecules with optical absorption or electronic resonance peaks in ultraviolet (UV) or a far-infrared range on gold and silver nanoparticles with visible plasmon resonance peaks, only the shift of the plasmon resonance peak was observed. However, in this chapter, metalloprotein Cytochrome c (Cyt c) was conjugated on a single 30-nm gold nanoparticle in order to observe the direct quantized plasmon resonance energy transfer from the nanoplasmonic particle to Cyt c (Fig. 14.1).

14.2 PRET-Enabled Biomolecular Absorption Spectroscopy

The intentional overlap of the absorption peak positions of desired biomolecules with the plasmon resonance peak of the metallic nanoparticle generates distinguishable spectral quenching dips on the Rayleigh scattering spectrum of a single nanoparticle; this also allows near-single molecular-level nanoscopic absorption spectroscopy (Figs. 14.1 and 14.2). Cyt c, a metalloprotein in the cellular mitochondria membrane, acts as the charge transfer mediator [16] and plays a crucial role in bioenergy generation, metabolism, and cell apoptosis [17]. Unlike many other proteins, Cyt c has several optical absorption peaks in a visible range of around 550 nm, coinciding with the 30-nm gold nanoparticle plasmon resonance, and more importantly it is a natural energy acceptor with electron tunneling channels [18]. Similar to the donor-acceptor energy matching in Fluorescent (or Förster) Resonance Energy Transfer (FRET) between two fluorophores, the critical matching of the localized resonating plasmon kinetic energy E_p in gold nanoparticles with the electron transition energy from ground to excited state $E_e - E_g$ in Cyt c molecules permits the PRET process (Fig. 14.1*b*).

The quantized energy is likely transferred through the dipole-dipole interaction between the artificial alternating dipole—resonating the plasmon in the nanoparticle and the biomolecular dipoles. Previous work on the surface plasmon-mediated FRET process [19, 20], superlens imaging [21], and surface plasmon resonance shift of redox molecules [22], and very recent work on bulk optical extinction spectroscopy of nanoplasmonic particle clusters with conjugated resonant molecules [23] also indicates the possibility of such dipole-dipole interaction. Distinctively in this experiment, the plasmon energy quenching of the nanoparticle due to PRET is represented as the "spectral dips" in the single-nanoparticle scattering spectrum (Figs. 14.1*c* and 14.2) and the position of the dips

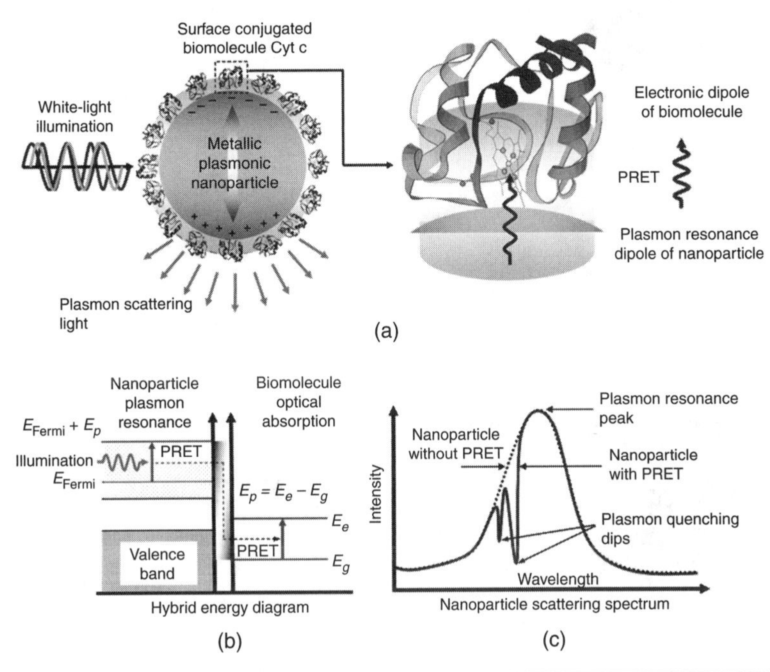

Figure 14.1 Schematic diagram of a PRET-enabled nanoscopic biomolecular absorption spectroscopy. (*a*) PRET from a single metallic nanoparticle to surface-conjugated biomolecules. The wavelength-specific plasmon resonance (collective free-electron oscillation) in a metallic nanoparticle is excited by white-light illumination. The plasmon resonance dipole can interact with the biomolecular dipole and transfer energy to biomolecules. (*b*) Hybrid energy diagram showing the quantized energy transfer process. With optical excitation, the free electrons in the conduction band of the metallic nanoparticle are elevated from Fermi to a higher energy level forming resonating plasmon. The plasmon resonance energy is transferred to the metalloprotein biomolecules (i.e., Cyt c) conjugated on the nanoparticle surface when matched with the electronic transition energy in biomolecule optical absorption. (*c*) Typical Rayleigh scattering spectrum of the single PRET probe. The energy transition in PRET is represented as quenching dips in the nanoparticle scattering spectrum, and the dip positions correspond to the biomolecule optical absorption peaks.

match with the molecular absorption peak positions (Figs. 14.1*c* and 14.2). PRET is a direct energy transfer process and thus much more efficient and faster than optical energy absorption [10], so the absorption spectral peaks of conjugated Cyt c molecules on single gold nanoparticles can be detected with a simple optical system, which is impossible using conventional visible absorption spectroscopic methods.

The Cyt c conjugated 30-nm gold nanoparticles are dispersedly tethered on the surface of a transparent glass slide. The glass slide is mounted on a white light dark-field microscopy system with a true-color camera and a spectrometer to characterize the scattering image

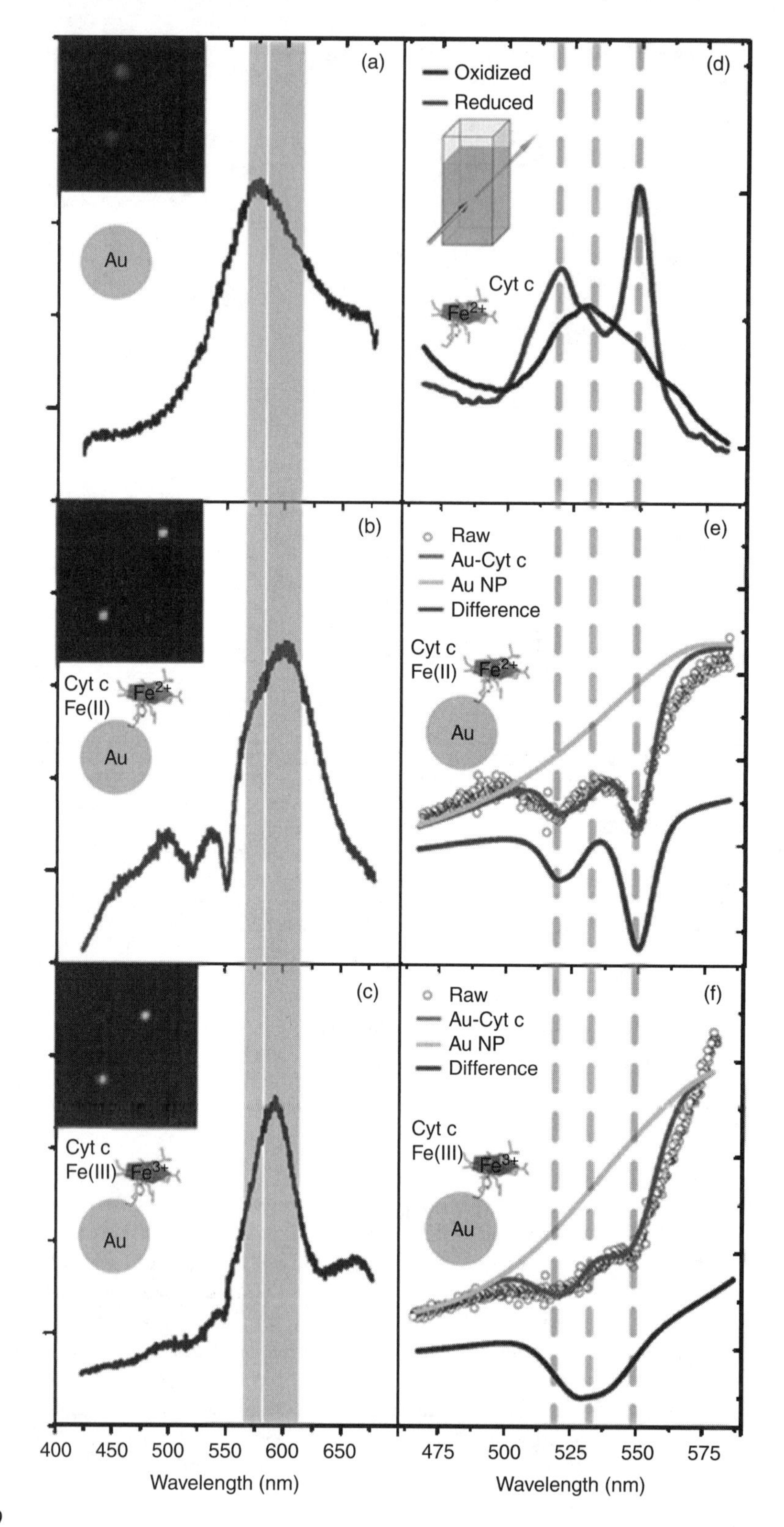
(a)
Au
(b)
Cyt c
Fe(II)
Fe^{2+}
Au
(c)
Cyt c
Fe(III)
Fe^{3+}
Au
400
450
500
550
600
650
Wavelength (nm)
(d)
Oxidized
Reduced
Cyt c
Fe^{2+}
(e)
Raw
Au-Cyt c
Au NP
Difference
Cyt c
Fe(II)
Fe^{2+}
Au
(f)
Raw
Au-Cyt c
Au NP
Difference
Cyt c
Fe(III)
Fe^{3+}
Au
475
500
525
550
575
Wavelength (nm)

FIGURE 14.2 Experimental results of PRET from single gold nanoparticle to conjugated Cyt c molecules. The Rayleigh scattering spectrum of a single gold nanoparticle coated with (*a*) only Cysteamine coating, (*b*) Cysteamine cross linker and reduced Cyt c (*c*) Cyteamine and oxidized Cyt c. The Rayleigh scattering spectrum was obtained using 1-second integration time. (*d*) The visible absorption spectra of the Cyt c bulk solution in reduction form (line with two peaks), and in oxidation form (line with one peak). (*e*) The fitting curve for the spectrum in (*b*). Black open circle, raw data; green solid line, fitting curve of the raw data; yellow solid line, Lorenzian scattering curve of bare gold nanoparticle; and red solid line, processed absorption spectra for the reduced conjugated Cyt c by subtracting the yellow curve from the green curve. (*f*) The fitting curve for the spectrum in (*b*). Open circle, raw data; line overlaid on the raw data, fitting curve of the raw data; lighter solid line, Lorenzian scattering curve of bare gold nanoparticle; and darker solid line, processed absorption spectra for the oxidized conjugated Cyt c by substrating the yellow curve for the green curve.

and spectrum of individual gold nanoparticles as well as the hybrid PRET probes (i.e., specific metallic nanoparticles with conjugated Cyt c molecules).

In comparison with the visible scattering spectrum of gold nanoparticles coated with only Cysteamine cross-linker molecules (Fig. 14.2*a*), the raw scattering spectra of gold nanoparticles conjugated with reduced (Fig. 14.2*b*) and oxidized Cyt c (Fig. 14.2*c*) show not only a scattering peak (plasmon resonance peak) but distinctive dips next to it. The spectral dips modulated on the nanoparticle scattering spectrum can be decoupled and converted into the visible absorption peaks of the reduced and oxidized Cyt c molecules (Fig. 14.2*e* and *f*). In accordance with the conventional visible absorption spectrum of bulk Cyt c solutions (Fig. 14.2*d*), the processed spectra have matched absorption peaks of reduced Cyt c around 525 and 550 nm and oxidized Cyt c at 530 nm [24]. Although the near-field optical excitation efficiency from the nanoparticle scattering light is much higher than the far-field optical excitation (i.e., the photon scattered from the nanoparticle is more likely to transmit through, and be absorbed by, the surface conjugated biomolecules than those far away from the nanoparticle), the optical absorption at 550 nm by the ferrocytochrome c molecule monolayer only accounts for 0.03% of the nanoparticle scattering light even for 100% excitation efficiency. (The absorption coefficient ε of horse heart Cyt c at 550 nm is 20.4 mM^{-1} cm^{-1}. For a completely packed monolayer of ~3 nm in diameter Cyt c molecules on a 30-nm gold nanoparticle surface, the local concentration C of Cyt c on the single nanoparticle is 47 mM. The optical path length L of a Cyt c monolayer is ~3 nm, so the optical absorbance is $A = \varepsilon CL$ $\sim= 3 \times 10^{-4}$ assuming 100% of scattering photon passing through the biomolecule monolayer.) Therefore, the dramatic spectral dips are not a result of the direct optical absorption of Cyt c molecules.

The energy-matching condition in PRET is further confirmed by three negative control experiments. For the first control experiment, synthesized peptides which have absorption peaks out of wavelength range of the plasmon resonance of a 30-nm gold nanoparticle are intentionally conjugated. As expected, the scattering spectrum of this hybrid system shows only the scattering peak because the absorption peaks of peptide do not coincide with the plasmon resonance spectrum of the nanoparticle (Fig. 14.3*a*). For the second control experiment, the importance of matching resonant frequency and molecular absorption peaks was tested by using a large gold nanoparticle cluster which has a plasmon resonance wavelength beyond 650 nm. As anticipated, the conjugated Cyt c absorption peaks can hardly be observed, including the 525 and 550-nm peaks for reduced Cyt c (Fig. 14.3*b*). For the third control experiment, dielectric polystyrene nanoparticles were conjugated with Cyt c and characterized. As expected, the plasmon-quenching spectral dips cannot be found on the scattering spectrum of the single dielectric polystyrene nanoparticle without plasmon resonance, even though the dielectric nanoparticle also scatters light; this indicates that the presence of excited free electrons is necessary for the PRET process (Fig. 14.3*c*).

The average surface density of Cyt c molecules on an individual gold nanoparticle is controlled by the molar concentration ratio used in the conjugation process. Considering the effective cross-sectional area of single Cyt c molecules and the surface area of a single 30-nm gold nanoparticle, maximally around 400 Cyt c, molecules can be tethered on a 30-nm gold nanoparticle. The scattering spectrum of many individual 30-nm nanoparticles was measured, and extracted the reduced Cyt c visible absorption peaks. Due to the nonuniformity of the surface molecule numbers on each particle and nanoparticle plasmon resonance wavelength, the plasmon-quenching or Cyt c absorption peak intensity shows variations from particle to particle (Fig. 14.4*a*), whereas the spectral measurement on each individual nanoparticle is repeatable and stable (Fig. 14.4*b*) and no photochemical changes are observed. An unpolarized white-light source is used in all of the above experiments.

Similar to the energy transfer process in FRET, the PRET efficiency is possibly dependent on the distance from the spectrally active moiety of biomolecules, for example, the Heme group for Cyt c, to the plasmonic nanoparticle surface as well as the relative orientations between the polarized plasmon resonance dipole and molecular dipoles (Fig. 14.5*a*). On the other hand, the simulated single nanoparticle PRET spectra (Fig. 14.5*b*) shows that the strongest plasmon resonance mode for a single Cyt c conjugated 30-nm gold nanoparticle occurs around 550 nm. This resonant frequency (i.e., energy-) matching condition potentially explains why the plasmon-quenching peak amplitude at 550 nm is relatively higher than at 525 nm (Fig. 14.2*d*) compared to the peak intensity ratio in a Cyt c bulk solution absorption measurement (Fig. 14.2*d*). The scattering peak wavelength of the gold nanoparticle in experiments is higher than the simulated results due

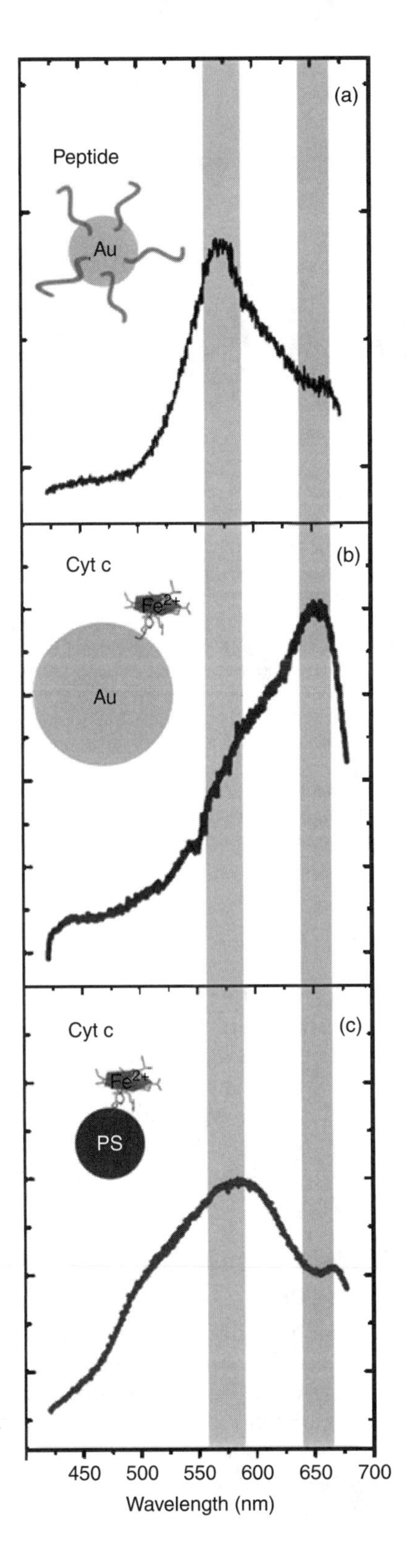

Figure 14.3 Negative control results showing the importance of critical-energy matching for PRET. (*a*) The scattering spectrum of a 30-nm gold nanoparticle coated with Cys-(Gly-Hyp-Pro)^6 peptides. (*b*) The scattering spectrum of a single large gold nanoparticle cluster conjugated with Cysteamine and Cyt c. (*c*) The scattering spectrum of a 40-nm amine-modified polystyrene bead conjugated with Cyt c.

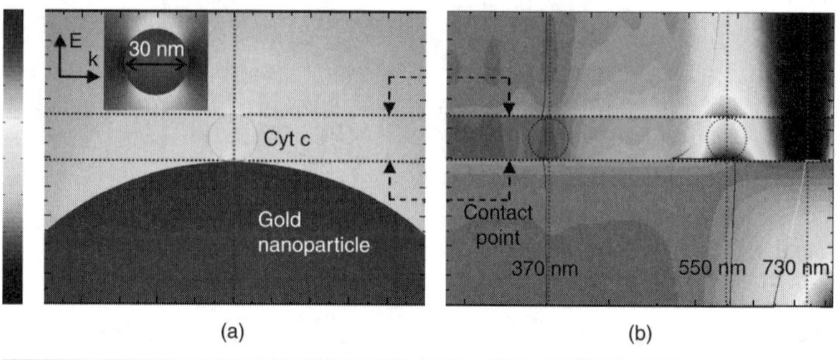

Figure 14.4 Simulation of nanoparticle plasmon resonance coupling to a single Cyt c molecule. (*a*) Time-averaged total electromagnetic (EM) energy at 550-nm vertically polarized light excitation around the interface of a single 30-nm gold nanoparticle and a single 3-nm spherical molecule. The dielectric nanosphere is used to simulate a single reduced Cyt c molecule with a wavelength-dependent complex refractive index [27]. The EM energy is transferred to the single molecule and forms the dipolar energy distribution across the molecule. The inset image of the whole nanoparticle shows that the energy coupling only occurs in the light polarization direction. (*b*) Time-averaged total energy profile at the cross-section line in (*a*) as the function of the excitation wavelength or energy. The energy distribution at each wavelength is normalized to the energy at the inner surface of the nanoparticle. The representative line plots of the energy profile at 370, 550, and 730 nm are superposed on the two-dimensional (2D) color-coded energy distribution at corresponding wavelength positions. The dipolar energy of nanoparticle plasmon resonance is electromagnetically coupled to the single molecule around 550 nm, while at other wavelengths, much less energy transfer is observed.

to larger numbers of Cyt c and Cysteamine molecules conjugated on the surface.

We also studied the PRET effect for hemoglobin molecules conjugated with single silver nanoparticles. The hemoglobin molecule has a distinctive absorption peak of around 407 nm which is the sorbet band of the Heme group. The absorption spectra of hemoglobin with and without binding to oxygen are shown in Fig. 14.6. Using the PRET nanosensor, we observed plasmon quenching dips corresponding to the Soret band (407 nm) of hemoglobin molecules on the surface of a single nanoparticle as shown in Fig. 14.7.

14.2.1 PRET Imaging of Nanoplasmonic Particles in Living Cells

The nanoplasmonic particles can be internalized into living cells in multiple ways, either by natural endocytosis of cells or by assisted endocytosis such as in the lipo-transfection technique. The nanoplasmonic particles can be effectively internalized in living cancer cells using Fugene 6 (Roche Diagnostics, Indianapolis, IN) transfection reagent [25]. The nanoplasmonic particles can be clearly imaged using our nanospectroscopy imaging system. Figure shows the dark-field

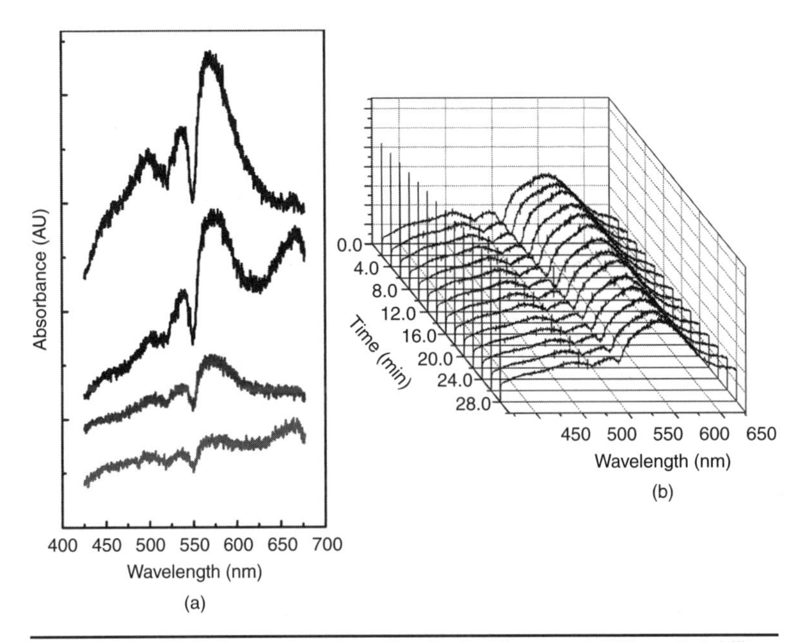

FIGURE 14.5 Multiple experiment results showing the repeatability and stability of PRET spectroscopic measurements. (*a*) Raw scattering spectra of four representative gold nanoparticles conjugated with reduced Cyt c molecules. The nanoparticle plasmon resonance peaks and PRET-induced plasmon-quenching dips have variable intensities from particle to particle due to the nonuniformity of the conjugated molecule number and particle geometry; however, the plasmon-quenching peak positions are consistent. More than 50 individual nanoparticles were tested, and PRET was consistently observed. (*b*) Time-lapse measurement of scattering spectra of a single gold nanoparticle conjugated with reduced Cyt c molecules. The plasmon-quenching spectral dips remain nearly constant during the whole time period of measurement. No photobleaching effect was observed.

scattering image of individual liver cancer cells with many internalized nanoplasmonic particles. A recent study also shows that intracellular Cyt c molecules can be imaged using PRET nanospectroscopy in a long-term continuous measurement as shown Fig. 14.8 [26].

14.2.2 Whole-Field Plasmon Resonance Imaging

All previous PRET-imaging works have been limited to the spectroscopy of several particles at most, and whole-field intracellular PRET imaging has not been implemented to date due to the limitations imposed by current spectral imaging systems. All existing imaging and spectroscopy systems for metallic nanostructures are based on a microscopy system with a white-light illumination source, a dark-field condenser, and a polychromator spectrograph as shown before. In such a system, the field of view is limited to several microns by the width of the entrance slit to the polychromator even though the field

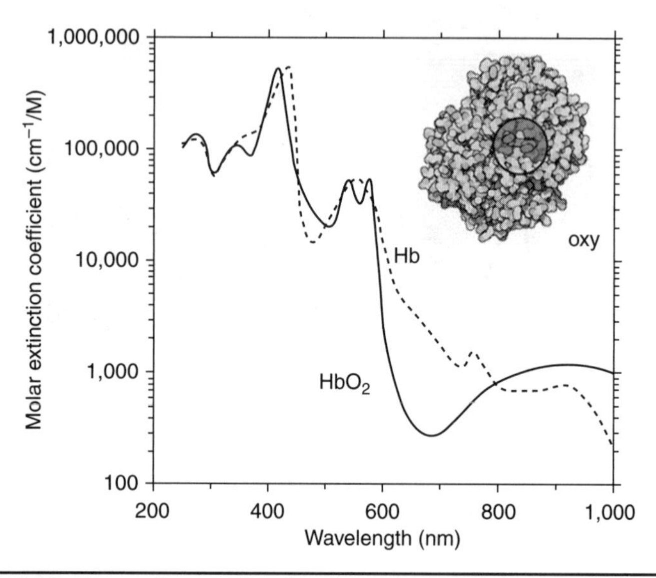

FIGURE 14.6 Absorption spectra of oxy-hemoglobin and deoxy-hemoglobin bulk high-concentration solutions.

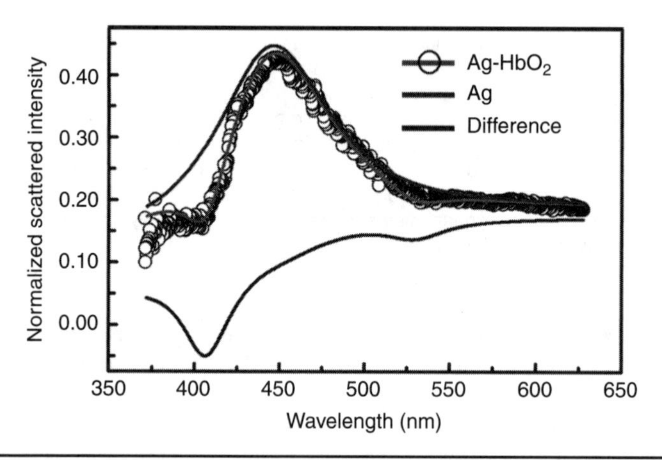

FIGURE 14.7 Absorption spectrum of hemoglobin molecules on single nanoparticle surface, and the spectrum is acquired and can only be acquired using our PRET nanomolecular sensor.

of view of the objective lens is on the order of several hundred microns; therefore only a few nanoparticles or nanostructures that are aligned in a column can be imaged spectrally at once.

The polychromator entrance slit is required to limit data pollution for the high-resolution spectroscopy of each individual nanoparticle. Hence in the case of spectrally imaging hundreds or even thousands of sparsely distributed nanostructures, the sample

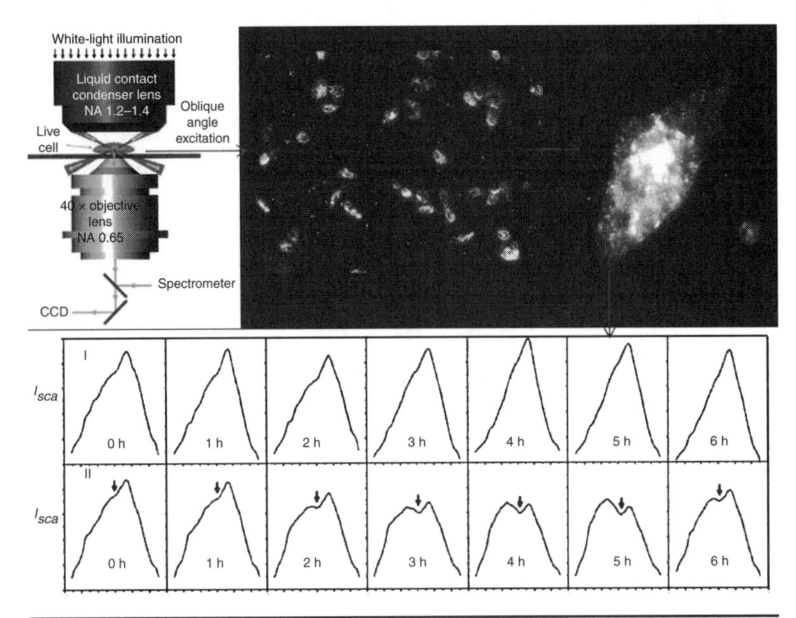

FIGURE 14.8 Dark-field imaging of PRET nanoparticles in living cancer cells and the time-lapse observation of intracellular Cyt c molecules using PRET nanospectroscopy [26].

stage of the microscopy system must be translated either manually or automatically; this has two disadvantages. First, the spectra of separated nanostructures cannot be simultaneously captured and the acquired data will be from different instants in time, for example, 20 to 30 minutes for a 100 × 100 array; this voids their applicability in the real-time monitoring of multiple biomolecular interactions. Second, the focal plane will tend to drift due to the stage translation which may change the condition of the immersion liquid between the sample and dark-field condenser lens, and thus manual refocusing is often required.

We constructed a new multispectral imaging system to simultaneously monitor the individual scattering spectra and plasmon resonance wavelength of large numbers of nanoplasmonic particles distributed within the field of view of a microscopy objective lens without mechanically scanning the sample. The multispectral imaging system currently supports frame rates as high as 2 seconds per frame (wavelength) that could potentially be increased by using a light source with higher power or an image detector with greater sensitivity, by which the image signal-to-noise ratio can also be increased. In contrast with the configurations of previous imaging systems, our whole-field plasmonic imaging system consists of a multispectral illumination source synchronized with an intensity-imaging camera rather than a polychromatic spectrograph. As shown in Fig. 14.9, the white light from a halogen lamp is

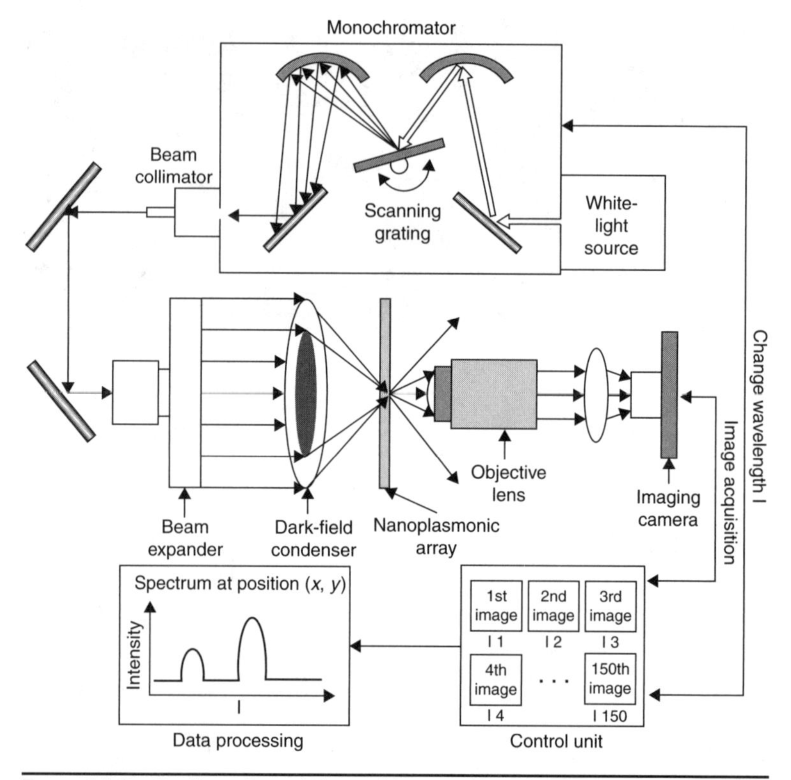

FIGURE 14.9 Configuration of the multispectral imaging system for the scattering spectra measurement of nanoplasmonic particles in the whole field of view.

coupled into a scanning monochromator which is controlled by a computer program to output a monochromatic light beam with varying wavelength and specified spectral step size. The monochromatic light beam is then focused and illuminated on the sample by a dark-field condenser (NA = 1.2 ~ 1.4). A 40 × microscopy objective lens (NA = 0.8) is used to collect the scattered light, which is captured by a Charge Couple Device (CCD) camera. The monochromator and image acquisition control software are integrated, and they are synchronized to capture a single image at each wavelength of interest. The image acquisition can be finished in a few minutes, and the acquisition time is dependent on the chosen spectral range and resolution.

All images are stored as gray-scale data files and are analyzed by an image-processing program. The bright-spot regions (typically 1 to 10 pixels) in each image are recognized by the analysis program as individual nanoplasmonic structures of interest. The mean intensity value of these small regions is extracted from the image at each wavelength as the raw scattering spectra data. The mean intensity value in a large empty (black) region is also measured at each wavelength as the background spectrum, which is subsequently subtracted from the raw scattering spectra. The difference spectra are then scaled according to the previously stored spectrum of the light output from the monochromator to yield the final scattering spectra. The process of the image analysis and the spectral data reconstruction is completely automated by the computer program.

As a demonstration, we use randomly dispersed Au colloidal nanoparticles on a glass slide as the imaging sample. The diameter of the Au nanoparticles vary from 20 to 80 nm, so their plasmon resonance wavelengths, and thus their scattering colors, are different. Figure 14.10*a* shows the true-color scattering image of the nanoparticles within ~1/10 of the whole view field of the objective lens. The true-color image is taken in the same dark-field microscopy system but with a white-light illumination source and color camera. Figures 14.10*b* and 14.2*c* show the scattering intensity images of the same nanoparticles within the same field of view taken by our system at 550 and 630 nm, respectively. Figure 14.10*d* shows the scattering spectra of three representative nanoparticles marked in Fig. 14.10*a*, *b*, and *c*. The plasmon resonance wavelengths (spectral peaks) of these three nanoparticles are respectively 560, 580, and 630 nm, which agree well with their scattering images in Fig. 14.10*a*, *b*, and *c*. The Au "particle" with red scattering color could be a cluster of a few Au nanoparticles, because the plasmon resonance wavelength of an individual Au nanoparticle is shorter than 600 nm according to Mie scattering theoretical predictions. Although only the scattering spectra of three typical particles are shown here, the spectral information for all the

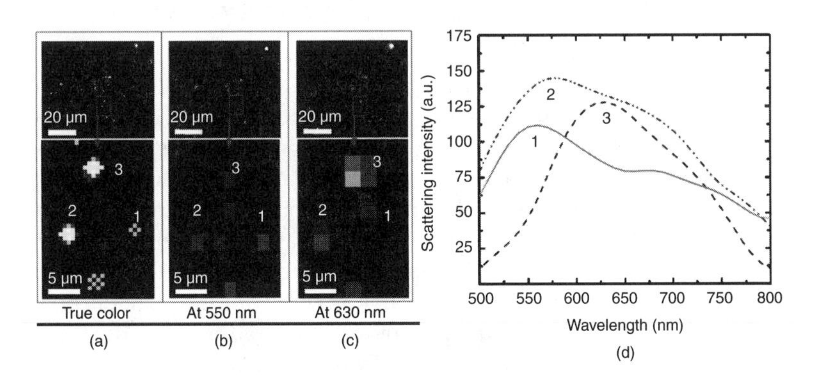

FIGURE 14.10 Multispectral images of nanoplasmonic particles and scattering spectra of selected particles. (*a*) True-color scattering image of thousands of dispersed Au nanoparticles. For the clarity of the image, a field of 100 × 100 μm is cropped from the whole view field of ~300 × 300 μm. The lower picture is the zoom-in image from the marked area (square) of the upper image. (*b*) and (*c*) The scattering intensity image of the same Au nanoparticles as in (*a*) with 550- and 630-nm monochromatic illumination, respectively. (*d*) Scattering spectra of three representative particles marked 1, 2, and 3, respectively, in the images.

other nanoparticles in the field of view are also stored at once and can be reconstructed in the same fashion.

14.3 Experimental Procedures

14.3.1 Preparation of Cyt c Conjugated Gold Nanoparticles on a Glass Slide

A cleaned glass slide was modified with 3-mercaptopropyl trimethoxy silane (MTS) by incubation in 1 mM MTS acetone for 24 hours. The glass slide was then rinsed with acetone, dried with clean nitrogen gas, and 30-nm spherical gold nanoparticles (Ted Pella, Inc., Redding, California) were cast on and wet the MTS functionalized glass surface. The gold nanoparticles were then immobilized by the free thiol groups. The surface was incubated in 0.1-mM Cysteamine solution for 2 hours. The resulting glass slide was thoroughly rinsed with Phosphate Buffer Solution (PBS) buffer to remove physically adsorbed Cysteamine, and then incubated in 10-μM horse heart Cyt c PBS solution (pH = 7.2) (Sigma, St. Louis, Missouri) for 40 minutes. Cysteamine has a thiol group at the one end to connect with gold, and an amino group at the other end to anchor the carboxyl groups in the peptide chain of Cyt c with 1-ethyl-3-[3-dimethyaminopropyl] carbodiimide hydrochloride (EDC) as the conjugation catalyst. The Cyt c molecules are in the oxidized form when purchased, and the reduced form of Cyt c is made by the addition of excess sodium dithionite ($Na_2S_2O_4$) in a deoxygenated PBS buffer solution.

14.3.2 Scattering Imaging and Spectroscopy of Single Gold Nanoparticles

The microscopy system consists of a Carl Zeiss Axiovert 200 inverted microscope (Carl Zeiss, New York, NY) equipped with a dark-field condenser (1.2 < NA < 1.4), a true-color digital camera (CoolSNAP cf, Roper Scientific, New Jersey), and a 300-mm focal-length and 300-grooves/mm monochromator (Acton Research, Massachusetts) with a 1024- × 256 pixel cooled spectrograph CCD camera (Roper Scientific). A few-micron-wide aperture was placed in front of the entrance slit of the monochromator to keep only a single nanoparticle in the region of interest at the grating dispersion direction. The true-color scattering images of gold nanoparticles were taken using a 40X objective lens (NA = 0.8) and the true-color camera with a white-light illumination from a 100-W halogen lamp. The scattering spectra of gold nanoparticles were taken using the same optics, but they were routed to the monochromator and spectrograph CCD. The immobilized nanoparticles were immersed in a drop of PBS buffer solution, which was deoxygenated by clean nitrogen gas; the buffer liquid also served as the contact fluid for the dark-field condenser. The distance between the condenser and nanoparticles was 1 to 2 mm. The microscopy system was completely covered by a dark shield, which prevents ambient-light interference and serious evaporation of the buffer solution.

14.3.3 Finite Element Simulation of Electromagnetic Energy Coupling in PRET

For the simulations of the electromagnetic (EM) energy distribution presented in Section 14.1, a commercial software package FEMLAB, available from Comsol Inc. (Los Angeles, California), was used, which numerically solves the Helmholtz equation for a set of predefined boundary conditions. The computation domain is a 1.2×3.0-μm square with all sides treated as matched low-reflection boundaries. The ambient refractive index of the domain was set to be the value for water in accordance with the experimental setup. The excitation source is a plane wave with its electric field oscillating in the plane of propagation. Although the simulated wave from the excitation source experiences diffraction over its propagation, its wavefront approximates that of a plane wave in the length scale of the nanoparticles under consideration. The refractive index of the gold nanoparticles is set to the values of bulk gold reported by P. B. Johnson and R. W. Christy [27]. In order to cope with sharp resonance peaks, interpolated values of the refractive index are used. A conjugated Cyt c molecule is simplified as a sphere, or a solid circle in 2D simulations. The real part of the refractive index of Cyt c molecules is assumed to be 1.6 as that of most of the other macromolecules. The imaginary part of the refractive index is calculated according the definition by Robin Pope and Edward Fry [28], $n''(\lambda) = \varepsilon\lambda/4\pi$, where ε is the linear absorption coefficient of Cyt c and λ is the wavelength. Triangular elements are used for the computation mesh. The built-in mesh generator is used to regulate the mesh size in simulating different geometries. The distribution of local EM energy distribution is obtained from the built-in plotting function of FEMLAB and MATLAB®.

14.4 Conclusions

Although only the PRET in the visible wavelength range is observed here due to the optical properties of gold nanoparticles and Cyt c molecules, the PRET process at UV and near-infrared range could be envisioned by using different properties (i.e., size, shape, free-electron density, etc.) of metallic nanoparticles with UV or near-infrared plasmon resonance wavelength. The PRET-based ultrasensitive biomolecular absorption spectroscopy on a single metallic nanoparticle could be used for molecular imaging such as genetic analysis of small copies of latent nucleotides, activity measurements of small numbers of functional cancer biomarker proteins, and rapid detection of little biological toxin, pathogen, and virus molecules. Additionally, PRET could be applied in intracellular biomolecule conjugated nanoparticle sensors to detect localized *in vivo* electron transfer, oxygen concentration, and pH-value changes in living cells with nanoscale spatial resolutions. Furthermore the optical energy in advanced plasmonic devices could be potentially tuned by functional biomolecules by taking advantage of the PRET process [29].

References

1. C. F. Bohren, and D. R. Huffman, *Absorption and Scattering of Light by Small Particles*, 335–336, Wiley, New York, 1998.
2. J. J. Mock, D. R. Smith, and S. Shultz, "Local Refractive Index Dependence of Plasmon Resonance Spectra from Individual Nanoparticles," *Nano Lett.* 3: 485–491, 2003.
3. A. Otto, "Excitation of Nonradiative Surface Plasma Waves in Silver by the Method of Frustrated Total Reflection," *Z. Phys.* 216: 398–410, 1968.
4. S. A. Maier, P. G. Kik et al., "Local Detection of Electromagnetic Energy Transport below the Diffraction Limit in Metal Nanoparticle Plasmon Waveguides," *Nat. Mater.* 2: 229–232, 2003.
5. S. A. Maier, and H. A. Atwater, "Plasmonics: Localization and Guiding of Electromagnetic Energy in Metal/Dielectric Structures," *J. Appl. Phys.* 98: 011101, 2005.
6. M. Moskovits, "Surface-Enhanced Spectroscopy," *Rev. Mod. Phys.* 57: 783–826, 1985.
7. J. R. Lombardi, R. L. Birke, T. H. Lu, and J. Xu, "Charge-Transfer Theory of Surface-Enhanced Raman Spectroscopy—Herzberg-Teller Contributions," *J. Chem. Phys.* 84: 4174–4180, 1986.
8. S. Nie, and S. R. Emory, "Probing Single Molecules and Single Nanoparticles by Surface-Enhanced Raman Scattering," *Science* 275: 1102–1106, 1997.
9. K. Kneipp et al., "Single Molecule Detection Using Surface-Enhanced Raman Scattering (SERS)," *Phys. Rev. Lett.* 78: 1667–1670, 1997.
10. P. Das, and H. Metiu, "Enhancement of Molecular Fluorescence and Photochemistry by Small Metal Particles," *J. Phys. Chem.* 89: 4680–4687, 1985.
11. G. L. Liu et al., "A Nanoplasmonic Molecular Ruler for Measuring Nuclease Activity and DNA Footprinting," *Nat. Nano.* 1: 47–52, 2006.
12. T. Endo, Kerman, K. N, Nagatani, Y. Takamura, and E. Tamiya, "Label-Free Detection of Peptide Nucleic Acid-DNA Hybridization Using Localized Surface Plasmon Resonance Based Optical Biosensor," *Anal. Chem.* 77: 6976–6984, 2005.
13. G. Rascheke et al., "Biomolecular Recognition Based on Single Gold Nanoparticle Light Scattering," *Nano Lett.* 3: 935–938, 2003.
14. C. Sönnichsen, "Plasmons in Metal Nanostructures," PhD Diss. University of Munich, Germany 2001.
15. M. A. van Dijk et al., "Absorption and Scattering Microscopy of Single Metal Nanoparticles," *Phys. Chem. Chem. Phys.* 8: 3486–3495, 2006.
16. D. S. Wuttke, M. J. Bjerrum, J. R. Winkler, and H. B. Gray, "Electron-Tunneling Pathways in Cytochrome-c," *Science* 256: 1007–1009, 1992.
17. J. Yang, X. S. Liu et al., "Prevention of Apoptosis by Bcl-2: Release of Cytochrome c from Mitochondria Blocked," *Science* 275: 1129–1132, 1997.
18. C. Lange, and C. Hunte, "Crystal Structure of the Yeast Cytochrome bc_1 Complex with Its Bound Substrate Cytochrome c," *Proc. Natl. Acad. Sci. USA* 99: 2800–2805, 2002.
19. P. Andrew, and W. L. Barnes, "Energy Transfer across a Metal Film Mediated by Surface Plasmon Polaritons," *Science* 306: 1002–1005, 2004.
20. R. P. van Duyne, "Molecular Plasmonics," *Science* 306: 985–986, 2004.
21. N. Fang, H. Lee, C. Sun, and X. Zhang, "Sub-Diffraction-Limited Optical Imaging with a Silver Superlens," *Science* 308: 534–537, 2005.
22. A. J. Bard, and L. R. Faulkner, *Electrochemical Methods: Fundamentals and Applications*, 685, Wiley, New York, 2nd ed., 2001.

23. E. Ozbay, "Plasmonics: Merging Photonics and Electronics at Nanoscale Dimensions," *Science* 311: 189–193, 2006.
24. S. Boussaad, J. Pean, and N. J. Tao, "High-Resolution Multiwavelength Surface Plasmon Resonance Spectroscopy for Probing Conformational and Electronic Changes in Redox Proteins," *Anal. Chem.* 72: 222–226, 2000.
25. E. S. Lee, G. L. Liu, F. Kim, and L. P. Lee, "Remote Optical Switch for Localized Control of Gene Interference," *Nano Lett.* 9: 562–570, 2009.
26. Y. Choi, T. Kang, and L. P. Lee, "Plasmon Resonance Energy Transfer (PRET)-Based Molecular Imaging of Cytochrome c in Living Cells," *Nano Lett.* 9: 85–90, 2009.
27. P. B. Johnson, and R. W. Christy, "Optical Constants of the Noble Metals," *Phys. Rev. B* 6: 4370–4379, 1972.
28. R. M. Pope, and E. S. Fry, "Absorption Spectrum (380–700 nm) of Pure Water. II. Integrating Cavity Measurements," *Appl. Opt.* 36: 8710–8723, 1997.
29. A. J. Haes, S. Zou, J. Zhao, G. C. Schatz, and R. P. van Duyne, "Localized Surface Plasmon Resonance Spectroscopy Near Molecular Resonances," *J. Am. Chem. Soc.* 128: 10905–10914, 2006.

CHAPTER 15

Erythrocyte Nanoscale Flickering: A Marker for Disease

Catherine A. Best

Department of Medical Cell and Structural Biology
College of Medicine
University of Illinois at Urbana–Champaign

15.1 Introduction to Hematology

15.1.1 Standard Blood Testing

Blood is the life-sustaining fluid that circulates through the heart and vasculature. Almost all patients have their blood tested because of the importance of detecting the presence of anemia and leukocyte (white blood cell, WBC) anomalies. Standard testing includes both quantitative and qualitative assessments. More specifically, routine analysis, by sophisticated automated analyzers, includes measurement of the complete blood count (CBC), which refers to the number of erythrocytes (red blood cells, RBCs), WBCs, and platelets per liter of blood; the hematocrit (the proportion of blood that is occupied by RBCs); the mean cell hemoglobin concentration (MCHC, the average concentration of hemoglobin in a given volume of RBCs); the mean cell hemoglobin (MCH, the average amount of hemoglobin in an average RBC); the average volume of the RBCs (the mean corpuscular value, MCV), and the red blood cell distribution width (RDW, which is a measure of RBC size variability). In addition, the analyzers generate a variety of histograms and scatter plots that give a visual representation of RBC population characteristics from which increased numbers of abnormal RBCs can be easily detected. Classifications of anemia and the identification of a number of other diseases are based on these parameters.

Microscopic examination of peripheral blood smears, a process whereby a thin layer of blood is spread on a glass slide to form a *blood film*, provides additional information regarding all the formed blood elements (RBCs, WBCs, and platelets). Blood smears are typically air-dried and stained with the dyes eosin and methylene blue (Wright), with or without Giemsa (Giemsa Wright), so that the detailed structure of individual cells can be examined. Wet preparations may also be prepared to ensure that the reported RBC abnormalities are not artifacts inadvertently introduced by the drying or staining procedures.

Peripheral blood smears are initiated by physicians in response to abnormalities detected in the CBC or in the context of other suggestive clinical features such as unexplained jaundice, sudden splenic enlargement, or unexpected severe infection, and also by laboratory staff in response to automated instrument "flags." A *flag* is reported whenever the hemoglobin concentration is unexpectedly low, for example.

Preparation and examination of peripheral blood smears is time- and labor-intensive, and is usually preformed by experienced and qualified laboratory technicians, hematologists, or pathologists. Staining accounts for more than 80% of the time and cost in smear preparation [1].

Microscopic examination of peripheral blood smears plays an important role in the rapid diagnosis of a number of disorders including infections, leukemia, and lymphoma and is essential in the differential diagnosis of anemia and thrombocytopenia (low platelet count). In addition, the examination of blood smears is particularly important in detecting diseases in which the CBC is normal. These diseases include compensated acquired hemolytic anemia, consumptive coagulopathy, hemoglobin C disease, hereditary spherocytosis, elliptocytosis, lead poisoning, malaria, severe infection, infectious mononucleosis, allergic reactions, multiple myeloma, chronic lymphocytic leukemia, and acute (early) leukemia [2]. The evaluation and diagnosis of these disorders are based on the observation of dysmorphic RBCs in peripheral blood smears.

15.1.2 RBC Morphology Assay

The RBC shape is dependent on a number of combined factors including the environment, its metabolic status, and age. The normal, resting mature RBC, the discocyte, appears as a smooth biconcave disk with an outer rim of dark hemoglobin and an inner pale central area. Normal RBCs are approximately 8 μm in diameter and have a volume of 90 fL and a surface area of 140 μm^2 (see Fig. 15.1*a*).

The abnormalities of RBC morphology that occur in disease include variations in RBC shape, size (dimensions), and staining

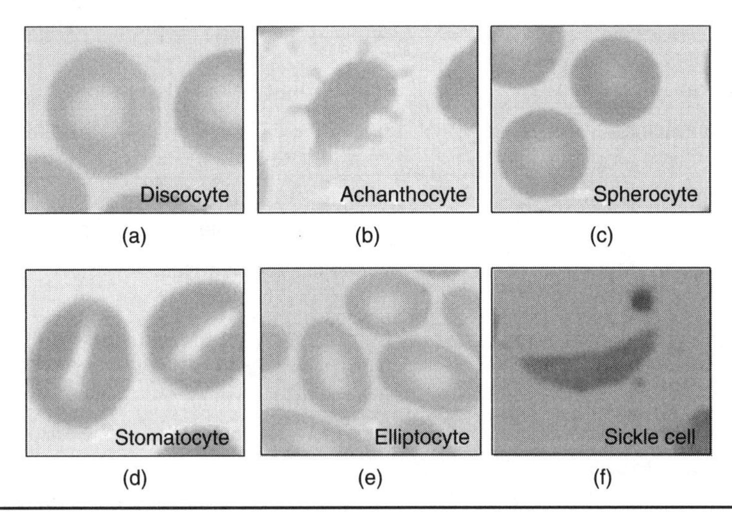

FIGURE 15.1 Red blood cell morphologies observed on a peripheral blood smear. (*a*) Discocyte. (*b*) Achanthocyte. (*c*) Spherocyte. (*d*) Stomatocyte. (*e*) Elliptocyte. (*f*) Drepanocyte (Sickle cell). Wright-Giemsa stain, 1000×. (Courtesy of Dr. Charles E. Hess. Division Hematology/Oncolgy, School of Medicine University of Virgina Charlottesville, Virginia.)

properties. Microcytic RBCs are small RBCs with MCVs less than 80 fL. Macrocytes are large RBCs that have MCVs greater than 100 fL. Hypochromic RBCs have a large, characteristic central pallor and a low MCHC (below 32 g per deciliter). Hyperchromic RBCs are defined as having an MCHC above 36 g per deciliter. RBCs may also contain inclusions such as Howell Jolly bodies (nuclear fragments), Pappenheimer bodies, or basophilic stippling.

RBC disorders are the most common inherited disorders; one in six people worldwide are affected by RBC abnormalities [6]. The most commonly observed abnormal RBC morphologies include *drepanocytes, acanthocytes*, and *spherocytes*, (see Fig. 15.1). *Drepanocytes* are crescent-shaped cells that are pathognomonic for sickle-cell disease (hemoglobin SS disease). *Acanthocytes* are irregularly shaped cells with 2 to 10 hemispherically tipped spicules of varied length and diameter. *Spherocytes* are thick RBCs with a greatly reduced central concavity. The spherocytic shape is not diagnostically specific and results from many processes (see Table 15.1). Normal RBC morphology and examples of abnormal RBC morphology are shown in Fig. 15.1.

A variety of environmental conditions, including pH, temperature, and presence or absence of albumin or a number of chemical agents can induce rapid reversible RBC shape transformations. Discocytes can readily change into two other forms: the echinocyte, characterized by 10 to 30 uniform short projections evenly spaced over the cell surface; and the stomatocyte, a cup-shaped cell.

Abnormality Identified on Peripheral Blood Smear	Associated Clinical Disorder
Specific Findings	
Burr cells (echinocytes)	Uremia, pyruvate kinase deficiency, stomach carcinoma, and peptic ulcers
Spur cells (acanthocytes)	Alcoholic liver disease, abetalipoproteinemia, and post-splenectomy
Keratocytes (bite cells)	Acute hemolysis induced by oxidant damage as seen in glucose-6-phosphate dehydrogenase (G6PD) deficiency
Schistocytes (RBC fragments), microspherocytes (hyperchromatic and small cells)	Microangiopathic hemolytic anemia (pregnancy-associated hypertension, disseminated cancer, disseminated intravascular coagulation (DIC), hemolytic-uremic syndrome (HUS), or thrombotic thrombocytopenic purpura (TTP), HELLP, and vasculitis), severe burns, and in heart-valve hemolysis
Codocyte (target cells)	Liver disease, thalassemia, and iron deficiency
Dacrocytes (teardrop cells)	Myelofibrosis/myeloid metaplasia, thalassemia, pernicious anemia, and some hemolytic anemias
Spherocytes	Autoimmune hemolytic anemia and hereditary spherocytosis
Stomatocytes	Hereditary stomatocytosis, spherocytosis, alcoholism, and cirrhosis
Elliptocytes	Hereditary elliptocytosis (hemolytic anemia)
Hypoproliferative, hypochromic (MCHC < 32 g/dL), microcytic (MCV < 80 fL)	Iron deficiency anemia (chronic blood loss, pregnancy, dietary deficiency, malabsorption, chronic low grade hemolysis), and thalassemias
Macrocytosis (MCV > 100 fL)/ megoblastic anemia	B12 or folate deficiency, alcoholism, liver disease, renal disease, hypothyroidism, and drug effect (chemotherapy)

Table 15.1 MCV, Mean Corpuscle Volume; HELLP, Hemolysis Elevated Liver Enzymes and Low Platelet Count; MCHC, Mean Cell Hemoglobin Concentration

Abnormality Identified on Peripheral Blood Smear	Associated Clinical Disorder
Macrocytes, oval macrocytes, poikilocytosis, RBC fragments, and hypersegmented neutrophils,	Macrocytic anemia
Basophilic stippling	Lead poisoning
Drepanocyte (speculated forms)	Sickle-cell disease (polymerized hemoglobin S)
Inclusions	
Howell Jolly bodies	Splenectomy and sickle-cell disease (functional asplenia)
Heinz bodies (α globin precipitates)	Glucose 6-phosphate dehydrogenase (G6PD), congenital hemolytic anemia, alpha thalassemia, and unstable hemoglobin variant
Nonspecific Findings	
Anisocytosis (RCDW > 15%)	Feature of many types of anemia

Table 15.1 RCDW, Red Cell Distribution Width (*Continued*)

Amphipathic compounds can preferentially intercalate into either the inner or outer leaflets of the membrane depending on their individual characteristics. Those that accumulate in the outer membrane are echinocytogenic (form membrane crenation). Echinocytogenic compounds include lysophosphatidylcholine and barbiturates; and echinocytogenic conditions include metabolic starvation, calcium loading, cholesterol enrichment, and high salt concentrations. Conversely, agents that preferentially concentrate in the inner leaflet are stomatocytogenic (induce cup formation). Chloropromazine is stomatocytogenic. Conditions including low pH, low salt, and cholesterol depletion are also stomatocytogenic [3].

Flow alters RBC morphology as well (see Fig. 15.2). This was first described by Antonie van Leeuwenhoek in 1675. "They [RBCs] should be soft and flexible, that they may be capable of passing through capillary veins and arteries by changing their round Figures into ovals and also reassuming their former roundness when they come into vessels where they find larger room" [4]. Today, microfluidic flow chambers produced by soft lithography allow for three-dimensional (3D) analysis of RBCs. They are used experimentally to investigate RBC material properties, shape changes, and flow dynamics [5]. Figure 15.2 shows the typical RBC morphologic changes that occur during flow. Note the RBC deformation as it enters and leaves the channel. General observations regarding RBC entrance

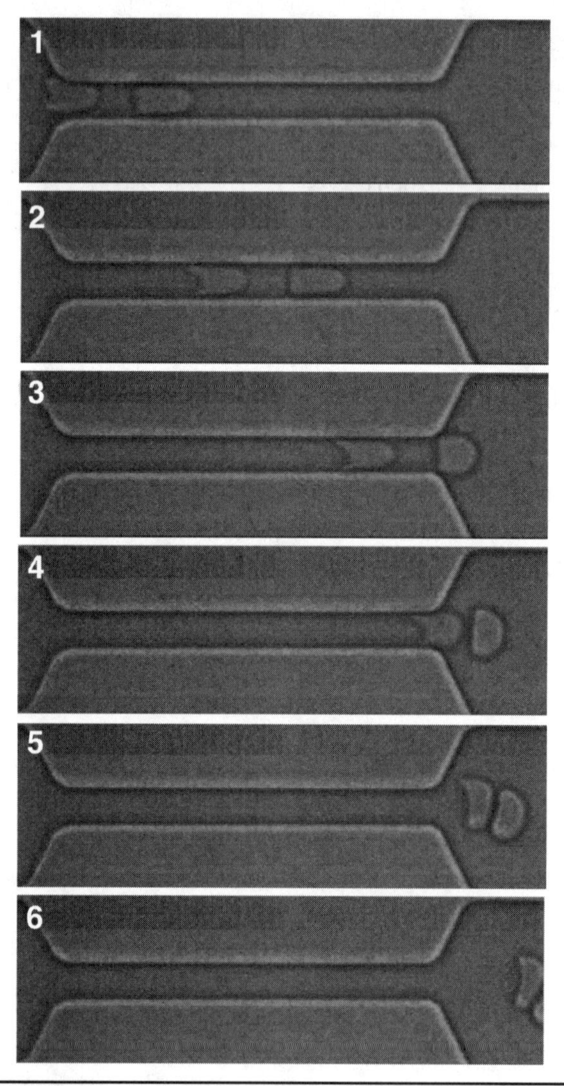

Figure 15.2 RBC deformability: Characteristic RBC morphologies during flow. Panels 1–6. Red blood cell (RBC) entrance, flow, and exit morphologies (deformability) through a 5 × 5 × 50 μm microchannel. The images were recorded by a photron high-speed video camera that was mounted onto a regular optical microscope (Leica DM IRB). The camera allowed 4 seconds of recording at 1000 frames per second with a 512 × 512-pixel resolution. Panels 1–6 were made using "snag it" software on a representative high-speed movie. Typically, normal RBC display a parachute- or jellyfish-like shape when exiting microchannels and then quickly revert to a biconcave disc shape.

length, time to reach steady shape, and exit shape can be quantified. The ability of RBCs to return to their original form following deformations illustrates the elastic nature of the RBC membranes.

15.1.3 Cell Deformability Assay

One of the most critical features of RBCs is their flexibility (reversible shape deformability). RBC deformability is essential for both RBC function and survival. Throughout the four-month life span of RBCs, they withstand many circulatory stresses. RBCs must repeatedly undergo rapid extensive deformations as they navigate through the microvasculature. RBCs flow continuously through 3-micron diameter capillaries and squeeze through tiny 1- to 2-micron slits in the sinusoidal walls of the spleen. Normal RBCs can withstand large linear extensions of up to 230% while maintaining a constant surface area. It is the biconcave disc shape and redundant membrane that allow the RBC to undergo these impressive deformations while maintaining a constant surface area. A decrease in RBC deformability results in either their removal by the reticuloendothelial system or lysis during circulation [6]. Loss of RBC deformability is a feature of senescent cells and is observed in a number of inherited and acquired RBC disorders.

Three factors regulate RBC deformability: (1) Cell shape and geometry determine the ratio of cell surface area to volume (higher ratios facilitate deformation). (2) Cytoplasmic viscosity is determined by hemoglobin concentration and degree of cell hydration. RBC viscosity is thus influenced by changes in cell volume. (The increase in RBC viscosity associated with elevated hemoglobin concentrations impairs deformability and flow characteristics.) (3) Membrane deformability and mechanical stability are regulated by the structural organization of the RBC membrane proteins and lipid bilayer (see Figs. 15.3 and 15.4). Reduced membrane deformability hinders shape-change generation which is necessary for entry into, and exit from, arteries and capillaries, as well as RBC escape from splenic fenestrations. A decreased mechanical stability leads to premature RBC removal from circulation because of RBC fragmentation and eventual lysis. Diverse pathologies differentially affect RBC deformability and mechanical stability. This suggests that these two properties are independently regulated by different membrane protein interactions [6].

15.2 Physical Properties of RBC Membranes

The highly elastic and resilient RBC membrane, a 4- to 5-nm thick lipid bilayer (see Fig. 15.5), encloses a concentrated hemoglobin solution. It has fluid-like properties with a finite bending modulus κ and a vanishing shear modulus $\mu \approx 0$. The resistance to shear, crucial for

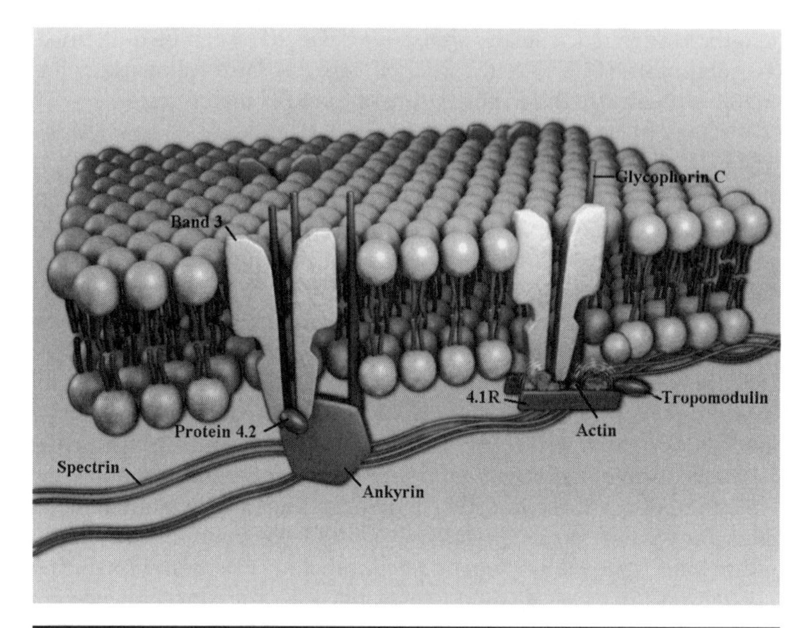

FIGURE 15.3 Schematic representation of the structural organization of the RBC membrane (the proteins and lipids are not drawn to scale). The cytoskeleton is made up of a spectrin tetramer base and a number of linking proteins. The cytoskeleton is linked to the lipid bilayer through b-spectrin-ankyrin-band 3 interactions and through b-spectrin-protein 4.1 R-glycophorin C interactions. b-spectrin also forms a multiprotein complex with actin, protein 4.1R, adducin tropomyosin, and tropomodulin. These interactions regulate cell shape and deformability. Figure by Zachery Johnson Imaging Technology Group Beckman Institute, University of Illinois.

normal RBC function, is provided by the underlying 2D cytoskeletal spectrin network which has a mesh size of ~80 nm. The RBC cytoskeleton helps maintain the RBC shape and plays a role in regulating RBC deformability and mechanical stability. The cytoskeletal filamentous network of proteins is tethered to the bilayer through multiple-anchoring protein mechanisms [7].

By mass, the RBC membrane contains 40% lipids (cholesterol, phospholipids, free fatty acids, and glycolipids), 52% proteins (over 100 RBC-associated proteins have been identified), and 8% carbohydrates. The major membrane lipid components, unesterified cholesterol and phospholipids, are present in equimolar quantities. The phospholipids [phosphotidly choline (PC), phosphatidyl ethanolamine (PE), phosphatidyl serine (PS), and sphingo myclin (SM)] are asymmetrically distributed in the membrane, with >75% of choline-containing uncharged phospholipids (PC and SM) localized to the outer monolayer of the lipid bilayer, while 80% of the charged phospholipids (phosphatidyl ethanolamine and phosphatidyl serine) are found in the inner monolayer. Free fatty acids and glycolipids are

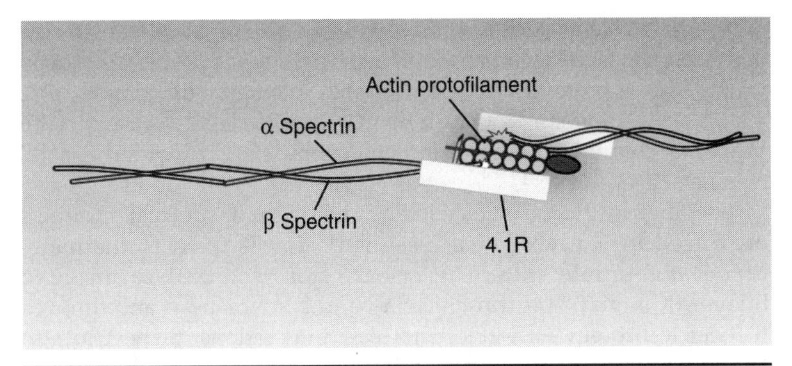

Figure 15.4 Schematic representation of the RBC cytoskeletal lateral linkage proteins (the proteins and lipids are not drawn to scale). The *lateral junctional complexes* are (1) a- and b-spectrin dimers self-associated to form spectrin tetramers and (2) interactions between the spectrin dimers and actin, protein 4.1R, adducin, tropomyosin, and tropomodulin. These lateral interactions regulate cell shape and deformability. Spectrin has to be phosphorylated by an ATP-dependent protein kinase in order to associate with actin.

present in small amounts [2]. The RBC lipid bilayer is shown in the scanning electron transmission micrograph in Fig. 15.5.

There are 10 major membrane proteins and over 200 minor proteins asymmetrically organized in the RBC membrane. RBC proteins are categorized as either integral or peripheral proteins. Integral proteins are embedded within the membrane through hydrophobic interactions with the lipids within the bilayer. Many carry RBC antigens and act as receptors or are transport proteins. Glycophorins are the major integral membrane proteins in the RBC.

Peripheral proteins are located on the cytoplasmic surface of the membrane. The major peripheral protein is spectrin. Spectrin is a flexible, filimentous 200-nm long protein that binds with other

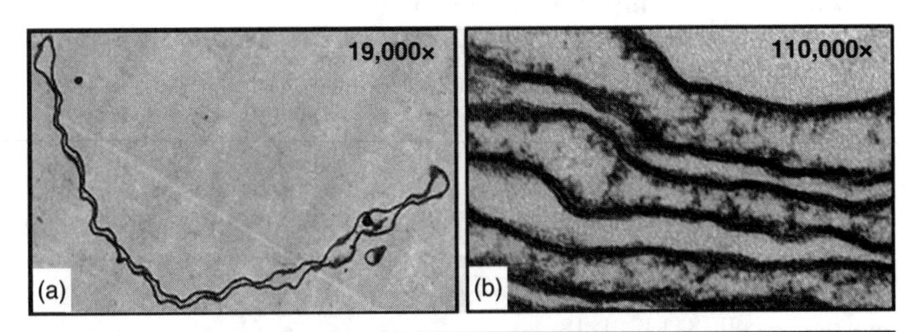

Figure 15.5 *Transmission electron micrograph images* of red blood cell membranes (ghosts) (*a*) Cross section of a normal ghost. (*b*) Continuous bilayered membranes of three RBCs in parallel. (Taken by Martin Karl Selig at Massachusetts General Hospital, Boston, Massachusetts.)

peripheral proteins such as actin to form a skeleton of microfilaments that strengthens the membrane and gives it its elastic properties. β-actin is made up of protofiliments 30- to 40-nm in length and consists of 12 to 14 actin monomers that form the cytoskeletal vertices. Actin filiments contain tropomyosin which binds to tropomodulin [2] (see Figs. 15.3 and 15.4).

Geometrically, the cytoskeleton is made up of spectrin tetramers organized into a hexagonal meshwork that is fixed to the membrane at multiple sites. The cytoskeleton is linked to the lipid bilayer through β-spectrin-ankyrin-band 3 interactions and through β-spectrin-protein 4.1R-glycophorin C interactions. β-spectrin also forms a multiprotein complex with actin, protein 4.1R, adducin (a calmodulin-binding protein that functions to cap and regulate the length of the actin filament), tropomyosin, and tropomodulin (see Fig. 15.5). Spectrin–actin junctions also contain dematin, an actin-bundling protein. The spectrin-based cytoskeletal connections are categorized as vertical or lateral. The cytoskeletal *vertical linkage* is made up of interactions between Band 3, RhAg, cd 47, and glycophorin A proteins which are linked to the spectrin-based cytoskeleton by ankyrin and protein 4.2. Loss of this vertical linkage as a result of a deficiency in either Band 3, ankyrin, or RhAg leads to a reduction in membrane surface area and the subsequent generation of spherocytic RBCs. The *lateral junctional complexes* are (1) α- and β-spectrin dimers that self-associate to form spectrin tetramers and (2) interactions between the spectrin dimers and actin, protein 4.1R, adducin, tropomyosin, and tropomodulin. These lateral interactions regulate cell shape and deformability. Defective lateral interactions among skeletal proteins lead to decreased membrane mechanical stability, elliptocytosis, and cell fragmentation [2, 6].

In addition to controlling RBC morphology, the protein-protein and protein-lipid interactions form the cellular basis of RBC membrane dynamics.

15.2.1 RBC Membrane Fluctuations

The RBC membrane has a characteristic "flicker" (surface undulations) or internal scintillation when observed by phase-contrast microscopy (PCM) or differential interference microscopy (DIC). This property and its effect on RBC overall structure, dynamics, and function of the whole cell have been widely studied [8–17]. These spontaneous motions have also been modeled theoretically under both static and dynamic conditions in an attempt to connect the statistical properties of the membrane displacements to relevant mechanical properties of the cell [8, 13, 14, 18, 19].

Measurements of the membrane mean-squared displacement versus the spatial wave vector, $\Delta u^2(q)$, revealed a q^{-4} dependence, which is indicative of fluid-like behavior [8, 20–23]. These results are

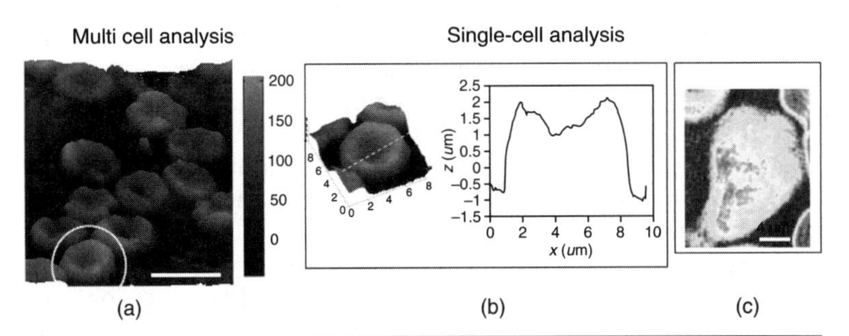

Figure 15.6 (*a*) Phase image of whole blood. The gray scale bar represents optical path length in nm: the scale bar is 10 μm. (*b*) Expanded analysis of the red blood cell circled in (*a*) and a thickness profile in microns along the dotted line indicated in (*b*). (*c*) Leukocyte (white blood cell, WBC). (This figure was modified with permission from the authors [58].)

in conflict with the static deformation measurements provided by micropipette aspiration [24, 25], high-frequency electric fields [26, 27], and more recently, optical tweezers [28]. This indicates an average value for the shear elasticity on the order of $\mu \sim 10^{-6} J/m^2$. It was also predicted that the RBC cytoskeleton confines the fluctuations of the bilayer, which has the net effect of an apparent increased superficial tension in the membrane [13]. This increase in tension was then quantified experimentally by analyzing the time-averaged (static) behavior of the RBC membrane displacements [16]. RBC fluctuations are modulated by the viscosity of the surrounding medium and by ATP hydrolysis, which was interpreted as an indication of nondeterministic processes [9, 12, 29]. Such nonthermal contributions have also been the subject of recent theoretical work [14, 17]. In spite of these efforts, the role of molecular interactions between the lipid bilayer and underlying cytoskeletal network on the overall mechanical properties of the cell membrane remains to be elucidated.

15.2.2 Existing Techniques for Measuring RBC Membrane Fluctuations

A number of different techniques have been used to study the dynamic mechanical properties of live cells [30]. Among these, pipette aspiration [24], electric field deformation [27], and optical tweezers [31, 32] provide quantitative information about the shear and bending moduli of RBC membranes under *static* conditions. Rheoscopy, ektacytometry, microfluidic flow channels, and optical-trapping strategies have also been used to investigate the material characteristics of RBCs [5, 6]. However, *dynamic*, frequency-dependent knowledge of the RBC mechanical response is limited to very recent developments [33].

Quantifying the RBC membrane motion is experimentally challenging. These motions typically occur at the nanometer and millisecond scales across the entire cell, and reliable methods for spatial and temporal data acquisition are currently limited. Optical techniques are appealing as they retrieve dynamic information without physical contact. Point dark-field microscopy has been developed as a technique that relies on focusing light in the vicinity of the rim of an RBC and sensing the submicron lateral motions of the membrane [9, 11, 12, 29, 34, 35]. Although this approach has provided novel insight into the cell membrane dynamics, it is a single-point measurement technique with the typical limitations associated with such techniques.

In contrast, full-field methods can measure both spatial and temporal features of the membrane fluctuations. Existing techniques include PCM [8, 36], reflection interference contrast microscopy (RICM) [20, 22], and fluorescence interference contrast (FLIC) [37]. These techniques are limited in their ability to measure cell membrane displacements. PCM provides quantitative phase shifts *only* for samples that are optically much thinner than the wavelength of light. Similarly, a single RICM measurement cannot provide the absolute cell thickness unless additional measurements or approximations are made [38]. FLIC relies on inferring the position of fluorescent dye molecules attached to the membrane from the absolute fluorescence intensity; this may limit both the sensitivity and acquisition rate of the technique [37]. Thus, these techniques are not suitable for making spatially resolved measurements of RBC membrane fluctuations.

15.2.3 Quantitative Phase Imaging

PC and DIC microscopy have been used extensively to infer morphometric features of live cells without the need for exogenous contrast agents [39]. These techniques transfer the information encoded in the phase of the imaging field into the intensity distribution of the final image. Thus, the optical phase shift through a given sample can be regarded as a powerful endogenous contrast agent, as it contains information about both the thickness and refractive index of the sample. However, both PCM and DIC are *qualitative* in terms of optical path-length measurement; that is, the relationship between the irradiance and phase of the image field is generally nonlinear [40, 41].

Quantifying the optical phase shifts associated with cells gives access to information about morphology and dynamics at the *nanometer scale*. Over the past decade, the development of quantitative phase-imaging techniques has received increased scientific interest. These are noncontact, nondestructive high-sensitivity methods that measure the optical path-length shifts associated with the sample of interest and can provide accurate topographic information. The technology can be divided into *single-point* and *full-field* measurements, according to the experimental geometry employed. Several point-measurement techniques have been applied for investigating the structure and dynamics of live cells [42–48]. This type of measurement allows for fiber-optic implementation and also high-speed punctual phase measurement by using a single, fast photodetector. Full-field phase measurement techniques, on the other hand, provide simultaneous information from a large number of points on the sample; this has the benefit of allowing the study of both the temporal and spatial behavior of the biological system under investigation [49–61]. With the recent advances in two-dimensional array detectors, full-field phase images can now be acquired at high speeds (i.e., thousands of frames per second).

Point Measurements

Various point-measurement techniques have been developed for quantifying phase shifts at a given point through biological samples. This class of techniques can be described as an extension of optical coherence tomography (OCT) to provide measurements of phase, phase dispersion, and birefringence associated with biological structures [62]. De Boer et al. demonstrated depth-resolved birefringence measurements with a polarization sensitive OCT system [63]. Differential phase-contrast OCT images have also been generated with a polarization-sensitive OCT instrument [64]. Recently, polarization-sensitive OCT was used to quantify phase retardation in the retinal nerve fiber [65]. An instantaneous quadrature technique based on using a $1 \times N$-fiber coupler and the inherent phase shift between different output fibers was proposed [66]. Electrokinetic [67] and thermo-refractive [68] properties of tissue and tissue phantoms have been measured by differential-phase OCT. Phase-sensitive OCT-type measurements have also been performed for studying static cells [48], for monitoring electric activity in nerves [46, 47], and for investigating spontaneous beating in cardiomyocytes [44]. However, these methods rely on single-point measurements, which, for imaging purposes, require faster scanning. This procedure is often time consuming, and this reduces the applicability of these techniques.

Full-Field QPI

Recently, new full-field phase-imaging techniques, which are suitable for spatially resolved investigation of biological structures, have been developed to overcome these limitations. Combining phase-shifting interferometry with Horn microscopy, DRIMAPS (digitally recorded interference microscopy with automatic phase shifting) has been proposed as a new technique for quantitative biology [49, 69]. This quantitative phase-imaging technique has been successfully used for measuring cell spreading [50], cell motility [51], cell growth, and dry mass [70]. A full-field quantitative phase microscopy method was developed also by using the transport-of-irradiance equation [71, 72]. The phase retrieval technique is inherently stable against phase noise because it does not require using two separate beams as in typical interferometry experiments. This approach requires, however, recording images of the sample displaced through the focus and subsequently solving numerically partial differential equations.

Digital holography was developed a few decades ago [73] as a technique that combines digital recording with traditional holography [74]. Typically, the phase and amplitude of the imaging field are measured at an out-of-focus plane. For optically thin objects, this method allows for the reconstruction of the in-focus field and, thus, retrieval of the phase map and the characterization of the sample under investigation. This method has been implemented in combination with phase-shifting interferometry [75]. More recently, digital holography was adapted for quantitative phase imaging of cells [56, 76, 77].

Significant advancements in a number of qualitative phase imaging (QPI) methods for studying live cells and tissues have been made (for a review see [89]). Fourier-phase microscopy (FPM) [58, 78], Hilbert-phase microscopy, (HPM) [59, 60], and diffraction-phase microscopy (DPM) [61, 79] have been developed in response to the need for high phase stability over broad temporal scales.

Currently, there is an effort to advance the QPI technology and expand its applicability to both basic science and clinical practice. The QPI technology has been generalized to include polarization imaging [80], speckle-free imaging [81], and high-throughput cytometry [82]. Furthermore, knowledge of the phase and amplitude of an image field allows for numerical reconstruction of the far field, i.e., the *angular scattering* associated with the specimen [83]. This approach, termed *Fourier Transform Light Scattering* (FTLS), provides unprecedented sensitivity to scattering measurements from single cells and tissues, in both static and dynamic conditions [84–86]. Compared to typical flow cytometers, where only forward and side scattering angles are measured, FTLS provides simultaneous multiangle information.

Quantitative Phase Imaging of Blood

Mature RBCs are unique cells in that they lack nuclei and are organelle free. Thus, they can be modeled as optically homogeneous objects; that is, they produce local optical phase shifts that are proportional to their thickness. Measuring quantitative phase images of RBCs provides cell thickness profiles with an accuracy that corresponds to a very small fraction of the optical wavelength. Using the refractive index of the cell and surrounding plasma of 1.40 and 1.34, respectively [87], the phase information associated with the RBCs can be translated into a nanometer-scale image of the cell topography (see Fig. 15.6). Thus, quantitative phase imaging provides RBC volumetry directly from an *unstained* blood smear from which blood cell abnormalities can be readily detected [82]. This represents a significant advance with respect to current techniques that require the cells to be sphered [88].

Imaging technologies for evaluating patients with hematologic disorders are changing rapidly. QPI is able to recover all the parameters reported by current blood cell analyzers directly from an *unstained* blood smear. The high-throughput, label-free method also reveals all the structural information provided by a regular *stained* smear. Individual cells, both RBCs and WBCs, are analyzed, and a set of parameters that quantifies cellular morphology in great detail is generated. Cell volume, cell surface area, and minimum capillary diameter (the smallest blood vessel that the cell can squeeze through) can be determined through static evaluation of the smear. Evaluation occurs in the order of thousands of cells per second. The rapid image acquisition permits the analysis of RBC mechanical properties of hundreds of cells per second, or a full smear in less than a minute. The dynamic parameters reported by the instrument are bilayer bending modulus, shear modulus of spectrin, and effective 3D shear modulus of the cell. Furthermore, the entire high-resolution [0.5 μm, over a very large field of view (1.5×1.5 mm^2)] image of a smear, or droplet of blood, containing approximately 3000 cells, is stored digitally. This eliminates the need for fixing the specimen for possible future analysis [82]. Thus, full-field qualitative and quantitative blood cell assessment is now possible to aid in the evaluation and diagnosis of hematologic disorders. Furthermore, this technology of automated phase imaging and analysis will permit remote examination of blood samples and remote expert second opinions (telehematology).

Typically, the preparation of blood samples for QPI analysis of RBCs requires blood collection in vacutainer tubes containing ethylenediaminetetraacetic acid (EDTA) to prevent clotting. Whole blood can be analyzed, or RBCs can be isolated and washed. Samples are centrifuged to isolate RBCs from plasma. The RBCs are then washed and resuspended in an isotonic solution of PBS at a concentration of 10% by volume. Droplets of the suspension are then sandwiched between two coverslips and imaged at room temperature.

15.3 Nanoscale Characterization of RBC Dynamics

15.3.1 Static (Spatial) Behavior of Membrane Displacements

Assessing full-field images of RBC fluctuations over time provides a means to investigate intrinsic RBC membrane dynamics. RBCs with the typical discocytic shape, as well as altered forms such as spiculated echinocytes and spherocytes, have been analyzed (see Figs. 15.6 to 15.13). By taking into account the free-energy contributions of the bilayer and the cytoskeletal network, these morphological forms have all been successfully modeled [90]. Figure 15.7*a*, *b*, and *c*

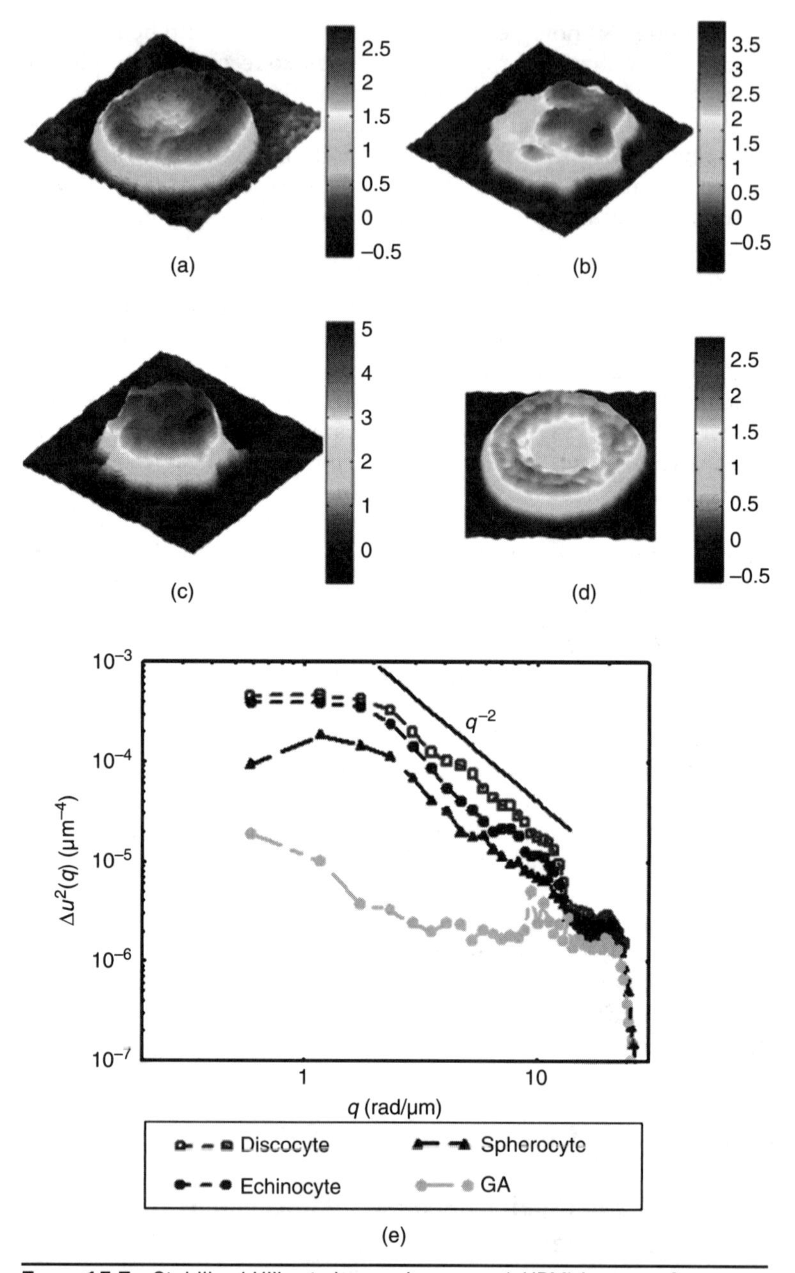

FIGURE 15.7 Stabilized Hilbert-phase microscopy (sHPM) images of a (*a*) discocyte, (*b*) echinocyte, (*c*) spherocyte, and (*d*) a gluteraldehyde (GA) fixed cell. The gray scale bar shows cell thickness in microns. (*e*) Mean-squared displacements (*y*-axis) plotted versus spatial wave vector *q* (*x*-axis) for the three RBC groups (top three curves) and for gluteraldehyde (GA)-fixed cells (bottom curve). (Adapted with permission from authors [98].)

show typical sHPM (stabilized Hilbert-phase microscopy) images of RBCs in these three morphological groups. As a control for background thermal motion of cells and the lipid bilayer, RBCs fixed with 40-μM gluteraldehyde were also analyzed in Fig. 15.7*d*. The resultant mean-squared displacements $\Delta u^2(q)$ for each group of four to five cells are summarized in Fig. 15.7*e* [98]. The gluteraldehyde-fixed cells show significantly diminished fluctuations as expected. The curves associated with the three untreated RBC groups exhibit a power-law behavior with an exponent $\alpha = 2$. As detailed elsewhere [16], this dependence is an indication of vesicle tension, which in the case of RBCs is modulated by the confinement of the lipid bilayer by the underlying cytoskeletal network [13, 91].

On the basis of this model, the data was fit to extract the tension coefficient for each individual cell. The average values obtained for the discocytes, echinocytes, and spherocytes are $\sigma = (1.5 \pm 0.2) \times 10^{-6}$ J/m^2, $\sigma = (4.05 \pm 1.1) \times 10^{-6}$ J/m^2, and $\sigma = (8.25 \pm 1.6) \times 10^{-6}$ J/m^2, respectively. The tension coefficient of RBCs is 4 to 24 times larger than that of giant artificial vesicles of the lipid bilayer [98]; this suggests that the membrane cytoskeleton might be responsible for this enhancement. Further, it is known that the cytoskeleton plays a role in the transition from a normal red blood cell shape to abnormal morphology, such as an echinocyte and spherocyte [90]. Therefore, the consistent increase in tension measured for the discocyte–echinocyte–spherocyte transition is likely explained by changes in the cytoskeletal architecture and/or membrane-cytoskeleton associations. Finally, these findings provide experimental support for the theoretical suggestion that the RBC membrane fluctuations are laterally confined by a characteristic length, $\xi_0 = 2\pi\sqrt{\kappa / \sigma}$, which is much smaller than the cell size [13].

15.3.2 Dynamic Behavior of Membrane Displacements: Spatial and Temporal Correlations

The spatial *and* temporal coherence of RBC membrane motion has also been measured [92]. Thus, RBC dynamic quantities can now be assessed in a spatially resolved manner. The quantitative phase image obtained is from a single recorded interferogram via a 2D Hilbert transform, which enables the study of rapid phenomena [93]. The physical profile of the cell (Fig. 15.8) is obtained by using the RBC refractive index information and the acquisition of a series of quantitative phase images for 4 seconds at a rate of 128 frames per second.

Figure 15.8*c* shows the instantaneous membrane displacement map $\Delta h(x,y,t)$ of the same RBC shown in Fig. 15.8*b*. Approximating the membrane with a sheet of entropic springs, the equivalent elastic constant $k_e = 1.90 \pm 0.25\, \mu N/m$ was measured as $k_e = k_B T / \langle \Delta h^2 \rangle$. The instantaneous restoring force $f_e(x,y) = -k_e \Delta u(x,y)$, ranges to $f_e \in (-0.2; 0.2) pN$.

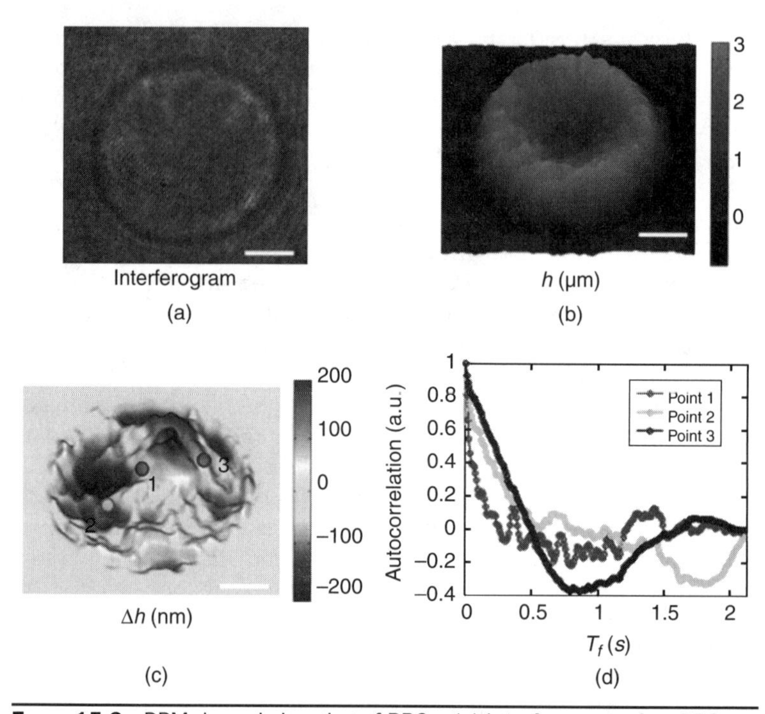

FIGURE 15.8 DPM dynamic imaging of RBCs. (*a*) Interferogram of an RBC. (*b*) DPM image (physical map) of the same cell. (*c*) Instantaneous displacement map of the cell. (*d*) Temporal autocorrelation of the displacements associated with the points shown in (*c*). (Adapted with permission from authors [96]. Copyrighted by the American Physical Society.)

In addition, diffraction phase microscopy (DPM) was used to quantify the flickering of RBC membranes and was used to extract the viscoelastic properties of RBC. RBCs were separated into three groups corresponding to their shapes: discocytes, echinocytes, and spherocytes. Data was collected from 18 discocytes, 30 echinocytes, and 6 spherocytes. The commonly occurring discocyte–echinocyte–spherocyte shape transition within the same cell was analyzed. During this transition, the root-mean-squared amplitude of the membrane fluctuations $\sqrt{\langle \Delta h^2 \rangle}$ decreases progressively, from 134 nm for discocytes to 92 nm for echinocytes and 35 nm for spherocytes; this indicates a progressive increase in cell stiffness during this shape transition (Fig. 15.9). Similarly, the long spatial wavelength fluctuations visible in the instantaneous displacement map of discocytes are extinguished in spherocytes; this is a manifestation of the loss in deformability.

It is thought that the cytoskeletal spectrin network provides the RBC with the necessary shear resistance [94]. However, little is known about

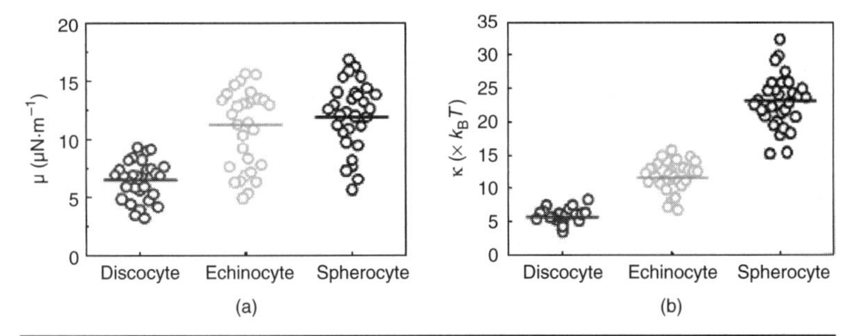

FIGURE 15.9 Shear modulus and bending modulus. (*a*) Shear moduli of discocytes (DSs), echinocytes (ECs), and spherocytes (SCs) with their mean values represented by the horizontal lines. *p*-values verify that the differences in the shear moduli between morphological groups are statistically significant: $p < 10^{-5}$ between DCs and ECs and between DCs and SCs; $p < 10^{-4}$ between ECs and SCs. (*b*) bending modulus of three groups. The means of the bending moduli (shown by horizontal lines) are 5.3 ± 1.1, 11.5 ± 2.4, and 23.5 ± 3.9, respectively. $p < 10^{-7}$ both between DCs and ECs and between ECs and SCs.

the molecular and structural transformations that take place in the membrane or the cytoskeleton as it loses its elasticity during shape changes. In order to characterize the time-averaged (static) behavior of the membrane elasticity, the cells were mapped in terms of an effective spring constant k_e, by assuming thermal equilibrium, $k_e = k_B T / \langle \Delta h^2 \rangle$ (Fig. 15.10).

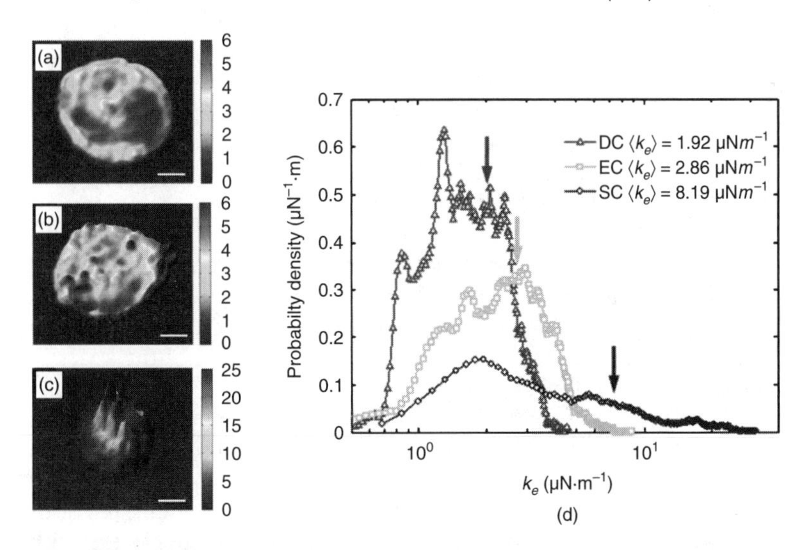

FIGURE 15.10 Maps of the effective spring constant for a (*a*) discocyte, (*b*) echinocyte, and (*c*) spherocyte. Gray scale bars are in $\mu N \cdot m^{-1}$ and the scale bar is 2 μm. (*d*) The corresponding histograms of spring constants have the abscissa in the logarithmic scale. The legend shows the respective mean values, $<k_e>$, which are also indicated by the arrows on the graph. (Adapted with permission from authors [103]).

This spatially resolved representation reveals material inhomogeneities especially in echinocytes and spherocytes. Remarkably, the center ("dimple") region of normal cells appears stiffer (higher k_e values) than the rest of the cell; this is likely due to a combination of the geometrical (i.e., high curvature) and structural properties of that membrane area. The highly inhomogeneous k_e map associated with echinocytes may indicate topological defects in the cytoskeleton mesh or correspond to the increased density of relatively rigid lipid rafts in these regions. On average, spherocytes are characterized by an elastic constant which is four times higher than that of discocytes. The average elastic constant measured for discocytes, $k_e = 1.9\mu N/m$, shown in Fig. 15.10, is a factor of 3.5 to 10 lower than what was measured by micropipette aspiration [95] and electric field deformation [27]. This difference can be explained by noting that these two techniques probe larger cell deformations ($\Delta h \geq 1\mu m$), while the DPM technique is sensitive to much smaller membrane displacements ($3\,nm \leq \Delta h \leq 200\,nm$) in the absence of external stress, and therefore explores the *linear* viscoelastic regime.

It is hoped that these intrinsic signals, which report on the cell mechanics, can be used as markers for disease [83, 96–101]. The quantitative measurements of membrane tension, for example, can be used as a diagnostic or therapeutic assay for pathological processes that result in increased RBC tension.

15.3.3 Effects of Adenosine Triphosphate (ATP)

Although RBC membrane dynamics have been explored extensively, no definitive experiment has determined whether flickering is a purely thermally driven phenomena or if it requires active contributions. Different interference microscopic techniques have been used to study membrane fluctuations and mechanical properties assuming Brownian dynamics [102, 103]. First observed a century ago, flickering was generally believed to stem from thermal forces [8, 104]. Theoretically, RBC membrane fluctuations have traditionally been studied using models of thermally driven equilibrium systems [8, 102]. However, a more recent theoretical model [105, 106], validated by simulation [107, 108], showed that local breaking and reforming of the spectrin network can result in enhanced fluctuations.

The direct full-field and quantitative measurements of ATP effect on RBC membrane morphology and fluctuations through DPM also reveal an ATP dependence factor [109]. This indicates that RBC flickering is not the sole product of thermal forces. By extracting the optical path-length shifts produced across the cell, cell thickness with nanometer sensitivity and millisecond temporal resolution was measured. RBC samples were prepared under four different conditions: healthy RBCs, an irreversibly ATP-depleted group, a metabolically ATP-depleted group, and an ATP-repleted group. After collection, the RBCs were minimally prepared. For RBCs in the irreversibly

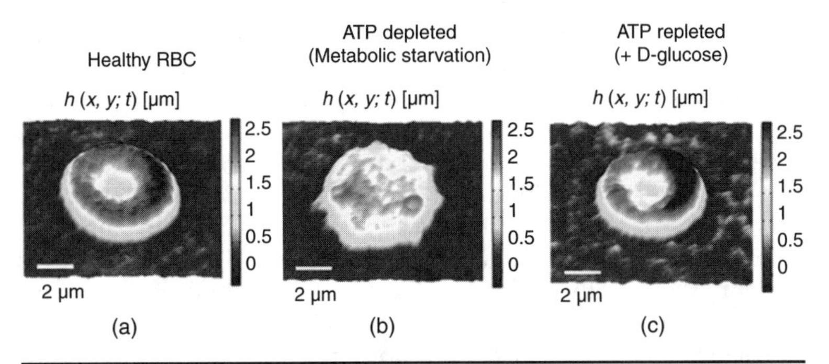

Figure 15.11 The effect of ATP on red blood cell morphology. (*a*) Height map and topography of a normal RBC. (*b*) Height map and topography of a metabolically starved RBC. (*c*) Height map and topography of a cell in which the ATP was first depleted for 24 hours, and then D-glucose was added to the cell medium. Adapted with permission from the authors [109].

ATP-depleted group, the cytoplasmic pool of ATP was depleted by inosine and iodoacetamide. For the metabolically ATP-depleted group, healthy RBCs were incubated in a glucose-free medium for 24 hours. For RBCs in the ATP-repleted group, cytoplasmic ATP was first metabolically depleted, and then regenerated through the addition of D-glucose. ATP affects RBC morphologies. From the measured cell thickness profiles at a given time t, $h(x, y, t)$, time-averaged heights $\langle h(x,y)\rangle$ were calculated and the RBCs from the four groups were analyzed. When ATP is depleted from RBCs, both irreversibly and metabolically, they lose their biconcave shape and are transformed into echinocytes. Furthermore, reintroducing ATP results in the recovery of the biconcave shape (see Fig. 15.11). This shows that ATP is crucial to maintaining the biconcave shape of RBCs [109, 110].

The membrane-displacement map was analyzed by subtracting the averaged shape from the cell-thickness map: $\Delta h(x,y,t) = h(x,y,t) - \langle h(x,y)\rangle$. ATP depletion also reduces RBC membrane fluctuation amplitudes. Compared to healthy RBCs, the fluctuation amplitudes were decreased in both ATP-depleted groups. Additionally, ATP repletion increased the fluctuation amplitudes to healthy RBC levels. The root-mean-squared (RMS) displacement of membrane fluctuations, $\sqrt{\langle \Delta h^2\rangle}$, of healthy RBCs is 41.5 ± 5.7 nm. In contrast, the fluctuations are significantly decreased to 32.0 ± 7.8 and 33.4 ± 8.7 nm in both the irreversibly and metabolically ATP-depleted groups, respectively. And the fluctuations in the ATP-repleted group returned to the level of healthy RBCs (48.4 ± 10.2 nm). This is in agreement with an earlier report using the point measurement technique that qualitatively measured the local fluctuations of RBC membranes and reported a correlation between the ATP concentration and the fluctuation amplitude [111].

Although the membrane fluctuations indeed decrease in the absence of ATP, this result does not yet answer the question of whether ATP drives "active" nonequilibrium dynamics or simply modifies membrane elastic properties. Of course, the two different situations can give rise to fundamentally different dynamics: (1) fluctuations exhibit out of equilibrium or (2) the equilibrium Gaussian statistics are preserved. In order to answer this question, the non-Gaussian parameter κ for the membrane fluctuations was measured. Theoretically, $\kappa = 2$ for purely thermally driven Gaussian motion and increases above 2 for active nonequilibrium dynamics [112]. For healthy RBCs, the average value of κ was 2.8; this shows that membrane fluctuations contain nonequilibrium dynamic components, particularly on short length and time scales ($q > 5$ rad/μm and $\Delta t < 0.5$ seconds). With depletion of ATP, κ decreased to 2, as expected in purely thermally driven dynamics (the average values of κ were 2.06 and 2.19 for the irreversibly depleted and metabolically depleted ATP groups, respectively). Reintroducing ATP increased κ to healthy RBC levels (average value $\kappa = 2.98$). The data clearly illustrate that active metabolic energy from ATP enhances the RMS displacements by 44.9%. This measured value is lower than predicted by a theoretical model, where an increase of at least 100% was expected [112]. However, this disparity can be explained by recognizing that the ATP effect is more significant at large q-values, comparable with the size of the spectrin network [105]. For example, the ATP-mediated RMS displacement at $q = 17 \pm 0.5$ rad/μm showed an increase of 143% compared with the thermal components. Thus, in the overall assessment that includes all spatial frequencies, ATP enhancement is likely to be underestimated.

In order to study further spatial aspects of active motions, the morphologies and fluctuations for RBCs were analyzed in a polar coordinate system with its origin at the cell center. Assuming cylindrical symmetry, the average height of the RBC membrane $\langle h(r) \rangle$ and the membrane mean-squared displacements $\langle h^2(r) \rangle$ are shown as functions of the radial distance r (Fig. 15.11). In healthy RBCs, the membrane fluctuations are enhanced and strongly localized at the outer region, while both ATP depletion groups show little variation in membrane fluctuations over the cell surface. Remarkably, reintroducing ATP not only restored the biconcave shape but also caused enhanced fluctuations in the outer area. This is striking because continuum models predict a stronger restoring force and a decreased fluctuation amplitude in regions of high membrane curvature [113]. This result shows that active contributions are spatially inhomogeneous and that they are correlated with the maintenance of the biconcave shape. It also helps explain different mechanistic inferences reported in the literature from prior measurements of membrane fluctuations [111, 114, 115]. Probing the edge shape of RBCs alone does not capture ATP-dependent enhanced fluctuations [114] since they are localized on the outer cell. Dark-field microscopy, which qualitatively measures the averaged dynamics of the RBC surface, can however measure ATP dependence [111, 115].

Other cytoskeleton models, incorporating actin, microtubules, and motor proteins such as myosin, have demonstrated active motion [116]. However, this cannot be the case for the ATP-enhanced fluctuations in RBC because motor proteins are absent here. The results were further analyzed in the context of RBC cytoskeletal structures. The non-Gaussian parameter, κ, at short-time delays, was plotted as a function of spatial distance $\Lambda = 2\pi/q$. Interestingly, in the presence of ATP, κ showed distinct peaks at specific distances ($\Lambda = 361, 512, 680, 860$, and 1030 nm); these peaks are equally spaced at 167 ± 10 nm. These results indicate that ATP-dependent enhanced fluctuations are correlated with the network structure of the underlying cytoskeleton. Considering the roughly hexagonal lattice of spectrin network, these peaks can be related to the dynamic remodeling of the spectrin network by ATP. Possible elements responsible for this remodeling are the junctional complexes of the spectrin network. It was proposed that this ATP-induced remodeling takes the form of local associations and dissociations of spectrin filaments within the network or between the cytoskeleton and the lipid membrane [105, 117]. Both processes result in the formation of a structural defect in the hexagonal network, of size $\sim 2\Delta$, where Δ is a distance between neighboring junctions. This remodeling of the cytoskeletal attachment causes a local release of the cytoskeleton-induced membrane tension and results in local bilayer deformation [105, 117]. The length scale of this local, ATP-induced, bilayer deformation will therefore be in multiples of the junction spacing, 2Δ. In this study, $\Delta \sim 83.5 \pm 5$ nm, which is in good agreement with the separation of the junctions complexes measured by electron microscopy [118].

This dynamic remodeling of the cytoskeleton may be related to protein phosphorylation powered by ATP. One possible candidate is the phosphorylation of spectrin. Spectrin has to be phosphorylated by an ATP-dependent kinase in order to associate with actin (see Fig. 15.4). The interactions between the spectrin dimers and actin, protein 4.1R, adducin, tropomyosin, and tropomodulin make up the lateral cytoskeletal junctional complexes. These interactions regulate RBC shape and deformability. Defective lateral interactions lead to decreased membrane mechanical stability and elliptocytosis. Interestingly, in the above study, the ATP-depleted cells were morphologically similar to those found in patients with hereditary elliptocytosis. In hereditary elliptocytosis Glycophorin C (GP) does not properly interact with protein 4.1; this results in the lack of biconcave shape and deformability [119, 120]. This dynamic remodeling of the spectrin network also offers a possible explanation for the observed metabolic dependence of RBC deformability [121]. The biconcave shape and nonequilibrium dynamics in the membrane are both consequences of the same biochemical activity: the dissociations of the cytoskeleton at the spectrin junctions, powered by ATP metabolism.

Nonphosphorylated spectrin is unbound from actin, and this may result in increased membrane fluctuations since tension applied to the bilayer by the spectrin network is locally and transiently released. In the absence of ATP, this dynamic remodeling may not occur, and thus RBCs exhibit only thermally driven membrane fluctuations.

These results provide further experimental evidence for the metabolism-dependent shape transformation; here the ATP-dependent transient binding of junctional complexes are localized over the cell outer area, and the spectrin network should therefore exert a lower tension on the membrane.

15.3.4 Effects of Osmotic Pressure on Red Blood Cell Mechanics

RBCs show dramatic changes in deformability in response to osmotic stress; however, the underlying mechanism of this mechanical transformation has not previously been explored. Park et al. investigated the dynamic membrane fluctuations in RBCs at different osmotic pressures, from which they retrieved cell mechanical properties [122]. It was shown that water influx increases the lipid-membrane bending modulus in hypotonic media and water efflux increases cytosolic viscosity under hypertonic conditions. In both cases, RBC deformability significantly increases, but due to entirely different physical phenomena. Interestingly, RBC membrane properties are optimized at physiological osmotic pressure (see Fig. 15.12).

15.4 Clinical Relevance of RBC Membrane Fluctuations: Malaria as the First Example

Malaria, an infection by the protozoan *Plasmodium falciparum* (*P. falciparum*) parasite, claims 1 to 2 million children every year. Following injection into the human host's bloodstream by mosquitoes motile sporozoites enter the liver. In the liver they multiply and redifferentiate to generate thousands of merozoites which then invade the host's RBCs to initiate the blood stage of the infection. The intraerythrocytic parasite morphs into the ring, trophozoite, and schizont stages and ultimately bursts the RBC with the release of 16 to 32 daughter merozoites [123].

In a recent study by Park et al., the nanoscale cell membrane fluctuations of *P. falciparum*–infected RBCs were analyzed by DPM [101]. The experiments covered all intraerythrocytic stages of parasite development under normal physiological and febrile temperatures. Figure 15.13 shows the DPM topographic images of healthy and

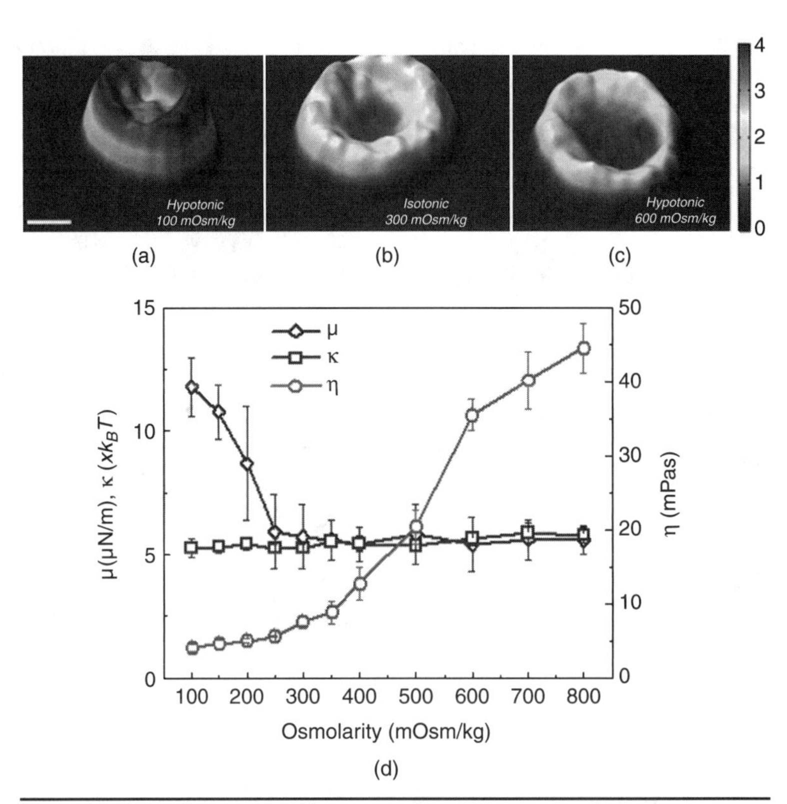

Figure 15.12 Shapes of RBCs exposed to different osmolarity. (*a*) Hyportonic (100 mOsm). (*b*) Isotonic (300 mOsm). (*c*) Hypertonic (800 mOsm). Topological images measured by DPM. Scale bar is 2 μm. Gray scale bar in μm. (*d*) Shear modulus μ, cytosol viscosity η, and bending modulus κ versus different osmotic pressure.

P. falciparum parasitized RBCs. The effective stiffness map of the cell, $K_e(x, y)$, was obtained at each point on the cell. Representative k_e maps of RBCs at the different stages of parasite development were shown. The map of instantaneous displacement of cell membrane fluctuation, $\Delta h(x, y, t)$, was obtained by subtracting the time-averaged cell shape from each thickness map in the series. This approach to studying the parasitized RBCs uniquely combined optical interferometry, biophysics, cell nanomechanics, and parasitology. The DPM technique was also validated by quantitative comparisons of the RBC elastic moduli over all parasite maturation stages with prior, independent experimental data obtained with laser tweezers [101, 124, 125]. Compared to other techniques for assessing RBC mechanical properties such as electric field deformation, micropipette aspiration, optical tweezers, and magnetic-bead excitation, DPM has distinct advantages of being spatially resolved and noncontact.

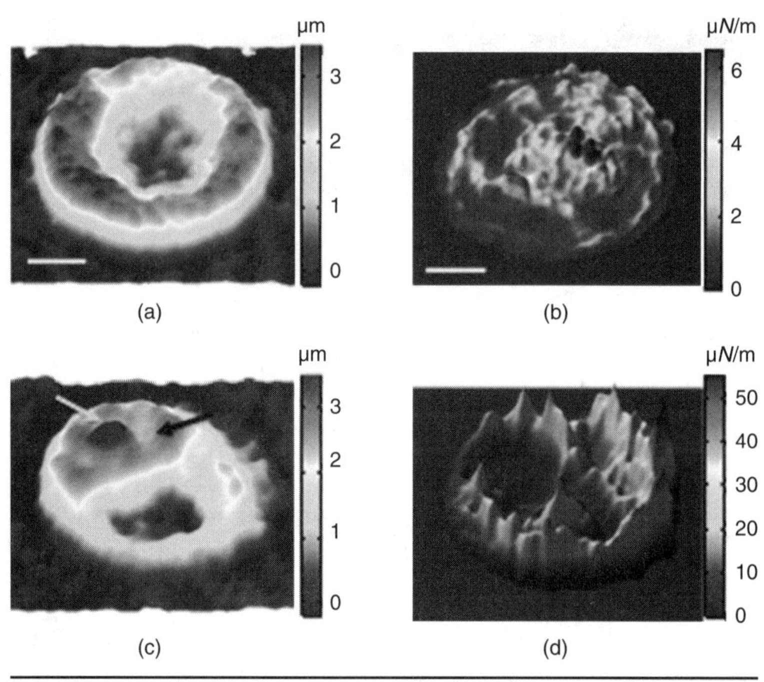

Figure 15.13 Topographic images (*a and c*) and effective elastic constant maps (*b and d*) of normal and *P. falciparum infected RBC (Pf-RBCs)*. (*a and b*) Healthy red blood cell. (*c and d*) Schizont stage of *Pf-RBCs*. The topographic images shown in (*a* and *c*) are the instant thickness map of normal and *Pf-RBCs*, respectively, measured by diffraction-phase micros copy (DPM). The effective elastic constant maps are calculated from the root-mean-square displacement of the thermal membrane fluctuations in the Pf-RBC membranes using the equipartition theorem. Black arrows in (*c*) indicate the location of *P. falciparum* merozoite, and the gray arrow indicates the location of hemozoin. (Bright-field and fluorescence micrographs provide information on locations of merozoites and hemozoin.) Scale bar is 1.5 µm. (Adapted with permission from authors [101]).

It is hoped that the nanoscale intrinsic signals measured by DPM, which report on the cell mechanics, can be used as markers of disease [83, 96–101]. The quantitative measurements of membrane tension for example, may be used as a diagnostic or therapeutic assay for pathological processes that result in increased RBC tension.

Acknowledgments

The author would like to thank colleagues and collaborators who contributed to these studies, most importantly M. Laposata, M. S. Feld, Y. K. Park, H. Stone, M. Abkarian, M. Faivre, M. K. Selig, and G. Popescu.

References

1. B. J. Bain, "Diagnosis from the Blood Smear," *N. Engl. J. Med.* 353: 498–507, 2005.
2. W. J. Williams, E. Beutler, A. J. Erslev and M. A. Lichtman, *Hematology*, McGraw-Hill, New York, 1990.
3. M. P. Sheetz and S. J. Singer, "Biological Membranes as Bilayer Couples: A Molecular Mechanism of Drug-Erythrocyte Interactions," *Proceedings Natl. Acad. Sci. USA* 71: 4457–4461, 1974.
4. A. van Leeuwenhoek, "Other Microscopical Observations Made by the Same, about the Texture of the Blood, the Sap of Some Plants, the Figures of Sugar and Salt and the Probable Cause of the Difference of Their Tastes," *Philos. Trans. R. Soc Lond.* 1675: 380–385.
5. M. Abkarian, M. Faivre, R. Horton, K. Smistrup, C. A. Best-Popescu and H. A. Stone, "Cellular-Scale Hydrodynamics," *Biomedical Materials* 3: 2008.
6. N. Mohandas and P. G. Gallagher, "Red Cell Membrane: Past, Present and Future," *Blood* 112: 3939–3948, 2008.
7. D. E. Discher and P. Carl, "New Insights into Red Cell Network Structure, Elasticity and Spectrin Unfolding—A Current Review," *Cell. Mol. Biol. Lett.* 6: 593–606, 2001.
8. F. Brochard and J. F. Lennon, "Frequency Spectrum of the Flicker Phenomenon in Erythrocytes," *J. Physique*, 36: 1035–1047, 1975.
9. S. Levin and R. Korenstein, "Membrane Fluctuations in Erythrocytes are Linked to Mgatp-Dependent Dynamic Assembly of the Membrane Skeleton," *Biophys. J.* 60: 733–737, 1991.
10. D. H. Boal, U. Seifert and A. Zilker, "Dual Network Model for Red-Blood Cell Membranes," *Phys. Rev. Lett.* 69: 3405–3408, 1992.
11. S. Tuvia, S. Levin and R. Korenstein, "Correlation between Local Cell-Membrane Displacements and Filterability of Human Red-Blood Cells," *FEBS Lett.* 304: 32–36, 1992.
12. S. Tuvia, A. Almagor, A. Bitler, S. Levin, R. Korenstein and S. Yedgar, "Cell Membrane Fluctuations Are Regulated by Medium Macroviscosity: Evidence for a Metabolic Driving Force," *Proc. Natl. Acad. Sci. USA* 94: 5045–5049, 1997.
13. N. Gov, A. G. Zilman and S. Safran, "Cytoskeleton Confinement and Tension of Red Blood Cell Membranes," *Phys. Rev. Lett.* 90: 228101, 2003.
14. N. Gov, "Membrane Undulations Driven by Force Fluctuations of Active Proteins," *Phys. Rev. Lett.* 93: 268104, 2004.
15. L. C. L. Lin and F. L. H. Brown, "Brownian Dynamics in Fourier Space: Membrane Simulations Over Long Length and Time Scales," *Phys. Rev. Lett.* 93: 2004.
16. G. Popescu, T. Ikeda, K. Goda, C. A. Best-Popescu, M. Laposata, S. Manley, R. R. Dasari, K. Badizadegan and M. S. Feld, "Optical Measurement of Cell Membrane Tension," *Phys. Rev. Lett.* 97: 218101, 2006.
17. L. C. L. Lin, N. Gov and F. L. H. Brown, "Nonequilibrium Membrane Fluctuations Driven by Active Proteins," *J. Chemical Physics* 124: 2006.
18. N. S. Gov and S. A. Safran, "Red Blood Cell Membrane Fluctuations and Shape Controlled by ATP-Induced Cytoskeletal Defects," *Biophys. J.* 88: 1859–1874, 2005.
19. R. Lipowski and M. Girardet, "Shape Fluctuations of Polymerized or Solid-Like Membranes," *Phys. Rev. Lett.* 65: 2893–2896, 1990.
20. A. Zilker, H. Engelhardt and E. Sackmann, "Dynamic Reflection Interference Contrast (Ric-) Microscopy—A New Method to Study Surface Excitations of Cells and to Measure Membrane Bending Elastic Moduli," *J. Physique* 48: 2139–2151, 1987.
21. K. Zeman, E. H. and E. Sackman, Bending Undulatios and Elasticity of the Erythrocyte Membrane: Effects of Cell Shape and Membrane Organization," *Eur. Biophys. J.* 18: 203, 1990.

22. A. Zilker, M. Ziegler and E. Sackmann, "Spectral-Analysis of Erythrocyte Flickering in the 0.3-4-Mu-M-1 Regime by Microinterferometry Combined with Fast Image Processing," *Phys. Rev. A* 46: 7998–8002, 1992.
23. H. Strey, M. Peterson and E. Sackmann, "Measurement of Erythrocyte Membrane Elasticity by Flicker Eigenmode Decomposition," *Biophys. J.* 69: 478–488, 1995.
24. D. E. Discher, N. Mohandas and E. A. Evans, "Molecular Maps of Red Cell Deformation: Hidden Elasticity and *In Situ* Connectivity," *Science* 266: 1032–1035, 1994.
25. R. M. Hochmuth, P. R. Worthy and E. A. Evans, "Red Cell Extensional Recovery and the Determination of Membrane Viscosity," *Biophys. J.* 26: 101–114, 1979.
26. H. Engelhardt and E. Sackmann, "On the Measurement of Shear Elastic Moduli and Viscosities of Erythrocyte Plasma Membranes by Transient Deformation in High Frequency Electric Fields," *Biophys. J.* 54: 495–508, 1988.
27. H. Engelhardt, H. Gaub and E. Sackmann, "Viscoelastic Properties of Erythrocyte Membranes in High-Frequency Electric Fields," *Nature,* 307: 378–380, 1984.
28. S. Suresh, J. Spatz, J. P. Mills, A. Micoulet, M. Dao, C. T. Lim, M. Beil and T. Seufferlein, "Connections between Single-Cell Biomechanics and Human Disease States: Gastrointestinal Cancer and Malaria," *Acta Biomaterialia,* 1: 15–30, 2005.
29. S. Tuvia, S. Levin, A. Bitler and R. Korenstein, "Mechanical Fluctuations of the Membrane-Skeleton are Dependent on F-Actin Atpase in Human Erythrocytes," *J. Cell Biol.* 141: 1551–1561, 1998.
30. G. Bao and S. Suresh, "Cell and Molecular Mechanics of Biological Materials," *Nat. Mat.* 2: 715–725, 2003.
31. M. Dao, C. T. Lim and S. Suresh, "Mechanics of the Human Red Blood Cell Deformed by Optical Tweezers," *J. Mechanics and Physics of Solids* 51: 2259–2280, 2003.
32. J. Sleep, D. Wilson, R. Simmons and W. Gratzer, "Elasticity of the Red Cell Membrane and Its Relation to Hemolytic Disorders: An Optical Tweezers Study," *Biophys. J.* 77: 3085–3095, 1999.
33. M. Puig-de-Morales, K. T. Turner, J. P. Butler, J. J. Fredberg and S. Suresh, "Viscoelasticity of the Human Red Blood Cell," *J. Appl. Physiol.* 293: 597–605, 2007.
34. S. Tuvia, S. Levin and R. Korenstein, "Oxygenation–Deoxygenation Cycle of Erythrocytes Modulates Submicron Cell-Membrane Fluctuations," *Biophys. J.* 63: 599–602, 1992.
35. Y. Alster, A. Loewenstein, S. Levin, M. Lazar and R. Korenstein, "Low-Frequency Submicron Fluctuations of Red Blood Cells in Diabetic Retinopathy," *Arch. Ophthalmol.* 116: 1321–1325, 1998.
36. K. Fricke and E. Sackmann, "Variation of Frequency Spectrum of the Erythrocyte Flickering Caused by Aging, Osmolarity, Temperature and Pathological Changes," *Biochim. Biophys. Acta* 803: 145–152, 1984.
37. Y. Kaizuka and J. T. Groves, "Hydrodynamic Damping of Membrane Thermal Fluctuations Near Surfaces Imaged by Fluorescence Interference Microscopy," *Phys. Rev. Lett.* 96: 118101, 2006.
38. A. Zidovska and E. Sackmann, "Brownian Motion of Nucleated Cell Envelopes Impedes Adhesion," *Phys. Rev. Lett.* 96: 048103, 2006.
39. D. J. Stephens and V. J. Allan, "Light Microscopy Techniques for Live Cell Imaging," *Science* 300: 82–86, 2003.
40. F. Zernike, "How I Discovered Phase Contrast," *Science* 121: 345, 1955.
41. F. H. Smith, "Microscopic Interferometry," *Research (London)* 8: 385, 1955.
42. C. Yang, A. Wax, M. S. Hahn, K. Badizadegan, R. R. Dasari and M. S. Feld, Phase-Referenced Interferometer with Subwavelength and Subhertz Sensitivity Applied to the Study of Cell Membrane Dynamics," *Opt. Lett.* 26: 1271–1273, 2001.
43. C. H. Yang, A. Wax, I. Georgakoudi, E. B. Hanlon, K. Badizadegan, R. R. Dasari and M. S. Feld, "Interferometric Phase-Dispersion Microscopy," *Opt. Lett.* 25: 1526–1528, 2000.

44. M. A. Choma, A. K. Ellerbee, C. H. Yang, T. L. Creazzo and J. A. Izatt, "Spectral-Domain Phase Microscopy," *Opt. Lett.* 30: 1162–1164, 2005.
45. C. Joo, T. Akkin, B. Cense, B. H. Park and J. E. de Boer, "Spectral-Domain Optical Coherence Phase Microscopy for Quantitative Phase-Contrast Imaging," *Opt. Lett.* 30: 2131–2133, 2005.
46. C. Fang-Yen, M. C. Chu, H. S. Seung, R. R. Dasari and M. S. Feld, "Noncontact Measurement of Nerve Displacement during Action Potential with a Dual-Beam Low-Coherence Interferometer," *Opt. Lett.* 29: 2028–2030, 2004.
47. T. Akkin, D. P. Dave, T. E. Milner and H. G. Rylander, "Detection of Neural Activity Using Phase-Sensitive Optical Low-Coherence Reflectometry," *Opt. Exp.* 12: 2377–2386, 2004.
48. C. G. Rylander, D. P. Dave, T. Akkin, T. E. Milner, K. R. Diller and A. J. Welch, "Quantitative Phase-Contrast Imaging of Cells with Phase-Sensitive Optical Coherence Microscopy," *Opt. Lett.* 29: 1509–1511, 2004.
49. D. Zicha and G. A. Dunn, "An Image-Processing System for Cell Behavior Studies in Subconfluent Cultures," *J. Microscopy.* 179: 11–21, 1995.
50. G. A. Dunn, D. Zicha and P. E. Fraylich, "Rapid, Microtubule-Dependent Fluctuations of the Cell Margin," *J. Cell Sci.* 110: 3091–3098, 1997.
51. D. Zicha, E. Genot, G. A. Dunn and I. M. Kramer, "TGF Beta 1 Induces a Cell Cycle-Dependent Increase in Motility of Epithelial Cells," *J. Cell Sci.* 112: 447–454, 1999.
52. D. Paganin and K. A. Nugent, "Noninterferometric Phase Imaging with Partially Coherent Light," *Phys. Rev. Lett.* 80: 2586–2589, 1998.
53. B. E. Allman, P. J. McMahon, J. B. Tiller, K. A. Nugent, D. Paganin, A. Barty, I. McNulty, S. P. Frigo, Y. X. Wang and C. C. Retsch, "Noninterferometric Quantitative Phase Imaging with Soft X-Rays," *J. Opt. Soc. Am. A-Opt. Image Sci. Vis.* 17: 1732–1743, 2000.
54. S. Bajt, A. Barty, K. A. Nugent, M. McCartney, M. Wall and D. Paganin, "Quantitative Phase-Sensitive Imaging in a Transmission Electron Microscope," *Ultramicroscopy* 83: 67–73, 2000.
55. C. J. Mann, L. F. Yu, C. M. Lo and M. K. Kim, "High-Resolution Quantitative Phase-Contrast Microscopy by Digital Holography," *Opt. Exp.* 13: 8693–8698, 2005.
56. P. Marquet, B. Rappaz, P. J. Magistretti, E. Cuche, Y. Emery, T. Colomb and C. Depeursinge, "Digital Holographic Microscopy: A Noninvasive Contrast Imaging Technique Allowing Quantitative Visualization of Living Cells with Subwavelength Axial Accuracy," *Opt. Lett.* 30: 468–470, 2005.
57. H. Iwai, C. Fang-Yen, G. Popescu, A. Wax, K. Badizadegan, R. R. Dasari and M. S. Feld, "Quantitative Phase Imaging Using Actively Stabilized Phase-Shifting Low-Coherence Interferometry," *Opt. Lett.* 29: 2399–2401, 2004.
58. G. Popescu, L. P. Deflores, J. C. Vaughan, K. Badizadegan, H. Iwai, R. R. Dasari and M. S. Feld, "Fourier Phase Microscopy for Investigation of Biological Structures and Dynamics," *Opt. Lett.* 29: 2503–2505, 2004.
59. T. Ikeda, G. Popescu, R. R. Dasari and M. S. Feld, "Hilbert Phase Microscopy for Investigating Fast Dynamics in Transparent Systems," *Opt. Lett.* 30: 1165–1168, 2005.
60. G. Popescu, T. Ikeda, C. A. Best, K. Badizadegan, R. R. Dasari and M. S. Feld, "Erythrocyte Structure and Dynamics Quantified by Hilbert Phase Microscopy," *J. Biomed. Opt. Lett.* 10: 060503, 2005.
61. G. Popescu, T. Ikeda, R. R. Dasari and M. S. Feld, "Diffraction-Phase Microscopy for Quantifying Cell Structure and Dynamics," *Opt. Lett.* 31: 775–777, 2006.
62. D. Huang, E. A. Swanson, C. P. Lin, J. S. Schuman, W. G. Stinson, W. Chang, M. R. Hee, T. Flotte, K. Gregory, C. A. Puliafito and J. G. Fujimoto, "Optical Coherence Tomography," *Science* 254: 1178–1181, 1991.
63. J. F. de Boer, T. E. Milner, M. J. C. van Gemert and J. S. Nelson, "Two-Dimensional Birefringence Imaging in Biological Tissue by Polarization-Sensitive Optical Coherence Tomography," *Opt. Lett.* 22: 934–936, 1997.
64. C. K. Hitzenberger and A. F. Fercher, "Differential-Phase Contrast in Optical Coherence Tomography," *Opt. Lett.* 24: 622–624, 1999.

65. J. Park, N. J. Kemp, T. E. Milner and H. G. Rylander, "Analysis of the Phase Retardation in the Retinal Nerve Fiber Layer of Cynomolus Monkey by Polarization Sensitive Optical Coherence Tomography," *Lasers Surg. Med.* 55–55, 2003.
66. M. A. Choma, C. H. Yang and J. A. Izatt, "Instantaneous Quadrature Low-Coherence Interferometry with 3 × 3 Fiber-Optic Couplers," *Opt. Lett.* 28: 2162–2164, 2003.
67. J. I. Youn, T. Akkin, B. J. F. Wong, G. M. Peavy and T. E. Milner, "Electrokinetic Measurements of Cartilage Measurements of Cartilage Using Differential Phase Optical Coherence Tomography," *Lasers Surg. Med.* 56–56, 2003.
68. J. Kim, S. A. Telenkov and T. E. Milner, "Measurement of Thermo-Refractive and Thermo-Elastic Changes in a Tissue Phantom Using Differential Phase Optical Coherence Tomography," *Lasers Surg. Med.* 8: 8, 2004.
69. G. A. Dunn and D. Zicha, eds., *Using DRIMAPS System of Transmission Interference Microscopy to Study Cell Behavior,* Academic Press, 1997.
70. G. A. Dunn and D. Zicha, "Dynamics of Fibroblast Spreading," *J. Cell Sci.* 108, 1239–1249, 1995.
71. T. E. Gureyev, A. Roberts and K. A. Nugent, "Phase Retrieval with the Transport-of-Intensity Equation—Matrix Solution With Use of Zernike Polynomials," *J. Opt. Soc. Am. A-Opt. Image Sci. Vis.* 12, 1932–1941, 1995.
72. T. E. Gureyev, A. Roberts and K. A. Nugent, "Partially Coherent Fields: The Transport-of-Intensity Equation and Phase Uniqueness," *J. Opt. Soc. Am. A-Opt. Image Sci. Vis.* 12: 1942–1946, 1995.
73. J. W. Goodman and R. W. Lawrence, "Digital Image Formation from Electronically Detected Holograms," *Appl. Phys. Lett.* 11: 77, 1967.
74. D. Gabor, "A New Microscopic Principle," *Nature.* 161: 777, 1948.
75. I. Yamaguchi and T. Zhang, "Phase-Shifting Digital Holography," *Opt. Lett.* 22: 1268–1270, 1997.
76. C. J. Mann, L. F. Yu, C. M. Lo and M. K. Kim, High-Resolution Quantitative Phase-Contrast Microscopy by Digital Holography, *Opt. Exp.* 13: 8693–8698, 2005.
77. D. Carl, B. Kemper, G. Wernicke and G. von Bally, "Parameter-Optimized Digital Holographic Microscope for High-Resolution Living-Cell Analysis," *Appl. Opt.* 43: 6536–6544, 2004.
78. N. Lue, W. Choi, G. Popescu, R. R. Dasari, K. Badizadegan and M. S. Feld, "Quantitative Phase Imaging of Live Cells using Fast Fourier Phase Microscopy," *Appl. Opt.* 46: 1836, 2007.
79. Y. K. Park, G. Popescu, K. Badizadegan, R. R. Dasari and M. S. Feld, "Diffraction Phase and Fluorescence Microscopy," *Opt. Exp.* 14: 8263, 2006.
80. Z. Wang, L. J. Millet, M. U. Gillette and G. Popescu, "Jones Phase Microscopy of Transparent and Anisotropic Samples," *Opt. Lett.* 33: 1270, 2008.
81. Z. Wang, L. J. Millet, H. Ding, M. Mir, S. Unarunotai, J. A. Rogers, M. U. Gillette and G. Popescu, "Spatial Light Interference Microscopy (SLIM)," *Nat. Methods* (under review).
82. M. Mir, Z. Wang, K. Tangella and G. Popescu, "Diffraction Phase Cytometry: Blood on a CD-ROM," *Opt. Exp.* 17: 2579, 2009.
83. H. Ding, Z. Wang, F. Nguyen, S. A. Boppart and G. Popescu, "Fourier Transform Light Scattering of Inhomogeneous and Dynamic Structures," *Phys. Rev. Lett.* 101: 238102, 2008.
84. H. Ding, F. Nguyen, S. A. Boppart and G. Popescu, "Optical Properties of Tissues Quantified by Fourier Transform Light Scattering," *Opt. Lett.* 34: 1372, 2009.
85. H. Ding, E. Berl, Z. Wang, L. J. Millet, M. U. Gillette, J. Liu, M. Boppart and G. Popescu, "Fourier Transform Light Scattering of Biological Structures and Dynamics," *IEEE J. Selected Topics in Quantum Electronics* (in press).
86. H. Ding, L. J. Millet, M. U. Gillette and G. Popescu, "Active Cell Membrane Fluctuations Measured Probed by Fourier Transform Light Scattering," *Phys. Rev. Lett.* (under review).

87. M. Hammer, D. Schweitzer, B. Michel, E. Thamm and A. Kolb, "Single Scattering by Red Blood Cells," *Appl. Opt.* 37: 7410–7418, 1998.
88. C. A. Best, "Fatty Acid Ethyl Esters and Erythrocytes: Metabolism and Membrane Effects," PhD Thesis. Northeastern University, Boston, 2005.
89. G. Popescu, in *Methods in Cell Biology*, ed., (B. P. Jena), Elsevier, 2008.
90. H. W. G. Lim, M. Wortis and R. Mukhopadhyay, "Stomatocyte–Discocyte–Echinocyte Sequence of the Human Red Blood Cell: Evidence for the Bilayer-Couple Hypothesis from Membrane Mechanics," *Proc. Natl. Acad. Sci. USA*, 99: 16766–16769, 2002.
91. N. Gov, A. Zilman and S. Safran, "Cytoskeleton Confinement of Red Blood Cell Membrane Fluctuations," *Biophys. J.* 84: 486A, 2003.
92. G. Popescu, K. Badizadegan, R. R. Dasari and M. S. Feld, "Observation of Dynamic Subdomains in Red Blood Cells," *J Biomed Opt.* 11: 059802, 2006.
93. T. Ikeda, G. Popescu, R. R. Dasari and M. S. Feld, "Hilbert Phase Microscopy for Investigating Fast Dynamics in Transparent Systems," *Opt. Lett.* 30: 1165–1167, 2005.
94. J. B. Fournier, D. Lacoste and E. Raphael, "Fluctuation Spectrum of Fluid Membranes Coupled to An Elastic Meshwork: Jump of the Effective Surface Tension at the Mesh Size," *Phys. Rev. Lett.* 92: 018102, 2004.
95. R. Waugh and E. A. Evans, "Thermoelasticity of Red Blood Cell Membrane," *Biophys. J.* 26: 115–131, 1979.
96. G. Popescu, Y. K. Park, R. R. Dasari, K. Badizadegan and M. S. Feld, Coherence Properties of Red Blood Cell Membrane Motions, *Phys. Rev. E.* 76: 031902, 2007.
97. M. S. Amin, Y. K. Park, N. Lue, R. R. Dasari, K. Badizadegan, M. S. Feld and G. Popescu, "Microrheology of Red Blood Cell Membranes Using Dynamic Scattering Microscopy," *Opt. Exp.* 15: 17001, 2007.
98. G. Popescu, T. Ikeda, K. Goda, C. A. Best-Popescu, M. Laposata, S. Manley, R. R. Dasari, K. Badizadegan and M. S. Feld, "Optical Measurement of Cell Membrane Tension," *Phys. Rev. Lett.* 97: 218101, 2006.
99. G. Popescu, K. Badizadegan, R. R. Dasari and M. S. Feld, "Observation of Dynamic Subdomains in Red Blood Cells," *J. Biomed. Opt. Lett.* 11: 040503, 2006.
100. G. Popescu, in *Methods in Cell Biology*, ed., (P. J. Bhanu), 87, Elsevier, 2008.
101. Y. K. Park, M. Diez-Silva, G. Popescu, G. Lykotrafitis, W. Choi, M. S. Feld and S. Suresh, "Refractive Index Maps and Membrane Dynamics of Human Red Blood Cells Parasitized by *Plasmodium falciparum*," *Proc. Natl. Acad. Sci. USA*, 105: 13730, 2008.
102. A. Zilker, H. Engelhardt and E. Sackmann, "Dynamic Reflection Interference Contrast (RIC-) Microscopy: A New Method to Study Surface Excitations of Cells and to Measure Membrane Bending Elastic Moduli, *J. Physique*, 48: 2139–2151, 1987.
103. Y-K. Park, C. A. Best, K. Badizadegan, R. Dasari, M. S. Feld, T. Kuriabova, M. L. Henle, A. J. Levine and G. Popescu, "Measurement of Red Blood Cell Mechanics During Morphological Changes," *Proc. Natl. Acad. Sci.* USA, 107: 6731–6736, 2010.
104. A. Parpart and J. Hoffman, "Flicker in Erythrocytes: Vibratory Movements in the Cytoplasm," *J. Cellular and Comparative Physiology* 47: 295–303, 1956.
105. N. S. Gov and S. A. Safran, "Red Blood Cell Membrane Fluctuations and Shape Controlled by ATP-Induced Cytoskeletal Defects," *Biophys. J.* 88: 1859, 2005.
106. N. S. Gov, "Active Elastic Network: Cytoskeleton of the Red Blood Cell," *Phys. Rev. E.* 75: 11921, 2007.
107. J. Li, G. Lykotrafitis, M. Dao and S. Suresh, "Cytoskeletal Dynamics of Human Erythrocyte," *Proc. Natl. Acad. Sci. USA* 104: 4937, 2007.
108. R. Zhang and F. Brown, "Cytoskeleton Mediated Effective Elastic Properties of Model Red Blood Cell Membranes," *J. Chemical Physics.* 129: 065101, 2008.
109. Y.-K. Park, C. A. Best, T. Auth, N. Gov, S. Safran, G. Popescu, S. Suresh and M. S. Feld, "Metabolic Remodeling of the Human Red Blood Cell Membrane," *Proc. Natl. Acad. Sci. USA*, 107: 1289–1294, 2010.
110. M. Sheetz and S. Singer, "On the Mechanism of ATP-Induced Shape Changes in Human Erythrocyte Membranes. I. The Role of the Spectrin Complex," *J. Cell Biol.* 73: 638–646, 1977.

111. S. Tuvia, S. Levin, A. Bitler and R. Korenstein, "Mechanical Fluctuations of the Membrane-Skeleton Are Dependent on F-Actin Atpase in Human Erythrocytes," *J. Cell Biol.* 141: 1551–1561, 1998.
112. C. Lawrence, N. Gov and F. Brown, "Nonequilibrium Membrane Fluctuations Driven by Active Proteins," *J. Chemical Phy.* 124: 074903, 2006.
113. T. Auth, S. Safran and N. Gov, "Fluctuations of Coupled Fluid and Solid Membranes with Application to Red Blood Cells," *Phys. Rev. E.* 76: 51910, 2007.
114. J. Evans, W. Gratzer, N. Mohandas, K. Parker and J. Sleep, "Fluctuations of the Red Blood Cell Membrane: Relation to Mechanical Properties and Lack of ATP Dependence," *Biophysical J.* 94: 4134, 2008.
115. S. Tuvia, A. Almagor, A. Bitler, S. Levin, R. Korenstein and S. Yedgar, "Cell Membrane Fluctuations Are Regulated by Medium Macroviscosity: Evidence for a Metabolic Driving Force," *Proc. Natl. Acad. Sci. USA* 94: 5045–5049, 1997.
116. D. Mizuno, C. Tardin, C. Schmidt and F. MacKintosh, "Nonequilibrium Mechanics of Active Cytoskeletal Networks," *Science* 315: 370, 2007.
117. N. Gov, A. G. Zilman and S. Safran, "Cytoskeleton Confinement and Tension of Red Blood Cell Membranes," *Phys. Rev. Lett.* 90: 228101, 2003.
118. F. Liu, H. Mizukami, S. Sarnaik and A. Ostafin, "Calcium-Dependent Human Erythrocyte Cytoskeleton Stability Analysis through Atomic Force Microscopy," *J. Structural Biology* 150: 200–210, 2005.
119. G. Tchernia, N. Mohandas and S. Shohet, "Deficiency of Skeletal Membrane Protein Band 4.1 in Homozygous Hereditary Elliptocytosis. Implications for Erythrocyte Membrane Stability," *J. Clinical Investigation* 68: 454, 1981.
120. S. Suresh, "Mechanical Response of Human Red Blood Cells in Health and Disease: Some Structure-Property Function Relationships," *J. Mater. Res.* 21: 1872, 2006.
121. M. Fred and M. Pickens, "Metabolic Dependence of Red Cell Deformability," *J. Clinical Investigation* 48: 795, 1969.
122. Y. K. Park, C. A. Best, T. Kuriabova, M. L. Henle, M. S. Feld, A. J. Levine and G. Popescu, "Optical Measurement of Red Blood Cell Mechanics during Osmotic Changes," unpublished.
123. A. G. Maier, B. M. Cooke, A. F. Cowman and L. Tilley, "Malaria Parasite Proteins That Remodel the Host Erythrocyte," *Nat. Rev. Microbiology* 7: 341–54, 2009.
124. J. P. Mills, M. Diez-Silva, D. J. Quinn, M. Dao, M. J. Lang, K. S. W. Tan, C. T. Lim, G. Milon, P. H. David, O. Mercereau-Puijalon, S. Bonnefoy and S. Suresh, "Effect of Plasmodial Resa Protein on Deformability of Human Red Blood Cells Harboring *Plasmodium falciparum*," *Proc. Natl. Acad. Sci. USA*, 104: 9213–9217, 2007.
125. S. Suresh, J. Spatz, J. P. Mills, A. Micoulet, M. Dao, C. T. Lim, M. Beil and T. Seufferlein, "Connections between Single-Cell Biomechanics and Human Disease States: Gastrointestinal Cancer and Malaria," *Acta. Biomater.* 1: 15–30, 2005.

CHAPTER 16

Superresolution Far-Field Fluorescence Microscopy

Manuel F. Juette

Department of Cell Biology
Yale University School of Medicine
New Haven, Connecticut
Department of Biophysical Chemistry
University of Heidelberg
Department of New Materials and Biosystems
Max Planck Institute for Metals Research
Stuttgart, Germany

Travis J. Gould

Department of Cell Biology
Yale University School of Medicine
New Haven, Connecticut

Joerg Bewersdorf

Department of Cell Biology
Yale University School of Medicine
New Haven, Connecticut

16.1 Introduction and Historical Perspective

The driving force to improve the resolution of fluorescence microscopy is biological research. Understanding biological mechanisms on the subcellular level requires the detailed investigation of the morphology,

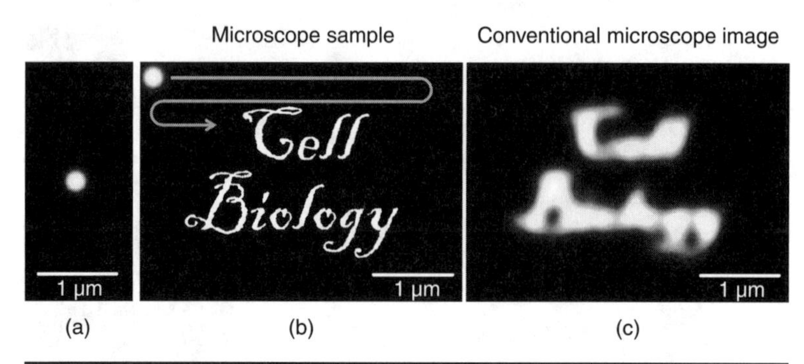

FIGURE 16.1 (*a*) The point-spread function (PSF) of a far-field microscope is diffraction-limited. The typical full width at half maximum is 200 to 250 nm. (*b–c*) Scanning the microscope sample (*b*) with the PSF in a confocal laser scanning microscope yields a diffraction-limited image (*c*). Similar results would be obtained in a wide-field microscope. (*a–c*) show simulated data with the feature sizes representing typical PSFs and subcellular structures of interest.

spatial distribution, and dynamics of the structures involved, ideally under physiologically relevant (live) conditions. Fluorescence microscopy is a tool well suited for such experiments (Lakowicz 2006; Pawley 2006). It relies on specimen labeling using fluorescent marker molecules which can be excited by light of a certain wavelength to emit light of a longer wavelength. Light microscopy is easily applicable to fixed or living cells and has continuously driven important biological discoveries ever since the publication of Robert Hooke's seminal *Micrographia* (1665), the first book concerning the use of microscopy in the life sciences.

However, the size or distances relevant to the structural details of many biological systems are on the order of tens of nanometers (Alberts 2008). Synaptic vesicles, the carriers of neurotransmitters in the brain's nerve cells, have, for example, a diameter of about 40 nm; single proteins, the nanomachines in our cells, are typically between 2 and 10 nm in size. Unfortunately, these length scales are not accessible to conventional far-field light microscopy,* whose resolution is limited by diffraction to about half the wavelength of visible light (i.e., 200 nm) in the focal plane of the objective lens (Fig. 16.1). This limitation can be understood intuitively: a light wave cannot carry spatial information about structures that are significantly smaller than the dimension given by its own spatial frequency (for a more thorough discussion of image formation in the frequency domain, see Section 16.2.4).

This fundamental limitation became apparent in the 19th century with the emergence of the electromagnetic wave theory of light

*The term *far field* indicates that the detector or light source is placed "far away" from the sample, that is, at a distance much longer than the wavelength. In contrast, near-field optical microscopy techniques (not discussed here) take images by scanning a probe tip of subdiffraction size across the surface of a sample and measuring electromagnetic waves in the immediate vicinity (Betzig 1991).

propagation. The first rigorous analysis of image formation in the microscope was carried out by Ernst Abbe in 1873 (Abbe 1873) and later expanded on by Rayleigh (Rayleigh 1879) based on G. B. Airy's calculation of the diffraction pattern created by a circular aperture (Airy 1835). Due to diffraction, even the image of an infinitesimal point-source emitter in the specimen appears as an extended spatial intensity distribution, the so-called *point-spread function* (PSF) of the instrument (Hecht 2001).* Hence, the image formed by a lens-based system is always blurred and resolution is limited (Fig. 16.1). According to the *Rayleigh criterion*, a well-established measure for resolution, two incoherent point sources can only be resolved if their lateral distance d_0 (distance perpendicular to the optical axis) is at least

$$d_0 \approx \frac{0.61\,\lambda}{n\sin\alpha} \tag{16.1}$$

where λ is the wavelength of the light emitted by the point sources, n is the refractive index of the medium surrounding the sample, and α is the maximum angle, as measured from the optical axis, under which the objective lens is able to gather light (Born and Wolf 1999). The quantity $n \sin \alpha$ is commonly referred to as the *numerical aperture* (NA) of the lens.

The theory developed by Abbe and Rayleigh is valid for wide-field microscopes, the most commonly used form of microscopy, in which an extended field of view is illuminated. Another widely distributed imaging mode is *confocal laser-scanning microscopy* (or simply confocal microscopy), which scans the sample point by point using a focused laser beam. The collected light emitted from the laser focus is imaged onto a pinhole which acts as a spatial filter to suppress light originating from out-of-focus planes (*optical sectioning*). The intensity modulation is recorded behind the pinhole using a sensitive detector (e.g., a photomultiplier tube) and digitally converted to an image. Confocal microscopy allows for three-dimensional imaging and offers a slight resolution improvement over Eq. (16.1) by up to a factor of $\sqrt{2}$ (Wilson and Sheppard 1984). Its basic principle was invented by Minsky in 1961 (U.S. Patent No. 3013467) and started to find an application in biology at the end of the 1980s (White et al. 1987). Today, it is a standard tool in biological imaging (Pawley 2006). Both wide-field and confocal microscopy mostly use fluorescence as their signal source (see Section 16.2.3).

A restricted depth of the focal plane can also be achieved in wide-field microscopy using *total internal reflection fluorescence* (TIRF) microscopy. TIRF microscopy (Axelrod 1981) relies on the evanescent electric field created by total internal reflection of the incident light at a cover glass–water interface to excite only those molecules that are very close (typically less than about 100 nm) to the cover glass. This geometry is

*Note that the superposition of point sources is a very good description of the image formation in fluorescence microscopy, where the specimen is labeled with many light emitters of subdiffraction size.

particularly useful to reduce the background in the imaging of membrane structures and their dynamics (Zenisek et al. 2000).

Despite the discouraging limit found by Abbe and Rayleigh, attempts at improving the resolution of light microscopes were made throughout the 20th century, eventually leading to the advent of superresolution microscopy—microscopy beyond the diffraction limit—in the early 1990s. This section is intended to give a brief overview of the large variety of concepts developed for improving the resolution of far-field microscopy. The remainder of the chapter will discuss in more detail the fundamentals of the two most successful approaches underlying genuine diffraction-unlimited imaging: PSF engineering and single-molecule localization.

Improving resolution by adjusting the variables in Eq. (16.1) is possible only to a limited extent. The denominator does not offer much room for optimization, since modern objective lenses feature numerical apertures close to the theoretical limits of their respective immersion media (e.g., for typical immersion oil with $n = 1.518$, numerical apertures of over 1.45 are now routinely achieved, corresponding to $\sin \alpha > 0.95$).

Using probes of significantly shorter wavelengths than those of visible light, such as X-ray radiation or matter beams (electron microscopy), can offer greater resolution improvements. However, important advantages of microscopy using visible light may have to be sacrificed: live-cell compatibility, the possibility to image deep inside cells or tissue, and the availability of a broad range of specific labeling strategies make fluorescence microscopy indispensable for biological research.

A logical step after the realization that Eq. (16.1) does not leave much room for resolution improvement is to ask which of the assumptions in the derivation of the Rayleigh criterion might be altered by implementing new experimental geometries. One of these assumptions is that of unobstructed, uniform circular apertures. Articles based on information theory from the early 1950s often emphasized that it is possible to make the central PSF maximum arbitrarily sharp by dropping this assumption (Di Francia 1952), for example, by using annular apertures (Sheppard and Choudhur 1977). However, this is of limited practical use as it leads to pronounced side-maxima even for generating a moderately narrower central peak.

A related technique called *structured illumination microscopy* (SIM) enhances the resolution by illuminating only two (or three) extreme points in the aperture of the objective, this creates a wide-field interference pattern in the sample. Here, the spatial information content of the micrographs is enhanced by a factor of 2 by recording a series of images at different positions and orientations of a spatially modulated excitation pattern (Heintzmann and Cremer 1999; Gustafsson 2000; Kner et al. 2009). SIM also improves the optical sectioning capabilities of wide-field microscopy.

Instead of trying to work within the limits of the aperture of the objective lens or to further increase it, *4Pi* microscopy (Hell and Steltzer 1992) and *I*^{5}*M* (Gustafsson et al. 1999) combine the apertures of two opposing objectives (Bewersdorf et al. 2006). The two lenses that focus into the same spot are illuminated coherently so that they act as one unit. While 4Pi microscopy is implemented in a confocal point-scanning geometry, I^{5}M uses wide-field illumination and detection. The resulting interference pattern created by the two opposing objectives improves the axial resolution (resolution along the optical axis) by a factor of up to 7 down to ~100 nm (full width at half-maximum, FWHM). To eliminate "ghost images" produced by side-maxima in the PSF, these methods utilize in most cases image deconvolution, a method of mathematical postprocessing. The increased resolution results from the combined apertures of the two lenses; thus fundamentally, both methods are still diffraction-limited.

In the last two decades, much effort has been put forth to develop methods that fundamentally break the diffraction barrier. In 1994, the concept of stimulated emission depletion (STED) microscopy (Hell and Wichmann 1994) showed for the first time that diffraction-*unlimited* resolution could be attained in a lens-based microscope. By forcing fluorescent molecules out of their excited states before they can fluoresce, focal volumes smaller than the diffraction limit can be produced in a laser-scanning geometry (see Section 16.2.1). The concept of STED has also been generalized to include other reversible saturable optical fluorescence transitions (RESOLFT) (Hell et al. 2003), including reversibly photoswitching, fluorescent proteins (Hofmann et al. 2005), ground-state depletion (Hell and Kroug 1995; Bretschneider et al. 2007), modulation of quantum dot fluorescence (Irvine et al. 2008), and molecular optical bistability (Bossi et al. 2006).

More recently, a new approach toward superresolution imaging that relies on localizing single fluorescent molecules has been developed.* Localizing the positions of isolated light emitters is a concept that has been used for many years in single-particle tracking experiments (Saxton and Jacobson 1997) in which localization precisions of down to ~ 2 nm have been demonstrated (Yildiz et al. 2003). To extend this method to resolve complex structures below the diffraction limit, fluorescence photoactivation localization microscopy (FPALM) (Hess et al. 2006), photoactivated localization microscopy (PALM) (Betzig et al. 2006), and stochastic optical reconstruction microscopy (STORM) (Rust et al. 2006) were developed simultaneously in 2006. All three methods localize optically distinguishable subsets of single photoactivatable fluorophores that are cyclically activated, their fluorescence read out, and bleached (see Section 16.2.2). In addition

***Localization* in the context of this chapter means "determining the position of a particle mathematically."

to generating images, these localization-based methods can also provide single-molecule information such as absolute numbers of molecules, diffusion properties (Hess et al. 2007; Manley et al. 2008), and anisotropies (Gould et al. 2008; Testa et al. 2008). This additional information is inaccessible to ensemble imaging techniques.

In this chapter we review the fundamental principles behind these two concepts toward superresolution imaging: PSF engineering as used in STED and RESOLFT and single-molecule localization and image reconstruction as used in FPALM and related methods. For simplicity of notation, for the remainder of this chapter we will refer to RESOLFT-type approaches as STED and localization-based approaches as FPALM.

16.2 Fundamentals of Superresolution Microscopy

16.2.1 PSF Engineering

The first successful attempts at superresolution in the far-field domain were based on confocal laser-scanning microscopy. Because the image is generated point by point, the resolution of a confocal microscope is dictated by the smallest achievable focal volume, whose lateral extent can be approximated by Eq. (16.1) (even for high NA lenses, it is about 2.5 times more prolonged in the axial direction). In order to turn confocal microscopy into superresolution microscopy, the shape of the focus needs to be modified, a concept known as *PSF engineering*.

STED was originally proposed in 1994 (Hell and Wichmann 1994) and realized experimentally in 1999 (Klar and Hell 1999). It overcomes the diffraction limit by exploiting the inherent photophysical properties of the probe molecules themselves to reduce the size of the effective focal volume.

In a standard confocal microscope, fluorophores within the excitation focus are transferred into an excited electronic state by the absorption of a photon. Within that electronic state, nonradiative relaxation processes dissipate some of the absorbed energy and put the molecule into a lower vibrational level. The molecule returns to its electronic ground state by emitting a photon of slightly lower energy than that of the originally absorbed photon. This loss in energy is equivalent to an increase (red shift) in wavelength and is known as the *Stokes shift*. The red-shifted fluorescence light can be separated from the excitation light with a filter and is detected to form the image.

In STED, molecules are forced back to the ground state before they can fluoresce. Following illumination of a diffraction-limited volume with the excitation laser, a second laser of a longer wavelength, the depletion laser, is focused into the same volume (see Fig. 16.2*a* for a schematic setup). A phase mask in its beam path gives the resulting

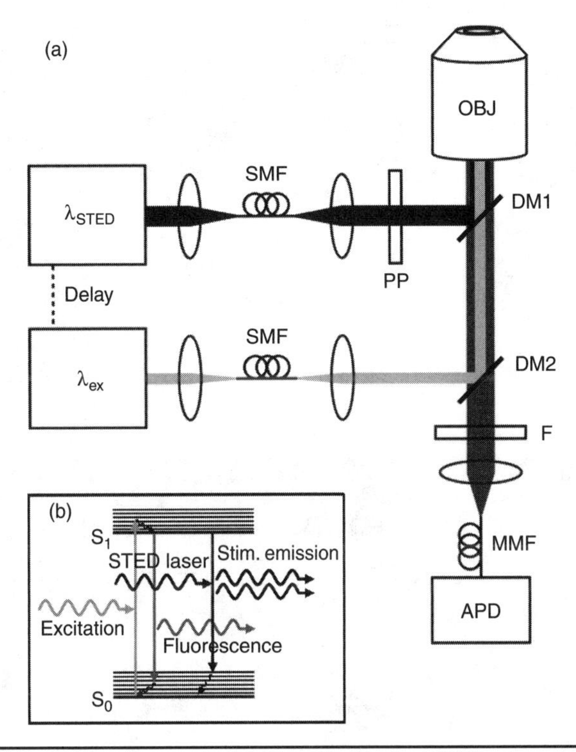

FIGURE 16.2 STED microscopy. (*a*) Simplified typical STED microscopy setup. Two pulsed lasers, one for the excitation wavelength (λ_{ex}) and the STED wavelength (λ_{STED}), each are focused into single-mode optical fibers (SMF) to stretch the pulses and collimated again. A phase plate (PP) shapes the STED pulse to form a donut after focusing into the sample by the objective lens (OBJ). The dichroic beam splitters (DM1 and DM2) coalign the two laser beams with the detection beam path. Fluorescence is transmitted by DM1 and DM2, the remaining laser light filtered out by the band-pass filter (F) and focused onto a multimode fiber (MMF) which acts as a pinhole and transfers the fluorescence to an avalanche photodiode (APD). (*b*) Jablonski diagram of a typical fluorophore, showing the transitions of excitation, spontaneous fluorescence, and stimulated emission.

focal spot a donut shape, featuring an intensity-minimum of essentially 0 which is aligned to overlap with the excitation intensity-maximum (see Fig. 16.3*a-c*). After excitation, fluorescent molecules outside the intensity 0 of the depletion beam are forced to return to the ground state by stimulated emission (Fig. 16.2*b*).* The photon generated in this process has the same wavelength as the stimulating photon (which is tuned to the tail of the probe's emission spectra) and can be filtered out from the rest of the collected fluorescence. Only molecules

*The same phenomenon is also the basis of the operation of lasers.

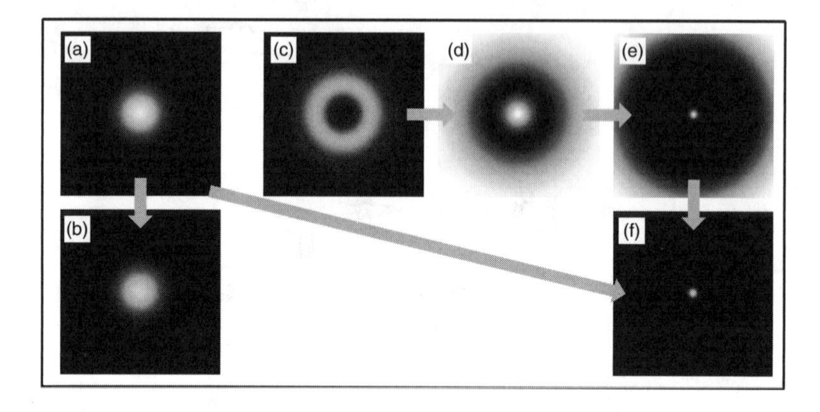

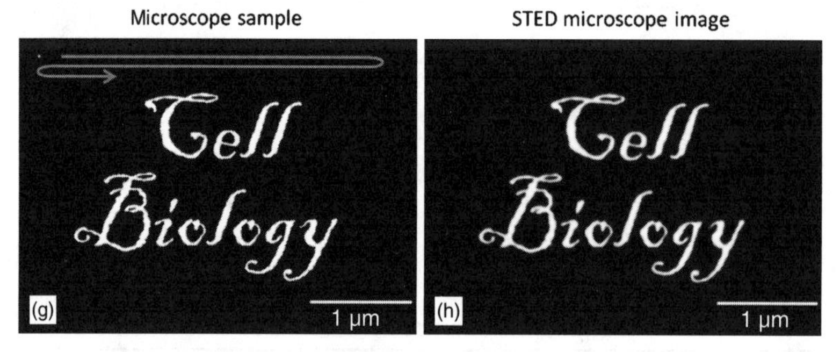

FIGURE 16.3 STED microscopy principle. (*a*) Excitation PSF. (*b*) Effective PSF of a conventional confocal microscope. (*c*) STED donut. (*d*) Depletion profile of the STED donut at low intensities. (*e*) Depletion profile of the STED donut at high intensities. (*f*) Effective STED-PSF. (*g–h*) Scanning the microscope sample (*g*) with the effective STED-PSF yields a subdiffraction-limited image (*h*).

near the center of the focal volume return to the ground state by spontaneous emission and contribute to the image (Fig. 16.3*d*).

The phenomenon of stimulated emission is insufficient to break the diffraction barrier since the width of the intensity minimum in the depletion beam is itself limited by diffraction. The additional requirement for superresolution is saturation of the stimulated emission process (Hell and Wichmann 1994). By increasing the depletion intensity, even relatively low amplitudes of the donut are sufficient to saturate the depletion process; that is, practically all fluorescence in these regions is quenched (Fig. 16.3*e*). The insides of the donut become "steeper" and fluorescence emission is confined to a volume smaller than the diffraction limit (see Fig. 16.3*e–h*). The inherent nonlinearity of the saturation process results in an effective PSF that can be made arbitrarily small, in principle, by increasing the depletion intensity. Thus the resolution in STED microscopy is fundamentally diffraction-unlimited.

Resolution in STED microscopy is determined by the FWHM of the effective PSF which can be approximated by (Hell 2009)

$$d_{\mathrm{STED}} \approx \frac{\lambda}{2n \sin\alpha \sqrt{1+I_{\max}/I_{\mathrm{sat}}}} \tag{16.2}$$

where I_{max} denotes the peak intensity of the depletion beam, I_{sat} the fluorophore-dependent saturation intensity in which the emitted fluorescence is reduced by a factor of $1/e$, and n, λ, and α are as defined in Eq. (16.1). For typical dyes, I_{sat} is on the order of tens of MW/cm^2 (Hell 2007); this requires I_{max} to be on the order of hundreds of MW/cm^2 to push the resolution significantly beyond the diffraction limit. While these large depletion intensities may at first appear incompatible with biological imaging, they are still several orders of magnitude less than those used in multiphoton microscopy (Hell et al. 2004). In recent implementations, STED has been reported to achieve resolutions of about 30 nm (Schmidt et al. 2009) down to less than 10 nm (Rittweger et al. 2009).

Typically, STED is implemented with complex pulsed-laser systems that require synchronization and pulse stretching (Klar et al. 2000; Donnert et al. 2006). However, the use of continuous wave lasers has been demonstrated more recently; this greatly simplifies the experimental setup, although three- to five-fold higher depletion intensities are required to produce similar resolution to that with pulsed lasers (Willig et al. 2007). The pulsed-laser setup can also be simplified using white light produced through supercontinuum generation in an optical fiber source in which the excitation and depletion beams originate from the same laser and thus are inherently synchronized (Wildanger et al. 2008).

16.2.2 Localization-Based Microscopy

STED, in brief, achieves superresolution by targeted sequential switching of molecules to ensure that at any given time, only fluorophores within a subdiffraction volume of the specimen contribute to the signal. It makes use of the nonlinearity of saturable transitions.

Localization-based microscopes like FPALM are based on the realization that a nontargeted, stochastic switching process can be used for the same purpose, as long as a minimum of three molecular states is available, two of which belong to the regular fluorescence of the molecule. The third state is a "dark" (i.e., nonfluorescent) state which has typically a photoinduced transition to the fluorescent ground state. Early attempts at using single-molecule localization in imaging applications relied on using probes with distinct emission spectra (Lacoste et al. 2000), blinking of quantum dots (Lidke et al. 2005; Lagerholm et al. 2006), or stepwise photobleaching of single probes (Gordon et al. 2004; Qu et al. 2004). However these approaches are limited so that only a few molecules can reside within a diffraction-limited area; this results in an effective resolution severely limited by the density of localized molecules (see Section 16.4). In another approach to improve resolution, diffusing molecules are localized as they become fluorescent on binding to a target species (Sharonov and Hochstrasser 2006), yet control over the number of visible molecules still requires controlling the concentration of probes.

The general principle of FPALM can be summarized in the following way: Initially, the entire population of fluorescent molecules resides in the dark state. The sample is illuminated with both the activation and the excitation wavelength in a wide-field geometry (Fig. 16.4). By illuminating with the activation laser (often with a wavelength close to the ultraviolet) at a low intensity, sparse subsets of molecules are stochastically converted into their active state, from which they can be repeatedly excited by the readout laser and their fluorescence detected until they are finally "deactivated" by spontaneous photobleaching or return to the dark state (reversible photoconversion) (Fig. 16.5). Thus control over the number of visible molecules is maintained *optically*.

As a possible simplification, the use of a separate activation beam is not generally required (Hess et al. 2006; Egner et al. 2007), provided that the probe features a sufficiently high activation cross section even at the excitation wavelength. A single laser can also be used to drive molecules into long-lived dark states, and molecules can be localized as they individually return to the fluorescent state (Fölling et al. 2008; Baddeley et al. 2009). However, independent adjustment of activation and bleaching rates to control the density of activated molecules is only possible if separate lasers are used. The activation and bleaching rates

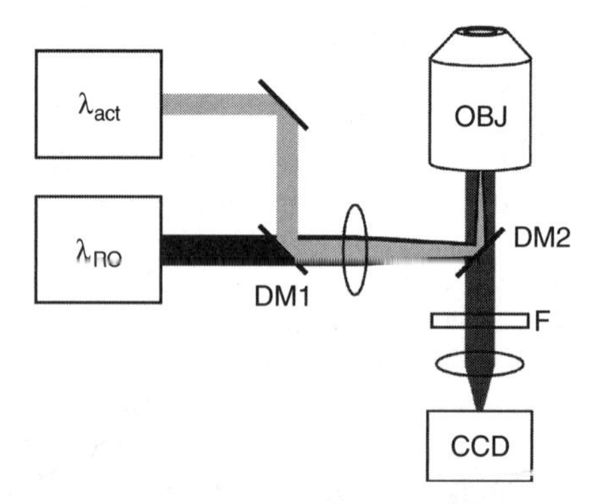

FIGURE 16.4 FPALM. Simplified typical FPALM setup. Two lasers, one for the readout wavelength (λ_{RO}) and the activation wavelength (λ_{act}), each are focused into the back aperture of the objective lens (OBJ) to create a homogeneous wide-field illumination in the sample. The dichroic beam splitters (DM1 and DM2) coalign the two laser beams with the detection beam path. Fluorescence is transmitted by DM2, the remaining laser light is filtered out by the band-pass filter (F) and the fluorescence imaged by a charge-coupled device (CCD) camera.

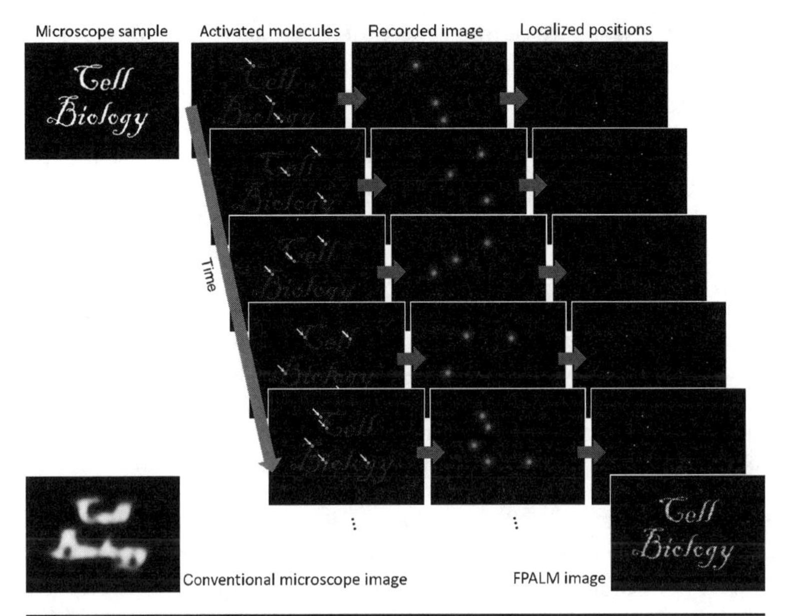

FIGURE 16.5 FPALM imaging method. Multiple images of different sparse sets of activated molecules are imaged and the recorded molecules localized. An FPALM image is assembled from all determined positions.

are proportional to the corresponding laser intensities. Appropriate adjustment of these intensities can be used to make sure that, at any given time, only a sparse distribution of probe molecules is visible, so that their average separation is greater than the Rayleigh limit (i.e., individual molecules are resolvable in the conventional sense). Using a sensor array with single-photon sensitivity such as an electron-multiplying charge-coupled device (EMCCD) camera, it is possible to record a series of frames, each of which contains on average not more than one active fluorescent molecule per diffraction-limited volume.

Assume a sufficiently large number N of photons emitted by an individually resolvable molecule is detected by the camera to image the molecule's PSF. The PSF can be interpreted as the probability density function for each single photon to hit a particular spot on the camera. The standard deviation σ_0 of this distribution (related to the quantity d_0 from Eq. 16.1) quantifies the localization precision of the molecule based on the detection of one photon. Detecting N photons means performing N identical measurements with a standard deviation σ_0; thus the resulting localization precision (in the absence of instrumentation-induced errors and background signal) is

$$\sigma_{\text{localisation}} = \frac{\sigma_0}{\sqrt{N}} \tag{16.3}$$

From a sequence of frames recorded by the camera, we can now calculate the position of each molecule with a precision determined by the number of detected photons and the achieved signal-to-noise ratio. Localization can be performed for example by a full least-squares fit to a two-dimensional Gaussian (Cheezum et al. 2001) or by faster methods like the Gaussian mask algorithm (Thompson et al. 2002).

All the determined particle positions are stored in a list, which is finally used to reconstruct a high-resolution image by plotting the stored position data in a coordinate grid. This can be done by simply binning the localized molecules into pixels of a size corresponding to the expected resolution or by plotting the position of each molecule as a two-dimensional Gaussian with a width corresponding to the localization precision. Figure 16.5 shows a schematic of this procedure. In practice, enough detected photons per molecule can be obtained to achieve lateral localization precisions less than a few tens of nanometers (Betzig et al. 2006; Hess et al. 2006; Rust et al. 2006). From Eq. (16.3) it is evident that the position of a single emitter can be localized with a precision much better than the diffraction limit. However, producing an image with enhanced resolution also requires that a high enough density of molecules is localized to represent the structure of interest in accordance with the Nyquist frequency (Betzig et al. 2006) (see Section 16.4).

To summarize, FPALM records sequences of individual frames (typically tens of thousands) in which the images of the well-separated single-molecule PSFs are captured as they are cyclically activated, read out, and bleached by appropriate adjustment of the readout and activation laser intensities. During analysis, single molecules are identified and their positions are calculated with a precision that is determined by the number of detected photons, the camera pixel size, and the noise in the background signal. The positions of all molecules are then plotted to generate an image with subdiffraction resolution. Detailed procedures for image acquisition and analysis can be found in Gould et al. (2009).

16.2.3 Fluorescent Probes

The fluorescent probes commonly used in fluorescence microscopy generally fall into one of the following categories: fluorescent proteins (FPs), organic dyes, or fluorescent nanoparticles. Figure 16.6 shows examples of fluorescent probes. FPs can be genetically encoded to label proteins of interest when expressed in cell cultures or organisms, whereas dyes and nanoparticles often have to be inserted into permeabilized cells during sample preparation and attached to the structure of interest using labeling strategies such as immunostaining.

FPs and organic dyes differ significantly in size with small dye molecules being made up of only tens of atoms; whereas proteins often have chain lengths of hundreds of peptides, corresponding to molecular

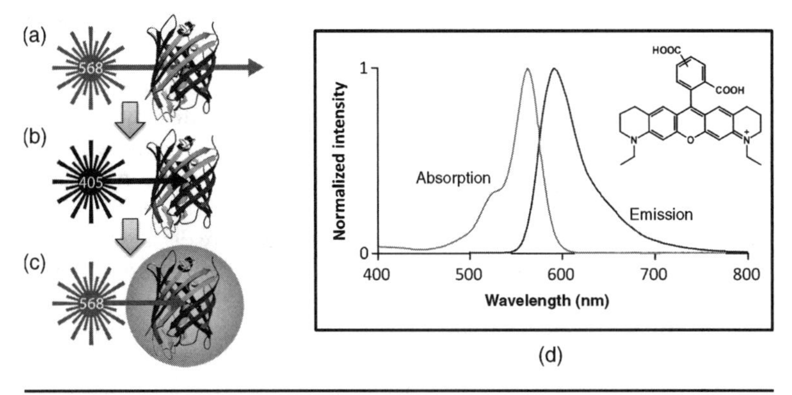

Figure 16.6 Examples of fluorescent probes used in superresolution microscopy. (*a–c*) Photoactivation of the photoactivatable fluorescent protein PAmCherry. (*a*) In its inactive state, PAmCherry shows almost no absorption at 568 nm. (*b*) Absorption of a 405-nm photon induces a conformational change which "activates" the molecule. (*c*) The active molecule absorbs at 568 nm and emits fluorescence with a peak around 595 nm (Subach et al. 2009). (*d*) Molecular structure (top right), absorption (gray curve) and emission spectra (black curve) of the STED-compatible fluorescent dye ATTO 565. The PAmCherry structure was taken from Protein Data Bank, http://www.rcsb.org/pdb/. The structure of ATTO 565 was downloaded from the manufacturer's website, http://www.atto-tec.com. Spectral data was obtained from the fluorophore database at Graz University of Technology, http://www.fluorophores.org.

weights in the kDa-range. It should be kept in mind that the definition of resolution on the molecular scale depends on factors such as label size and density. Immunofluorescence techniques may also require the length of linker molecules to be taken into account.

In their initial implementations, localization-based superresolution techniques made use of specialized fluorescent probes that possess at least two optically distinct states (an activated and an inactivated state). This subset of probes includes variants of FPs, organic dyes, and organic dye pairs that can all be switched between states using light of an appropriate wavelength. Such variants of FPs include photoactivatable FPs (PAFPs), which can be converted from an inactive dark (essentially nonfluorescent) state to an active (potentially fluorescent) state (see Fig. 16.6*a–c*), and photoswitchable FPs, which can be converted between states that emit at different wavelengths. Often both types of FPs are collectively referred to as PAFPs. Some examples of PAFPs include PA-GFP (Patterson and Lippincott-Schwartz 2002), the photoactivatable variant of green fluorescent protein, and PAmCherry (Subach et al. 2009), the photoactivatable version of the red-emitting mCherry (Shaner et al. 2004); examples of photoswitchable FPs include the green-to-orange switching EosFP (Wiedenmann et al. 2004), and Dendra2 (Chudakov et al. 2007). The photophysical properties of many PAFPs can be found in several reviews (Chudakov et al. 2005; Shaner et al. 2007; Day and Davidson 2009).

The preceding examples exhibit mostly irreversible conversions between states; that is, each PAFP can only be activated or switched and imaged once. However, Dronpa (Ando et al. 2004) and variants (Ando et al. 2007; Stiel et al. 2007), a family of green-emitting PAFPs, exhibit reversible switching and can be imaged over many switching cycles. A reversible red emitter has also been recently developed (Stiel et al. 2008). It has also been demonstrated that localization microscopy is possible using the conventional-enhanced yellow fluorescent protein (Biteen et al. 2008), which was observed to exhibit switching behavior more than a decade ago (Dickson et al. 1997).

Aside from PAFPs, photoswitchable organic dyes such as caged-fluorescein (Invitrogen) and caged-rhodamine (Fölling et al. 2007) can also be used. STORM has typically made use of reporter-activator pairs of the cyanine dyes which have been shown to exhibit photoswitching behavior when used in buffers with reduced levels of oxygen (Bates et al. 2005), a condition that may not be desirable for live-cell imaging. Recently, the use of conventional (i.e., nonswitching) dyes has been demonstrated in localization-based microscopy, based on exploiting intrinsic dark states of such molecules (Fölling et al. 2008; Heilemann et al. 2008; Lemmer et al. 2008; Baddeley et al. 2009).

In principle, any fluorescent probe is capable of stimulated emission and should be compatible with STED microscopy. However, in practice the use of many dyes is limited by photobleaching caused by high intensities of the depletion laser. While using reduced laser repetition rates (1 MHz) to image with a triplet-state relaxation scheme has been shown to reduce photobleaching and thus increase the fluorescence signal in many cases (Donnert et al. 2006), probes typically used in STED have high quantum efficiencies and low levels of triplet-state buildup such as the ATTO family of dyes (ATTO-TEC, Germany). However, the successful use of certain FPs in STED has been reported (Willig et al. 2006; Hein et al. 2008).

16.2.4 The Optical Transfer Function (OTF)

This section covers some aspects of the theoretical background of superresolution microscopy. It is not required in order to understand the rest of the chapter but is intended as a reference for readers with an interest in the mathematics behind image formation.

Image Formation in the Frequency Domain

Two attributes inherent in most far-field fluorescence microscopes provide the basis for a sample-independent definition of their resolution: first, due to careful optical design, the imaging performance of a microscope is spatially invariant; that is, a microscope's imaging performance is independent of the position of the observed object within a reasonably large field of view. Second, the system is linear; that is, an image of a complex object can be interpreted as the sum of the images of object components. This latter attribute is a consequence of the incoherence of fluorescence. The phase of the electromagnetic wave emitted by one fluorescent probe is not correlated to that of any other probe molecule and therefore intensities rather than field amplitudes are superimposed. Mathematically, the microscope image $i(\mathbf{x})$ of a structure $s(\mathbf{x})$ can therefore be described as the convolution with a spatially invariant PSF, $h(\mathbf{x})$.

$$i(\mathbf{x}) = s(\mathbf{x}) \otimes h(\mathbf{x}) \tag{16.4}$$

$h(\mathbf{x})$ describes the image of an infinitesimal fluorescent probe, and $s(\mathbf{x})$ describes the spatial distribution of these fluorescent probes, $\mathbf{x} = (x, y, z)$ is the three-dimensional position vector.

The convolution theorem of the Fourier theory (Goodman 2004) provides an alternative to PSFs to describe a microscope's imaging performance. Equation (16.4) can be transferred into the frequency domain by a Fourier transform, replacing the convolution by simple multiplication:

$$I(\mathbf{k}) = S(\mathbf{k})\, H(\mathbf{k}) \tag{16.5}$$

Here, I, S, and H are the Fourier transforms of the functions i, s, and h, respectively, $\mathbf{k} = (k_x, k_y, k_z)$ represents the three-dimensional vector in spatial frequency space. The image, structure, and PSF are now described by their spatial frequency content (i.e., the degree to which this function can be described by a wave represented by $\exp(i\mathbf{kx})$ with spatial frequency $\mathbf{k}$ instead of their distribution in real space. Usually, they carry next to their real component an imaginary part. The function H is called the *optical transfer function* (OTF) and describes how the different spatial frequency components of the observed structure are transmitted by the microscope. For a point-symmetric PSF, which is a reasonable approximation for this discussion, the properties of a Fourier transform tell us that the imaginary component of H vanishes and H is real.

The OTF is of great value in the discussion of spatial resolution, its fundamental limitation, and how to overcome it. The shape of the OTF can be directly derived in frequency space from the electromagnetic field distribution analog to the PSF which is calculated from the focused electromagnetic wave. In a confocal laser-scanning microscope, the effective PSF of the system is the product of the excitation and detection probability distributions. The first one is represented by the focused laser intensity; the second one is determined by the constraints of detecting through the confocal pinhole. For simplicity, we invert the detection beam path and approximate the probability to detect a photon emitted at a position $\mathbf{x}$ in the sample by the diffraction-limited image of an infinitesimal pinhole in the sample. We further use here the scalar approximation of electrodynamics by omitting the vectorial nature of the E-field and also approximate excitation as well as fluorescence by monochromatic spectra. The excitation focus as well as the detection can each be represented by a traveling, focused light wave in the focal region of the sample. Their real components are displayed in Fig. 16.7*a* and *b* for the example of an objective lens with a high numerical aperture. To obtain the excitation and detection probability, that is, the field intensity, the electromagnetic waves have to be multiplied by their complex conjugates (Fig. 16.7*c* and *d*). For the final effective probability to detect a molecule at a certain position, these two probability distributions are multiplied and yield the effective PSF (Fig. 16.7*e*).

In frequency space, the electromagnetic wave of the focused excitation laser beam resembles a very simple shape; Fraunhofer diffraction shows that the focused wave can be interpreted as the superposition of planar waves (Hecht 2001). All nonzero contributions therefore have to lie on a spherical shell with radius of the wave number $k = 2\pi n/\lambda$ with λ/n being the wavelength of light in the medium with refractive index n.

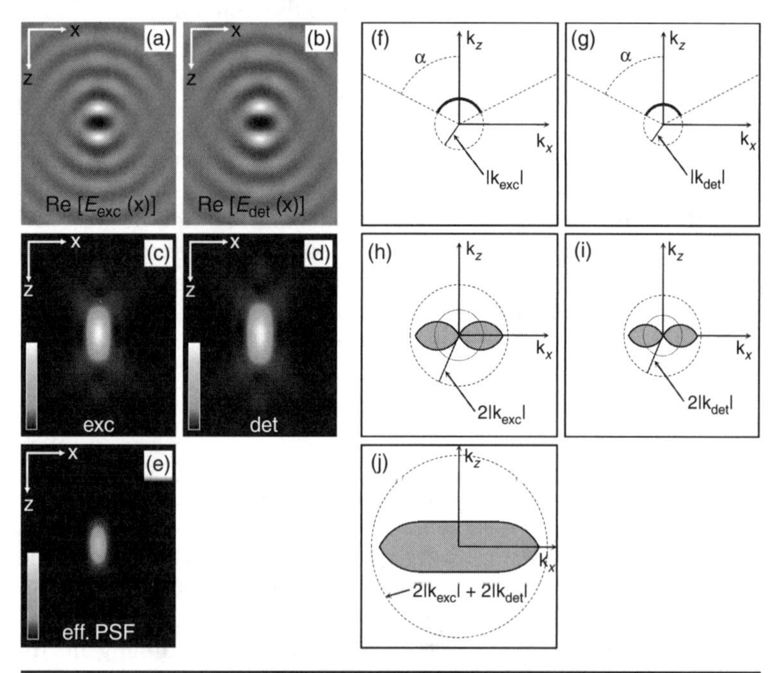

Figure 16.7 PSF and OTF assembly for a confocal microscope. (*a*) Real part of the focused electromagnetic wave of the excitation laser. (*b*) Equivalent for the fluorescence detection. (*c–d*) Corresponding intensity distributions. (*e*) Effective confocal PSF obtained by multiplying the excitation and detection PSFs. (*f–j*) corresponding supports in frequency space. The dashed circles represent the diffraction limit at every step, α denotes the half-opening angle of the objective aperture. The lookup table in (*c*), (*d*), and (*e*) has been modified from a linear scale to highlight low intensities in the PSFs.

The aperture angle of the objective lens used determines which wave vector directions contribute to the focus. This angle limits the nonzero contributions in frequency space to the cap of the spherical shell with the same opening angle (see Fig. 16.7*f*). The longer wavelength of fluorescence manifests itself by a smaller sphere diameter (Fig. 16.7*g*). To obtain the OTFs of excitation and detection, the multiplication of the E-field with its complex conjugate in real space translates into the convolution of the structures of Fig. 16.7*f* and g with their point-reflected counterpart. The result is a three-dimensional shape similar to a donut but with a singularity in the center. The support of the resulting function is displayed as a cross section in Fig. 16.7*h* and *i*.* Analogous to the multiplication of excitation and detection PSFs, these shapes have to be convolved with each other to obtain the effective OTF of a confocal laser-scanning microscope. The support of the OTF (see Fig. 16.7*j*) demonstrates that higher spatial frequencies are transmitted in the focal plane of the microscope (x, y) than in the axial direction (z), a phenomenon that also manifests itself in the larger axial extent of the PSF compared to its lateral dimensions (Fig. 16.7*e*).

The most important observation related to the OTF in the context of this chapter is, however, that its support is finite. Spatial frequencies outside the support are not transmitted. As can be seen from the above derivation of the OTF's shape, the size of the support depends on the aperture angle and wavelengths involved. A larger aperture angle can increase the support, but its size is ultimately limited to a sphere with a radius of $2k_{exc} + 2k_{det}$ with $k_{exc} = 2\pi n/\lambda_{exc}$ and $k_{det} = 2\pi n/\lambda_{det}$. This sphere represents the diffraction limit of conventional confocal microscopy.

Similar limits can be found for all other conventional microscopy techniques: for regular wide-field microscopy, where the whole field of view is illuminated homogeneously, Fig. 16.7*f* reduces to a single point at the top of the shell. Consequently, the excitation OTF does not contain any spatial frequency beyond the zero-frequency and the effective OTF is identical to the detection OTF. Structured illumination microscopy creates not a single but several points in the cap shown for the confocal microscopy in Fig. 16.7*f*. This creates a final OTF support very similar to that of confocal microscopy (Gustafsson 2000).† 4Pi microscopy (using one-photon excitation) mirrors the cap of E-field vectors shown in Fig. 16.7*f* to the bottom of the same shell, and the resulting effective OTF can extend to the top and bottom of the $2k_{exc} + 2k_{det}$ shell but cannot extend beyond it (Bewersdorf et al. 2006). I^5M and I^5S microscopy are combinations of the 4Pi microscopy approach with wide-field and structured illumination microscopy, respectively (Gustafsson et al. 1999; Bewersdorf et al. 2006; Shao et al. 2008).

---

*The support is the set of points over which a function is not zero, which in the case of OTFs denotes the generally transferred spatial frequencies.

†Its typically enhanced resolution compared to confocal microscopy is due to the more effective transmission of high spatial frequencies.

STED exploit means to create spatial frequencies beyond the diffraction-limiting sphere. FPALM circumvents the problem of the diffraction limit by replacing the problem of transmitting higher spatial frequencies with determining the position of a PSF-like object which is encoded in the argument of its complex Fourier transform.

STED in the Context of OTFs

To create spatial frequencies beyond the limiting sphere shown in Fig. 16.7*j*, nonlinearities of the excitation (or theoretically the detection) PSF have to be exploited. In two-photon excitation (2PE), for example, the excitation probability is proportional to the square of the excitation-light intensity (Denk et al. 1990). The 2PE PSF therefore has to be squared compared to the regular one-photon excitation (1PE) PSF. For the excitation OTF, this in turn means that the 1PE OTF has to be convolved with itself; this doubles the corresponding frequency range and leads to an effective limiting sphere of a 2PE confocal microscope of $4k_{exc} + 2k_{det}$ instead of $2k_{exc} + 2k_{det}$ as reported for the 1PE case. Unfortunately, each absorbed photon in 2PE carries only approximately half the required excitation energy resulting in double the wavelength, or half the k_{exc} value of 1PE. Although 2PE microscopy has its benefits over 1PE microscopy due to the longer wavelength and the associated reduction in scattering and spatial confinement of absorption, this long wavelength eliminates the resolution advantage. The same is true for higher-order multiphoton absorption effects.

Other nonlinear processes, however, which do not require longer wavelengths carry the potential to enhance the resolution of fluorescence microscopy. An unlimited OTF support can be created by nonlinear effects which are represented by a power series. In particular, exponential functions of the PSF fall into this category; for example,

$$h'(\mathbf{x}) = \exp[a\, h(\mathbf{x})] = \sum_{n=0}^{\infty} \frac{a^n}{n!} h^n(\mathbf{x}) \tag{16.6}$$

They appear in practice, in the form of saturation effects, which can be approximated by an exponential curve. A fluorophore cannot be excited to emit another fluorescence photon while it is still in its excited electronic state from the previous excitation. Fluorescence emission therefore has to increase less and less with increasing excitation intensities. For high-laser powers, the excitation PSF saturates in its peak, and so creates a plateau with steeper edges. These steeper edges are responsible for the higher spatial frequencies contained in the OTF.

STED microscopy utilizes the same mechanism in a more elegant scheme; here, the saturation effect is not created in excitation but in depletion of the excited fluorophores. Thus it narrows the remaining excitation volume with increasing depletion intensity. This creates higher spatial frequencies in a more favorable manner where the superresolution effect is represented directly by the sharper effective

PSF and not just sharper edges of a broad PSF which requires extensive data processing to create a useful image.

FPALM in the Context of OTFs

In FPALM, single fluorophores are imaged separately. The analysis of every molecule can ideally be separated from all other molecules. Since the molecule is essentially a point object, its image $I_{\text{fluorophore}}$ resembles a PSF shifted to the location of the molecule. This shift can be described mathematically by the convolution of the effective PSF with the δ-distribution $\delta(\mathbf{x} - \mathbf{x}_0)$, where $\mathbf{x}_0$ denotes the three-dimensional particle position. This convolution is translated into multiplication of the OTF with the Fourier transform of the δ-distribution to describe the image of the single fluorophore. Assuming that the PSF is symmetrical, Fourier theory tells us that the OTF is a real function. The Fourier transform of the δ-distribution is essentially the phase factor exp $(-i\mathbf{k}\mathbf{x}_0)$. The fluorophore's position is therefore encoded in the phase of the Fourier-transformed image while the OTF describes the shape of its image (see Fig. 16.8).

$$I_{\text{fluorophors}}(\mathbf{k}) = H(\mathbf{k})\, e^{-i\mathrm{x}_0\mathbf{k}} \tag{16.7}$$

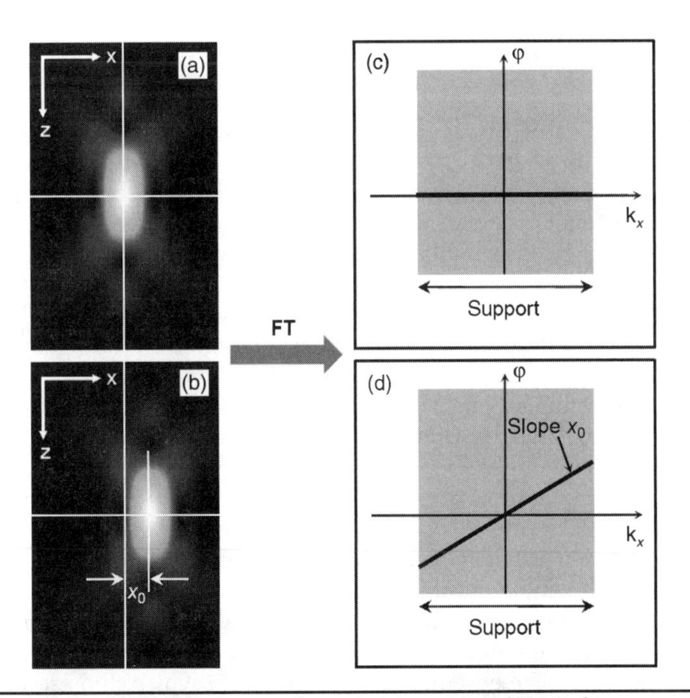

Figure 16.8 Shift of the image of a molecule and its effect in frequency space. (*a* and *b*) Centered and shifted molecules, respectively. (*c* and *d*) Their respective phase in the k_x-direction of frequency space. The gray-shaded area denotes the support of the OTF (compare to Fig. 16.5). The slope of the phase is proportional to the molecule's position.

To determine the position, one can fit $\mathbf{x}_0\mathbf{k}$ to the observed phase term with $\mathbf{x}_0$ as the slope being the fit parameter. This fit is dependent on the size of the support of the OTF, since the slope can be determined more accurately if values at large **k**-values are known, but it is not fundamentally limited by it. What limits the precision of the fit is noise, such as shot noise stemming from the limited number N of detected photons. This noise is present at all spatial frequencies. The randomness of the noise introduces asymmetries in $I_{\text{fluorophore}}(\mathbf{x})$ which result in additional imaginary parts in its Fourier transform even if it is centered at the origin. Not only the position shift but also the noise is therefore changing the phase function. The noise contribution becomes dominant and practically destroys the interpretability of the phase values if its amplitude gets close to or even exceeds the amplitude of the Fourier-transformed signal. Since the OTF amplitude decreases at higher spatial frequencies, this reduces the frequency range over which the slope can be fit with increasing noise. The resolution in FPALM is, hence, dependent on the noise level of the images which is mostly determined by the signal and the associated shot noise.

16.3 Applications

16.3.1 Multicolor Imaging

The ability to visualize multiple fluorescent labels within the same specimen is highly desirable in fluorescence microscopy. Unfortunately, the acquisition of multicolor images by superresolution techniques has been less than trivial. Localization methods using PAFPs have been troubled by the fact that many PAFPs with red-shifted emission emit green light prior to photoconversion (e.g., EosFP and Dendra2) and thus make it impossible to distinguish the signal from that of a single green-emitting PAFP such as PAGFP.

Initially, this problem was overcome by taking advantage of the reversible switching of Dronpa which allows sequential imaging (first the red probe and then the green probe) of each probe (Shroff et al. 2007). However, this approach requires that the red emitter be completely bleached away before imaging the green emitter, and thus requires long acquisition times that are undesirable for live-cell imaging.

Alternatively, a spectroscopic approach has been employed in which individual probes can be distinguished by the relative amount of fluorescence emitted in separate spectral detection channels (Bossi et al. 2008). Unlike ensemble measurements in which fluorescence is simultaneously collected from many fluorophores, the single-molecule nature of localization-based microscopy allows probes with differing emission spectra to be individually identified. This approach is also not limited to only two colors (Bossi et al. 2008).

The recent development of PAmCherry (Subach et al. 2009), a bright dark-to-red switching fluorescent protein, has allowed for a more conventional approach for dual-color imaging in combination with a green-emitting PAFP. The use of such proteins is also attractive for dual-color live-cell imaging in which changes in local cellular environments can cause slight changes in fluorophore emission spectra and complicate analysis for a spectroscopic approach.

Yet another mode of multicolor imaging is accessible to localization microscopy using the photoswitching of activator-reporter dye pairs (Bates et al. 2005) used in STORM (Rust et al. 2006). By combining different activator dyes with a given reporter dye, multicolor imaging is possible by selective activation, that is, using a specific activation laser to image each dye pair (Bates et al. 2007).

Dual-color imaging has also presented challenges in STED microscopy. Thus far, to image two spectrally separated dyes has required complex laser systems and the precise alignment of four laser beams (Donnert et al. 2007; Meyer et al. 2008). These setups have also required sequential imaging of each dye due to excitation of the red-shifted dye by the depletion beam needed for the blue-shifted dye (Donnert et al. 2007; Meyer et al. 2008). Alternatively, a dye with a long stokes shift has been used so that a single STED beam can be used to deplete both dyes (Schmidt et al. 2008); this greatly simplifies the dual-color implementation. However, this approach also requires sequential imaging since both dyes emit in the same detection range.

16.3.2 Temporal Resolution and Live-Cell Imaging

In comparison to electron microscopy and near-field techniques, the relatively noninvasive nature of fluorescence imaging has facilitated its compatibility with live-cell imaging. In most cases, observing dynamics requires high-speed imaging. However, the inherent trade-off between spatial and temporal resolution in fluorescence microscopy becomes obvious as the spatial resolution approaches the nanoscale.

FPALM is typically implemented in wide-field illumination and detection geometries and does not require scanning the laser or sample, yet it is limited in temporal resolution by the number of frames required to localize a sufficient number of molecules to observe a structure of interest at the desired resolution. Since the density of localized molecules is often the limiting factor in achieved resolution (see Section 16.4), increasing the number of localized molecules per frame is necessary for increasing temporal resolution. Precise localization of molecular positions requires that only one molecule can be localized per diffraction-limited area per frame, yet as the average number of molecules visible per frame increases so does the probability that visible molecules will have overlapping images (Gould et al. 2009). Although the stepwise photobleaching of single molecules can in principle be used to identify multiple emitters within a diffraction-limited area

(Gordon et al. 2004; Qu et al. 2004), the diffusion of molecules in living cells additionally complicates such analysis. Also, a balance must be established between camera-frame rates and excitation rates (i.e., laser intensity) so that images of single molecules are not significantly blurred by diffusion, yet enough photons can be collected per molecule to minimize localization uncertainty.

Rearrangement of the structure of interest during the overall image acquisition time results in a blurred image and loss of resolution. The effect is a resolution that is limited by the motion of the structure and the image acquisition time (Shroff et al. 2008). However, motion of individual molecules in consecutive frames allows for the characterization of diffusion properties of labeled molecules (Hess et al. 2007; Manley et al. 2008).

In contrast to FPALM, STED microscopy is implemented in a laser-scanning geometry in which image pixels are generated by scanning either the beams or the sample. In accordance with the Shannon-Nyquist sampling frequency, a working rule of thumb is that the pixel size in confocal microscopy should not be more than half the Abbe resolution limit (Pawley 2006). The reduced size of the effective PSF used in STED microscopy then requires smaller pixels which in turn result in a reduced pixel dwell time for a given scan speed. Thus, maintaining fluorescence signal strength may require increased pixel dwell times which will reduce imaging speed. Despite these trade-offs, video-rate acquisition of images in STED microscopy has been demonstrated (Westphal et al. 2008).

Using organic fluorophores, the application of STED to labeled membranes (Klar et al. 2000) or proteins in synaptic vesicles using secondary antibodies (Westphal et al. 2008) has been demonstrated. However, the use of fluorescent proteins in STED microscopy provides a genetically encodable tool for imaging the interior of living cells (Hein et al. 2008; Nagerl et al. 2008).

High laser intensities are typically required in STED to achieve subdiffraction-limited resolution but may also be necessary to generate complete images in a short period of time in FPALM. Although certain cell lines have been demonstrated to be relatively photon-tolerant in superresolution imaging conditions (Shroff et al. 2008), the effects of phototoxicity due to intense laser exposure should be assessed on a case-by-case basis.

16.3.3 Superresolution in the Axial Direction

The extension of superresolution imaging to three dimensions is of great importance since the extent of most biological structures is fundamentally three-dimensional. While STED, like any confocal method, inherently provides optical sectioning, the resolution enhancement achieved by the typical donut-shaped depletion beam is limited to the lateral dimensions. This is particularly unfortunate as axial resolution is inferior to lateral resolution by approximately 2.5- to 3-fold. Wide-field–based localization methods, on the other hand, have no inherent optical sectioning ability and usually show plane projections of the imaged structures. They must rely on three-dimensional localization methods to extend their scope into the axial dimension.

A variety of approaches have now been implemented to provide subdiffraction-limited resolution in the axial direction, even without the use of two objective lenses as in 4Pi and I^5M microscopy. By introducing a π phase shift into the central region of the depletion laser, it is possible to generate an STED beam profile that features a central intensity 0 with high intensity above and below the focal plane capable of reducing the axial extent of the effective PSF to subdiffraction size (Klar et al. 2000). This depletion pattern can be combined with an additional depletion beam with the typical donut shape to efficiently reduce the size of the effective PSF below the diffraction limit in all directions (Harke et al. 2008).

The extension of STED to 3D is also accomplished by implementing STED in a 4Pi geometry. This approach has demonstrated axial resolution of $\sim\lambda/23$ (Dyba and Hell 2002), and has been recently optimized to generate spherical focal volumes of diameter < 45 nm in the isoSTED configuration (Schmidt et al. 2008; Schmidt et al. 2009). Figure 16.9 shows an example of an isoSTED image.

Three-dimensional localization of single molecules has been realized in several ways. Although a standard wide-field image of a single point source does contain information about the emitter's axial distance from the geometrical focal plane (encoded in the width of the PSF cross section), it cannot usually be inferred whether the emitter is

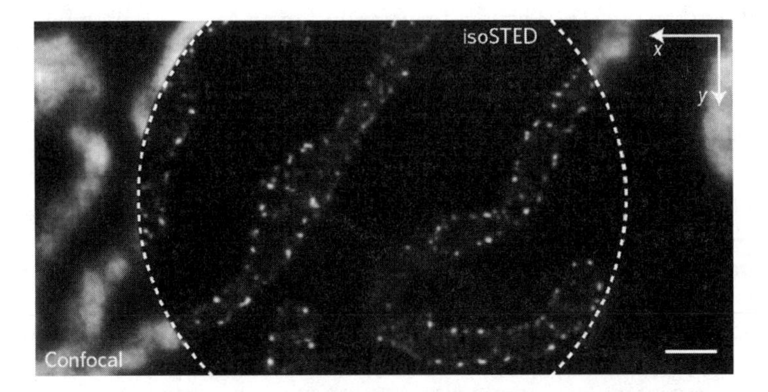

Figure 16.9 Application example of STED microscopy. Optical section taken with a 3D isoSTED microscope of the outer mitochondrial membrane of a mammalian cell. The mitochondria are labeled by an antibody against the protein Tom20. The area in the circle shows the STED image, the area around the conventional confocal counterpart. Scale bar 500 nm. (Adapted by permission from Macmillan Publishers Ltd., Nature Photonics (Hell 2009), Copyright 2009.)

located above or below that plane. In order to overcome this ambiguity, position-dependent asymmetry needs to be introduced into the detection PSF (Holtzer et al. 2007). This has been implemented either by using cylindrical lenses resulting in an astigmatically distorted PSF (Huang et al. 2008) or by a "biplane" detection geometry imaging two planes in the sample simultaneously at an axial distance of several hundred nanometers (Juette et al. 2008) (Fig. 16.10*d*). Both methods have demonstrated axial resolution below 100 nm and provide comparable resolution (Mlodzianoski et al. 2009). The photoswitchable dye pairs used in STORM have also been able to provide multicolor 3D imaging (Huang et al. 2008). Alternatively, a modified detection PSF in the shape of a double helix has been used to determine the axial position of single molecules from the varying relative orientation of the two resulting spots at different axial positions (Pavani et al. 2009).

Most recently, localization microscopy has been implemented in a 4Pi geometry to localize single molecules with an axial resolution on the order of 10 nm (Shtengel et al. 2009), in agreement with the expected sevenfold axial resolution improvement of such a geometry over single-lens setups (von Middendorff et al. 2008; Shtengel et al. 2009).

16.4 Discussion

Overcoming the diffraction limit and achieving a resolution in the far-field of well below 100 nm is not realizable by optics alone. With over 100 years of optics development since Abbe's seminal work, the diffraction barrier could not be broken. Optical manufacturing and design has been optimized by impressive engineering efforts since then, but the resolution achievable with visible light came to a halt above 100 nm. The key to higher resolution and in fact removing diffraction as the resolution-limiting factor lies in the utilization of the photophysical properties of the observed probe molecules. As in the general case of nonlinear optics, the interaction between light and matter is essential to create effects beyond those of classical optics.

All successful superresolution methods in biomedical optics today rely on the optical switching of fluorescent probes. In FPALM, PALM, STORM, and related techniques, single fluorescent molecules are activated, so that they can be imaged separately, and superresolution is obtained from localizing large numbers of molecules with precision well below the diffraction limit. In PSF engineering techniques such as STED microscopy, a structured beam of light is used to drive fluorescent molecules into one state in a saturating manner. This process is used in a highly nonlinear fashion which can be idealized by a binary switching process between an "on" and "off" state.

Although both FPALM and STED utilize switching of fluorescent probes, they differ quite significantly in their realization. Therefore they feature different strengths but also face different limitations and constraints.

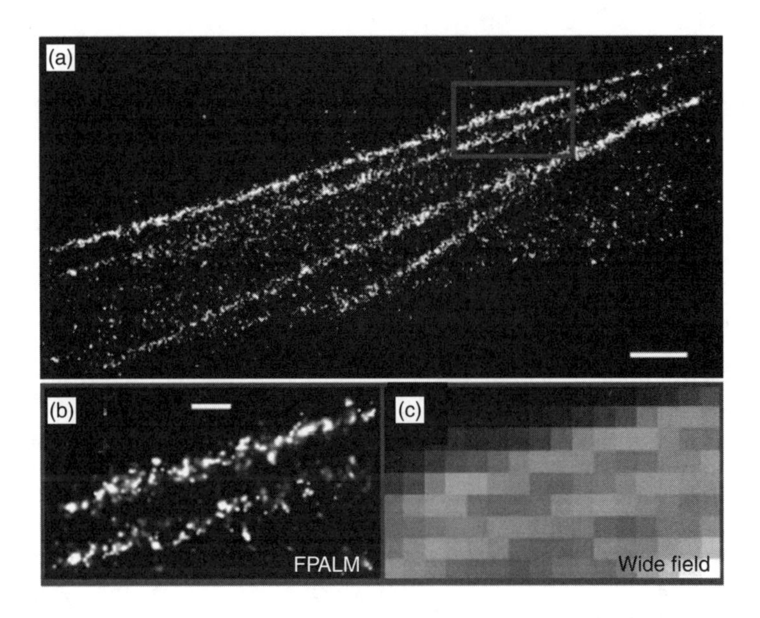

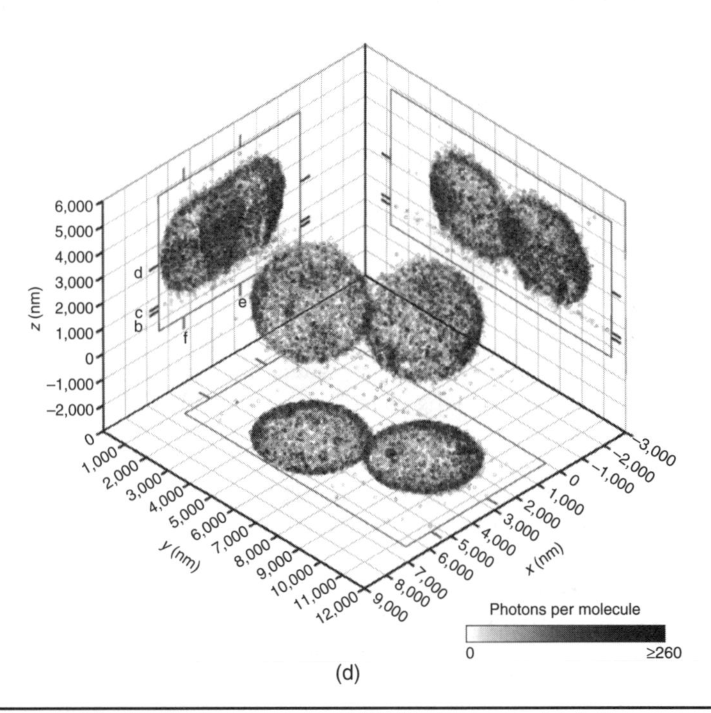

FIGURE 16.10 Application example of FPALM microscopy. (*a*) Actin cytoskeleton imaged by FPALM. (*b*) Detail from (*a*) as denoted by the gray box. (*c*) Comparison to a conventional wide-field image. (*d*) Example of biplane FPALM imaging showing the 3D reconstruction of two four-micron diameter beads surface-labeled with caged fluorescein. Scale bar in (*a*) 2 μm, in (*b*) 500 nm. (*a*–*c*) adapted by permission from Macmillan Publishers Ltd., Nature *Protocols* (Gould 2009), Copyright 2009.

The most obvious difference from an instrumental perspective is that STED microscopy is a (confocal) laser scanning technique whereas FPALM is based on conventional wide-field microscopy. The stimulated emission process utilized by STED microscopy competes with the spontaneously occurring fluorescence with a lifetime of typically 2 to 3 nanoseconds. Therefore, STED intensity levels need to be high enough to deplete the fluorophores within the first few hundred picoseconds after excitation. This high laser intensity cannot be maintained simultaneously across large fields of view and therefore favors a point-scanning approach. Additionally, the success of an STED microscope setup depends critically on the quality of the illumination pattern. If the central minimum of the STED beam is too high (greater than a few percent of the intensity maxima), its intensity will be high enough to deplete fluorescence at higher STED powers (Hell and Schönle 2006). This depletion at the center of the focus counteracts the resolution improvement effect which relies on steep inner edges of the depletion donut. A high beam quality can be realized most easily with a point-scanning system, where cross talk between neighboring illuminated areas is avoided.

In contrast, FPALM relies on a switching mechanism which is not as instantaneous and reversible as stimulated emission and therefore does not pose a minimum requirement on the illumination intensity. Typically, switching in FPALM requires conformational changes of a fluorophore. These are often irreversible and slow. A scanning approach works best with reversible switching since the imaged molecules need to be switched on and off multiple times during the scanning process depending on the relative position of the focus at any time. Since the recording time that the system requires to image each pixel has to be longer than the switching time, scanning would also slow down the imaging speed dramatically. Wide-field imaging is therefore the chosen approach for all successful FPALM setups.

These particular forms of realization provide practical limitations to resolution and recording speed in STED microscopy and FPALM. FPALM requires cameras fast enough to keep up with the appearance and disappearance of fluorescing molecules. The wide-field approach is prone to background which especially becomes a problem for thicker samples where background fluorescence originating at many depths can accumulate. STED microscopy, on the other hand, is limited by the dead times of the typically used photon-counting detectors which limit the maximum detectable signal per unit of time.

Which are the fundamental factors that determine the resolution and imaging speed in STED and FPALM? The resolution of an STED microscope can be expressed by Eq. (16.2).

The saturation intensity I_{sat} is inversely proportional to the stimulated emission cross section of the fluorophore. From this fact, it follows that d_{STED} is proportional to the diffraction-limited resolution and, for $I_{max} >> I_{sat}$, inversely proportional to the square root of the

Resolution Dependency	STED	FPALM
Switching intensity	$I^{-1/2}$	–
Switching cross section	$\sigma^{-1/2}$	–
Number of detected photons per molecule	–	$N^{-1/2}$
PSF size	d	d

TABLE 16.1 Resolution Dependencies of STED and FPALM

STED intensity and the stimulated emission cross section. To enhance the resolution of an STED microscope by a factor of 2, the STED intensity (or alternatively the cross section) has to be quadrupled.

In FPALM, a fourfold brighter molecule can be localized twice as precisely according to Eq. (16.3). In contrast to STED microscopy, the resolution in FPALM is independent of the intensity of the laser light (see Table 16.1 for an overview). For lower intensities, FPALM imaging takes longer, but provides essentially the same image.

This dependency of the recording time on the excitation intensity is not really a limitation in STED microscopy. Typically, even a relatively weak laser source provides enough laser power to excite fluorescence close to the excitation saturation limit in a laser-scanning microscope. Raising the laser intensity therefore provides no further advantage (but contributes to increased photobleaching). In FPALM, high laser intensities have to be maintained not only in a small focus but over the entire field of view to read out and bleach (or deactivate) activated molecules fast enough. Often the limits of available laser power are reached before excitation saturation, and thus, the recording time, which is defined by the average time it takes to switch off an activated molecule, depends inversely on the available laser power. Alternatively, the laser power can be concentrated to a smaller field of view in FPALM to achieve higher intensities leading to a linear relationship between recording time and imaging area. This same relationship is observed in STED microscopy because of the time it takes to scan a field of view with a laser beam while keeping the pixel dwell time constant. While this relationship is independent of any laser powers, in FPALM, a larger imaging area can be compensated for by an equally increased laser power to maintain the same recording time (see Table 16.2 for an overview).

Recording Time Dependency	STED	FPALM
Excitation intensity	–	I^{-1}
Imaging area if limited by power	A	A
Imaging area if not limited by power	A	–

TABLE 16.2 Recording Time Dependencies of STED and FPALM

The different methods to achieve superresolution also fundamentally influence the data processing after image-recording. STED creates superresolution data directly; after scanning of the field of view, no further postprocessing is required. In fact, as in conventional confocal microscopy the creation of the superresolution image can be observed in real time as the image is sequentially built up during the scan process. Basic postprocessing to reduce the noise in the images or a mathematical deconvolution step to further improve the resolution slightly can be performed optionally, but are not required and in most cases not used.

In FPALM, data processing is a relatively complex procedure whereas imaging itself is fairly simple. Typically, thousands of recorded camera frames have to be analyzed to identify the signal of single molecules. For every identified particle, the center of the image has to be determined with subpixel accuracy to get an estimate for the molecule's position. The determined positions are collected and combined to finally create the image representing the recorded structure.

This method of assembling an image points to a fundamental limitation of superresolution fluorescence microscopy: Fluorescent probes are used as representatives of the labeled structure. The recorded image displays the distribution of the subset of probes which had emitted enough photons to significantly contribute to the image. Some of the probe molecules could have been bleached before producing a sufficient signal, or they might not have been activated over the course of imaging. Moreover, the probes are typically not the target structure of interest. The subpopulation of target molecules labeled by antibodies can be small, whereas FPs linked to proteins coexist within the endogenous pool of unlabeled proteins. Some of the probe molecules might not function because they have been bleached before imaging or proteins have not folded correctly. The fluorescent label is slightly offset in position compared to the target structure since linkers and the probes themselves have dimensions of several up to about 10 nm. These problems become even more apparent if the structure of interest is not the target protein but this protein is a probe itself, for example, representing a certain cell organelle such as mitochondria or the Golgi apparatus. A low density of probe molecules cannot represent complex organelle structures on the tens of nanometers size scale.

In fact, in FPALM it becomes apparent that "resolution" is a combination of localization precision and probe density. A complex structure can only be resolved in FPALM if the density of localized molecules is high enough so that the structure is "sampled" sufficiently. It has been suggested that the effective resolution in localization-based techniques can be approximated as

$$d = \sqrt{\sigma^2 + a^2} \qquad (16.8)$$

where σ is the localization precision and *a* the average nearest neighbor distance between localized molecules (Gould et al. 2008).

Although not as apparent, STED microscopy also suffers from low probe densities: the number of molecules that fluoresce at any point in time becomes smaller as the resolution is increased. "Holes" appear in the imaged structures due to fluctuations in the probe distribution. As in FPALM, it is difficult to distinguish between these statistical fluctuations and real structural features.

Therefore, probe development and novel labeling strategies experience a renaissance in superresolution microscopy. Image quality can only be as good as the sample and labeling allow it to be. Next to brighter and more photostable fluorescent markers, new labeling strategies are required that allow denser labeling without disturbing (live) samples.

Nonetheless, superresolution microscopy is capable of resolving objects 10 times smaller than the diffraction limit. This holds an enormous potential as a rapidly increasing number of published applications show. A growing number of labs is actively developing in this field, which promises more exciting developments in the next years. In particular, multicolor, live-cell, and three-dimensional imaging still hold room for improvement and are at the same time of particular importance in biological applications.

Acknowledgments

We thank Joachim Spatz for support, Tobias Hartwich and Michael Mlodzianoski for careful reading of the manuscript. M. J. is supported by a fellowship from the German Academic Exchange Service (DAAD Doktorandenstipendium).

References

Abbe, E. (1873). "Beiträge zur Theorie des Mikroskops und der mikroskopischen Wahrnehmung, *Archiv für Mikroskopische Anatomie* 9: 413–468.

Airy, G. B. (1835). "On the Diffraction of an Object-Glass with Circular Aperture," *Transactions of the Cambridge Philosophical Society* 5: 283–291.

Alberts, B. (2008). *Molecular Biology of the Cell*. Garland Science, New York.

Ando, R., C. Flors et al. (2007). "Highlighted Generation of Fluorescence Signals Using Simultaneous Two-Color Irradiation on Dronpa Mutants," *Biophys. J.* 92(12): L97–L99.

Ando, R. H. Mizuno, et al. (2004). "Regulated Fast Nucleocytoplasmic Shuttling Observed by Reversible Protein Highlighting," *Science* 306(5700): 1370–1373.

Axelrod, D. (1981). "Cell-Substrate Contacts Illuminated by Total Internal Reflection Fluorescence," *J. Cell Biol.* 89(1): 141–145.

Baddeley, D., I. D. Jayasinghe et al. (2009). "Light-Induced Dark States of Organic Fluochromes Enable 30-nm Resolution Imaging in Standard Media," *Biophys. J.* 96(2): L22–L24.

Bates, M., T. R. Blosser et al. (2005). "Short-Range Spectroscopic Ruler Based on a Single-Molecule Optical Switch," *Phys. Rev. Lett.* 94(10): 108101.

Bates, M., B. Huang et al. (2007). "Multicolor Super-Resolution Imaging with Photo-Switchable Fluorescent Probes," *Science* 317(5845): 1749–1753.

Betzig, E., G. H. Patterson et al. (2006). "Imaging Intracellular Fluorescent Proteins at Nanometer Resolution," *Science* 313(5793): 1642–1645.
Betzig, E., J. K. Trautman et al. (1991). "Beating the Diffraction Barrier: Optical Microscopy on a Nanometer Scale," *Science* 251: 1468–1470.
Bewersdorf, J., R. Schmidt et al. (2006). "Comparison of I^{5}M and 4Pi-Microscopy," *J. Microsc.* 222(Pt 2): 105–117.
Biteen, J. S., M. A. Thompson et al. (2008). "Super-Resolution Imaging in Live *Caulobacter crescentus* Cells Using Photoswitchable EYFP," *Nat. Methods* 5(11): 947–949.
Born, M. and E. Wolf (1999). *Principles of Optics: Electromagnetic Theory of Propagation, Interference, and Diffraction of Light*, Cambridge University Press, New York.
Bossi, M., J. Fölling et al. (2008). "Multicolor Far-Field Fluorescence Nanoscopy through Isolated Detection of Distinct Molecular Species," *Nano Lett.* 8(8): 2463–2468.
Bossi, M., J. Fölling et al. (2006). "Breaking the Diffraction Resolution Barrier in Far-Field Microscopy by Molecular Optical Bistability," *New J. Physics* 8: 275.
Bretschneider, S., C. Eggeling et al. (2007). "Breaking the Diffraction Barrier in Fluorescence Microscopy by Optical Shelving," *Phys. Rev. Lett.* 98(21): 218103.
Cheezum, M. K., W. F. Walker et al. (2001). "Quantitative Comparison of Algorithms for Tracking Single Fluorescent Particles," *Biophys. J.* 81(4): 2378–2388.
Chudakov, D. M., S. Lukyanov et al. (2005). "Fluorescent Proteins as a Toolkit for *in Vivo* Imaging," *Trends Biotechnol.* 23(12): 605–613.
Chudakov, D. M., S. Lukyanov et al. (2007). "Tracking Intracellular Protein Movements Using Photoswitchable Fluorescent Proteins PS-CFP2 and Dendra2," *Nat. Protoc.* 2(8): 2024–2032.
Day, R. N. and M. W. Davidson (2009). "The Fluorescent Protein Palette: Tools for Cellular Imaging," *Chem. Soc. Rev.* 38(10): 2887–2921.
Denk, W., J. H. Strickler et al. (1990). "Two-Photon Laser Scanning Fluorescence Microscopy," *Science* 248(4951): 73–76.
Di Francia, G. (1952). "Super-Gain Antennas and Optical Resolving Power," *Il Nuovo Cimento (1943–1954)* 9: 426–438.
Dickson, R. M., A. B. Cubitt et al. (1997). "On/off Blinking and Switching Behaviour of Single Molecules of Green Fluorescent Protein," *Nature* 388(6640): 355–358.
Donnert, G., J. Keller et al. (2006). "Macromolecular-Scale Resolution in Biological Fluorescence Microscopy," *Proc. Natl. Acad. Sci. USA* 103(31): 11440–11445.
Donnert, G., J. Keller et al. (2007). "Two-Color Far-Field Fluorescence Nanoscopy," *Biophys. J.* 92(8): L67–L69.
Dyba, M. and S. W. Hell (2002). "Focal Spots of Size $\lambda/23$ Open Up Far-Field Fluorescence Microscopy at 33-nm Axial Resolution," *Phys. Rev. Lett.* 88(16): 163901.
Egner, A., C. Geisler et al. (2007). "Fluorescence Nanoscopy in Whole Cells by Asynchronous Localization of Photoswitching Emitters," *Biophys. J.* 93(9): 3285–3290.
Fölling, J., V. Belov et al. (2007). "Photochromic Rhodamines Provide Nanoscopy with Optical Sectioning," *Angew. Chem. Int. Ed. Engl.* 46(33): 6266–6270.
Fölling, J., M. Bossi et al. (2008). "Fluorescence Nanoscopy by Ground-State Depletion and Single-Molecule Return," *Nat. Methods* 5(11): 943–945.
Goodman, J. W. (2004). *Introduction to Fourier Optics*, Roberts & Company, Greenwood Village, CO.
Gordon, M. P., T. Ha et al. (2004). "Single-Molecule High-Resolution Imaging with Photobleaching," *Proc. Natl. Acad. Sci. USA* 101(17): 6462–6465.
Gould, T. J., M. S. Gunewardene et al. (2008). "Nanoscale Imaging of Molecular Positions and Anisotropies," *Nat. Methods* 5: 1027–1030.
Gould, T. J., V. V. Verkhusha et al. (2009). "Imaging Biological Structures with Fluorescence Photoactivation Localization Microscopy," *Nat. Protoc.* 4(3): 291–308.
Gustafsson, M. G. (2000). "Surpassing the Lateral Resolution Limit by a Factor of 2 Using Structured Illumination Microscopy," *J. Microsc.* 198(Pt 2): 82–87.
Gustafsson, M. G. L., D. A. Agard et al. (1999) "I^{5}M: 3D Wide-Field Light Microscopy with Better Than 100-nm Axial Resolution," *J. Microsc.* 195: 10–16.
Harke, B., C. K. Ullal et al. (2008). "Three-Dimensional Nanoscopy of Colloidal Crystals," *Nano Lett.* 8(5): 1309–1313.

Hecht, E. (2001). *Optics*, Addison Wesley, New York.
Heilemann, M., S. van de Linde et al. (2008). "Subdiffraction-Resolution Fluorescence Imaging with Conventional Fluorescent Probes," *Angew. Chem. Int. Ed. Engl.* 47(33): 6172–6176.
Hein, B., K. I. Willig et al. (2008). "Stimulated Emission Depletion (STED) Nanoscopy of a Fluorescent Protein-Labeled Organelle Inside a Living Cell," *Proc. Natl. Acad. Sci. USA* 105(38): 14271–14276.
Heintzmann, R. and C. Cremer (1999). "Laterally Modulated Excitation Microscopy: Improvement of Resolution by Using a Diffraction Grating," *Optical Biopsies and Microscopic Techniques III, Proceedings* 3568: 185–196.
Hell S., and E. H. K. Steltzer (1992). "Properties of a 4Pi Confocal Fluorescence Microscope," *J. Opt. Soc. Am. A.* 9(12): 2159–2167.
Hell, S. W. (2007). "Far-Field Optical Nanoscopy," *Science* 316(5828): 1153–1158.
Hell, S. W. (2009). "Microscopy and Its Focal Switch," *Nat. Methods* 6(1): 24–32.
Hell, S. W., M. Dyba et al. (2004). "Concepts for Nanoscale Resolution in Fluorescence Microscopy," *Curr. Opin. Neurobiol.* 14(5): 599–609.
Hell, S. W., S. Jakobs et al. (2003). "Imaging and Writing at the Nanoscale with Focused Visible Light through Saturable Optical Transitions," *Appl. Phys. A* 77: 859–860.
Hell, S. W., and M. Kroug (1995). "Ground-State-Depletion Fluorescence Microscopy: A Concept for Breaking the Diffraction-Resolution Limit," *Appl. Phys. B-Lasers and Optics* 60: 495–497.
Hell, S. W. and A. Schönle (2006). "Nanoscale Resolution in Far-Field Fluorescence Microscopy," *Science of Microscopy*, ed. P. W. Hawkes, and J. C. H. Spence, Springer, New York.
Hell, S. W. and J. Wichmann (1994). "Breaking the Diffraction-Resolution Limit by Stimulated Emission: Stimulated-Emission Depletion Fluorescence Microscopy," *Opt. Lett.* 19(11): 780–782.
Hess, S. T., T. P. Girirajan et al. (2006). "Ultra-High Resolution Imaging by Fluorescence Photoactivation Localization Microscopy," *Biophys. J.* 91(11): 4258–4272.
Hess, S. T., T. J. Gould et al. (2007). "Dynamic Clustered Distribution of Hemagglutinin Resolved at 40 nm in Living Cell Membranes Discriminates between Raft Theories," *Proc. Natl. Acad. Sci. USA* 104(44): 17370–17375.
Hofmann, M., C. Eggeling et al. (2005). "Breaking the Diffraction Barrier in Fluorescence Microscopy at Low Light Intensities by Using Reversibly Photoswitchable Proteins," *Proc. Natl. Acad. Sci. USA* 102(49): 17565–17569.
Holtzer, L., T. Meckel et al. (2007). "Nanometric Three-Dimensional Tracking of Individual Quantum Dots in Cells," *Appl. Phys. Lett.* 90(5): 053902-3.
Huang, B., S. A. Jones, et al. (2008). "Whole-Cell 3D STORM Reveals Interactions between Cellular Structures with Nanometer-Scale Resolution," *Nat. Methods* 5: 1047–1052.
Huang, B., W. Wang et al. (2008). "Three-Dimensional Super-Resolution Imaging by Stochastic Optical Reconstruction Microscopy," *Science* 319(5864): 810–813.
Irvine, S. E., T. Staudt et al. (2008). "Direct Light-Driven Modulation of Luminescence from Mn-Doped ZnSe Quantum Dots," *Angew. Chem. Int. Ed. Engl.* 47(14): 2685–2688.
Juette, M. F., T. J. Gould et al. (2008). "Three-Dimensional Sub-100 nm Resolution Fluorescence Microscopy of Thick Samples," *Nat. Methods* 5(6): 527–529.
Klar, T. A. and S. W. Hell (1999). "Subdiffraction Resolution in Far-Field Fluorescence Microscopy," *Opt. Lett.* 24(14): 954–956.
Klar, T. A., S. Jakobs et al. (2000). "Fluorescence Microscopy with Diffraction Resolution Barrier Broken by Stimulated Emission," *Proc. Natl. Acad. Sci. USA*, 97(15): 8206–8210.
Kner, P., B. B. Chhun et al. (2009). "Super-Resolution Video Microscopy of Live Cells by Structured Illumination," *Nat. Methods* 6(5): 339–342.
Lacoste, T. D., X. Michalet et al. (2000). "Ultrahigh-Resolution Multicolor Colocalization of Single Fluorescent Probes," *Proc. Natl. Acad. Sci. USA* 97(17): 9461–9466.
Lagerholm, B. C., L. Averett et al. (2006). "Analysis Method for Measuring Submicroscopic Distances with Blinking Quantum Dots," *Biophys. J.* 91(8): 3050–3060.

Lakowicz, J. R. (2006). *Principles of Fluorescence Spectroscopy*, Springer Science, New York.

Lemmer, P., M. Gunkel et al. (2008). "SPDM: Light Microscopy with Single-Molecule Resolution at the Nanoscale," *Appl. Phys. B–Lasers and Optics* 93(1): 1–12.

Lidke, K. A., B. Rieger et al. (2005). "Superresolution by Localization of Quantum Dots Using Blinking Statistics," *Opt. Exp.* 13(18): 7052–7062.

Manley, S., J. M. Gillette et al. (2008). "High-Density Mapping of Single-Molecule Trajectories with Photoactivated Localization Microscopy," *Nat. Methods* 5(2): 155–157.

Meyer, L., D. Wildanger et al. (2008). "Dual-Color STED Microscopy at 30-nm Focal-Plane Resolution," *Small* 4(8): 1095–1100.

Mlodzianoski, M. J., M. F. Juette et al. (2009). "Experimental Characterization of 3D Localization Techniques for Particle-Tracking and Super-Resolution Microscopy," *Opt. Exp.* 17(10): 8264–8277.

Nagerl, U. V., K. I. Willig et al. (2008). "Live-Cell Imaging of Dendritic Spines by STED Microscopy," *Proc. Natl. Acad. Sci. USA* 105(48): 18982–18987.

Patterson, G. H. and J. Lippincott-Schwartz (2002). "A Photoactivatable GFP for Selective Photolabeling of Proteins and Cells," *Science* 297(5588): 1873–1877.

Pavani, S. R., M. A. Thompson et al. (2009). "Three-Dimensional Single-Molecule Fluorescence Imaging beyond the Diffraction Limit by Using a Double-Helix Point Spread Function," *Proc. Natl. Acad. Sci. USA* 106(9): 2995–2999.

Pawley, J. B. (2006). *Handbook of Biological Confocal Microscopy*. Springer Science, New York.

Qu, X., D. Wu et al. (2004). "Nanometer-Localized Multiple Single-Molecule Fluorescence Microscopy," *Proc. Natl. Acad. Sci. USA* 101(31): 11298–11303.

Rayleigh, L. (1879). "Investigations in Optics with Special Reference to the Spectroscope," *Philos. Mag.* 5(8): 261–274.

Rittweger, E., K. Y. Han et al. (2009). "STED Microscopy Reveals Crystal Colour Centres with Nanometric Resolution," *Nat. Phot.* 3: 144–147.

Rust, M. J., M. Bates et al. (2006). Sub-Diffraction Limit Imaging by Stochastic Optical Reconstruction Microscopy (STORM)," *Nat. Methods* 3(10): 793–796.

Saxton, M. J. and K. Jacobson (1997). "Single-Particle Tracking: Applications to Membrane Dynamics," *Annu. Rev. Biophys. Biomol. Struct.* 26: 373–399.

Schmidt, R., C. A. Wurm et al. (2008). "Spherical Nanosized Focal Spot Unravels the Interior of Cells," *Nat. Methods* 5(6): 539–544.

Schmidt, R., C. A. Wurm et al. (2009). "Mitochondrial Cristae Revealed with Focused Light," *Nano Lett.* 9(6): 2508–2510.

Shaner, N. C., R. E. Campbell et al. (2004). "Improved Monomeric Red, Orange and Yellow Fluorescent Proteins Derived from *Discosoma* sp. Red Fluorescent Protein," *Nat. Biotechnol.* 22(12): 1567–1572.

Shaner, N. C., G. H. Patterson et al. (2007). "Advances in Fluorescent Protein Technology," *J. Cell Sci.* 120(Pt 24): 4247–4260.

Shao, L., B. Isaac et al. (2008). "I^5S: Wide-Field Light Microscopy with 100-nm Scale Resolution in Three Dimensions," *Biophys. J.* 94(12): 4971–4983.

Sharonov, A. and R. M. Hochstrasser (2006). "Wide-Field Subdiffraction Imaging by Accumulated Binding of Diffusing Probes," *Proc. Natl. Acad. Sci. USA* 103(50): 18911–18916.

Sheppard, C. J. R. and A. Choudhur (1977). "Image Formation in the Scanning Microscope," *J. Modern Optics*, 24(10): 1051–1073.

Shroff, H., C. G. Galbraith et al. (2008). "Live-Cell Photoactivated Localization Microscopy of Nanoscale Adhesion Dynamics," *Nat. Methods* 5(5): 417–423.

Shroff, H., C. G. Galbraith et al. (2007). "Dual-Color Superresolution Imaging of Genetically Expressed Probes within Individual Adhesion Complexes," *Proc. Natl. Acad. Sci. USA* 104(51): 20308–20313.

Shtengel, G., J. A. Galbraith et al. (2009). "Interferometric Fluorescent Super-Resolution Microscopy Resolves 3D Cellular Ultrastructure," *Proc. Natl. Acad. Sci. USA* 106(9): 3125–3130.

Stiel, A. C., M. Andresen et al. (2008). "Generation of Monomeric Reversibly Switchable Red Fluorescent Proteins or Far-Field Fluorescence Nanoscopy," *Biophys. J.* 95(6): 2989–2997.

Stiel, A. C., S. Trowitzsch et al. (2007). "1.8 A Bright-State Structure of the Reversibly Switchable Fluorescent Protein Dronpa Guides the Generation of Fast-Switching Variants," *Biochem. J.* 402(1): 35-42.
Subach, F. V., G. H. Patterson et al. (2009). "Photoactivatable mCherry for High-Resolution Two-Color Fluorescence Microscopy," *Nat. Methods* 6(2): 153–159.
Testa, I., A. Schonle et al. (2008). "Nanoscale Separation of Molecular Species Based on Their Rotational Mobility," *Opt. Exp.* 16(25): 21093–21104.
Thompson, R. E., D. R. Larson et al. (2002). "Precise Nanometer Localization Analysis for Individual Fluorescent Probes," *Biophys. J.* 82(5): 2775–2783.
Middendorff, C., A. Egner et al. (2008). "Isotropic 3D Nanoscopy Based on Single Emitter Switching," *Opt. Exp.* 16(25): 20774–20788.
Westphal, V., S. O. Rizzoli et al. (2008). "Video-Rate Far-Field Optical Nanoscopy Dissects Synaptic Vesicle Movement," *Science* 320(5873): 246–249.
White, J. G., W. B. Amos et al. (1987). "An Evaluation of Confocal versus Conventional Imaging of Biological Structures by Fluorescence Light Microscopy," *J. Cell Biol.* 105(1): 41–48.
Wiedenmann, J., S. Ivanchenko et al. (2004). "EosFP, a Fluorescent Marker Protein with UV-Inducible Green-to-Red Fluorescence Conversion," *Proc. Natl. Acad. Sci. USA* 101(45): 15905–15910.
Wildanger, D., E. Rittweger et al. (2008). "STED Microscopy with a Supercontinuum Laser Source," *Opt. Exp.* 16(13): 9614–9621.
Willig, K. I., B. Harke et al. (2007). "STED Microscopy with Continuous Wave Beams," *Nat. Methods* 4(11): 915–918.
Willig, K. I., R. R. Kellner et al. (2006). "Nanoscale Resolution in GFP-Based Microscopy," *Nat. Methods* 3(9): 721–723.
Wilson, T. and C. J. R. Sheppard (1984). *Theory and Practice of Scanning Optical Microscopy*, Academic Press, New York.
Yildiz, A., J. N. Forkey et al. (2003). "Myosin V Walks Hand-over-Hand: Single Fluorophore Imaging with 1.5-nm Localization," *Science* 300(5628): 2061–2065.
Zenisek, D., J. A. Steyer et al. (2000). "Transport, Capture and Exocytosis of Single Synaptic Vesicles at Active Zones," *Nature* 406(6798): 849–854.

Index

Note: Page numbers referencing figures are followed by an "*f*"; page numbers referencing tables are followed by a "*t*".

A

B

D

E

N

O

P